新生物学丛书

间充质干细胞基础与临床
（第二版）

韩忠朝　李宗金　韩之波　主编

科 学 出 版 社
北　京

内 容 简 介

近十来年,国内外间充质干细胞临床应用与药物开发取得了极大的进展,使得间充质干细胞成为最具有临床应用价值的细胞治疗制品之一。自 2012 年出版《间充质干细胞基础与临床》一书以来,该书得到读者的好评,期间多次加印。这几年国内外科研工作者在间充质干细胞研究与应用方面又陆续出现不少新成果,我们在第一版的基础上,更新和补充了近几年关于间充质干细胞的基础与临床研究进展,并新增了间充质干细胞治疗的分子影像学研究、无标记检测、间充质干细胞外泌体和药物开发策略等方面的内容。

本书系统地对间充质干细胞的国内外研究成果进行总结和综述,可供相关专业的研究人员、干细胞或组织工程药物开发人员、临床医生,以及在校研究生阅读参考。

图书在版编目(CIP)数据

间充质干细胞基础与临床/韩忠朝,李宗金,韩之波主编. —2 版. —北京:科学出版社,2019.6

(新生物学丛书)

ISBN 978-7-03-061518-3

Ⅰ. ①间… Ⅱ. ①韩… ②李… ③韩… Ⅲ. ①干细胞–临床应用–研究 Ⅳ. ①Q24

中国版本图书馆 CIP 数据核字(2019)第 106428 号

责任编辑:罗 静 岳漫宇 闫小敏/责任校对:张怡君
责任印制:赵 博/封面设计:刘新新

科学出版社 出版

北京东黄城根北街 16 号
邮政编码:100717
http://www.sciencep.com

北京凌奇印刷有限责任公司印刷
科学出版社发行 各地新华书店经销

*

2012 年 3 月第 一 版 开本:787×1092 1/16
2019 年 6 月第 二 版 印张:20 1/4
2025 年 1 月第四次印刷 字数:477 000

定价:150.00 元

(如有印装质量问题,我社负责调换)

《新生物学丛书》专家委员会

主　　任：蒲慕明

副 主 任：吴家睿

专家委员会成员（按姓氏汉语拼音排序）

昌增益	陈洛南	陈晔光	邓兴旺	高　福
韩忠朝	贺福初	黄大昉	蒋华良	金　力
康　乐	李家洋	林其谁	马克平	孟安明
裴　钢	饶　毅	饶子和	施一公	舒红兵
王　琛	王梅祥	王小宁	吴仲义	徐安龙
许智宏	薛红卫	詹启敏	张先恩	赵国屏
赵立平	钟　扬	周　琪	周忠和	朱　祯

《间充质干细胞基础与临床》(第二版)编委会

主　编

韩忠朝　细胞产品国家工程研究中心
李宗金　南开大学
韩之波　细胞产品国家工程研究中心

编　委（按姓氏汉语拼音排名）

白　华　天津工业大学电子与信息工程学院
曹晓沧　天津医科大学总医院
陈旭义　武警后勤学院附属医院
耿　洁　围产期干细胞北京市工程实验室
郝好杰　北京恒峰铭成生物科技有限公司
何　睿　复旦大学基础医学院
黄平平　中国医学科学院血液病医院（血液学研究所）
李美蓉　中国人民解放军总医院
梁　璐　天津百恩生物科技有限公司
马步鹏　北京协和医院
王　斌　天津百恩生物科技有限公司
王伟强　江西省细胞产品工程技术研究中心
王有为　俄亥俄州立大学詹姆斯肿瘤医院
武开宏　南京医科大学附属儿童医院
吴志宏　北京协和医院
杨舟鑫　浙江医院老年医学研究所
张　颢　济宁医学院附属医院
张　磊　博雅干细胞科技有限公司
竺晓凡　中国医学科学院血液病医院（血液学研究所）

参编人员（按姓氏汉语拼音排名）

陈惠华　北京恒峰铭成生物科技有限公司

杜　为	南开大学
高弘烨	中国医学科学院血液病医院（血液学研究所）
侯慧星	天津医科大学总医院
李丽丽	天津市妇女儿童保健中心
林美光	汉氏联合（天津）干细胞研究院
牛学刚	武警后勤学院附属医院
孙　军	北京市第六医院
陶红燕	南开大学
易　军	北京恒峰铭成生物科技有限公司
张帅强	南开大学
赵钦军	江西省干细胞工程技术研究中心
赵向男	南开大学

《新生物学丛书》丛书序

当前，一场新的生物学革命正在展开。为此，美国国家科学院研究理事会于2009年发布了一份战略研究报告，提出一个"新生物学"（New Biology）时代即将来临。这个"新生物学"，一方面是生物学内部各种分支学科的重组与融合，另一方面是化学、物理、信息科学、材料科学等众多非生命学科与生物学的紧密交叉与整合。

在这样一个全球生命科学发展变革的时代，我国的生命科学研究也正在高速发展，并进入了一个充满机遇和挑战的黄金期。在这个时期，将会产生许多具有影响力、推动力的科研成果。因此，有必要通过系统性集成和出版相关主题的国内外优秀图书，为后人留下一笔宝贵的"新生物学"时代精神财富。

科学出版社联合国内一批有志于推进生命科学发展的专家与学者，联合打造了一个21世纪中国生命科学的传播平台——《新生物学丛书》。希望通过这套丛书的出版，记录生命科学的进步，传递对生物技术发展的梦想。

《新生物学丛书》下设三个子系列：科学风向标，着重收集科学发展战略和态势分析报告，为科学管理者和科研人员展示科学的最新动向；科学百家园，重点收录国内外专家与学者的科研专著，为专业工作者提供新思想和新方法；科学新视窗，主要发表高级科普著作，为不同领域的研究人员和科学爱好者普及生命科学的前沿知识。

如果说科学出版社是一个"支点"，这套丛书就像一根"杠杆"，那么读者就能够借助这根"杠杆"成为撬动"地球"的人。编委会相信，不同类型的读者都能够从这套丛书中得到新的知识信息，获得思考与启迪。

《新生物学丛书》专家委员会
主　任：蒲慕明
副主任：吴家睿
2012年3月

第二版前言

自 2011 年出版《间充质干细胞基础与临床》一书以来，近几年国内外间充质干细胞临床应用与药物开发取得了极大的进展。我们在第一版的基础上，补充了近几年关于间充质干细胞的基础与临床研究进展，并新增了间充质干细胞治疗的分子影像学研究、无标记检测、外泌体和药物开发策略等方面内容。

我们一直坚信间充质干细胞是目前最具临床应用前景的干细胞，其在免疫调节、促血管新生及抗炎等方面均表现出良好的应用前景及临床疗效。在本书中，除了对前一版本中已有的临床应用相关章节进行了更新，如间充质干细胞用于治疗自身免疫系统疾病、血液系统疾病、糖尿病，以及在治疗神经系统疾病中的应用等，也新增了间充质干细胞在治疗骨关节炎、肝硬化、炎性肠病和慢性肺疾病等中应用的内容，希望本书内容对各位读者有所帮助。

本书出版过程中，在内容设置方面得到了众多专家的指导与帮助，在文字校对与排版方面喻昊博士做了大量工作，在此一并表示感谢。由于撰稿时间仓促，在内容方面难免存在不足之处，欢迎批评指正。

我们希望本书能对我国间充质干细胞临床研究与间充质干细胞药物开发起到积极的推动作用。

韩忠朝
2018 年 11 月 18 日

第一版前言

干细胞是一类具有自我复制和多向分化能力的细胞，它们可以不断地自我更新，并在特定条件下转分化成为一种或多种构成人体组织或器官的细胞。干细胞是机体的起源细胞，是形成人体各种组织的祖宗细胞。干细胞技术是一门尚处于试验阶段的医疗新技术。干细胞的临床治疗目前存在着许多争议，这是由于干细胞疗法是目前临床使用的最复杂的生物疗法；许多疾病的动物模型不能准确反映人类所患的疾病，有时在动物中的毒理学研究很难预示在人体内的毒性；干细胞可能作用几个靶标部位，发挥有利或有害的作用，因此对临床前试验的安全性要求更高；不同来源的干细胞具有不同的分子特征，这方面尚未形成国际公认的标准；干细胞技术范围很广，不同的技术成熟度也不一致。造血干细胞移植等已被证明是安全有效的干细胞疗法，而胚胎干细胞以及诱导多能干细胞（induced pluripotent stem cell，iPSC）技术仍处于实验室阶段，其安全性和分化可控性问题仍未解决，技术尚远未成熟至可用于临床。

间充质干细胞（mesenchymal stem cell，MSC）是干细胞家族的重要成员，来源于发育早期的中胚层和外胚层。MSC最初在骨髓中发现，因其具有多向分化潜能、造血支持和促进干细胞植入、免疫调控和自我复制等特点而日益受到人们的关注。例如，间充质干细胞在体内或体外特定的诱导条件下，可分化为脂肪、骨、软骨、肌肉、肌腱、韧带、神经、肝、心肌、内皮等多种组织细胞，连续传代培养和冷冻保存后仍具有多向分化潜能，可作为理想的种子细胞用于衰老和病变引起的组织器官损伤修复。

间充质干细胞的临床研究已经在许多国家开展，美国批准了数十项临床试验，中国批准了2项临床试验。目前，间充质干细胞除了用来促进恢复造血，与造血干细胞共移植治疗白血病和难治性贫血等以外，还用于心血管疾病、肝硬化、骨和肌肉衰退性疾病、脑及脊髓神经损伤、老年痴呆、红斑狼疮和硬皮病等自身免疫系统疾病的临床研究，已经取得的部分临床结果令人鼓舞。

迄今的研究表明，脐带、胎盘来源的间充质干细胞不但能够成为骨髓间充质干细胞的理想替代物，而且具有更大的应用潜能。脐带、胎盘间充质干细胞表达多种胚胎干细胞的特有分子标志，具有分化潜能大、增殖能力强、免疫原性低、取材方便、无道德伦理问题的限制、易于工业化制备等特征，因此有可能成为最具临床应用前景的多能干细胞。

本书涉及间充质干细胞的基础研究与国内外临床研究进展、间充质干细胞库、间充质干细胞制剂的质量标准，以及临床医生十分关注的间充质干细胞安全性研究进展、间充质干细胞与肿瘤的关系等；同时将干细胞应用所涉及的伦理学问题及相关知识产权保护进行了适当的介绍。各章节除介绍该领域的最新进展外，还适当介绍了有关的背景资料。由于全书各章节所属领域研究工作的深度不同，不同作者的写作风格也有差别，全书在文字上难免有较大的差异。由于干细胞研究进展很快，难免会有不同意见，甚至观点相悖，敬请读者谅解。另外，由于撰稿时间短促，在内容上难免出现不足之处，欢迎批评指正。

本书能够得以出版需要感谢各位作者的辛勤工作。同时，还需要感谢中国医学科学院血液病医院（血液学研究所）徐茂强主任的前期准备工作，以及细胞产品国家工程研究中心的郑重在文字处理方面所做的工作。

世界范围内干细胞技术的迅猛发展，给人类健康带来了巨大福音。2007年，美国食品药品监督管理局已经批准间充质干细胞作为治疗造血干细胞移植排斥的一线药物进入Ⅲ期临床。尽管对于间充质干细胞的临床治疗机制还有许多未解之谜，但这并不能阻碍间充质干细胞的临床应用之路，因为越来越多的临床证据显示应用间充质干细胞治疗多种疾病的确安全有效。正如阿司匹林的临床应用过程，作为一种最有效的解热、镇痛、消炎药物在临床上使用已经有100余年，但它的作用机制直到其进入市场70年后才得以阐明。带着疑问和目标上路的人最终不会迷路，我们希望并相信我们可以成为间充质干细胞探索之路上清醒的推动者。

细胞产品国家工程研究中心　韩忠朝
二〇一一年十月一日

目　　录

第一章　间充质干细胞概述 ··· 1
　　参考文献 ·· 8
第二章　不同组织来源间充质干细胞的生物学特性 ·· 11
　　第一节　不同组织来源间充质干细胞表面标志的比较 ·· 11
　　第二节　不同组织来源间充质干细胞增殖能力的比较 ·· 16
　　第三节　不同组织来源间充质干细胞分化能力的比较 ·· 16
　　第四节　不同组织来源间充质干细胞造血支持能力的比较 ································ 17
　　第五节　不同组织来源的间充质干细胞免疫调节能力的比较 ···························· 18
　　第六节　不同组织来源间充质干细胞促血管生成能力的比较 ···························· 19
　　第七节　不同组织来源间充质干细胞迁移能力的比较 ·· 19
　　第八节　小结 ··· 20
　　参考文献 ·· 20
第三章　间充质干细胞制剂的质量控制 ·· 22
　　第一节　间充质干细胞质量研究的必要性 ··· 23
　　第二节　干细胞治疗监管政策的演进 ··· 23
　　第三节　间充质干细胞质量研究的主要内容 ··· 25
　　第四节　小结 ··· 32
　　参考文献 ·· 33
第四章　间充质干细胞的安全性研究进展 ·· 34
　　第一节　间充质干细胞的体外安全性研究进展 ··· 34
　　第二节　间充质干细胞的体内安全性研究进展 ··· 39
　　第三节　目前临床上间充质干细胞治疗的安全性问题 ·· 41
　　第四节　小结 ··· 43
　　参考文献 ·· 43
第五章　间充质干细胞的免疫调节机制 ·· 48
　　第一节　免疫系统的构成和功能 ·· 48
　　第二节　间充质干细胞的免疫学特性 ··· 50
　　第三节　间充质干细胞的免疫调节机制研究 ··· 51
　　第四节　间充质干细胞免疫调节能力的可塑性 ··· 56
　　第五节　间充质干细胞治疗免疫相关疾病的临床应用研究 ································ 58
　　参考文献 ·· 59

第六章　间充质干细胞与治疗性血管新生 ························· 65
第一节　血管新生的概念 ··· 65
第二节　间充质干细胞在血管新生中的角色 ······················· 67
第三节　间充质干细胞的内皮分化 ····································· 71
第四节　间充质干细胞作为载体搭载血管新生因子 ············· 73
第五节　间充质干细胞促进血管新生的进展与展望 ············· 73
参考文献 ··· 76

第七章　间充质干细胞的生物学特性及在造血干细胞移植中的应用 ············· 79
第一节　间充质干细胞的生物学特性 ································· 79
第二节　间充质干细胞在造血干细胞移植中的应用 ············· 80
第三节　小结 ·· 85
参考文献 ··· 85

第八章　间充质干细胞在心血管疾病治疗中的应用 ············· 90
第一节　心血管疾病的概述 ·· 90
第二节　干细胞移植的方法 ·· 91
第三节　干细胞移植的作用机制 ··· 92
第四节　干细胞移植存在的问题 ··· 93
第五节　干细胞构建组织工程心肌 ····································· 95
第六节　问题与展望 ··· 99
参考文献 ··· 99

第九章　间充质干细胞在慢性肺疾病治疗中的应用 ············· 105
第一节　慢性肺疾病治疗现状 ·· 105
第二节　间充质干细胞治疗慢性肺疾病的临床前研究 ········· 106
第三节　间充质干细胞治疗慢性肺疾病的临床研究 ············· 110
第四节　间充质干细胞治疗慢性肺疾病的注册临床试验 ······ 112
参考文献 ··· 113

第十章　间充质干细胞在自身免疫系统疾病治疗中的应用 ············ 116
第一节　自身免疫系统疾病 ·· 116
第二节　间充质干细胞驱动的免疫调节 ······························ 116
第三节　间充质干细胞治疗自身免疫系统疾病的临床前模型 ············· 118
第四节　间充质干细胞在自身免疫系统疾病治疗中的临床应用 ·········· 120
第五节　间充质干细胞在自身免疫系统疾病治疗中面临的问题 ·········· 126
参考文献 ··· 127

第十一章　间充质干细胞在下肢缺血性疾病治疗中的应用 ············· 133
第一节　下肢缺血性疾病的概述 ··· 133
第二节　间充质干细胞治疗下肢缺血性疾病的研究进展 ······ 134
参考文献 ··· 142

第十二章	间充质干细胞在神经损伤修复中的应用	144
第一节	间充质干细胞与神经损伤修复的概述	144
第二节	间充质干细胞与神经损伤修复的基础研究	145
第三节	间充质干细胞与神经损伤修复的临床研究	149
参考文献		151
第十三章	**间充质干细胞在炎性肠病治疗中的应用**	**157**
第一节	炎性肠病	157
第二节	间充质干细胞治疗炎性肠病的可能机制	157
第三节	间充质干细胞治疗炎性肠病的基础及临床研究	159
第四节	间充质干细胞对炎性肠病相关瘘管的局部治疗	160
参考文献		163
第十四章	**间充质干细胞在肝硬化治疗中的应用**	**168**
第一节	肝硬化治疗现状	168
第二节	间充质干细胞治疗肝硬化的临床前研究	169
第三节	间充质干细胞治疗肝硬化的临床研究	171
第四节	间充质干细胞治疗肝硬化的临床试验	173
参考文献		174
第十五章	**间充质干细胞在糖尿病治疗中的应用**	**176**
第一节	糖尿病的概述	176
第二节	间充质干细胞治疗糖尿病的机制	177
第三节	间充质干细胞治疗糖尿病的临床试验进展	179
第四节	问题与展望	181
参考文献		183
第十六章	**间充质干细胞在骨关节炎疾病治疗中的应用**	**187**
第一节	间充质干细胞与骨关节炎	187
第二节	间充质干细胞治疗骨关节炎的研究进展	189
第三节	间充干细胞治疗骨关节炎的机制	192
第四节	问题与展望	194
参考文献		194
第十七章	**间充质干细胞在皮肤修复与再生中的应用**	**198**
第一节	皮肤创面修复的一般过程	198
第二节	间充质干细胞参与皮肤修复的机制	200
第三节	影响间充质干细胞促修复能力的因素	204
第四节	间充质干细胞治疗皮肤损伤的临床研究	206
参考文献		207
第十八章	**间充质干细胞组织工程**	**216**
第一节	间充质干细胞组织工程的概述	216

第二节　间充质干细胞在组织工程中的应用 218
　　第三节　发展趋势预测与展望 223
　　参考文献 225

第十九章　间充质干细胞基因治疗 229
　　第一节　干细胞的来源 229
　　第二节　间充质干细胞的归巢能力 230
　　第三节　对干细胞进行基因工程改造的方法 231
　　第四节　研究进展 234
　　第五节　总结与展望 238
　　参考文献 239

第二十章　间充质干细胞的分子影像学 243
　　第一节　干细胞的分子影像学 243
　　第二节　分子影像学方法 243
　　第三节　间充质干细胞的标记方法 245
　　第四节　间充质干细胞治疗的分子影像学 248
　　第五节　间充质干细胞治疗的分子影像技术应用前景 251
　　参考文献 251

第二十一章　基于拉曼散射的间充质干细胞无标记检测 255
　　第一节　引言 255
　　第二节　拉曼散射的基本概念 255
　　第三节　间充质干细胞的典型拉曼光谱 260
　　第四节　拉曼光谱在间充质干细胞检测中的应用 261
　　参考文献 267

第二十二章　间充质干细胞外泌体 269
　　第一节　间充质干细胞和外泌体 269
　　第二节　外泌体的研究 269
　　第三节　间充质干细胞来源外泌体的治疗作用 274
　　第四节　提高间充质干细胞来源外泌体生物活性的研究 277
　　参考文献 282

第二十三章　间充质干细胞药物开发策略 286
　　第一节　干细胞临床转化探索之路 286
　　第二节　间充质干细胞药品开发流程 292
　　第三节　制约干细胞药物开发的技术瓶颈 296
　　第四节　国外已上市的干细胞药物及生物制品概要 300
　　参考文献 305

第一章　间充质干细胞概述

间充质干细胞（mesenchymal stem cell，MSC）是中胚层来源的具有高度自我更新能力和多向分化潜能的多能干细胞，广泛存在于全身多种组织中，可在体外培养扩增，并能在特定条件下分化为神经细胞、成骨细胞、软骨细胞、肌肉细胞、脂肪细胞等[1]。MSC是多能干细胞，具有"横向分化"或"跨系分化"的能力，不仅支持造血干细胞（hematopoietic stem cell，HSC）的生长，还可以在地塞米松、抗坏血酸、胰岛素、异丁基甲基黄嘌呤等不同诱导条件下，在体外分化为多种组织细胞如神经细胞、成骨细胞、软骨细胞、脂肪细胞、心肌细胞等。MSC具有广阔的临床应用前景，是细胞替代治疗和组织工程的首选种子细胞，是移植和自身免疫系统疾病治疗领域的研究热点。

一、间充质干细胞的发现

间充质干细胞的发现最早可以追溯到130多年前。1867年，德国病理学家Cohnheim在研究伤口愈合时，首次提出骨髓中存在非HSC的观点，指出成纤维细胞可能来源于骨髓。直到1976年，Friedenstein等[2]从骨髓细胞中培养出这种成纤维细胞，并证实其可在体外大量扩增，易贴壁，呈集落样生长，且具有定向分化的特点。1988年这类干细胞被命名为"骨髓基质干细胞"[3]。随后，Prockop[4]和Pittenger等[1]分别证明了这种成纤维细胞包含间充质干细胞，其具有形成多种间质组织（如骨、软骨、脂肪和平滑肌）的能力。Caplan教授在1992年进一步把这类干细胞命名为"间充质干细胞"[5]。"间充质干细胞"的命名逐渐被广泛接受和使用。有学者认为间充质干细胞的英文名称应为mesenchymal stromal cell，但是更多学者习惯采用mesenchymal stem cell。

骨髓是MSC的主要来源，对骨髓间充质干细胞（bone marrow-mesenchymal stem cell，BM-MSC）的研究也最为深入。但骨髓中MSC的含量极少，仅占骨髓单个核细胞（bone marrow mononuclear cell，BMMC）的0.002%~0.005%。近年来，研究人员陆续从其他组织如脐血、脐带、胎盘、脂肪、肌肉、头皮、牙周质等中分离得到MSC。da Silva等[6]报道，MSC几乎存在于所有的成体组织器官中，他认为血管周围是MSC的体内"干细胞龛"。Bruno Peault实验室[7]也从成体的肺、心脏、脑、小肠等中分离出MSC。

二、间充质干细胞的定义

一直以来MSC被作为一种具有特征性形态、表型及功能性质的细胞群体来研究，由于没有找到特异性的表面标志物，对MSC的定义一直存在争议。MSC高表达CD13、CD29、CD105、CD44，低表达CD106，不表达CD14、CD34、CD11a、CD31、CD45[1]及人白细胞抗原Ⅱ（human leukocyte antigen-Ⅱ，HLA-Ⅱ），不表达或低表达HLA-Ⅰ抗原[8]。其中CD29属于整合素家族，CD105是间充质相关抗原，CD14是单核巨噬细胞表面标记，CD11a是淋巴细胞功能相关抗原-1，CD34和CD45是HSC阳性标记，CD31是内皮细胞特异性抗原标记。同时 MSC 也不表达或低表达移植免疫排斥相关表面标志 CD80（B7-1）、CD86（B7-2）、CD40、CD40L[9]，表明细胞高表达 MSC 相关标志而不表达 HSC、内皮细胞特异性抗原及移植排斥相关表面标志。Pittenger 等[1]提出的有关 MSC 的定义是目前公认的"金标准"，即 BM-MSC 可以在体外适宜的刺激下，向骨细胞、脂肪细胞及软骨细胞分化。2006 年，国际细胞治疗协会（International Society for Cellular Therapy，ISCT）确定 MSC 的鉴定标准[10]为：①在塑料培养皿内，贴壁生长，在含血清的培养基内，高度增殖；②表达 CD105、CD73、CD90，不表达 CD45、CD34、CD14 或 CD11β、CD79α 或 CD19；③在体外可以分化为成骨细胞、脂肪细胞、软骨细胞。

人们通过高密度培养和多次传代获得的形态均一和表型一致的 MSC 的这一培养方法可能已经丢失了一些 MSC 的特有特性。通过单细胞来源克隆培养 MSC 发现人 MSC（human MSC，hMSC）至少包括三种不同形态的细胞：①一种体积小但自我更新很快的细胞；②一种细长的纺锤状的成纤维样的细胞；③一种大而扁平的增殖速度慢的细胞。其中体积小的 MSC 具有更好的增殖能力和更强的分化、迁移和定植能力，而连续传代后这类细胞会不断减少[11]。这也说明连续传代后 MSC 出现了不对称分裂和分化。除了表型，早期的 MSC 的蛋白质表达也是多样的，一些对细胞活性十分重要的受体蛋白只表达于特定的亚群中。MSC 的异质性会造成得到不同的研究结果，也会影响细胞治疗的疗效。因此研究不同 MSC 亚群的增殖、分化和生物学特性具有深远的意义。将具有不同生物学活性的特定 MSC 亚群应用到相对应的疾病治疗中，可使其更有效地发挥治疗作用。

Stro-1 高表达于增殖能力高的 MSC 细胞表面，是评价 MSC 功能特性的一个重要的表面标志物。在贴壁培养的骨髓细胞中，表达 Stro-1 的细胞大约占 6%。Stro-1$^+$ MSC 具有更强的迁移和组织修复能力[12]。CD271（低亲和力神经生长因子受体）被认为是更原始的 MSC 的一个表面标志物，它在不同组织来源 MSC 中的表达量差异大，CD271$^+$ MSC 表达成软骨的基因水平比其他 MSC 亚群要高，具有很强的成软骨能力，适合应用于软骨修复[13,14]。CD105 是转化生长因子-β（transforming growth factor-β，TGF-β）受体复合物的一部分，对血管形成和重塑至关重要。脂肪和骨髓来源 MSC 表达 CD105 的水平较高，超过 90%[14]。CD105$^+$ MSC 具有很强的成心肌能力，研究表明 CD105$^+$ MSC 可以减少心肌梗死面积和改善心功能[15]。CD106 [血管细胞黏附分子-1（vascular cell adhesion molecule-1，VCAM-1）]在胎盘组织中表达量最高，大约为 75%。CD106$^+$ MSC 表达免疫调节因子的水平要高于其他亚群，具有很强的免疫抑制能力，适合应用于自身免疫系统疾病[16]。CD146 是一种膜糖蛋白，在 BM-MSC 中占 40%~70%，在 UC-MSC 中占 16%~40%，在脂肪间充质干细胞（adipose-derived mesenchymal stem cell，AD-MSC）中约占 20%[17]。CD146$^+$ MSC 同时还表达 NG2 和 PDGFRβ，CD146$^+$ MSC 表达骨相关因子的水平要高于其他亚群，具有很强的成骨分化能力，适合应用于骨折后的骨愈合和骨重建[18]。PDGFRα 是血小板源生长因子（platelet derived growth factor，PDGF）家族成员的酪氨酸激酶受体，PDGFRα$^+$ MSC 对皮肤疾病具有很好的治疗作用，可以向损伤皮肤趋化并分化成表皮细胞，促进皮肤损伤修复和重建[19]。另外，其他表型如 Nestin、CXCR4、SSEA-4、Stro-3、GD2、Leptin R 等都被报道过单独表达或者联合表达在 MSC 的表面，表达这些标志物的细胞或是具有更强的集落形成能力、造血支持作用，或是具有更好的迁移或者分化能力[20]。随着研究者的深入研究，更纯的 MSC 的生物学功能将被确定，并特定地应用于治疗相应的疾病。

三、间充质干细胞的作用机制

MSC 具有很强的分化能力，这一特性曾经被认为是其参与组织损伤修复的主要作用机制。然而目前许多的研究结果表明，MSC 在动物疾病模型中确实能发挥治疗作用，但是在免疫健全的动物体内定植和分化的细胞数量甚微[21,22]。MSC 具有强大的旁分泌功能，通过分泌一系列的细胞因子、化学分子和生长因子来调节微环境，激活内源性的干细胞发挥组织损伤修复作用[23]。通过已有的蛋白质检测方法可以检测有一定表达量的分泌蛋白。应用质谱仪定量检测方法检测到鼠 MSC 表达 258 种蛋白，其中分泌蛋白为 54 种[24]，但许多细胞因子和生长因子的含量要低于质谱仪的检测限度；蛋白质芯片法可以检测到低浓度的蛋白质，但是只有已知的蛋白质可以通过这个方法检测。有研究人员通过质谱法确定了 hMSC 表达的 247 种蛋白，包括 29 个未知蛋白，通过芯片法确定了低表达的 72 种蛋白，其中只有 3 个与质谱法检测到的蛋白相同。通过生物信息学的方法分析了编码这些蛋白的 201 种基因，这些基因涉及 58 种生物过程和 30 条信号通路，58 种生物过程主要分成 3 个部分，即代谢、防御应答和组织分化；30 条信号通路涉及的范围很广，包括受体结合、信号转导、细胞直接接触、细胞迁移、免疫应答和代谢[25]。MSC 通过分泌血管内皮生长因子（vascular endothelial growth factor，VEGF）、肝细胞生长因子（hepatocyte growth factor，HGF）、成纤维细胞生长因子（fibroblast growth factor，FGF）、胰岛素样生长因子-1（insulin-like growth factor-1，IGF-1）和 TB4 等来保护缺血性心肌细胞，促进心肌再生[26]。MSC 通过分泌一系列的生长因子如 HGF、VEGF、基质细胞衍生因子-1（stromal cell derived factor-1，SDF-1）、角质细胞生长因子（keratinocyte growth factor，KGF）、FGF、胎盘生长因子（placental growth factor，PLGF）、单核细胞趋化蛋白-1（monocyte chemoattractant protein-1，MCP-1）和 IGF-1 等来促进细胞增殖和组织损伤修复[27]。MSC 通过分泌抗炎因子[白介素-10（interleukin，IL-10）、肿瘤坏死因子刺激基因-6（tumor necrosis factor stimulated gene 6，TSG-6)]和抑制促炎因子[IL-1a、IL-6、IL-17、干扰素-γ（interferon，IFN-γ）、粒细胞集落刺激因子（granulocyte colony stimulating factor，G-CSF）、粒细胞-巨噬细胞集落刺激因子（granulocyte-macrophage colony stimulating factor，GM-CSF）、巨噬细胞炎性蛋白-2α（macrophage inflammatory protein-2α，MIP-2α）、MCP-1]来发挥抗炎作用[28]。

外泌体（exosome）是由细胞内多泡体（multivesicular body，MVB）与细胞膜融合后，释放到细胞外基质（extracellular matrix，ECM）中的一种直径为 30~120nm 的膜性囊泡。外泌体所携带的"货物"具有重要的生物学意义，包括具有多种生物学特性的脂质、蛋白质、RNA 和 DNA。其中脂质包括胆固醇、鞘磷脂和己糖基神经酰胺。蛋白质包括溶酶体相关蛋白、脂筏相关蛋白、四跨膜蛋白和离散蛋白等。RNA 包括 mRNA、miRNA、tRNA、rRNA、siRNA、lncRNA，DNA 包括 mtDNA、ssDNA、dsDNA。MSC 的外泌体含有的脂筏相关蛋白大约是其产生的全部脂筏相关蛋白的 1/3[29]。越来越多的研究表明，MSC 通过外泌体来进行细胞间的交流，通过循环系统到达其他细胞与组织，进而发挥其组织损伤修复作用。MSC 释放的外泌体通过增加Ⅱ型胶原和硫酸化糖胺聚糖（sulfated-glycosaminoglycan，s-GAG）的合成来填充新生组织进行软骨损伤修复[30]，通过促进心肌细胞存活和抑制其凋亡来进行心肌修复，恢复心血管功能[31]。

线粒体转移是 MSC 另一种可能的作用机制。MSC 是非常好的线粒体供者，MSC 高表达 RHOT1（一个关键的 Rho GTP 酶），从而支持 MSC 对其他细胞进行线粒体转移[32]。MSC 与损伤的组织细胞共培养时，会通过线粒体转移主动修复存在 mtDNA 缺陷的细胞，从而挽救受损的细胞，促进组织的再生[33]。增强 MSC 的线粒体转移能力将得到更好的治疗效果[34]。

四、间充质干细胞的应用

MSC 具有许多独特的性质，如归巢迁移能力、造血支持能力、免疫调节能力、多向分化能力等，MSC 针对不同的适应证能发挥不同的治疗作用。

MSC 是造血微环境的主要细胞成分，在 HSC 的生长、增殖、分化中起重要作用。MSC 通过与 HSC 相互作用及分泌多种细胞因子，或表达与 HSC 黏附、归巢有关的黏附分子及 ECM 蛋白[35]，支持和促进了造血。体外实验证明，MSC 对 CD34$^+$骨髓细胞的维持和扩增起了很重要的作用[36]。目前有关 MSC 促进造血干细胞移植（hematopoietic stem cell transplantation，HSCT）后造血重建的研究已取得了很大进展。研究人员发现 MSC 与 HSC 共移植可以支持 HSC 生长，明显促进巨核细胞和血小板形成[37]。Sang 等[38]发现 MSC 可能在 HSC 分化过程的表观遗传调控中起了一定的作用。国内外已有 HSC 和 MSC 共移植用于临床研究的报道[39]，并取得了良好的临床效果。无论是自体移植还是同种异体移植，细胞的输入均没有引起不良反应。2016 年日本发表了 BM-MSC 治疗 25 例难治性Ⅲ期或者Ⅳ期急性移植物抗宿主病（graft-versus-host disease，GVHD）的Ⅱ/Ⅲ期临床试验，细胞治疗期间不使用免疫抑制剂治疗，在第一次细胞治疗 4 周后，6 位患者（24%）达到完全缓解，9 位患者（36%）达到部分缓解，总有效率达到 60%。24 周后单独细胞治疗达到完全缓解后的存活率是 48%（12/25），52 周后另有 6 位患者在细胞联合其他治疗后达到完全缓解[40]。2010 年欧洲进行了 MSC 治疗儿童急性 GVHD（acute GVHD，aGVHD）的Ⅱ/Ⅲ期双盲安慰剂对照临床试验，完全缓解率为 77%，达到完全缓解后的存活率达 88%[41]。

MSC 是一类免疫缺陷细胞，具有低免疫原性和免疫调节特性，MSC 通常低表达 HLA-Ⅰ抗原，不表达 HLA-Ⅱ抗原[8]和共刺激分子 CD80（B7-1）、CD86（B7-2）与 CD40 等。MSC 可以抑制 T 淋巴细胞的增殖和激活，这种抑制作用并不依赖于主要组织相容性复合体（major histocompatibility complex，MHC），不需要细胞之间的相互接触；MSC 通过可溶性因子的介导来发挥免疫调节功能，实验发现 TGF-β、前列腺素 E2（prostaglandin

E2，PGE2)、吲哚胺 2,3-双加氧酶（indoleamine 2,3-dioxygenase，IDO）[42]等因子可能参与了 MSC 的免疫调节功能。在自身免疫系统疾病动物实验[43]及临床试验[44]中，MSC 可以有效地缓解疾病，抑制免疫反应，减少炎症发生。2016 年在欧洲多中心进行的克罗恩病的 MSC 的随机双盲对照Ⅲ期临床试验，共有 212 例患者入组，和安慰剂组相比，接受 MSC 细胞治疗的患者取得了显著的效果，总有效率为 50%（53/107），总体而言，细胞治疗免疫性肠炎是安全和有效的[45]。

MSC 具有多向分化潜能，在特定的诱导条件下，它不仅能分化为骨、软骨、脂肪、肌腱等中胚层组织细胞，而且能够向内胚层组织细胞如心肌细胞[46]、肝细胞[47]和外胚层组织细胞如神经细胞[48]分化。MSC 因这一特性而成为组织工程研究的首选种子细胞。Wakitanti 等[49]将体外扩增的 MSC 输注给 24 位进行了高位胫骨切开的膝关节炎患者 42 周后，观察到缺损部位已被软组织填充并有软骨样组织形成。这一结果提示 MSC 可以作为骨组织工程的种子细胞参与组织修复。MSC 可以分化为神经元和神经胶质细胞，因此可以用于外周神经的修复[50]。另外，其可以分泌神经营养因子来提高神经细胞的存活率和促进其再生，这提示 MSC 用于治疗神经系统疾病的潜力很大。

MSC 不仅在人医临床上有很好的应用，在动物临床上的应用也有了不小的发展。最开始是从赛马开始，主要是用于治疗运动系统方面的疾病，包括骨关节疾病和肌腱损伤等。对 141 例赛马的浅屈肌肌腱损伤进行干细胞治疗和至少 2 年的随访的研究结果表明，干细胞治疗赛马的肌腱损伤是安全有效的，与传统方法比较能够提高治疗后的恢复能力[51]。而后干细胞治疗逐渐应用到伴侣动物（猫和狗），从运动系统疾病、外伤疾病发展到眼科疾病和内科疾病[52]，通过关节腔注射干细胞治疗犬关节炎能有效减轻疼痛，改善临床症状[53]。伴侣动物和实验动物有所不同，由于经过了上万年的进化演变和与人类相近的生活习惯，伴侣动物所患疾病更接近人所患疾病，用伴侣动物进行干细胞治疗研究能更好地为人医临床提供很好的动物实验数据。

五、间充质干细胞的前景与展望

MSC 具有独特的生物学功能，干细胞治疗是继药物治疗、手术治疗之后的又一场医疗革命。MSC 具有在体外容易扩增的特性，在短时间内可以扩增上千倍，这是 MSC 广泛应用于实验研究和临床的重要前提条件。相对于骨髓，从脐带中分离 MSC 具有组织来源丰富、细胞原始、增殖能力强和安全无病毒感染风险等优点，因此 UC-MSC 可能会为今后临床应用需求提供更为理想的种子细胞。目前，发达国家已经开始利用干细胞治疗技术用于临床治疗自身免疫系统疾病、心肌梗死、脊髓损伤、心力衰竭等疾病，其发展速度十分惊人，已从基础研究过渡到临床应用研究阶段。中国在干细胞研究方面有很好的基础，与国外同行的差距不大。正因如此，我们更应该在干细胞技术向临床转化方面抢占制高点。

截至 2017 年 6 月，累计在 *Pubmed* 上发表的 MSC 相关的各类论文数量达到了 46 023 篇，从 2001 年开始论文数量呈现暴发式增长，2001 年发表了 202 篇，而 2016 年发表了 5442 篇（图 1-1）。关于 MSC 的鉴定、分化、生物学特性、作用机制、临床应用各个方

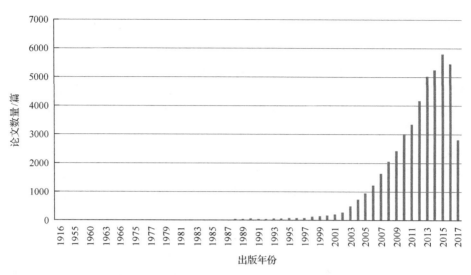

图 1-1 Pubmed 上发表的 MSC 相关的论文数量

面都有大量的研究。截至 2017 年 6 月，在 ClincalTrials.gov 上注册的应用 MSC 进行临床研究的项目达到 730 项，其中 GVHD 治疗研究 37 项，免疫系统疾病 101 项，肾脏疾病 19 项，肝脏疾病 39 项，肺部疾病 38 项，肠道疾病 74 项，骨相关疾病 21 项，中枢神经系统（central nervous system，CNS）疾病 94 项，心血管疾病 58 项，糖尿病 43 项，糖尿病足 11 项，骨关节炎（osteoarthritis，OA）40 项等。已注册的临床研究在美国开展的有 140 项，在欧洲开展的有 145 项，而在中国大陆地区开展的有 144 项，在中国台湾地区开展的有 13 项。临床试验所处阶段反映了干细胞产品的成熟程度，由图 1-2 可知，59.6% 的临床研究还处在 Ⅰ/Ⅱ 期和 Ⅱ 期，只有 11 个进入了 Ⅱ/Ⅲ 期，8 个进入了 Ⅲ 期临床研究阶段，其中在消化系统疾病、心血管疾病、免疫系统疾病和神经系统疾病方面的研究最为成熟[54]。已经有 14 个获准进入临床的干细胞药物，其中多数为异体 MSC，这也说明 MSC 具有广阔的临床应用前景。

要实现将干细胞变成药物还有许多的关键技术需要解决，如如何保持干细胞经过体外大规模培养后的分化能力、遗传特征、致瘤性的稳定性问题；建立与药效作用相关的体外生物学评价方法；筛选保持活细胞稳定的药物制剂配方；选择剂型及包装材料；规范储存、冷链运输条件；建立干细胞产品的技术标准及质量评价体系；摸索临床前药物有效性评价体系、安全性评价体系等。另外，建立健全规范的法律法规和统一的生产与质量标准也是发展干细胞技术的关键。只有通过严格的临床试验审查，经过严格监管，才能保证干细胞技术应用的安全性和有效性。通过建立统一的生产和质量标准，对组织来源、细胞筛选、制备过程和最后的产品进行严格的检验，获得符合标准的干细胞才具有质量上的可靠性，才可以用于临床使用，对没有达到标准的干细胞严格禁止使用，这样才可以遏制干细胞滥用的趋势，防止损害公众利益的事件发生，让干细胞的研究和应用健康发展。

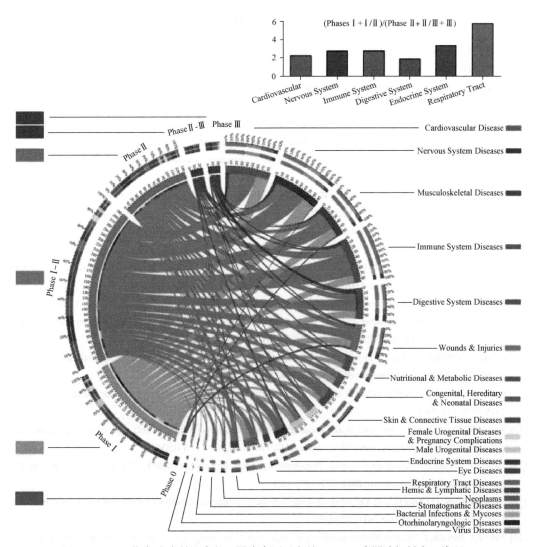

图 1-2 MSC 临床试验所处阶段（图片来源于文献[54]）（彩图请扫封底二维码）

Cardiovascular Disease. 心血管疾病，Nervous System Diseases. 神经系统疾病，Musculoskeletal Diseases. 肌肉骨骼疾病，Immune System Diseases. 免疫系统疾病，Digestive System Diseases. 消化系统疾病，Wounds & Injuries. 创伤与损伤，Nutritional & Metabolic Diseases. 营养与代谢疾病，Congenital, Hereditary & Neonatal Diseases. 先天性、遗传性和新生儿疾病，Skin & Connective Tissue Diseases. 皮肤和结缔组织疾病，Female Urogenital Diseases & Pregnancy Complications. 女性泌尿生殖系统疾病与妊娠并发症，Male Urogenital Diseases. 男性泌尿生殖系统疾病，Endocrine System Diseases. 内分泌系统疾病，Eye Diseases. 眼病，Respiratory Tract Diseases. 呼吸道疾病，Hemic & Lymphatic Diseases. 血液和淋巴系统疾病，Neoplasms. 肿瘤，Stomatognathic Diseases. 口腔疾病，Bacterial Infections & Mycoses. 细菌感染和真菌感染疾病，Otorhinolaryngologic Diseases. 耳鼻喉科疾病，Virus Diseases. 病毒感染疾病，Phase Ⅰ. 临床Ⅰ期，Phase Ⅰ-Ⅱ. 临床Ⅰ-Ⅱ期，Phase Ⅱ. 临床Ⅱ期，Phase Ⅲ. 临床Ⅲ期

（梁　璐　韩忠朝）

参 考 文 献

[1] Pittenger MF, Mackay AM, Beck SC, et al. Multilineage potential of adult human mesenchymal stem cells. Science, 1999, 284:143-147.

[2] Friedenstein AJ, Deriglasova UF, Kulagina NN, et al. Fibrolast precursors in normal and irradiate mouse hematopoietic organs. Exp Hematol, 1976, 2:83-92.

[3] Owen M, Friedenstein AJ. Stromal stem cells: marrow-derived osteogenic precursors. Ciba Found Symp, 1988, 136(136):42-60.

[4] Prockop DJ. Marrow stromal cells are stem cells for nohematopoietic tissues. Science, 1997, 276:71-74.

[5] Haynesworth SE, Baber MA, Caplan AI. Cell surface antigens on human marrow-derived mesenchymal cells are detected by monoclonal antibodies. Bone, 1992, 13(1):69-80.

[6] da Silva Meirelles L, Chagastelles PC, Nardi NB. Mesenchymal stem cells reside in virtually all post-natal organs and tissues. Journal of Cell Science, 2006, 119 (Pt 11):2204-2213.

[7] Crisan M, Yap S, Casteilla L, et al. A perivascular origin for mesenchymal stem cells in multiple human organs. Cell Stem Cell, 2008, 3 (3):301-313.

[8] Lu LL, Liu YJ, Yang SG, et al. Isolation and characterization of human umbilical cord mesenchymal stem cells with hematopoiesis-supportive function and other potentials. Haematologica, 2006, 91:1017-1026.

[9] Deans RJ, Moseley AB. Mesenchymal stem cells: biology and potential clinical uses. Exp Hematol, 2000, 28:875-884.

[10] Dominici M, Le Blanc K, Mueller I, et al. Minimal criteria for defining multipotent mesenchymal stromal cells. The International Society for Cellular Therapy position statement. Cytotherapy, 2006, 8(4):315-317.

[11] Colter DC, Sekiya I, Prockop DJ. Identification of a subpopulation of rapidly self-renewing and multipotential adult stem cells in colonies of human marrow stromal cells. Proc Natl Acad Sci USA, 2001, 98(14):7841-7845.

[12] Samsonraj RM, Rai B, Sathiyanathan P, et al. Establishing criteria for human mesenchymal stem cell potency. Stem Cells, 2015, 33(6):1878-1891.

[13] Arufe MC, De la Fuente A, Fuentes I, et al. Chondrogenic potential of subpopulations of cells expressing mesenchymal stem cell markers derived from human synovial membranes. J Cell Biochem, 2010, 111(4):834-845.

[14] Mifune Y, Matsumoto T, Murasawa S, et al. Therapeutic superiority for cartilage repair by CD271-positive marrow stromal cell transplantation. Cell Transplant, 2013, 22(7):1201-1211.

[15] Gaebel R, Furlani D, Sorg H, et al. Cell origin of human mesenchymal stem cells determines a different healing performance in cardiac regeneration. PLoS One, 2011, 6(2):e15652.

[16] Yang ZX, Han ZB, Ji YR, et al. CD106 identifies a subpopulation of mesenchymal stem cells with unique immunomodulatory properties. PLoS One, 2013, 8(3):e59354.

[17] Rider DA, Dombrowski C, Sawyer AA, et al. Autocrine fibroblast growth factor 2 increases the multipotentiality of human adipose-derived mesenchymal stem cells. Stem Cells, 2008, 26(6):1598-1608.

[18] Sacchetti B, Funari A, Michienzi S, et al. Self-renewing osteoprogenitors in bone marrow sinusoids can organize a hematopoietic microenvironment. Cell, 2007, 131(2):324-336.

[19] Iinuma S, Aikawa E, Tamai K, et al. Transplanted bone marrow-derived circulating PDGFRalpha+ cells restore type VII collagen in recessive dystrophic epidermolysis bullosa mouse skin graft. J Immunol, 2015, 194(4):1996-2003.

[20] Mo M, Wang S, Zhou Y, Li H, et al. Mesenchymal stem cell subpopulations: phenotype, property and therapeutic potential. Cell Mol Life Sci, 2016, 73(17):3311-3321.

[21] Liang L, Li Z, Ma T, et al. Transplantation of human placenta-derived mesenchymal stem cells alleviates critical limb ischemia in diabetic nude rats. Cell Transplant, 2017, 26(1):45-61.

[22] Creane M, Howard L, O'Brien T, et al. Biodistribution and retention of locally administered human mesenchymal stromal cells: quantitative polymerase chain reaction-based detection of human DNA in murine organs. Cytotherapy, 2017, 19(3):384-394.

[23] Gnecchi M, Danieli P, Malpasso G, et al. Paracrine mechanisms of mesenchymal stem cells in tissue repair. Methods Mol Biol, 2016, 1416:123-146.

[24] Estrada R, Li N, Sarojini H, et al. Secretome from mesenchymal stem cells induces angiogenesis via Cyr61. J Cell Physiol, 2009, 219:563-571.

[25] Sze SK, de Kleijn DP, Lai RC, et al. Elucidating the secretion proteome of human embryonic stem cell-derived mesenchymal stem cells. Mol Cell Proteomics, 2007, 6:1680-1689.

[26] Gnecchi M, Zhang Z, Ni A, et al. 2008. Paracrine mechanisms in adult stem cell signaling and therapy. Circ Res, 103:1204-1219.

[27] Kuraitis D, Giordano C, Ruel M, et al. Exploiting extracellular matrix-stem cell interactions: a review of natural materials for therapeutic muscle regeneration. Biomaterials, 2012, 33:428-443.

[28] Khubutiya MS, Vagabov AV, Temnov AA, et al. Paracrine mechanisms of proliferative, anti-apoptotic and anti-inflammatory effects of mesenchymal stromal cells in models of acute organ injury. Cytotherapy, 2014, 16:579-585.

[29] Carayon K, Chaoui K, Ronzier E, et al. Proteolipidic composition of exosomes changes during reticulocyte maturation. J Biol Chem, 2011, 286:34426-34439.

[30] Zhang S, Chu WC, Lai RC, et al. Exosomes derived from human embryonic mesenchymal stem cells promote osteochondral regeneration. Osteoarthr Cartil, 2016, 24:6-11.

[31] Yu B, Kim HW, Gong M, et al. Exosomes secreted from GATA-4 overexpressing mesenchymal stem cells serve as a reservoir of anti-apoptotic microRNAs for cardioprotection. Int J Cardiol, 2015, 182:349-360.

[32] Ahmad T, Mukherjee S, Pattnaik B, et al. Miro1 regulates intercellular mitochondrial transport & enhances mesenchymal stem cell rescue efficacy. EMBO J, 2014, 33:994-1010.

[33] Hsu YC, Wu YT, Yu TH, et al. Mitochondria in mesenchymal stem cell biology and cell therapy: from cellular differentiation to mitochondrial transfer. Semin Cell Dev Biol, 2016, 52:119-131.

[34] Islam MN, Das SR, Emin MT, et al. Mitochondrial transfer from bone-marrow-derived stromal cells to pulmonary alveoli protects against acute lung injury. Nat Med, 2012, 18:759-765.

[35] Deans RJ, Moseley AB. Mesenchymal stem cells: biology and potential clinical uses. Exp Hematol, 2000, 28:875-884.

[36] Cheng L, Qasba P, Vanguri P, et al. Human mesenchymal stem cells support megakaryocyte and pro-platelet formation from CD34(+) hematopoietic progenitor cells. J Cell Physiol, 2000, 184:58-69.

[37] Angelopoulou M, Novelli E, Grove JE, et al. Contransplantation of human mesenchymal stem cells enhances human myelopoiesis and megakaryocytopoiesis in NOD/SCID mice. Exp Hematol, 2003, 31(5): 413-420.

[38] Sang HK, Hyoung SC, Eun SP, et al. Co-culture of human CD34[+] cells with mesenchymal stem cells increase the survival of CD34[+] cells against the 5-aza-deoxycytidine or trichostatin A induced cell death. Biochemical and Biophysical Research Communications, 2005, 329(3):1039-1045.

[39] Lazarus HM, Koc ON, Devine SM, et al. Cotransplantation of HLA-identical sibling culture-expanded mesenchymal stem cells and hematopoietic stem cells in hematologic malignancy patients. Biol Blood Marrow Transplant, 2005, 11:389-398.

[40] Muroi K, Miyamura K, Okada M, et al. Bone marrow-derived mesenchymal stem cells (JR-031) for steroid-refractory grade III or IV acute graft-versus-host disease: a phase Ⅱ/Ⅲ study. Int J Hematol, 2016, 103(2):243-250.

[41] Szabolcs P, Visani G, Locatelli F, et al. Treatment of steroid-refractory acute GVHD With mesenchymal stem cells improves outcomes in pediatric patients; results of the pediatric subset in A phase Ⅲ randomized, placebo-controlled study. Biology of Blood & Marrow Transplantation, 2010, 16(2):S298.

[42] Munn DH, Sharma MD, Lee JR, et al. Potential regulatory function of human dendritic cells expressing indoleamine 2,3-dioxygenase. Science, 2002, 297:1867-1870.

[43] Karussis D, Kassis I, Kurkalli BG, et al. Immunomodulation and neuroprotection with mesenchymal bone marrow stem cells (MSCs): a proposed treatment for multiple sclerosis and other neuroimmunological/neurodegenerative diseases. J Neurol Sci, 2008, 265:131-135.

[44] Le Blanc K, Rasmusson I, Sundberg B, et al. Treatment of severe acute graft-versus-host disease with third party haploidentical mesenchymal stem cells. Lancet, 2004, 363:1439-1441.

[45] Panés J, García-Olmo D, Van Assche G, et al. Expanded allogeneic adipose-derived mesenchymal stem cells (Cx601) for complex perianal fistulas in Crohn's disease: a phase 3 randomised, double-blind controlled trial. Lancet, 2016, 388(10051):1281-1290.

[46] Xu W, Zhang X, Qian H, et al. Mesenchymal stem cells from adult human bone marrow differentiate into a cardiomyocyte phenotype *in vitro*. Experimental Biology and Medicine Maywood NJ, 2004, 229 (7):623-631.

[47] Lee KD, Kuo TK, Whang-Peng J, et al. *In vitro* hepatic differentiation of human mesenchymal stem cells. Hepatology Baltimore Md, 2004, 40 (6):1275-1284.

[48] Jin K, Mao XO, Batteur S, et al. Induction of neuronal markers in bone marrow cells: differential effects of growth factors and patterns of intracellular expression. Experimental Neurology, 2003, 184 (1):78-89.

[49] Wakitani S, Imoto K, Yamamoto T, et al. Human autologous culture expanded bone marrow mesenchymal cell transplantation for repair of cartilage defects in osteoarthritic knees. Osteoarthritis and Cartilage, 2002, 10(3):199-206.

[50] Cuevas P, Carceller F, Garcia-Gomezl, et al. Bone marrow stromal cell implantation for peripheral nerve repair. Neurol Res, 2004, 26(2):230-232.

[51] Godwin EE, Smith RKW. Implantation of bone marrow-derived mesenchymal progenitor cells demonstrates improved outcome in horses with over-strain injury of the superficial digital flexor tendon. Equine Vet J, 2012, 44:25-33.

[52] Lisa AF, Alexander JT. Stem cells in veterinary medicine. Fortier and Travis Stem Cell Research & Therapy, 2011, 2:9.

[53] Belen C, Monica R, Joaquin S, et al. Hip osteoarthritis in dogs: a randomized study using mesenchymal stem cells from adipose tissue and plasma rich in growth factors. Int J Mol Sci, 2014, 15:13437-13460.

[54] Monsarrat P, Vergnes JN, Planat-Bénard V, et al. An innovative, comprehensive mapping and multiscale analysis of registered trials for stem cell-based regenerative medicine. Stem Cells Transl Med, 2016, 5(6):826-835.

第二章 不同组织来源间充质干细胞的生物学特性

间充质干细胞（MSC）作为机体的一种成体干细胞，最早在骨髓基质中发现，为造血干细胞（HSC）的发育分化提供必需的微环境。目前对骨髓间充质干细胞（BM-MSC）的研究已经颇为清楚，对 MSC 的鉴定主要集中于表面标记和多向分化能力，同时，MSC 除具有造血支持外，其免疫抑制功能和组织修复能力也越来越多地应用于临床。近些年来随着研究的深入，在脂肪、脐血、脐带、胎盘和胎儿肺等组织中均可以培养出间充质干细胞。MSC 并不是一群均一的群体，并且不同来源的 MSC 的表面标志及生物学特性（造血支持，多向分化为脂肪、骨、软骨，免疫抑制及组织修复等）也不尽相同。不同组织来源的 MSC 在生物学特性方面存在着较大差异。本章就不同来源 MSC 生物学特性做一简略概述。

第一节 不同组织来源间充质干细胞表面标志的比较

根据国际细胞治疗协会（ISCT）定义 MSC 的最低标准，人 MSC（hMSC）表达 CD73、CD90 和 CD105，不表达血液系统或内皮等标志如 CD45、CD34、CD14、CD11b、CD79α 和 CD19，正常培养情况下也不表达主要组织相容性复合体Ⅱ（MHCⅡ）类分子人白细胞抗原-DR（HLA-DR）。这一系列指标定义了 MSC 的基本表面标志。除了上面提到的一些表面标志外，MSC 还表达一些干细胞祖细胞、内皮或表皮细胞、T 淋巴细胞、B 淋巴细胞、自然杀伤细胞（natural killer cell，NK 细胞）、单核/巨噬细胞、粒细胞、树突细胞（dendritic cell，DC）、血小板和红细胞等细胞上表达的一些表面分子。这些标志包括细胞黏附分子（CD44、CD50、CD54、CD102、CD106、CD146、CD166）、整合素（CD11a、CD18、CD29、CD49a-f、CD51、CD61、CD104）、选择素（CD62E、CD62L、CD62P）、细胞因子受体（CD117、CD119、CD121a、CD123、CD124、CD126、CD127、CD140a、CD140b），或者凋亡或坏死相关受体（CD95、CD178、CD120a、CD120b）。

造成 MSC 表达不同的表面标志这个情况的原因有很多，包括 MSC 的来源、供者的年龄、提取的方法和扩增的条件等都会影响 MSC 的表面标志。其中 MSC 的来源不同是其中很重要的一个原因。目前围产期干细胞如胎盘及脐带来源的 MSC 由于其增殖迅速且材料容易获得、伦理限制少及不容易被病毒污染等特点，越来越受到研究人员的关注，它们的表面标志物与成人来源的 MSC 有明显的差异。例如，Oct-4、Nanog、Rex-1、SSEA-3、SSEA-4、Tra-1-6、Tra-1-81 表达于前三个月的胎儿血、肝和骨髓来源的 MSC，但是并不表达于成人骨髓来源的 MSC。另外，未经处理的骨髓细胞存在一小部分细胞表达 CD271，这些细胞体外扩增以后具有干细胞的高度分化潜能。但是目前没有从脐血中成功提取这类细胞的例子。Soncini 等[1]报道了从骨髓、羊膜、绒毛膜中提取 MSC，所有这些 MSC 的表达谱相似，但是在传代 4 次以上之后，表达的表面标志发生了变化，主要是 CD73、

CD105 和 CD166。BM-MSC 相对于其他两种细胞，表达更高比例的 CD73。而羊膜来源的 MSC 相对于另两种细胞表达更低比例的 CD105 和 CD166 阳性细胞。CD271 在骨髓中比例少于 1%，但是在羊膜中大约有 20%，在绒毛膜提取物中大约有 20%。CD271 的表达在细胞扩增过程中消失。胎盘绒毛膜、BM-MSC 和 BM-MSC 一样，不表达 CD3、CD14、CD19、CD34、CD45、CD31、CXCR4、Stro-1、HLA-DR、CD80 和 CD86，表达 CD13、CD44、CD73、CD90、CD105、CD106、CD29、CD49e 和 HLA-ABC [2]。UC-MSC 的 Nestin、CD54、SSEA-4 和 OCT-4 表达量要高于 BM-MSC[2]。胎盘绒毛膜来源的 MSC 高表达 CD106，BM-MSC 中度表达 CD106，而 CD106 在 UC-MSC 低表达，在脂肪间充质干细胞（AD-MSC）中没有检测到明显的 CD106 的表达[3]。

  MSC 的表面标志往往与特定的功能相关。例如，MSC CD106 表达量的高低可能与其免疫调节功能及血管再生的能力密切相关[3,4]。MSC 表面除表达一些成体干细胞的表面标志外，还表达一些与细胞的黏附、趋化及免疫调节功能相关的免疫标志。不同组织来源 MSC 的表面标志存在或多或少的差异，说明不同来源 MSC 的相关功能可能存在差异。不同来源 MSC 的表型差异也提示它们生物学特性之间可能存在更多的差异。

  不同来源 MSC 的免疫表型见表 2-1。

表 2-1 不同来源间充质干细胞（MSC）的免疫表型

标志	来源	表达	标志	来源	表达
CD1a	BM	−	CD29	BM	+
	AM	−		FB	+
	CM	−		UCB	+
CD3	BM	−		UC*	+
	AM	−		AM	+
	UC*	−		CM	+
	CM	−		ENDO	+
CD4	BM	−		ATSVF	+
CD5	BM	−		PMSC	+
CD9	BM	+		DPSC	+
	PLA	+	CD31	BM	−
CD10	BM	+		PLA	−
	BM	−		CM	−
	PLA	varies		CD31	
CD11a	BM	−		CV	−
CD13	BM	+		ATSVF	−
	UCB	+		AT	−
	PLA	+	CD33	BM	−
	UC*	+	CD34	BM	+
	PLA	varies	CD34	BM	−
	AM	+	CD34	mPB	−
	CM	+		FB	−
	CV	+		UCB	−

续表

标志	来源	表达	标志	来源	表达
	ATSVF	+		PLA	−
	PMSC	+		AM	−
CD14	BM	−		CM	−
	mPB	−		UC*	−
	FB	−		CV	−
	UC*	−		ENDO	−
	UCB	−		ATSVF	−
	AM	−		PMSC	−
	CM	−		DPSC	−
	DPSC	−	CD38	BM	−
	AT	−		UCB	−
CD15	BM	−		ATSVF	−
CD18	BM	−	CD44	BM	+
CD19	UC*	−		FB	+
CD25	BM	−		UCB	+
CD26	PLA	varies	CD54	BM	−
	PLA	+		BM	+
	UC*	+		AM	+
	AM	+		CM	+
	CM	+	CD56	BM	−
	CV	+	CD58	BM	+
	ENDO	+	CD59	BM	+
	ATSVF	+	CD61	BM	weak
	AT	+		BM	+
	PMSC	+	CD62E	BM	−
	DPSC	+	CD62L	BM	+
CD44H	BM	+	CD62P	BM	−
CD45	BM	+	CD63	PLA	+
CD45	BM	low	CD68	FB	−
	BM	−	CD71	BM	−
	UC*	-		UCB	−
	mPB	−	CD73	BM	+
	FB	−		FB	+
	UCB	−		UCB	+
	PLA	−		UC*	+
	AM	−		PLA	+
	CM	−		AM	+
	CV	−		CM	+
	ENDO	−		CV	+
	ATSVF	−		ENDO	+
	AT	−		ATSVF	+

续表

标志	来源	表达	标志	来源	表达
	DPSC	−	CD80	UC*	−
CD49a	BM	+	CD86	UC*	−
CD49b	BM	weak	CD90	BM	+
	BM	+		mPB	+
CD49c	BM	weak		UCB	+
	BM	+		PLA	+
CD49d	BM	−		UC*	+
	BM	+		AM	+
CD49e	BM	+		CM	+
	UC*	+		CV	+
CD49f	BM	weak		ENDO	+
CD50	BM	−		ATSVF	+
CD51	BM	weak		AT	+
	BM	+		CM	−
CD95	UCB	+		ATSVF	−
CD102	BM	+	CD140a	BM	+
	BM	−	CD140b	BM	+
CD104	BM	weak		ENDO	+
	BM	+	CD144	BM	−
CD105	BM	+	CD146	BM	+
	UC*	+		ENDO	+
	mPB	+		AT	+
	FB	+	CD166	BM	varies
	CB	+		BM	+
	PLA	+		UCB	+
	AM	+		UC*	+
	CM	+		AM	+
	CV	+		CM	+
	ENDO	+		ATSVF	+
	ATSVF	+		AT	+
	AT	+	CD178	BM	−
CD106	BM	varies	CD271	BM	+
	BM	+		PLA	low
	UCB	weak		ATSVF	−
	UC*	weak	CD340	BM	+
	AT#	−	CD349	BM	+
	CM#	+		PLA	+
CD109	BM	subset	3G5	AT	weak
CD117	BM	−	α-smooth muscle actin	FB	+
	BM	50%	actin	BM	+
CD119	BM	+	alkaline phosphatase	BM	+

续表

标志	来源	表达	标志	来源	表达
CD120a	BM	+		AT	+
CD120b	BM	+		DPSC	+
CD121a	BM	+	β-tubulin	BM	+
CD123	BM	−	bFGFR	BM	+
CD124	BM	+	BMPRIA	BM	+
CD126	BM	+	BS1 lectin	BM	+
CD127	BM	+	CCR2	BM	+
CD133	BM	−	CCR8	BM	+
	mPB	+	Collagen I	BM	+
	UC*	−		PMSC	+
	UCB	+		DPSC	+
	AM	−	vimentin	mPB	+
collagen II	BM	−	MyoD	BM	−
	BM	+		DPSC	−
	DPSC	−	neurofilament	BM	+
collagen III	BM	+		DPSC	−
	PMSC	+	osteocalcin	BM	+
	DPSC	+		DPSC	+
collagen IV	BM	+	PPAR-γ	BM	−
CXCR1	BM	+		DPSC	+
CXCR2	BM	+	prolyl-4-hydrolase	FB	+
CXCR3	BM	+	SSEA-4	BM	+
cytokeratin (pan)	BM	+		UC*	+
cytokeratin 18	BM	+		BM	+
cytokeratin 19	BM	+		PLA	+
D7-FIB	BM	+	STRO-1	BM	+
desmin	BM	+		AT	−
EGFR3	BM	−		ENDO	−
FGF-1	BM	+	TGFβIIR	BM	+
FGF-2	BM	+	TGFβIR	BM	+
	DPSC	+	VEGFR2	BM	−
FGF-3	BM	+	vimentin	FB	+
FGF-4	BM	+	vWF	BM	−
fibronectin	BM	+		BM	+
	mPB	+		FB	−
	FB	+		PMSC	−
HLA-DR	BM	−	W8B2	BM	+
	mPB	−	HLA-AB	BM	+
	UC*	−	HLA-ABC	BM	+
	FB	−		UC*	+

续表

标志	来源	表达	标志	来源	表达
	UCB	−		UCB	+
	PLA	−		PLA	+
	PMSC	−		CV	+
laminin	FB	+			

注：不同来源 MSC 的免疫表型见参考文献[5]，其中*见参考文献[2]，#见参考文献[3]；BM. 骨髓，mPB. 粒细胞集落刺激因子动员后的外周血，FB. 胎儿血，UCB. 脐血，PLA. 胎盘，AM. 羊膜，CM. 绒毛膜，CV. 绒毛膜绒毛，ENDO. 子宫内膜，ATSVF. 脂肪组织基质血管层，AT. 脂肪组织，PMSC. 胰间充质干细胞，DPSC. 牙髓干细胞，UC. 脐带

第二节　不同组织来源间充质干细胞增殖能力的比较

研究发现，不同组织来源 MSC 的体外扩增能力不完全相同。不同的微环境可能是差异产生的原因。Trivanovic 等[6]的研究表明，脐带和外周血来源的 MSC 相对于牙周韧带与脂肪来源的 MSC，具有更强的增殖能力和端粒长度。Li 等[7]比较了 BM-MSC 和 AD-MSC 的增殖能力，发现 AD-MSC 具有更强的增殖潜能，但两者的克隆效率相似。Kern 等[8]从每一份骨髓和脂肪样本中均能成功提取出 MSC，但脐血不能做到每一份都成功提取；尽管脐血中 MSC 的含量较少，但是它的扩增能力要比从骨髓和脂肪中提取的要强得多。

MSC 具有年龄特性，随着年龄的增长，BM-MSC 的数量和增殖能力出现明显的下降。围产期 MSC 如从胎儿脐带、胎盘绒毛膜和羊膜中成功提取出的 MSC 的增殖能力比 BM-MSC 强得多。胎肺来源的 MSC 在传代 40 多代以后没有明显的表型和增殖能力的改变[9]。将从脐带组织中分离得到的 MSC 与 BM-MSC 相比，脐带组织提取物具有更高的 MSC 量，并且脐带间充质干细胞（umbilical cord-derived mesenchymal stem cell，UC-MSC）具有更高的增殖能力，在传代 30 代之后，UC-MSC 增殖能力未发生明显改变。而 BM-MSC 在传代 6 次以后就表现出增殖能力减弱，倍增时间延长。Baksh 等[10]的研究也证明了 UC-MSC 具有比 BM-MSC 更强的增殖能力。Trivanovic 等[11]的研究也表明，UC-MSC 具有比外周血来源的 MSC 更强的成纤维细胞集落生成单位（colony-forming unit-fibroblast，CFU-F）形成能力。围产期 MSC 具有更强的扩增能力，而且经过扩增后 MSC 的基本性质没有发生变化，并且没有伦理学的限制，这样就可以从同一样本中得到大量的 MSC 而满足临床细胞治疗的需要。因而相对于增殖较慢的成体来源的干细胞，围产期 MSC 具有其独特的优势。

第三节　不同组织来源间充质干细胞分化能力的比较

MSC 向多种中胚层细胞的分化，包括向骨细胞、脂肪细胞、软骨细胞的分化，被认为是定义 MSC 的功能性指标。最近的研究表明，MSC 可能不仅具有向中胚层细胞分化的能力，也具有向内胚层和神经外胚层分化的能力，包括向神经细胞、肝细胞和内皮细

胞分化等。Baksh 等[10]发现 UC-MSC 也具有向脂肪细胞、骨细胞、软骨细胞分化的能力，并且在相同的成骨条件下，UC-MSC 具有更快的成骨速度，具有更多的碱性磷酸酶（alkaline phosphatase，ALP^+）细胞和骨节的形成。推测可能是由于在最开始提取时，UC-MSC 相对于 BM-MSC 具有更多的骨祖细胞。UC-MSC 与 BM-MSC 一样，能够向软骨细胞分化，因而可用于修复软骨的细胞。因此，UC-MSC 在诱导条件下也具有多向分化的能力，是可用于组织修复的理想细胞。

但不同组织来源的 MSC，由于其生理功能的差异，分化能力会有所不同。Kern 等[8]比较了骨髓、脐血和脂肪来源 MSC 的基本性质。骨髓、脐血和脂肪来源的 MSC 均表现出典型的 MSC 特征，即成纤维细胞样的形态、能形成 CFU-F、多向分化的能力并表达特定的表面分子。但这些 MSC 的分化能力是有差别的。BM-MSC 和 AD-MSC 具有向脂肪、骨、软骨细胞分化的能力，脐血间充质干细胞（umbilical cord blood-derived mesenchymal stem cell，UCB-MSC）虽然也具有向骨和软骨细胞分化的能力，但缺乏向脂肪细胞分化的能力。同样的，Wagner 等[12]也证明骨髓、脂肪和脐血来源的 MSC 在向骨细胞分化方面没有明显差异，但是 UCB-MSC 向脂肪细胞分化的能力很弱，而其他两种细胞向脂肪细胞的分化能力较强。Sacchetti 等[13]也证明在体内骨髓、脐带和骨膜来源的 MSC 可以向成骨分化，但肌肉来源的 MSC 在体内不具备成骨分化能力；而骨髓、脐带和骨膜来源的 MSC 不能向肌肉细胞分化，只有肌肉来源的 MSC 可以分化为类似于肌肉的细胞。Li 等[7]的研究表明，BM-MSC 与 AD-MSC 相比，有更强的成骨和成软骨分化能力，但两者成脂分化能力相似。Trivanovic 等[6]也证明脱落乳牙和牙周韧带来源的 MSC 表现出很强的成骨分化能力。Davies 等[14]比较了骨髓、脂肪和牙髓来源 MSC 的成骨分化能力，发现牙髓 MSC 的成骨分化能力要强于骨髓、脂肪来源的 MSC；并且牙髓来源的 MSC 与其他两种 MSC 不一样，它们成骨分化并不形成大的矿化结节，而是形成一层分散的矿化层。这些研究表明 MSC 的分化能力与组织来源相关。

第四节 不同组织来源间充质干细胞造血支持能力的比较

造血支持是 MSC 的又一重要的生理功能。体内正常造血是 HSC 和 BM-MSC 等细胞相互作用的结果。BM-MSC 通过直接接触和分泌大量的可溶性细胞因子维持体内正常造血。MSC 是造血微环境的细胞，除可以为 HSC 提供支持作用外，还分泌大量的可溶性细胞因子来调控 HSC 的增殖、分化、发育、迁移、凋亡。近几年来的研究表明：不同组织来源的 MSC 均可以对造血起重要的支持作用，只是不同来源的 MSC 由于其来源不同，膜表面的受配体和分泌的细胞因子不同，提供造血支持的能力不完全相同。我们实验室比较了成人骨髓、胎儿骨髓、胎儿胰腺和脐带来源 MSC 的体外造血支持能力，发现 4 种不同来源的 MSC 都能在体外支持造血，造血支持能力的强弱为成人 BM-MSC＞胎儿 BM-MSC＞胎儿胰腺 MSC＞UC-MSC。同时，我们比较了 4 种来源 MSC 对脐血 $CD34^+$ 细胞来源的巨核祖细胞向巨核细胞分化的影响，发现 4 种 MSC 均可以促进向巨核细胞的分化，其中成人 BM-MSC 的作用最为明显[15]。Nishikawa 等[16]将胎盘或脐带来源的 MSC 与 $CD34^+$ 细胞共培养，比较培养 5 天后 $CD34^+CD45^-$ 细胞比例及粒细胞-巨噬细胞集落生成单位（colony forming unit-granulocyte macrophage，CFU-GM）的形成能力，发现胎盘间充质干细胞（placenta mesenchymal stem cell，P-MSC）的造血支持能力要强于 UC-MSC。这可能与胎盘具有造血支持的能力有关。由于 MSC 在体内本身就是造血支持作用的关键细胞，因此这一结果也说明 MSC 的特性与在其来源组织的功能密切相关。

第五节　不同组织来源的间充质干细胞免疫调节能力的比较

MSC 主要的免疫学特性是低免疫原性和免疫调节能力。MSC 的低免疫原性表现为不表达 MHC II 类分子，不表达共刺激分子，以及不引起同种异源淋巴细胞的活化和增殖。在特定情况下，如一定干扰素-γ（IFN-γ）等细胞因子的处理下，MSC 会一定程度表达 MHC II 类分子；但如果炎症较强，MSC 会分泌免疫调节相关分子如前列腺素 E2（PGE2）和吲哚胺 2,3-双加氧酶 1（IDO1），主要表现为免疫调节功能。MSC 对多种免疫细胞都具有免疫调节功能，主要表现在抑制活化 T 淋巴细胞的增殖和 IFN-γ 的分泌、抑制 B 淋巴细胞的增殖和免疫球蛋白的分泌、抑制 NK 细胞的杀伤活性、抑制巨噬细胞促炎细胞因子的分泌、抑制单核细胞向 DC 的分化。MSC 的免疫调节活性主要是通过可溶性细胞因子和细胞间相互作用来发挥的，同时组织来源的 MSC 均具有免疫调节功能。MSC 对多种免疫细胞的调节如图 2-1 所示。

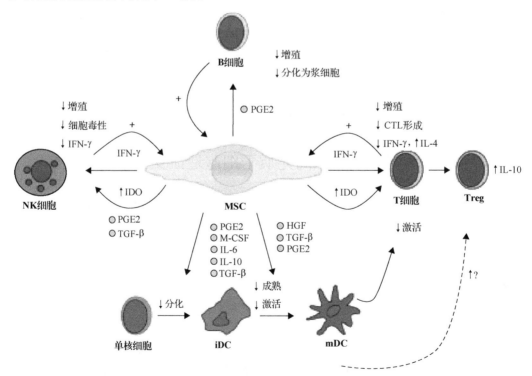

图 2-1　MSC 对多种免疫细胞的调节[17]（彩图请扫封底二维码）

IFN-γ. 干扰素-γ，IDO. 吲哚胺 2,3-双加氧酶，PGE2. 前列腺素 E2，TGF-β. 转化生长因子-β，MSC. 间充质干细胞，M-CSF. 巨噬细胞集落刺激因子，IL-6. 白介素-6，IL-10. 白介素-10，HGF. 肝细胞生长因子，CTL. 细胞毒性 T 细胞，IL-4. 白介素-4，Treg. 调节性 T 细胞，iDC. 不成熟树突细胞，mDC. 成熟树突细胞

Yoo 等[18]将 AD-MSC、UCB-MSC、脐带华通氏胶来源的间充质干细胞（Wharton's jelly-derived mesenchymal stem cell，WJ-MSC）和 BM-MSC 对 T 淋巴细胞的免疫调节功能进行了比较，发现 AD-MSC、UCB-MSC、WJ-MSC、BM-MSC 均可以抑制活化 T 淋巴细胞的增殖和 IFN-γ 与肿瘤坏死因子-α（tumor necrosis factor-α，TNF-α）的分泌。但不同来源的 MSC 其免疫调节功能并不完全相同，Bochev 等[19]发现 BM-MSC 和 AD-MSC 均可以抑制由佛波醇酯（phorbol-12- myristate-13-acetate，PMA）活化的 B 淋巴细胞中免疫球蛋白的分泌，但是 AD-MSC 对免疫球蛋白分泌的抑制作用比 BM-MSC 更加明显。Li 等[7]的研究也发现 AD-MSC 与 BM-MSC 相比，具有高的 IDO 活性。Bárcia 等[20]认为 UC-MSC 具有比 BM-MSC 更强的免疫调节能力，可以更有效地抑制 CD3 和 CD28 诱导的淋巴细胞增殖，并且能更有效地诱导 $CD3^+CD4^+CD25^+Foxp3^+$ 调节性 T 细胞（regulatory T cell，Treg）的产生。我们的研究并未发现 UC-MSC 和 BM-MSC 之间存在差异，但发现绒毛膜绒毛 MSC（chorionic villi-derived mesenchymal stem cell）与 UC-MSC 相比，免疫调节分子如环加氧酶 2（cyclo-oxygenase 2，COX2）和白介素-6（IL-6）等的表达量较高，并且能更有效地抑制 $CD4^+$ T 淋巴细胞中 IFN-γ 和 TNF-α 的分泌[3]。不同组织来源 MSC 的免疫活性可能由于 MSC 在不同组织中的活化状态不同而不完全相同，但是目前研究显示，不同组织来源 MSC 的免疫活性不同主要是表现为"强"或"弱"的差异，而非"有"或"无"的差异。

第六节 不同组织来源间充质干细胞促血管生成能力的比较

MSC 可以促进血管生成，主要通过两个可能的方式，即通过直接向内皮细胞分化和通过分泌内皮再生相关细胞因子来促进血管的生成。MSC 具有向内皮细胞风化的能力，体外实验也表明，在特定的培养条件下，MSC 可以分化为内皮样的细胞，并且具有形成血管样管状结构的能力。MSC 还可以分泌血管再生所需的一些重要细胞因子，如血管内皮生长因子（VEGF）、PGE2 和肝细胞生长因子（HGF）等，促进内皮祖细胞（endothelial precursor cell，EPC）分化的功能，参与内皮的形成。

不同组织来源的 MSC 具有不同的促血管生成能力。我们比较了骨髓、脂肪、脐带和胎盘来源 MSC 的促血管生成能力。体外诱导它们向内皮分化后，这 4 种 MSC 均表现出内皮的一样相关性质。与脂肪和脐带来源的 MSC 不同，骨髓和胎盘来源的 MSC 在体外可以直接形成管状结构；而体外实验也表明，骨髓和胎盘来源的 MSC 对内皮细胞血管形成有较强的促进作用。骨髓和胎盘 MSC 高表达多种促血管生成的细胞因子，如 BM-MSC 高表达 VEGF，而 P-MSC 高表达 PGE2 和 HGF[21]。因此，在将 MSC 用于促血管再生时，使用骨髓和胎盘来源的 MSC 可能会有更好的效果。

第七节 不同组织来源间充质干细胞迁移能力的比较

研究表明，MSC 可以向炎症部位或是损伤部位迁移，到达炎症、损伤部位后抑制炎症反应，并对损伤部位进行修复。Li 等[22]对骨髓、胎盘、脐带来源的 MSC 与迁移相关蛋白进行蛋白质组学比较，发现 UC-MSC 的 PAI-1 表达上升（PAI-1 抑制 MSC 的迁移能力），并且 UC-MSC 的迁移能力要弱于骨髓和胎盘来源的 MSC（分别为 UC-MSC 的 5.9 倍和 3.2 倍）。表明不同来源 MSC 的迁移能力是不同的，不同来源 MSC 在临床应用中的靶向治疗能力也是不同的。

第八节 小　　结

不同来源 MSC 在表面标志和功能上存在一些差异，表 2-3 比较了骨髓、脐带、胎盘和脂肪来源 MSC 的一些异同。除此之外，我们必须考虑到其提取的可行性，可提取的细胞数量及伦理学上的限制。BM-MSC 一般从成人中提取，其提取受年龄的限制，容易污染，可提取的细胞数量有限。胎肺来源的 MSC 受到伦理学的限制。从这个角度考虑，脐带和胎盘来源的 MSC 是一个理想的选择。

表 2-3 4 种间充质干细胞（MSC）在增殖、分化、造血支持、免疫调节等方面的差异比较

来源	增殖	分化			造血支持	免疫调节	其他
		成脂	成骨	成软骨			
骨髓	+	++	++	+++	+++	++	研究较为清楚，但是可提取的细胞数量相对较少，传代能力较弱（6~10 代）
脐带	+++	+	+++	++	++	+++	可提取的细胞数量明显多于骨髓，并且传代能力强，且无伦理学限制
胎盘	+++	++	++	++	+++	+++	可提取的细胞数量多于脐带和骨髓，无伦理学限制
脂肪	++	+++	+	++	++	++	来源较易、细胞数量相对较多，但是经体外培养优化后是否还能适应体内的环境有待进一步研究

总之，MSC 作为目前研究最多的一种成体干细胞，可以从多种组织中提取获得，但不同组织来源的 MSC 必须满足 MSC 的共性要求。与此同时，MSC 是一群混合的细胞，不同来源的 MSC 由于其来源不同，以及所用培养试剂和方法存在差异，因此其"干"性并不完全相同，这就导致了 MSC 的表面标志和增殖、分化、造血支持、免疫调节等功能并不完全相同，我们可以根据不同来源 MSC 的一些功能差异选择性地将其应用于临床，进而更好地为临床服务。

（杨舟鑫　韩之波）

参 考 文 献

[1] Soncini M, Vertua E, Gibelli L, et al. Isolation and characterization of mesenchymal cells from human fetal membranes. J Tissue Eng Regen Med, 2007, 1(4):296-305.

[2] Chen K, Wang D, Du WT, et al. Human umbilical cord mesenchymal stem cells hUC-MSCs exert immunosuppressive activities through a PGE2-dependent mechanism. Clin Immunol, 2010, 135(3):448-458.

[3] Yang ZX, Han ZB, Ji YR, et al. CD106 identifies a subpopulation of mesenchymal stem cells with unique immunomodulatory properties. PLoS One, 2013, 8(3):e59354.

[4] Du W, Xue L, Ying C, et al. VCAM-1$^+$ placenta chorionic villi-derived mesenchymal stem cells display potent pro-angiogenic activity. Stem Cell Research & Therapy, 2016, 7(1):1-13.

[5] Rojewski M T, Weber B M, Schrezenmeier H. Phenotypic characterization of mesenchymal stem cells from various tissues. Transfus Med Hemother, 2008, 35(3):168-184.

[6] Trivanović D, Jauković A, Popović B, et al. Mesenchymal stem cells of different origin: comparative evaluation of proliferative capacity, telomere length and pluripotency marker expression. Life Sciences, 2015,141:61-73.

[7] Li CY, Wu XY, Tong JB, et al. Comparative analysis of human mesenchymal stem cells from bone marrow and adipose tissue under xeno-free conditions for cell therapy. Stem Cell Research & Therapy, 2015, 6(1):1-13.

[8] Kern S, Eichler H, Stoeve J, et al. Comparative analysis of mesenchymal stem cells from bone marrow, umbilical cord blood, or adipose tissue. Stem Cells, 2006, 24(5):1294-1301.

[9] Fan CG, Tang FW, Zhang QJ, et al. Characterization and neural differentiation of fetal lung mesenchymal stem cells. Cell Transplantation, 2005, 14(5):311-321.

[10] Baksh D, Yao R, Tuan RS. Comparison of proliferative and multilineage differentiation potential of human mesenchymal stem cells derived from umbilical cord and bone marrow. Stem Cells, 2007, 25(6):1384-1392.

[11] Trivanović D, Kocić J, Mojsilović S, et al. Mesenchymal stem cells isolated from peripheral blood and umbilical cord Wharton's jelly. Srpski Arhiv Za Celokupno Lekarstvo, 2013, 141(3-4):178-186.

[12] Wagner W, Wein F, Seckinger A, et al. Comparative characteristics of mesenchymal stem cells from human bone marrow, adipose tissue, and umbilical cord blood. Exp Hematol, 2005, 33(11):1402-1416.

[13] Sacchetti B, Funari A, Remoli C, et al. No identical "mesenchymal stem cells" at different times and sites: human committed progenitors of distinct origin and differentiation potential are incorporated as adventitial cells in microvessels. Stem Cell Reports, 2016, 6(6):897-913.

[14] Davies OG, Cooper PR, Shelton RM, et al. A comparison of the in vitro mineralisation and dentinogenic potential of mesenchymal stem cells derived from adipose tissue, bone marrow and dental pulp. Journal of Bone and Mineral Metabolism, 2015, 33(4):371-382.

[15] Liu M, Yang SG, Liu PX, et al. Comparative study of *in vitro* hematopoietic supportive capability of human mesenchymal stem cells derived from bone marrow and umbilical cord. Zhongguo Shi Yan Xue Ye Xue Za Zhi, 2009, 17(5):1294-1300.

[16] Nishikawa E, Matsumoto T, Isige M, et al. Comparison of capacities to maintain hematopoietic stem cells among different types of stem cells derived from the placenta and umbilical cord. Regenerative Therapy, 2016, 4:48-61.

[17] Nauta AJ, Fibbe WE. Immunomodulatory properties of mesenchymal stromal cells. Blood, 2007, 110(10):3499-3506.

[18] Yoo KH, Jang IK,Lee MW, et al. Comparison of immunomodulatory properties of mesenchymal stem cells derived from adult human tissues. Cellular Immunology, 2009, 259(2):150-156.

[19] Bochev I, Elmadjian G, Kyurkchiev D, et al. Mesenchymal stem cells from human bone marrow or adipose tissue differently modulate mitogen-stimulated B-cell immunoglobulin production *in vitro*. Cell Biology International, 2008, 32(4):384-393.

[20] Bárcia RN. What makes umbilical cord tissue-derived mesenchymal stromal cells superior immunomodulators when compared to bone marrow derived mesenchymal stromal cells? Stem Cell International, 2015,(4-5):1-14.

[21] Du WJ, Chi Y, Yang ZX, et al. Heterogeneity of proangiogenic features in mesenchymal stem cells derived from bone marrow, adipose tissue, umbilical cord, and placenta.Stem Cell Res Ther, 2016, 7(1):163.

[22] Li G, Zhang XA, Wang H, et al. Comparative proteomic analysis of mesenchymal stem cells derived from human bone marrow, umbilical cord and placenta: implication in the migration. Proteomics, 2009, 720(1):51-68.

第三章 间充质干细胞制剂的质量控制

干细胞研究是当今生物医学界的热点,其成果广泛应用于再生医学、组织工程和替代医学领域。干细胞以其独特的生物学特性成为治疗某些重大疑难疾病的理想细胞来源。利用干细胞治疗某些疑难危重疾病,如恶性血液病、神经系统疾病、心血管疾病、自身免疫系统疾病等,其中一些确已初步取得令人满意的结果。近年来国际上干细胞研究取得了许多可喜的成果,大批的临床试验阐述了间充质干细胞(MSC)应用于临床的安全性和可行性。位于哥伦比亚的奥西里斯诊疗公司(Osiris)于2009年1月15日宣布获得美国食品药品监督管理局(FDA)对其干细胞治疗药物PROCHYMAL®的上市批准,用于治疗类固醇激素抵抗型急性移植物抗宿主病(aGVHD)和克罗恩病(Crohn's disease,CD)。迄今,澳大利亚、新西兰、加拿大、韩国、日本、印度、欧盟等国家或地区陆续批准了十多个干细胞药物上市。

MSC是属于中胚层的一类多能干细胞,主要存在于结缔组织和器官间质中,如骨髓、胎盘组织、脐带组织、脂肪组织、羊膜组织、牙髓等,MSC具有强大的增殖能力、多向分化潜能和免疫调节功能,具有分化为成骨细胞、软骨细胞、脂肪细胞、肌细胞、肝细胞、基质细胞等的能力。同时具备向炎症部位归巢、修复组织、调节免疫或炎症反应、造血支持等生物功能。MSC多次传代扩增后仍具有干细胞特性,还可以通过细胞间的相互作用与产生细胞因子抑制T淋巴细胞的增殖及其免疫反应,正是由于MSC具备这些免疫学特性,因此其在自身免疫系统疾病及各种替代治疗等方面具有广阔的临床应用前景。

MSC已被广泛用于血液疾病的治疗,尤其是造血干细胞移植(HSCT)。作为造血微环境主要细胞成分——基质细胞系的干细胞,MSC通过细胞对细胞的直接作用、分泌细胞外基质(ECM)及多种细胞因子,实现对造血的精细调控,支持造血干细胞(HSC)的生长,与HSC共同移植可促进HSC的植入。在HSCT中,MSC的作用主要包括促进HSC定植、治疗定植失败和移植功能不良,以及预防和缓解GVHD和再生障碍性贫血(aplastic anemia,AA)。

MSC具有低免疫原性并且能够逃避同种异体反应性T淋巴细胞和NK细胞的识别。MSC及其谱系的衍生物,表达中等水平的人白细胞抗原(HLA)主要组织相容性复合体Ⅰ(MHCⅠ),不表达共刺激分子B7-1、B7-2、CD40或是CD40配体。这些特征支持了MSC在宿主体内不易被免疫识别,使得MSC成为细胞治疗中理想的候选者,MSC在体内和体外被证实有免疫调节作用,能够抑制同种异体反应性T淋巴细胞的增殖。基于此特性,MSC已经用于GVHD的治疗,特别是应用于骨髓和HSCT治疗,除此之外还用于治疗炎症和自身免疫系统的功能障碍,如CD。

近年来,随着业界对MSC研究的深入,发现MSC对心肌再生[1]、重症肝病、卵巢早衰、胰岛细胞功能修复、糖尿病下肢缺血[2]、创伤愈合难等显示了良好的应用前景。

第一节 间充质干细胞质量研究的必要性

近年来我国干细胞的研究进展可谓突飞猛进。众多的研究机构开展了干细胞研究，一些医疗机构也纷纷涉足这一领域。但是，未经管理机构许可将干细胞直接应用于临床治疗的案例也不乏其数。某些医疗机构通过宣传干细胞可治疗糖尿病、心脏病、肝病等吸引患者，用于收取高额治疗费用，而仅仅是把人体血细胞抽出再回输，无法达到治疗目的，甚至很可能对健康造成危害。还有一些研究机构将实验室结果直接应用于临床，其细胞质量是否可控并未得到评估。由于研究机构间没有统一的规范，对细胞来源和细胞治疗产品没有统一的标准，这样，一则造成研究结果无法形成公认的结论，阻碍干细胞研究的进展，二则使得缺乏安全性和有效性评估的细胞进入临床，对人体造成伤害甚至危及生命，进而影响干细胞在医学领域的开发应用和其产业化进程。这些乱象的出现严重危害了公众的健康。国家卫生、药品行政管理部门于2012年1月发起了为期一年的干细胞临床研究和应用规范整顿，至2012年7月1日前禁止在治疗和临床应用中使用任何未经批准的干细胞，就是针对上述现象开展的规范性治理。只有建立起统一的标准，对组织来源、细胞筛选、制备过程和最后的产品进行严格的检验，获得符合标准的干细胞才具有质量上的可靠性，才可以用于临床使用，对没有达到标准的干细胞应严格禁止使用，这样才可以遏制干细胞滥用的趋势，防止损害公众利益的事件发生，让干细胞的研究和应用健康发展。

临床应用干细胞治疗必须保证细胞制剂的安全性、有效性和质量可控，其中质量可控性是安全和有效的基础与前提。本章针对干细胞中最具应用潜力的MSC，对其组织来源、细胞制备过程和产品进行质量研究，建立了一系列检验方法和检测标准来控制细胞产品的质量，建立供者细胞库及供临床用干细胞制剂质量标准，让人们在规范有序的环境下享受前沿科技带来的福祉。

第二节 干细胞治疗监管政策的演进

从我国卫生管理部门和药品管理部门对细胞治疗新药的管理演进过程中，不难看出我国对细胞治疗产品质量监管重视的一贯性。

1993年，负责药品注册审批的原卫生部药政管理局（1998年职能并入国家药品监督管理局）颁布了《人的体细胞治疗及基因治疗临床研究质控要点》，将人的体细胞治疗及基因治疗的临床研究纳入《中华人民共和国药品管理法》进行法制化管理。通知要求"今后，凡以人的体细胞治疗和基因治疗的单位，首先需按此质控要点要求向卫生部新药审评办公室申请，经专家委员会审查，卫生部批准后方可实施临床试验或临床验证"。这是我国卫生药监行政部门首次提出人体细胞治疗相关制剂按照新药管理，进行评审审批。

1999年，原国家药品监督管理局颁布了《新生物制品审批办法》，其中附件八《人的体细胞治疗申报临床试验指导原则》中明确了体细胞制剂按生物制品新药管理，并规定了体细胞的质量控制和质量检定相应的条款。

2003 年，原国家药品监督管理局颁布了《人体细胞治疗研究和制剂质量控制技术指导原则》，该指导原则较之前的两个文件，对细胞治疗产品的质量研究和制剂质量控制提出了较为系统的研究原则，这个技术文件也是国内研究机构在 2003 年之后开发干细胞新药、细胞治疗产品唯一可以参考的官方技术指导原则。我国已经申报的数个干细胞新药临床试验，在参考国际技术指导原则的同时，无不以此作为重要的研发依据，这份技术指导原则沿用至今。

以上历史沿革中的三份有关体细胞新药研究开发的技术指导原则，给予"体细胞治疗"一个相同的定义：指应用人的自体、异体或异种（非人体）的体细胞，经体外操作后回输（或植入）人体的治疗方法。尽管定义中未提及"干细胞"的概念，但是解读该定义，应涵盖目前所说的干细胞（人自体、异体的体细胞）治疗。

在 2003 年以后的十几年间，必须承认的是某些特殊情况下，干细胞技术的临床应用可能先于严格的新药开发程序完成，就是所谓的"医疗新技术应用"。但是对于公众而言，在干细胞科学研究尚在进行、临床应用条件尚未成熟及管理规则还有待制定的情况下，如何采取必要的措施保护自身的健康与经济利益，如何配合科学界与管理机构监督干细胞治疗的规范性，避免被不正当的商业手段所蒙骗，是一个值得认真考虑的问题。同时，我们必须清醒地认识到，从干细胞的研究到实际的应用还要经过漫长艰苦的研究发展过程。面对具有无限潜力的干细胞研究和应用前景，科学界、临床工作者、产业界及广大消费公众，都需要有一个现实的态度，需要耐心与理解，承担各自的责任，互相配合，以保证科学研究与临床应用能够在最有利的环境中进行。同时，亟待制定行业标准，改变不规范无秩序的现状，引导干细胞研究和应用走上健康发展的道路。

2015 年 8 月，国家卫生和计划生育委员会与国家食品药品监督管理总局联合颁布了《干细胞制剂质量控制及临床前研究指导原则（试行）》[3]，同时发布了《干细胞临床研究管理办法（试行）》。国家行政主管部门首次提出了"干细胞治疗"的定义，是指应用人自体或异体来源的干细胞经体外操作后输入（或植入）人体，用于疾病治疗的过程。明确了干细胞的范围：用于细胞治疗的干细胞主要包括成体干细胞、胚胎干细胞（embryonic stem cell，ESC）及诱导多能干细胞（induced pluripotent stem cell，iPSC）。

该指导原则立场鲜明地指出了从干细胞制剂的制备、体外实验、体内动物实验，到植入人体的临床研究及临床治疗过程的每一阶段，都需对所使用的干细胞制剂在细胞质量、安全性和生物学效力方面进行相关的研究与质量控制。干细胞研发机构应建立质量管理体系并在符合《药品生产质量管理规范》（*Good Manufacture Practice of Medical Products*，GMP）要求的基础上严格执行，为今后我国干细胞新药开发提供了现阶段应遵照的研究开发技术质量规范。

2017 年 12 月，国家食品药品监督管理总局发布了《细胞治疗产品研究与评价技术指导原则（试行）》，该指导原则针对当前我国细胞治疗产品研究迅猛发展和日趋激烈的态势，为更好地给相关科研机构和企业创造细胞产品研发环境并提供技术支持，经过广泛调研国外相关指南并充分征求业界意见，提出了细胞治疗产品在药学研究、非临床研究和临床研究方面应遵循的一般原则与基本要求。同时明确了本指导原则的适用范围，主要适用于按照药品进行研发与注册申报的人体来源的活细胞产品。对于按照医疗技术或其他管理路径研发的细胞治疗技术/产品，应执行其他相应管理规定及技术要求。该指导原则涵盖了多种细胞类型的产品，基于风险特性、风险程度的不同，在遵守传统药物安全性、有效性、质量可控性的原则下，充分考虑细胞治疗产品的特殊性，有效地指导细胞治疗产品的研发和注册申报。

第三节 间充质干细胞质量研究的主要内容

根据国家食品药品监督管理总局颁布的《人体细胞治疗研究和制剂质量控制技术指导原则》[4]、《多潜能间充质干细胞最低标准》（国际干细胞治疗学会，2006）[5]、《基于人类细胞的治疗产品指南》（欧洲药品审评委员会，2007）[6]、《21CFR1271 人用细胞组织和细胞的组织培养产品》（美国食品药品监督管理局）[7]、《药品生产质量管理规范（2015 年修订）》、《药物非临床研究质量管理规范》、《干细胞临床转化指南》（国际干细胞研究会 2009 年）、《干细胞制剂质量控制及临床前研究指导原则（试行）》（2015 年）、《干细胞临床研究管理办法（试行）》（2015 年）[8]、《细胞治疗产品研究与评价技术指导原则（试行）》（2017 年）[9]指导原则和文件，结合 MSC 的特点，研究建立其准确可靠、切实可行、能够保证产品安全性和有效性的质量标准和检定方法。

MSC 组织来源多样，供者的既往病史、家族史、性别、年龄、血型、组织相容性抗原等遗传背景复杂，体外分离、纯化、扩增和细胞制剂制备过程漫长且要引入抗生素、生长因子、抗体、胶原酶、蛋白酶、动物血清等多种外来成分，还可能污染细菌、支原体、病毒等外源因子，特别是制备新鲜细胞，一般要求 24h 内回输，这导致细胞制剂无法依照现行版《中国药典》进行制品放行检验，带来极大的急性毒性、异常毒性和病原体感染风险；体外操作过程可能会导致 MSC 的生物学效力、端粒酶、核型、组织相容性抗原、原癌抑癌基因、增殖能力、分化潜能等发生改变而带来免疫毒性和致瘤风险；在保存、运输、复苏、配制过程中，细胞制剂的存活率、生物学效力、均一性无时不在变化，难以保证 MSC 在临床应用中的安全性和有效性。

MSC 制剂的质量控制应通过对从细胞采集到输入受者体内全过程的质量控制来实现。要得到稳定均一的细胞产品，必须保证制备过程的稳定均一，因此必须建立严格的制造检定规程，加强操作过程关键点的质量控制，获得可靠的产品。

一、供者筛查

开发 MSC 制剂，首先应建立"供者的筛查标准"，根据供者来源，可分为异体供者和自体供者，供者的健康状况应记录并以书面报告的形式保存。对于异体干细胞的供者，必须严格排除乙型肝炎病毒（hepatitis B virus，HBV）、丙型肝炎病毒（hepatitis C virus，HCV）、人类免疫缺陷病毒（human immunodeficiency virus，HIV）、EB 病毒（Epstein-Barr virus，EBV）、巨细胞病毒（cytomegalovirus，CMV）、人类嗜 T 淋巴细胞病毒（human T-lymphotropic virus，HTLV）携带者和梅毒阳性者；排除易感染 HIV 的高危人群，如有吸毒史者、同性恋者、多个性伴侣者；排除各种结核病患者，如肺结核、肾结核、淋巴结核和骨结核患者等。对于自体干细胞供者，可根据具体情况具体分析，分析风险受益比，在具备有效控制或已存在的外源因子传播风险的措施后，可对供者的筛查标准进行调整。

2017年国家食品药品监督管理总局发布的《细胞治疗产品研究与评价技术指导原则（试行）》中，把供者细胞纳入主要生产用原材料的范围，这样参照生产用原材料进行质控和管理，监管程序清晰，便于执行。同时着重强调了供者细胞在整个工艺质量控制中的重要地位。

二、建立MSC供者细胞库技术标准

在种子细胞分离后，应尽早建立供者细胞库。生产用细胞的检验应满足安全性、质量可控性或有效性的基本要求。建立细胞库对细胞的构成成分进行系统的检测可以解决差异性的问题，使每批产品都有一个经过检定的共同起源。建立细胞库的优点，从制备的角度出发，可以由同一个种子细胞制备多批临床用细胞制剂，避免由供者的个体差异引起的产品间差异，影响产品有效性的评价。从安全的角度出发，建立细胞库可以有充裕的时间对种子细胞进行充分的检测，确保其安全性，并及时剔除不合格的种子细胞，避免造成更大的质量风险。因此这些细胞必须进行全面的质量鉴定，在细胞类型、纯度、活性、功能方面达到一定标准，才可以作为种子细胞，以保证扩增后的细胞的均一性和有效性。同时应兼顾种子细胞的数量，经过各个项目的质量检定，需要耗费大量细胞，检定后的种子细胞应保证有足够的数量以备后续制备之用。供者细胞库的标准应符合ICHQ5D"用于生物技术/生物制品生产的细胞基质的来源和鉴定"的一般要求[10]，并结合干细胞的特殊性，进行基因型分析，多向分化潜能、端粒酶、原癌基因的检测。供者细胞库的层级可根据细胞自身特性、扩增能力、体外操作风险和临床应用剂量综合考虑，设置合适的细胞库层级，既保证有足够的细胞数量，又能减少体外操作频次和时间，最大程度降低因体外操作导致的细胞状态、生长特性、生物功能等发生变化的风险。

（一）MSC分类鉴别

MSC分类鉴别是建立质量标准的基础，鉴别技术是制约MSC临床应用的关键因素，不能对细胞进行鉴别，从根本上无法确证所使用的细胞是否正确。MSC是一种在多种组织中存在的干细胞，目前采用多指标联合判断方式，这为MSC的鉴别带来了难度，也增加了临床使用MSC的质量不可控性。作为临床使用的MSC制品，首先需要明确的就是细胞的类型和纯度，这直接关系着用于治疗的MSC是否安全有效，因此MSC的分类鉴别是十分重要的基础检定项目。

目前比较通用的是国际细胞治疗协会（ISCT）于2006年推荐的一套定义MSC的最低标准。但对于不同来源MSC的特异性鉴别，不同实验室分别尝试应用各种不同的表面标记来定义MSC。例如，选取UC-MSC，采用流式细胞术[5]，研究结果表明UC-MSC的CD73、CD90、CD105呈阳性，阳性率不低于95%；CD11b、CD19、CD34、CD45、HLA-DR呈阴性，阳性率不高于2%。流式细胞术以其高灵敏度、高速度和在多参数测量、获取形态学信息等方面的优势在血液学、微生物学、分子生物学等领域中得到广泛的应用。应用这一技术，不仅可以分析细胞的表型，还可以研究细胞传代后的表型稳定性。

但是表型分析并不一定是唯一的干细胞鉴别方法，还可以通过细胞形态、遗传学、代谢酶亚型谱分析及特定基因表达产物检测等，对不同供体及不同类型的干细胞进行综合的细胞鉴别。

（二）核型分析

人体来源的正常细胞都具有 46 条正常染色体，但是这些细胞在体外培养的环境下其染色体核型通常具有不稳定性，而染色体的畸变必然导致细胞生物学特征的改变，最危险的改变就是细胞类型发生转化，从而导致细胞的生长不再受到调控，而具备了类似肿瘤细胞的无限增殖能力，这就是通常所说的恶性转化。通过对肿瘤细胞的核型分析，可以发现几乎所有肿瘤细胞中都存在非整倍体、染色体异位等染色体畸变现象，而这种现象的发生严重威胁着体外扩增的 MSC 的临床应用安全性。因此，迫切需要建立 MSC 染色体核型检测技术，并积累相关数据，以证明体外培养扩增的 MSC 的安全性。

关于核型分析，以筛查供者是否有遗传缺陷和染色体畸变，以及监测体外传代过程中 MSC 是否发生染色体畸变为目的。根据"人类细胞遗传学国际命名体制"（An International System for Human Cytogenomic Nomenclature，ISCN，2005）中的 G 显带法，获取不同世代数的 MSC，通过秋水仙碱处理，使之停留于分裂中期后，进行染色体数目检查和 G 显带分析，观察 20 个分裂相，染色体数目、核型均应无异常。

（三）端粒酶检测

人正常体细胞无端粒酶活性，而 85%～90%的恶性肿瘤可检出端粒酶活性。人细胞在胚胎发育早期，端粒酶呈活化状态，随后被抑制并一直处于失活状态，因而体细胞无端粒酶活性。当体细胞发生癌变时端粒酶再度活化，以逃避正常的细胞衰老过程，获得无限增殖潜能。因此，端粒酶再活化是恶性肿瘤一个显著的生物学特征[11]，需要检测粒酶活性以保证 MSC 的安全性。

（四）组织相容性抗原检测

MSC 临床使用属于同种异体细胞移植，虽然 MSC 本身免疫原性较低，在目前的临床使用中并不需要配型，但是通过分析以往的各种组织或细胞移植可以发现，移植治疗的效果直接受到 HLA 是否相合的影响。因此可以推测，HLA 分型对 MSC 移植的效果可能存在一定的影响。随着临床使用 MSC 的增加，病例数据的积累，使得分析 HLA 相合、半相合、不相合对移植效果的影响成为可能。因此建立 MSC 的 HLA 分型检测技术十分必要，这对日后提高 MSC 移植的安全性和有效性有着重要的意义。

（五）原癌抑癌基因检测

原癌基因（proto-oncogene）是细胞内与细胞增殖相关的基因，是机体维持正常生命活动所必需的，在进化上高度保守。当原癌基因的结构或调控区发生变异，基因产物增多或活性增强时，细胞过度增殖，从而形成肿瘤。体外培养的正常组织细胞也是一类具有增殖能力的细胞,其原癌基因也有一定程度的表达。确定正常组织原癌基因的表达水平,以此区分并预警发生恶性转化的 MSC 是一项十分重要的工作。

抑癌基因(antioncogene)也称为抗癌基因。早在 20 世纪 60 年代就有人将癌细胞与同种正常成纤维细胞融合,所获杂种细胞的后代只要保留某些正常亲本染色体就可表现为正常表型,但是随着染色体的丢失又可重新出现恶变细胞。这一现象表明,正常染色体内可能存在某些抑制肿瘤发生的基因,它们的丢失、突变或功能丧失,导致激活的癌基因发挥作用而致癌。抑癌基因正常时起抑制细胞增殖和肿瘤发生的作用。许多肿瘤细胞中均发现抑癌基因的两个等位基因缺失或失活,失去细胞增生的阴性调节因素,从而对肿瘤细胞的转化和异常增生失去抑制作用。检测抑癌基因正常与否,对确保 MSC 临床应用的安全性十分重要。

(六)群体倍增时间检测

与正常的体外培养细胞相比,发生恶性转化的细胞往往具有更强的增殖能力、更短的增殖时间、更快的增殖速度。由于内外因素的变化,体外培养的细胞也有可能发生衰老,衰老的细胞往往伴随着增殖速度的减缓,或失去增殖能力。

因此,有必要确定正常的 MSC 体外增殖的群体倍增时间范围,以此筛查出有可能发生恶性转化或衰老的 MSC。

(七)生物学效力实验

MSC 是一群具自我更新能力和多向分化潜能的细胞,但是在体外培养过程中,有可能随着传代次数的增加、培养条件的改变等而逐渐失去其多向分化能力。只有检测到 MSC 具多向分化能力,才能保证其具有干细胞的基本特征。无论何种来源的 MSC,都应进行体外其向多种类型细胞(如脂肪细胞、软骨细胞、成骨细胞等)分化能力的检测,以判断其细胞分化的多能性(multipotency)。这是判断干细胞具备生物学效力的基本特征。

除此以外,特定生物学效力试验应进行与 MSC 治疗适应证相关的生物学效力检验。例如,Du[12]等在研究不同组织来源 MSC 促进血管新生能力时发现,MSC 表达碱性成纤维细胞生长因子(basic fibroblast growth factor,bFGF)、血管内皮生长因子(VEGF)、肝细胞生长因子(HGF)和前列腺素 E2(PGE2),可促进内皮细胞增殖。当 MSC 作为药物治疗血管损伤性适应证时,可将这些细胞因子的表达作为该适应证生物学效力的检验指标。同理,在治疗免疫性疾病、神经性损伤疾病、皮肤损伤创伤等时,需要揭示干细胞治疗这些疾病的作用靶点和药理机制,开发与其适应证相关联的体外生物学效力检测方法,以定量考查 MSC 的生物活性。

(八)外源因子检测

临床应用的 MSC 制品,其纯净程度是十分重要的指标,这直接关系到 MSC 临床应用的安全性,而影响到纯净度的主要因素是外源因子,如病毒感染、微生物污染等。发生病毒感染的主要原因是作为供者的生物体在自然界暴露于各种感染因素之中,可能受到病毒的感染。此外，供者组织在手术获取、储运、分离培养过程中可能受到诸多外在因素的威胁。因此，对供者组织和 MSC 成品制剂进行病毒检测就显得尤为重要，这可以尽可能地避免由细胞治疗造成的感染性疾病的发生。

除采用不同细胞传代培养法检测病毒因子外，还要按照《中国药典》进行红细胞吸附实验，结果应为阴性。细胞制剂的支原体检查应采用培养法和指示细胞培养法两种方法，结果应均无支原体生长。

三、建立临床治疗用 MSC 制剂放行标准

作为一种回输于人体的治疗产品，其安全性和有效性必须得到保障。体外操作如采集、筛选、测试、扩增、储存、标记、包装和分发过程中都必须遵循对过程进行监控、对产品进行检定放行的原则，产品纯度、活力、活性和安全性指标都要求达到可应用的标准，才能达到治疗目的而不造成危害。

质量研究、方法的建立、质量标准的制定，其关键问题是要保证细胞的均一性和稳定性，为产品的安全性和有效性提供质量保证。

（一）细胞的纯度与均一性

制备基于 MSC 的产品的目的就是能够在一种具有充分的安全防范措施、无菌、可追溯的洁净环境下生产出数量充足的、均一的细胞。为了实现这一目标，就需要开发一种可以获得充足细胞数量的扩增方法，一种精确的诱导分化方法和经过验证的分析方法，以此来保证所生产的产品批与批之间具有很好的均一性。只要在符合 GMP 规范的环境条件下使用标准操作方案，以及开发有效的监测整个生产环节均一性的检测方法就可以大体上达到这种均一性要求。但是，对"均一性"的确认具有很大的困难，因为与小分子不同，细胞是一种有生命的有机体，并会随着时间的变化而发生改变。为了保证均一性，需要特别注意减少起始材料、分化调控方法及生产流程的差异，并且要对终产品进行严格的功能和成分检查。

在体细胞回输前，应该证明其纯度和均一性已达到可应用的水平。

1. MSC 的基因型分析

对于特定供者来说，其遗传基因型是特异的，而不同供者的脐带组织遗传基因型是不同的。所以，在组织分离、细胞培养过程中，同一供者来源组织应该独立分离、传代和保存，应严格防止不同产妇脐带组织在操作过程中发生交叉污染。在细胞原代培养、传代、制剂分装的不同阶段检测其基因图谱，证明细胞制备全过程中就是该细胞本身，来源单一，不存在任何与其他细胞的交叉污染。

采用聚合酶链反应（polymerase chain reaction，PCR）-毛细管电泳分析方法，检测 16 个人的短串联重复（short tandem repeat，STR）位点，将 STR 分型数据与 ATCC、JCRB 数据库进行对比分析[13]，该方法灵敏度高、特异性强，可用于生产用人源细胞株/系间的鉴别。在 MSC 的质量研究中，已将基因型分析列为常规的检测项目[14]。

2. MSC 的鉴别或特征性表面标志

对于经过数代培养的干细胞,在回输前,仍然应该进行细胞的鉴别实验,以保证在传代过程中细胞表型的稳定性及未被其他类型细胞污染或取代,确保同一株细胞在回输前与细胞库细胞的一致性。

(二)细胞制剂的生物学效力

细胞制剂在回输前,需要保证经过数代的体外培养,细胞生物学活性仍保持在一个稳定水平,以保证细胞产品的体内活性。生物学测定指采用生物学方法检测,以反映被测物的生物学特性的测定方法。根据具体情况,在产品进行常规质控时可采用其中的一种或几种。鉴于一种测定方法仅能反映制品某一方面的特性,且方法的变异一般较大,为更好控制产品质量,必要时需同时采用多种方法进行测定。

1. 细胞活率

细胞活率是反映细胞是否具有生物学功能的最直观指标,细胞治疗是将活细胞回输于人体,细胞总计数和细胞活率直接关系到治疗效果。细胞制剂是将培养收获的新鲜细胞与细胞稳定剂或细胞冻存保护剂配制成细胞悬液进行保存,新鲜保存的细胞一般在制备后 24h 内回输,而液氮冷冻保存的细胞需经过细胞复苏过程、一定的稀释再输入人体。一般采用台盼蓝染色法,也可以采用细胞快速分析仪自动检测细胞数目、浓度、活率、直径、体积、细胞碎片、细胞团等关于细胞状态的详细信息。对于细胞制剂,要特别注意细胞的标示量应为活细胞总数。

新鲜培养的细胞制剂的活率应不低于 90%,液氮冻存的细胞制剂复苏后的活率应不低于 80%。

2. 细胞生物学效力

根据开发的细胞制剂的药理作用研究结果及临床适应证,选择与药理作用相适应的特异性受体细胞测定细胞制剂生物学效力。细胞生物学效力实验一般能较好地反映制品的生物学活性,常用于各种生物制品的活性(效力)测定。

通过体外实验评价 MSC 的生物活性。MSC 进行临床治疗主要是针对 GVHD、系统性红斑狼疮(systemic lupus erythematosus,SLE)等免疫系统疾病,文献报道了 MSC 抑制免疫细胞的可能机制:激活的 T 淋巴细胞分泌包括干扰素-γ(IFN-γ)、肿瘤坏死因子-α(TNF-α)和白介素-1(IL-1α、IL-1β)等促炎性细胞因子,这些因子协同作用于 MSC 诱导其表达多种免疫抑制因子、趋化因子和一氧化氮(nitric oxide,NO),从而诱导 T 淋巴细胞和其他淋巴细胞趋向 MSC,局部高浓度免疫抑制因子、NO 可抑制免疫细胞的功能。

在生物学活性测定项目的选择上,可以选择 PGE2,激活的 MSC 分泌 PGE2,是介导 MSC 发生免疫抑制作用的主要因子。吲哚胺 2,3-双加氧酶(IDO)是促炎性细胞因子诱导产生的酶,其通过消耗免疫细胞激活所必需的色氨酸来介导 MSC 的免疫抑制作用。IFN-γ 是 T 淋巴细胞分泌的与 GVHD 发生过程相关的细胞因子,MSC 可以抑制 IFN-γ 的分泌从而有利于 GVHD 的治疗。sTNFRI 是 MSC 表达的肿瘤坏死因子受体(tumor necrosis factor receptor，TNFR），TNF-α 通过 TNFR 上调 PGE2 的分泌，诱导 IDO 的表达，促进 MSC 的迁移。TNF-α 是该机制中重要的细胞因子，通过结合于 MSC 细胞表面表达的受体（TNFRⅠ、TNFRⅡ）发挥功能[15]。

作为一种细胞学检测方法，使用 T 淋巴细胞增殖抑制实验检测 MSC 的生物学效力存在着系统误差较大、需要使用人外周血、结果重现性差的问题。因此，同时采用几种方法综合评价细胞制剂的生物学效力是必要的，以保证临床输注的 MSC 的有效性。

（三）动物实验

动物实验是指以整体动物为实验材料检测制品生物学活性（或效力）的实验方法，对于某些治疗用制品，由于其作用机制的原因，难以建立体外测量活性的方法，可以采用动物实验方法测定，但由于这类方法的变异一般相对较大，在进一步的研究中应尽可能以体外法代替。由于动物实验的成本高、周期长和变异性大，因此一般其仅用于成品检定。

细胞稳定表达特征性表面标志的水平也可以反映细胞制剂的生物学效力。

（四）添加物质的残留量

在当前的细胞分离技术手段下，不可避免地会将众多非人体来源的物质加入到细胞分离培养的过程中，这就为 MSC 的临床使用安全性埋下了隐患。目前最常使用的引入成分有抗生素、生长因子、抗体、胶原酶、蛋白酶、动物血清等。当前众多机构都在开发无血清培养基，也取得了较大的进展，特别是一些常用的工程细胞（CHO 细胞等）已经实现了无血清培养，但是由于 MSC 的特殊性，适用于 MSC 的无血清培养基尚未取得突破性进展，这就使得牛血清的使用成为必然。因此在用于临床治疗的 MSC 中，牛血清残留量直接影响 MSC 输注之后的安全性。另外，胰蛋白酶作为贴壁生长细胞培养过程中传代所必需的成分，也是无法替代的，由此对成品中胰蛋白酶含量进行检测同样重要。

（五）细胞制剂的理化性质

细胞制剂在输注前，理化性质与药品的注射剂检查要求一致，是不可或缺的控制项目，包括可见异物检查、装量检查、渗透压摩尔浓度、不溶性微粒检查。这些项目的控制是保证细胞制剂作为注射剂药品必须达到的基本安全性指标。

（六）细胞制剂的放行

细胞制剂放行前要保证放行检验项目全部完成，包括鉴别、纯度、无菌、支原体、外源病毒、细胞数和存活率、生物学效力、添加物质残留量、理化性质等。细胞制剂一般分为冻存制剂和新鲜制剂。冻存制剂是将细胞与一定浓度的细胞冻存保护剂混合配制，保存在 –130℃ 以下环境中，细胞可以稳定保存一年以上。故冻存制剂放行前有较充足的时间进行各项检验。

新鲜制剂一般以一定量的新鲜细胞与一定浓度的细胞保护剂配制形成，保存在4~8℃，这样的环境和条件下，细胞保持90%以上存活率的时间是非常有限的，一般要求在24h内输注。对于新鲜制剂，输注前没有足够的时间进行放行检验，这就给制备过程中的质量控制提出了更高的要求，制备的每个步骤需经过工艺验证，确保制备过程可控。无菌检查项目由终产品控制前移到过程控制，在细胞传代的每个过程，甚至在输入前24h需留取样品进行无菌检查。其结果虽然滞后，但仍然要作为质量记录保存。

（七）细胞制剂的稳定性

药物稳定性是指药物在生产制备后，在运输、储存、周转直至临床应用前一系列过程中质量变化的程度。稳定性研究的目的在于查明一种药品在各种不同环境因素影响下的不同表现，为药品辅料的筛选，生产工艺、生产条件、包装材料的选择和有效期的制定提供必不可少的科学依据。干细胞冻存制剂是将一定浓度的干细胞悬液，按比例与细胞冻存保护液混合配制，分装于适宜冷冻的内包装材料中，放于液氮中保存的制剂。作为一种不同于传统药物制剂的全新药物制剂形式，其具有非常强的特殊性，更需要通过细致全面的稳定性研究来指导制剂的开发、制备、储存、运输、使用等。耿洁等[16]的研究显示，MSC冻存制剂以市售包装形式在−196℃储存环境下经过36个月的储存后，包括理化性质、安全性、生物学活性三个方面的各项检测指标均保持稳定，未出现明显异常，也未表现出显著的变化趋势。

另外，干细胞冻存制剂在使用前需要进行细胞复苏，将冷冻状态的细胞复苏至活细胞状态，这个活细胞状态的稳定性对制剂的安全性、有效性非常关键，所以在稳定性研究中，既要研究干细胞制剂运输过程、长期储存过程中的质量变化情况，又要研究干细胞制剂在复苏后到输入人体前这段时间的质量变化情况，确保整个工艺链条中细胞质量的完整性和连续性。

第四节 小 结

随着我国干细胞治疗一系列规范化文件的相继出台，对干细胞药物的开发和临床使用提出了法制化、规范化要求；中国也正逐步加快与国际接轨的进程，将有条件加入"人用药物注册技术要求国际协调会议（International Council for Harmonization，ICH）"。在医药领域鼓励创新的大背景下，相信干细胞药物研究与应用将迎来前所未有的历史性机遇。同时随着药审改革的深入，如何保证药品质量与国际接轨是审评、审批、监管机构更加重视的问题。质量源于设计，在干细胞新药研究开发及临床转化过程中，着重研究干细胞的质量控制要点、研究方法，建立质量控制标准和完备质量体系，将我国的干细胞制剂推向科学、规范、健康的国内外市场，造福人类健康。

（耿 洁）

参 考 文 献

[1] Wu KH, Mo XM, Zhou B, et al. Cardiac potential of stem cells from whole human umbilical cord tissue. J Cell Biochem, 2009, 107(5):926-932.

[2] Liang L, Li Z, Ma T, et al. Transplantation of human placenta derived mesenchymal stem cell alleviates critical limb ischemia in diabetic nude rat. Cell Transplant, 2016, 26(1):45-61.

[3] 国家卫生计生委, 国家食品药品监管总局. 干细胞制剂质量控制及临床前研究指导原则（试行）. 2015 年 8 月 21 日发布. http://www.sda.gov.cn/WS01/CL1616/127242.html [2019-5-20]

[4] 国家食品药品监督管理总局. 人体细胞治疗研究和制剂质量控制技术指导原则. 2003 年 3 月 20 日颁布. http://www.cde.org.cn/zdyz.do?method=largePage&id=38 [2019-5-20]

[5] Dominici M, Le Blanc K, Mueller I, et al. Minimal criteria for defining multipotent mesenchymal stromal cells. The International Society for Cellular Therapy position statement. Cytotherapy, 2006, 8(4):315-317.

[6] 欧洲药品审评委员会. 基于人类细胞的治疗产品指南. 2008 年 9 月生效. http://www.ema.europa.eu/docs/en_GB/document_library/Scientific_guideline/2009/09/WC500003894.pdf [2019-5-20]

[7] 美国食品药品监督管理局. 21CFR1271 人用细胞组织和细胞的组织培养产品. 2007 年 8 月生效. http://www.fda.gov/downloads/BiologicsBloodVaccines/GuidanceComplianceRegulatoryInformation/Guidances/Tissue/ucm062592.pdf [2019-5-20]

[8] 国家卫生计生委, 国家食品药品监管总局. 干细胞临床研究管理办法（试行）. 2015 年 7 月 20 日发布. http://www.sda.gov.cn/WS01/CL0053/127243.html [2019-5-20]

[9] 国家食品药品监督管理总局. 细胞治疗产品研究与评价技术指导原则（试行）. 2017 年 12 月 22 日发布. http://www.sda.gov.cn/WS01/CL0087/220082.html [2019-5-20]

[10] ICH 指导委员会. 药品注册的国际技术要求. 质量部分. 周海钧主译. 北京: 人民卫生出版社, 2007:205-237.

[11] 龙军, 付曲波, 黄才斌, 等. 端粒、端粒酶及其基因转录调控与肿瘤. 赣南医学院学报, 2008, 3:467-468.

[12] Du WJ, Chi Y, Yang ZX, et al. Heterogeneity of proangiogenic features in mesenchymal stem cells derived from bone marrow, adipose tissue, umbilical cord, and placenta. Stem Cell Res Ther, 2016, 7(1):163.

[13] 孟淑芳, 吴瑜, 冯建平, 等. STR 图谱用于生物制品生产用人源细胞鉴别的研究. 中华微生物学和免疫学杂志, 2009, 7:636-641.

[14] 耿洁, 张磊, 韩之波, 等. 临床用间充质干细胞的质量控制研究. 中国医药生物技术, 2013, 3:225-230.

[15] Richard Y, Liu CF, Steven M, The role of type Ⅰ and type Ⅱ tumor necrosis factor (TNF) receptors in the ability of TNF-α to transduce a proliferative signal in the human megakaryoblastic leukemic cell line Mo7e. Cancer Res, 1998, 58:2217-2223.

[16] 耿洁, 张磊, 王斌, 等. 人脐带间充质干细胞注射液的长期稳定性研究. 中国医药生物技术, 2013, 3:230-234.

第四章 间充质干细胞的安全性研究进展

间充质干细胞（MSC）因具有多向分化潜能和造血支持、免疫调节与抗纤维化等功能[1]，已被广泛应用于再生医学、自身免疫系统疾病的临床研究之中。和造血干细胞（HSC）不同，MSC 是一类可以在体外扩增的干细胞。因此，虽然 MSC 在体内含量极少，但可以通过体外扩增制备大量的 MSC，以满足临床使用的需要。随着 2010 年美国食品药品监督管理局（FDA）授权人骨髓间充质干细胞（human BM-MSC，hBM-MSC）药物 Prochymal 作为"孤儿药"用于 1 型糖尿病的临床治疗；2011 年韩国批准自体 BM-MSC 上市，用于心梗和心衰治疗，2012 年批准异体脐血间充质干细胞（UCB-MSC）用于植入治疗关节软骨缺损；2012 年加拿大批准 hBM-MSC 药物 Prochymal 上市，用于治疗儿童移植物抗宿主病（GVHD）；2016 年 TiGenix 公司公布了脂肪间充质干细胞（AD-MSC）常规治疗或生物治疗无应答的克罗恩病（CD）患者复杂肛瘘Ⅲ期临床试验结果，显示 MSC 能够显著改善 CD 患者肛瘘的愈合[1]，2018 年 3 月，TiGenix 公司的 AD-MSC 药物以 Alofisel 的新名称被欧盟批准上市，用于瘘管对至少一种传统或生物疗法应答不良的非活动性/轻度活动性管内 CD 成人患者复杂性肛周瘘（complex perianal fistula，CPF）的治疗。随着多种 MSC 药物的上市，其已成为临床上最常用的"通用干细胞药物"。但干细胞研究者、临床医生和患者不仅仅关注其治疗效果，而且也关注其安全性。

第一节 间充质干细胞的体外安全性研究进展

早期多采用自体 MSC 治疗疾病[3-5]，但 MSC 具有低免疫原性，因此异体来源的 MSC 可以用于临床治疗。同时越来越多的异体 MSC 的临床研究与应用正在进行[6,7]。异体 MSC 已经成为临床使用的首选，这主要是因为 MSC 使用需要一定数量的细胞，异体 MSC 可以筛选优质的供者细胞，同时异体 MSC 易于大规模扩增和进行质量控制。由于每个人差异很大，自体 MSC 质量不易标准化，培养扩增需要的时间也不确定，有些甚至不能获得足够的治疗所需细胞。2008 年，Ning 等[8]对 10 位恶性血液病患者进行 MSC 与 HSC 共移植，MSC 的输入剂量大大低于其他研究报道，只有 2 例达到了预先设定的（1~2）×10^6/kg，有 4 例仅接受了（0.03~0.09）×10^6/kg。许多临床试验也已经证实同种异体 MSC 临床应用的有效性和安全性[2,6,9]。不管是自体或异体 MSC 用于临床治疗，都需要进行体外分离、传代扩增，以及制备成符合临床使用的制剂等体外操作程序。这些操作程序是否会导致 MSC 发生特性改变、自发转化成肿瘤细胞等安全问题，都是研究人员、临床医生和患者所关心的。

一、生物学特性

衰老的细胞会出现细胞体积增大、细胞内脂褐素颗粒沉积增多等形态学上的改变，会出现衰老相关β-半乳糖苷酶（senescence-associated β-galactosidase，SA-β-gal）活性[10]。长期传代扩增MSC，早期细胞可以保持一定的细胞形态，而随着代次的增加，细胞会表现出衰老的特征[11]，细胞体积会变大，并且出现SA-β-gal活性。虽然长期传代会导致细胞形态大小的变化，但不会影响MSC的贴壁特性、流式表型和多向分化潜能，即保持了MSC的基本特征[12]。虽然在体外MSC可以传代30代以上[13]，但体外培养的MSC并不是永生化的细胞，它会像正常的人体细胞一样，逐渐出现衰老的迹象。

二、端粒酶活性

端粒酶能延长缩短的端粒，提高体外细胞的增殖能力。肿瘤细胞通常表现为端粒酶高活性，端粒酶可能参与细胞恶性转化。*hTERT*是端粒酶的重要限速因子，对端粒酶的活性十分重要。Hahn等[14]的研究表明，转染*hTERT*和另外两个原癌基因（SV40大T抗原和*H-ras*癌基因）就可以使正常人的上皮细胞和成纤维细胞转变成肿瘤细胞。在体外长期传代的过程中，外源表达*hTERT*的MSC的基因组会变得更加不稳定，容易发生自发性基因组突变，进而具有能在免疫缺陷动物体内形成肿瘤的能力[15]。因此，检测供临床使用的MSC的端粒酶活性，能在一定程度上反映其在体内的致瘤风险，是反映其安全性的重要指标。我们在脐带间充质干细胞（UC-MSC）中检测到了*hTERT*的表达[13]。Jeon等[16]也以端粒重复序列扩增技术（telomeric repeat amplification protocol，TRAP）证实MSC确实具有微弱的端粒酶活性。虽然我们检测到了UC-MSC低表达*hTERT*，但是UC-MSC的端粒酶活性会随着传代次数的增加而逐渐减弱，甚至降到检测不出来的程度。MSC的端粒酶活性和肿瘤细胞或胚胎干细胞（ESC）相比存在很大差异。

三、染色体核型

研究表明，体外长期传代有可能导致ESC和MSC出现染色体核型的不稳定[17,18]，而核型异常往往是肿瘤细胞的重要特征[19]。因此，染色体核型稳定性是反映细胞制剂临床使用安全性的重要指标，有必要对供临床使用的MSC进行核型分析。我们对9株高代次的UC-MSC进行了核型分析[13]，对发现染色体异常的样本再取其低代次进行核型分析。8株UC-MSC在高代次时都表现出了正常的核型，而只有1株UC-MSC在高代次时检测到了10号染色体3体的现象（图4-1B），而这种染色体的异常在该株细胞的低代次中（图4-1A）未检测到。这说明该染色体数目的异常出现于体外长期培养过程中。因此，UC-MSC在体外培养的过程中，有可能会发生染色体数目异常，但其发生的频率比较低。

Bochkov和Nikitina等分别在脂肪和骨髓来源的MSC中也观察到了类似的结果[20-22]。虽然MSC在体外培养过程中发生染色体异常的频率很低，但还是应该在制备临床使用的细胞制剂时，通过定期的核型分析，排除体外培养过程中核型不稳定的样本，选择核型稳定的样本用于临床试验。

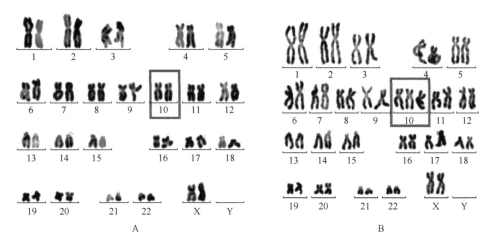

图 4-1 脐带间充质干细胞（UC-MSC）核型分析[13]

A. 样本在低代次（P3）进行核型分析时，是完全正常的核型，10 号染色体（方框）正常；B. 当经过了长期的体外培养，在高代次（P30）对样本进行核型分析时，10 号染色体三体（方框）

四、培养对细胞的影响

MSC 能广泛用于临床研究并开发为干细胞药物，这与其能大规模扩增特性是分不开的，既然需要扩增才可以满足临床使用，那么扩增过程的处理及培养添加物或培养方式对其安全性有何影响？

添加物对细胞培养过程中的细胞生长是必需的，最主要的是动物血清、自体血清[23]、人 AB 血清[24]、富含血小板血浆[25]和血小板裂解物[26]等。最常用的是胎牛血清（fetal bovine serum，FBS），含有丰富的贴壁因子和生长因子，对 MSC 的贴壁生长非常重要。目前，许多为 MSC 设计的无血清培养基含有牛血清白蛋白等动物来源成分。另外，无血清培养基促细胞贴壁生长的能力较差。因此，一些无血清培养基需要对培养表面进行包被，而这些包被材料大多为动物源或人源的材料。这些包被材料的成分是难以完全确定的，其有较大的批间差异。人源性的包被材料可能含有致病微生物，因其更容易在人与人之间传播，可能比动物源的材料具有更大的临床使用危险性。目前，我国规定同种异体人血清或血浆是不能用于干细胞制备的[27]。虽然无血清培养的 MSC 与添加牛血清培养的 MSC 具有类似的治疗作用[11,28]，但对无血清培养的 MSC 安全性研究尚少，其配方中通常会添加动物源或人源的蛋白质组分，在未证明其安全性的情况下，不建议过早用于临床治疗。建议选择安全的及质量可控的牛血清用于干细胞培养、制备疫苗等[29]。在目前阶段，建议优选质量可控的牛血清作为培养添加物。后续，无动物源组分、成分明确的无血清培养基应该是发展的方向。

现在许多研究人员为了提高 MSC 的治疗效应，开始对 MSC 进行相应预处理[30]，最常用的是低氧、细胞因子刺激[31-34]。2004 年 Xu 等用粒细胞-巨噬细胞集落刺激因子（GM-CSF）和白介素-4（IL-4）诱导体外培养的人胎儿 MSC 出现肿瘤样细胞群[35]，这也是我们不建议目前采用无血清培养基扩增 MSC 用于临床的原因之一，在无血清环境下需要细胞因子支持 MSC 的增殖，对 MSC 的安全性需要进一步研究。Ueyama 等[20]通过对 40 例体外扩增的 BM-MSC 的核型分析发现,在低氧条件下培养的 MSC 虽然可以获得更快的增殖速度,但会导致更多的染色体异常,包括染色体数目的异常和染色体结构的突变。因此,从安全性的角度考虑,有必要对供临床使用的 MSC 的核型稳定性进行定期、连续检测;而长期低氧培养或许不适合制备临床使用的 MSC。

在细胞制剂制备过程中可能用到的消化酶、抗生素等都可能影响制剂的安全,我们建议制备过程中所用消化酶最好为非动物源,如果是动物源,应该进行相关病毒检测;在细胞制剂制备中要严格遵守操作规程,尽量避免使用抗生素,例如,β-内酰胺类抗生素,抗生素残留可能导致极少数患者发生严重的过敏反应。如果细胞培养过程中使用过抗生素,一定要检测细胞制剂抗生素残留。

五、污染(微生物、细胞)

长期传代是一个长时间和多次操作的过程,其间不遵守操作规程、使用不合格耗材或试剂等都可能导致细胞发生微生物污染,这些污染对产品的安全性是致命的。细菌、真菌和酵母等污染相对容易辨别,多数情况下可以肉眼识别。而病毒、支原体等污染可能需要更复杂的检测才能确认,如通过定量聚合酶链反应(quantitative PCR,qPCR)检测。但污染不单单只有细菌、真菌和支原体等微生物,如果操作过程不规范,细胞间交叉污染也可能发生在 MSC 的长期传代之中[36-38]。这种污染可能是不同供体之间的细胞污染,或者是不同类型的细胞发生污染,甚至是肿瘤细胞污染。这个提示在制备临床级的 MSC 时需要严格遵守操作规则,不同个体细胞要独立处理,避免发生交叉污染。

六、恶性转化

MSC 在组织修复中起到重要作用,研究显示 MSC 也与癌症发生有关[39-41],Houghton 等[40]认为胃癌可能起源于骨髓来源细胞,也有研究证据显示骨肉瘤可能起源于 MSC。关于体外培养的原代细胞是否会发生转化,到目前还有很大的争议。1988 年,Boukamp 等[42]第一次报道 HaCAT 细胞系是由正常角质细胞自发转化而来的。2005 年,Wang 等[43]报道体外培养的 hBM-MSC 也可以出现自发转化。Miura 等[44]报道小鼠 MSC 会自发恶性转化。Serakinci 等[39]用端粒酶逆转录酶基因转染 MSC,转染后的细胞在体外长期培养也会发生恶性转化。这些认为 MSC 会发生恶性转发的研究多集中在早期。

细胞转化(自发或人为诱导)的发生至少需要 2 个遗传事件(突变),使得细胞获得永生化。早先有研究报道只有小鼠的 MSC 会发生自发转化,而人的 MSC 不会发生自发转化[45]。也有人认为向恶性转化的敏感性依赖人 MSC(hMSC)来源组织,认为来源于缺乏干细胞的脂肪组织的 MSC 比富含干细胞的 BM-MSC 更容易发生转化[46,47]。当然也有报道指出 hBM-MSC 也会自发转化,不过这一结论后来被否定[48]。Rubio 等[49]认为 MSC 自发转化时总是要先经历增殖停止阶段。但 Wang 等[43]认为自发转化可能早在分离 hMSC 时就已经出现。

报道称[50]自发转化的 hMSC 在形态上与低代次的 MSC 和衰老的 MSC 有明显区别,转化后的 BM-MSC 会增殖活跃,致密生长,显示出更高的核质比,有突出的核仁、更多的桥粒[51]。来源于骨髓的转化后 MSC 经常变成短梭形细胞,缺乏接触抑制生长特性,形态上会聚集呈悬浮样生长[43]。转化的 MSC 下调典型的早期膜标记 CD34、CD90、CD105 的表达，表型变成 CD133$^+$、CD34$^-$、CD45$^-$、CD105$^-$ 和血管内皮生长因子（VEGF）受体阳性[43,50]。自发转化总是伴随明显的转录组学改变。细胞通过衰老的"危机期"，发生转化，通过上调 c-myc 表达，抑制 p16 水平，获得端粒酶活性，同时 Ink4a/Arf 基因缺失，Rb 基因高度磷酸化[49]。转化后的 MSC 上调 Cdk1、Cdk2、Cdk4、Cdk6、周期蛋白（cyclin B1 和 D2）、RAD51a、DNA-PKcs、ERCC4、DNA 连接酶Ⅳ、DNA 聚合酶的转录和翻译及 VEGF 的表达，相反，转化后细胞会下调 cyclin D1、钙黏着蛋白 6（cadherin 6）、Snai1、ACTA2、基质金属蛋白酶-2（matrix metalloproteinase-2，MMP-2）、胶原蛋白（collagen Ⅰa2 和 Ⅴa2）、黏结蛋白聚糖 2（syndecan 2）、纤蛋白 2（fibulin 2）、结缔组织生长因子和成纤维细胞生长因子-2（FGF-2）的表达。与 hMSC 相比，转化后 MSC 表达更低水平的转化生长因子-β（TGF-β），这是由于在转化后细胞过度表达 Smad7、端粒酶和 c-myc，而导致表达 TGF-β 的信号相对不活跃。虽然转化后细胞有更高水平的端粒酶活性，但是它们经常为非整倍体、有不同的染色体移位。不同的转化后细胞有一致的核型变化，通常 3 号、11 号染色体发生移位，5 号染色体发生重排，偶尔出现一条等臂 8 号染色体。这些染色体的改变无论是由个体遗传因素导致，还是由体外培养引起，都需要进一步研究。因为这些抗原表型及转化后 MSC 形态明显发生改变，因此自发转化在早期就可以被检测出来。

虽然目前报道 MSC 发生自发恶性转化的文章不少，但也有人并不这么认为。有几个研究组已经报道 hMSC 经体外长期培养不会发生转化[13,46,47]。Bernardo 等[46]报道 hBM-MSC 不会发生自发转化，不表现出染色体核型或端粒酶活性异常。有报道认为培养条件是成体干细胞自发恶性转化的关键因素，适当的培养条件可以避免成人干细胞发生恶性转化的趋势。中山大学的 Guan 等[52]进行了人皮肤来源 MSC 体外长期培养的生物学安全性研究，对培养到 21 代之前的细胞进行了表型、核型、人白细胞抗原（HLA）及体内致瘤性实验检测，认为人皮肤来源 MSC 经过体外长期培养后用于体内移植是安全的。我们的研究也显示 UC-MSC 经体外长期传代后会发生细胞衰老、基因组不稳定，但不会发生恶性转化[13]。Gou 等[53]认为体外扩增的小鼠 BM-MSC 用于研究使用是安全的，他们研究发现 C57 小鼠来源的 BM-MSC 在体外稳定培养 50 代后，BM-MSC 并没有显示发生自发转化的标志。但长期培养的小鼠 BM-MSC 会失去其成骨和成软骨分化能力。不论是用成脂肪、成骨、成软骨的诱导培养基还是用非诱导培养基培养，它们都可以分化成脂肪细胞。

非常有趣的是，Rosland 等[54]之前报道 hBM-MSC 可发生自发转化，而且分离扩增在两个独立实验室进行。由于有报道显示 18%~36%的细胞系有交叉污染[55]，他们就用 DNA 指纹图谱和/或短串联重复（STR）分析比较正常 MSC 和它们的对应转化细胞。分析结果显示在一个实验室，转化后的 MSC 是与人纤维肉瘤或骨肉瘤细胞系发生交叉污染，而在另一个实验室是与两个胶质瘤细胞系发生交叉污染。他们的意见是培养原代细胞时要遵循严格的操作规程，包括培养 MSC。

在最近又有一篇有类似研究结果的文章发表。Garcia-Castro 实验室在 2005 年发表[50]关于 AD-MSC 发生自发恶性转化的原创文章后，他们继续鉴定 hMSC 自发转化后的特征，意想不到的是,他们没有再获得一个新的 hMSC 转化事件(大约 15 个新脂肪组织样本,包括成人和儿童来源,经过 25 次独立传代培养)。所有的长期传代细胞都在经历了一个明显的衰老过程后衰竭了,这与我们的研究结果类似[13]。他们回过头去分析原来的转化样本,通过 STR 分析转化后 MSC 和转化之前的 MSC,令人惊讶的是没有一个转化后 MSC 群落与原始 hMSC 样品具有相同的 STR 表型,而转化后 MSC 群落之间的 STR 十分相似。另外,当他们将转化后 MSC 的 STR 表型与人肿瘤细胞系相比较时,发现它们与一个对照细胞系 HT1080 十分相似,HLA 基因表型分析证实这个现象。最后他们重复了新、老样品的核型分析,发现虽然在共培养后筛选的克隆及样品中 MSC 存在高度的异质性,但是转化后 MSC 中存在的最多特征基因变化[der(5) and t(3;11)]与他们实验室拥有的 HT1080 细胞系相同。这些结果对他们实验室之前报道的 MSC 出现自发转化现象的最合理解释是人为因素导致培养的 MSC 被 HT1080 细胞交叉感染。

一般来说肿瘤细胞增殖能力强,但为什么发生潜在污染的肿瘤细胞在早期培养过程中没有明显增殖,而 3~4 个月后开始增殖,这个现象很难解释。Garcia-Castro 实验室又研究了长期体外培养过程中 hMSC 和 HT1080 之间的功能干扰。他们设计了一组实验,将 HT1080-增强型绿色荧光蛋白(enhanced green fluorescent protein,EGFP)细胞和来源于不同供者的 hMSC 混合(3 种稀释度:1/10 000、1/100 000 和 1/500 000)培养至结束或出现明显的 HT1080 样细胞。每周使用流式细胞术、PCR 和荧光缩微胶片检测两次 EGFP。在大多数共培养过程中,他们检测到了 HT1080-EGFP 细胞出现的生长延迟现象。结果表明 HT1080-EGFP 细胞的出现与理论上的时间比较延迟了,并且在某些例子中 HT1080 在生长之前是无法检测的。因为极少量的 HT1080 污染在培养的早期可以被 hMSC 培养所抑制,并且可能在几个月中都无法检测到。目前的研究结果也显示,MSC 在体外的确会抑制一些肿瘤细胞的增殖[56,57]。随着培养时间延长,一直到 hMSC 开始衰老并开始凋亡,HT1080 细胞才能够按照其正常的增殖速度生长[37]。

第二节 间充质干细胞的体内安全性研究进展

目前在动物体内研究 MSC 应用的文章很多,但在动物体内研究其安全性的文章却很少。Bartholomew 等[58]报道,给狒狒注射与供者匹配的 MSC,可以延迟对第三方供者皮肤移植物的排斥反应的发生。来自临床试验的数据和病例报道均提示 MSC 能治疗 GVHD。但 MSC 在产生这些治疗效应的同时在体内能存留多久?Devine 等[59]给未预处理过的狒狒静脉输注同种异体 MSC,9 个月后可以在不同组织中检测到异体 MSC。类似的在未预处理过的啮齿类动物模型中,小鼠 BM-MSC 植入大鼠骨髓和心肌中至少存活 13 周[60]。不过其他一些研究结果显示 MSC 在动物体内留存的时间并没有那么长[61]。在期待 MSC 在体内应用时起到最大的治疗效应的同时,我们也会担心,这些细胞进入体内时是否会发生不必要的分化,是否会导致肿瘤发生;在使用异体 MSC 时这些细胞在体内是否会长期存在,是否会导致严重的免疫反应。

Bartholomew 等[58]将异体 MSC 移植入未预处理过的狒狒体内,研究 MSC 在狒狒体内的存活情况,并评价初次和再次输注细胞后引发的细胞免疫与体液免疫反应。同时检测主要免疫参数，如丝裂原反应、外周血单个核细胞（peripheral blood mononuclear cell，PBMC）表型、免疫球蛋白水平，确定多次注射临床应用剂量的异体MSC的安全性。采用静脉输注和肌肉注射给药途径，后者可最大限度地诱发免疫反应。他们发现，在注射MSC之后，大部分受者的异体反应性T淋巴细胞对供者同种异体抗原反应减弱，虽然诱发了同种抗体反应。多次注射MSC不影响整体免疫系统的水平，因为反映整体免疫水平的参数包括T淋巴细胞对丝裂原的反应、PBMC分类、免疫球蛋白水平及异常抗体（抗核抗体）的产生。在长达6个月的研究期间内，MSC对细胞和体液免疫比较特异的效应不会导致明显的健康问题，说明本项研究中使用的治疗药物是安全的。有半数狒狒在注射后一个月仍然能够发现MSC，比预期的分化MSC会在1~2周被排斥的时间要长许多。注射6个月后进行尸检，所有的肌肉标本均未检测到供者MSC。

注射MSC的狒狒会产生同种抗体。本项研究中，大多数狒狒针对注射的同种异体MSC产生了同种抗体。唯一的例外是其中一只狒狒没有产生任何针对来自异体供者同种抗原的抗体。大多数动物在第二次注射MSC时没有发生抗体滴度的暴发效应（即抗体滴度保持稳定）。这种效应可能是种系相关的，因为静脉输注同种异体MSC的大鼠，在注射后4周并未产生同种抗体[62]。

总之，Bartholomew等[58]认为静脉输注临床级别的、高剂量的、同种异体MSC，随后再肌肉注射同一种同种异体MSC或第三方同种异体MSC，不会影响受体动物整体的健康和免疫水平。在半数正常受体动物的肌肉活检中发现了移植的同种异体MSC，时间持续至少一个月。MSC的免疫调节活性是保持异体MSC存活的原因。实验中，长时间后未能在肌肉活检中发现MSC，可能是因为细胞迁移，荧光标记丢失，细胞免疫或抗体导致了慢性排斥反应，或者是在正常的无损伤的组织环境中植入较差。选用的狒狒并不是组织修复模型，可能在损伤模型中，MSC能够被损伤部位环境吸引而利用并长期存在。

赵春华教授研究组[63]体外扩增Flk-1$^+$CD31$^-$CD34$^-$MSC后静脉输注，进行安全性的临床前和I期临床试验。恒河猴和人的Flk-1$^+$CD31$^-$CD34$^-$MSC被分离并最高传到6代。Flk-1$^+$CD31$^-$CD34$^-$MSC有正常二倍体核型。扩增的细胞被移植入恒河猴或志愿者体内，受者的生命体征正常。在注射细胞前后分析血、骨髓、肾及肝功能发现没有明显变化。通过对雌猴体内Y染色体性别决定区（sex-determining region of Y-chromosome，SRY）序列分析，显示MSC可以归巢并定植到骨髓。注射前后淋巴细胞群没有发生明显变化。这些结果显示静脉注射Flk-1$^+$CD31$^-$CD34$^-$MSC后没有明显不良反应。

对于供临床使用的干细胞来说，进行来源控制是必需的。对病毒、细菌等微生物控制大家基本都可以认识到，但对细胞遗传学稳定性方面相对关注较少。就以P53基因为例，其对肿瘤发生是关键的，大约50%的人类癌症有P53突变[64]。P53基因敲除鼠的BM-MSC具有更强的增殖能力，更短的倍增时间，更强的克隆形成能力，P53不仅仅影响MSC的增殖，还影响其分化。同时，MSC基因组不稳定，可以表达更高水平的c-myc mRNA。P53基因敲除鼠的BM-MSC在体外培养300天后，对免疫缺陷鼠移植可100%成瘤，而对照鼠的BM-MSC在体外培养350天后，也不能在免疫缺陷鼠体内成瘤[65]。

我们[61]也进行了hMSC在动物体内安全性的研究。组织分布和定植研究表明，UC-MSC单次静脉注射NOD/SCID小鼠后，24h后各组动物的肺和股骨（骨髓）中有

MSC 分布；3 周后可检测到股骨和心脏中有 MSC 分布；而 8 周后各组织中均未检出 MSC。同时，我们考察了长期传代 MSC 的动物体内安全性[13]，研究的 5 株高代次的 UC-MSC，不论其在体外长期传代过程中是否存在基因组的变异，即使在基质胶（matrigel）的支持下，也不能在 SCID 鼠体内形成肿瘤。而接受 ESC 皮下注射的 SCID 鼠，在注射部位皮下形成了肿瘤，病理分析为典型的畸胎瘤（图 4-2B~F）。

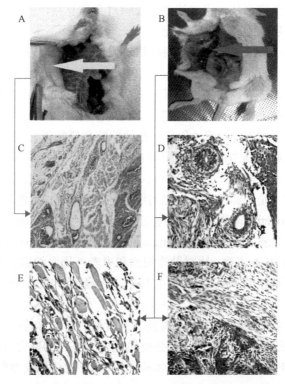

图 4-2　长期传代后的脐带间充质干细胞（UC-MSC）在 SCID 鼠体内的致瘤性实验
（彩图请扫封底二维码）

高代次人脐带间充质干细胞（hUC-MSC）皮下注射 SCID 鼠，在其注射部位和全身其他地方都未发现肿瘤形成迹象（A、C）；而接受胚胎干细胞（ESC）皮下注射的 SCID 鼠，在其注射部位皮下形成明显的肿瘤（B），病理分析该肿瘤为明显的畸胎瘤；D. 肠（内胚层）；E. 肌肉（中胚层）；F. 神经纤维（外胚层）

从临床角度来看，以上这些结果均支持 MSC 作为一种安全的"通用细胞"使用，即来自一名供者的 MSC 扩增后，可用于多名患者进行组织修复，特别是当自体分离和扩增的 MSC 不可用时尤为有利。

第三节　目前临床上间充质干细胞治疗的安全性问题

MSC 用于临床疗效是临床医生及患者所关心的，但其安全性也是他们关心的重点之一。在治疗过程中，是否会发生与治疗目的无关的不良反应，是否会引起肿瘤的发生等。

一位 46 岁女性，被诊断为系统性红斑狼疮（SLE）20 多年，后面发展成终末期肾病，2006 年她选择到私人干细胞诊所接受治疗。医生将由粒细胞集落刺激因子（G-CSF）动员后的外周血干细胞直接注入其肾中。根据发表在《美国肾脏病学会》杂志上的病例分析报告显示[66]，患者在接受治疗后肾注射部位出现了奇怪的肿块。来自泰国 Chulalongkorn 大学医院的 Duangpen Thirabanjasak 的报告指出，这些肿块几乎可以肯定来自于注射的干细胞。一年后她的一个肾坏死，两年后她因此而死亡。对于这些肿块的其中一个解释是：该患者肾已经严重受损并缺氧，干细胞被注入这个缺氧区域后，似乎开始构建新血管，从而为身体重建新的供应系统，也许这些新血管对身体有益，但它出现在了不合适的地方，是否最终会变成癌细胞我们还不得而知。

PLoS Medicine 期刊上一篇病例报告[67]描述了人 ESC 移植出现的一个少见的不良反应。以色列 Sheba 医学中心的 Ninette Amariglio 和 Gideon Rechavi 及他们的同事报告了一个患有一种罕见基因病共济失调性毛细血管扩张症（ataxia telangiectasia，AT）的男孩病例，他在莫斯科接受了人胚胎来源神经干细胞（neural stem cell，NSC）的移植，这些细胞来源于 8~12 周的胚胎[68]。NSC 治疗 4 年后，在他的脑和脊髓里有异常增生物。

以色列男孩大脑和脊髓中的肿瘤是否是由注射入的人胚胎来源 NSC 引起的呢？以色列研究人员对切除的肿瘤组织进行了检测，肿瘤类似于胶质瘤。作者分析了这个肿瘤并且证实了这个肿瘤不是来源于患者自己的组织。运用 X、Y 染色体探查荧光原位杂交，运用 PCR 检测 Amelogenin 基因 X 和 Y 特定等位基因，运用 MassArray 检测 ATM 患者特定的突变和严重的单核苷酸多态性（single nucleotide polymorphism，SNP），运用 PCR 检测小随体多态性和 HLA 表型，比对肿瘤细胞和患者外周血细胞。分子和细胞基因研究显示，肿瘤来自于所移植的 NSC 而非宿主本身。小随体和 HLA 分析显示，肿瘤至少来源于 2 名供者。

从上述两个例子我们可以看出，不规范的干细胞治疗蕴含着巨大的风险，在开展干细胞治疗之前，系统的安全性评价是十分必要的。关于 MSC 的研究报道国内外已有很多，在此不一一列举，以下我们各给出一个自体和异体 MSC 治疗的典型临床研究结果供大家参考。

日本大阪市立大学医学研究生院骨科的 Wakitani 等[69]随访了 1998 年 1 月至 2008 年 11 月共 41 位患者，在他们那接收了共 45 个关节用自体 BM-MSC 移植修复软骨的病例，随访时间最长的病例长达 11 年零 5 个月。他们检查了患者是否有肿瘤和感染，发现既没有局部感染也没有肿瘤发生，这强有力地证明了使用自体 BM-MSC 移植是安全的。

研究发现自体和异体 MSC 可用于治疗急性 GVHD（aGVHD），但个体化的产品在制造上很麻烦，也很难标准化。美国杜克大学医学中心儿科的 Prasad 等[70]牵头了一个多中心临床研究，从 2005 年 7 月到 2007 年 6 月期间，5 个移植中心协作使用预制的、通用型的 MSC 制剂，治疗了 12 例儿童 3 级和 4 级的 aGVHD。所有患者均有 3 级或 4 级的肠道综合征，一半有肝脏和/或皮肤损伤。所有患者使用激素治疗无效，其中 3 例还使用其他免疫抑制治疗。2 例患者采用 8×10^6cells/kg，其他采用 2×10^6cells/kg，静脉输注，每周 2 次，共 4 周。随后几周出现局部和混合反应是可以接受的。该 MSC 制剂不需要 HLA 配型，hMSC 开始使用的中位时间是在移植后 98 天（范围 45~237 天），总共用了 124 个剂量，平均每位患者用了 8 个剂量（每名患者 2~21 个剂量）。总体而言，7 例（58%）患者完全缓解，2 例（17%）部分缓解，3 例（25%）有混合反应。胃肠道症状完全解决的有 9 例（75%）。2 例患者在有初步反应后复发，重新治疗后显示部分反应。开始 Prochymal 治疗后，58%的患者生存超过 100 天，5 例（42%）12 人还活过中位数 730 天（存活天数范围为 527~1211 天）。所有患者没有出现可见的急性毒性反应。多次给药显示儿童患者是耐受的，也是安全的。

随着研究的进展，越来越多的自体或异体 MSC 进入临床研究，用于治疗 GVHD、关节损伤、肾移植、脊髓损伤、脑卒中（stroke）、视神经脊髓炎等不同的适应证，但未见明显的不良反应[7,71-74]。目前，成人骨髓[75]、脂肪[2]、脐血和胎盘来源 MSC 已经批准或正在进行临床研究，用于治疗 GVHD、克罗恩病（CD）、1 型糖尿病、慢性阻塞性肺炎、肢体缺血、关节损伤和心肌梗死。已报道的临床研究结果尚未见安全性问题。

第四节 小 结

体外培养的 MSC 是否真的会发生恶性转化，移植的细胞到体内转归如何？我们认为在保证供者细胞来源可控（调查遗传背景、检查病原微生物），细胞分离、培养环境符合必要条件，严格执行标准化操作及所用材料、设备符合相关要求的情况下，培养、传代的 MSC 发生恶性转化是小概率事件。但在干细胞治疗方面我们还是需要谨慎小心，如果需要使用的细胞是培养扩增后的细胞，我们需要严格控制供者主细胞库细胞的质量，严格控制传代次数。随着 MSC 临床应用的增加，或许各种不良反应会陆续出现，当然这并不意味着干细胞治疗研究应该禁止，任何药品都有可能出现这样或那样的不良反应，但我们需要评估这种风险是否可控。我们建议在进行干细胞治疗前需要大量的干细胞生物学研究和规范的前期实验，特别是安全性的研究。《涉及人体医学科学技术研究管理办法》《干细胞临床研究管理办法（试行）》和《干细胞制剂质量控制和临床前研究指导原则（试行）》等文件的陆续出台，会规范我国的干细胞治疗，保护患者的利益，促进干细胞行业健康发展。

（王有为 韩之波）

参 考 文 献

[1] Meirelles Lda S, Fontes AM, Covas DT, et al. Mechanisms involved in the therapeutic properties of mesenchymal stem cells. Cytokine Growth Factor Rev, 2009, 20:419-427.

[2] Panés J, García-Olmo D, Van Assche G, et al. Expanded allogeneic adipose-derived mesenchymal stem cells (Cx601) for complex perianal fistulas in Crohn's disease: a phase 3 randomised, double-blind controlled trial. Lancet, 2016, 388(10051):1281-1290.

[3] Koç ON, Gerson SL, Cooper BW, et al. Rapid hematopoietic recovery after coinfusion of autologous-blood stem cells and culture-expanded marrow mesenchymal stem cells in advanced breast cancer patients receiving high-dose chemotherapy. J Clin Oncol, 2000, 18(2):307-316.

[4] Bang OY, Lee JS, Lee PH, et al. Autologous mesenchymal stem cell transplantation in stroke patients. Ann Neurol, 2005, 57(6):874-882.

[5] Chen SL, Fang WW, Qian J, et al. Improvement of cardiac function after transplantation of autologous bone marrow mesenchymal stem cells in patients with acute myocardial infarction. Chin Med J (Engl), 2004, 117(10):1443-1448.

[6] Wu Y, Cao Y, Li X, et al. Cotransplantation of haploidentical hematopoietic and umbilical cord mesenchymal stem cells for severe aplastic anemia: Successful engraftment and mild GvHD. Stem Cell Res, 2014, 12(1):132-138.

[7] Reinders ME, Dreyer GJ, Bank JR, et al. Safety of allogeneic bone marrow derived mesenchymal stromal cell therapy in renal transplant recipients: the neptune study. J Transl Med, 2015, 13:344.

[8] Ning H, Yang F, Jiang M, et al. The correlation between cotransplantation of mesenchymal stem cells and higher recurrence rate in hematologic malignancy patients: outcome of a pilot clinical study. Leukemia, 2008, 22:593-599.

[9] Cai J, Wu Z, Xu X, et al. Umbilical cord mesenchymal stromal cell with autologous bone marrow cell transplantation in established type 1 diabetes: a pilot randomized controlled open-label clinical study to assess safety and impact on insulin secretion. Diabetes Care, 2016, 39(1):149-157.

[10] Dimri GP, Lee X, Basile G, et al. A biomarker that identifies senescent human cells in culture and in aging skin *in vivo*. Proc Natl Acad Sci USA, 1995, 92(20):9363-9367.

[11] Wang Y, Wu H, Yang Z, et al. Human mesenchymal stem cells possess different biological characteristics but do not change their therapeutic potential when cultured in serum free medium. Stem Cell Res Ther, 2014, 5(6):132.

[12] Dominici M, Le Blanc K, Mueller I, et al. Minimal criteria for defining multipotent mesenchymal stromal cells. The International Society for Cellular Therapy position statement. Cytotherapy, 2006, 8(4):315-317.

[13] Wang Y, Zhang Z, Chi Y, et al. Long-term cultured mesenchymal stem cells frequently develop genomic mutations but do not undergo malignant transformation. Cell Death Dis, 2013, 4:e950.

[14] Hahn WC, Counter CM, Lundberg AS, et al. Creation of human tumour cells with defined genetic elements. Nature, 1999, 400(6743):464-468.

[15] Burns JS, Abdallah BM, Guldberg P, et al. Tumorigenic heterogeneity in cancer stem cells evolved from long-term cultures of telomerase-immortalized human mesenchymal stem cells. Cancer Res, 2005, 65(8):3126-3135.

[16] Jeon BG, Kumar BM, Kang EJ, et al. Characterization and comparison of telomere length, telomerase and reverse transcriptase activity and gene expression in human mesenchymal stem cells and cancer cells of various origins. Cell Tissue Res, 2011, 345(1):149-161.

[17] Spits C, Mateizel I, Geens M, et al. Recurrent chromosomal abnormalities in human embryonic stem cells. Nat Biotechnol, 2008, 26(12):1361-1363.

[18] Tarte K, Gaillard J, Lataillade JJ, et al. Clinical-grade production of human mesenchymal stromal cells: occurrence of aneuploidy without transformation. Blood, 2010, 115(8):1549-1553.

[19] Roberts P, Burchill SA, Brownhill S, et al. Ploidy and karyotype complexity are powerful prognostic indicators in the Ewing's sarcoma family of tumors: a study by the United Kingdom Cancer Cytogenetics and the Children's Cancer and Leukaemia Group. Genes Chromosomes Cancer, 2008, 47(3):207-220.

[20] Ueyama H, Horibe T, Hinotsu S, et al. Chromosomal variability of human mesenchymal stem cells cultured under hypoxic conditions. J Cell Mol Med, 2012, 16(1):72-82.

[21] Nikitina VA, Osipova EY, Katosova LD, et al. Study of genetic stability of human bone marrow multipotent mesenchymal stromal cells. Bull Exp Biol Med, 2011, 150(5):627-631.

[22] Bochkov NP, Nikitina VA, Buyanovskaya OA, et al. Aneuploidy of stem cells isolated from human adipose tissue. Bull Exp Biol Med, 2008, 146(3):344-347.

[23] Stute N, Holtz K, Bubenheim M, et al. Autologous serum for isolation and expansion of human mesenchymal stem cells for clinical use. Exp Hematol, 2004, 32(12):1212-1225.

[24] Poloni A, Maurizi G, Serrani F, et al. Human AB serum for generation of mesenchymal stem cells from human chorionic villi: comparison with other source and other media including platelet lysate. Cell Prolif, 2012, 45(1):66-75.

[25] Murphy MB, Blashki D, Buchanan RM, et al. Adult and umbilical cord blood-derived platelet-rich plasma for mesenchymal stem cell proliferation, chemotaxis, and cryo-preservation. Biomaterials, 2012, 33(21):5308-5316.

[26] Fekete N, Gadelorge M, Fürst D, et al. Platelet lysate from whole blood-derived pooled platelet concentrates and apheresis-derived platelet concentrates for the isolation and expansion of human bone marrow mesenchymal stromal cells: production process, content and identification of active components. Cytotherapy, 2012, 14(5):540-554.

[27] 国家食品药品监督管理总局. 干细胞制剂质量控制及临床前研究指导原则（试行）. 2015 年 08 月 21 日发布. http://samr.cfda.gov.cn/WS01/CL1616/127242.html [2019-5-20]

[28] Wu M, Han ZB, Liu JF, et al. Serum-free media and the immunoregulatory properties of mesenchymal stem cells *in vivo* and *in vitro*. Cell Physiol Biochem, 2014, 33(3):569-580.

[29] Reinhardt J, Stühler A, Blümel J. Safety of bovine sera for production of mesenchymal stem cells for therapeutic use. Hum Gene Ther, 2011, 22(6):775.

[30] Hu X, Xu Y, Zhong Z, et al. A Large-scale investigation of hypoxia-preconditioned allogeneic mesenchymal stem cells for myocardial repair in nonhuman primates: paracrine activity without remuscularization. Circ Res, 2016, 118(6):970-983.

[31] Chang KC, Hung SC. Hypoxia-preconditioned allogeneic mesenchymal stem cells can be used for myocardial repair in non-human primates. J Thorac Dis, 2016, 8(7):E593-E595.

[32] Cruz FF, Rocco PR. Hypoxic preconditioning enhances mesenchymal stromal cell lung repair capacity. Stem Cell Res Ther, 2015, 6:130.

[33] Duijvestein M, Wildenberg ME, Welling MM, et al. Pretreatment with interferon-γ enhances the therapeutic activity of mesenchymal stromal cells in animal models of colitis. Stem Cells, 2011, 29(10):1549-1558.

[34] Fan H, Zhao G, Liu L, et al. Pre-treatment with IL-1β enhances the efficacy of MSC transplantation in DSS-induced colitis. Cell Mol Immunol, 2012, 9(6):473-481.

[35] Xu W, Qian H, Zhu W, et al. A novel tumor cell line cloned from mutated human embryonic bone marrow mesenchymal stem cells. Oncol Rep, 2004, 12:501-508.

[36] de la Fuente R, Bernad A, Garcia-Castro J, et al. Traction: spontaneous human adult stem cell transformation. Cancer Res, 2010, 70(16):6682.

[37] Garcia S, Bernad A, Martin MC, et al. Pitfalls in spontaneous *in vitro* transformation of human mesenchymal stem cells. Exp Cell Res, 2010, 316(9):1648-1650.

[38] Torsvik A, Rosland GV, Bjerkvig R. Comment to: "Spontaneous transformation of adult mesenchymal stem cells from cynomolgus macaques *in vitro*" by Ren Z, Wang J, Zhu W, et al. Exp. Cell Res, 2011, 317:2950-2957. In Press Spontaneous Transformation of Mesenchymal Stem Cells in Culture: Facts or Fiction? Exp Cell Res, 2012, 10318:441-443.

[39] Serakinci N, Guldberg P, Burns JS, et al. Adult human mesenchymal stem cell as a target for neoplastic transformation. Oncogene, 2004, 23:5095-5098.

[40] Houghton J, Stoicov C, Nomura S, et al. Gastric cancer originating from bone marrow-derived cells. Science, 2004, 306:1568-1571.

[41] Mohseny AB, Szuhai K, Romeo S, et al. Osteosarcoma originates from mesenchymal stem cells in consequence of aneuploidization and genomic loss of Cdkn2. J Pathol, 2009, 219:294-305.

[42] Boukamp P, Petrussevska RT, Breitkreutz D, et al. Normal keratinization in a spontaneously immortalized aneuploid human keratinocyte cell line. Cell Biol, 1988, 106:761-771.

[43] Wang Y, Huso DL, Harrington J, et al. Outgrowth of a transformed cell population derived from normal human BM mesenchymal stem cell culture. Cytotherapy, 2005, 7:509-519.

[44] Miura Y, Padilla-Nash HM, Molinolo AA, et al. Accumulated chromosomal instability in murine bone marrow mesenchymal stem cells leads to malignant transformation. Stem Cells, 2006, 24:1095-1103.

[45] Lepperdinger G, Brunauer R, Jamnig A, et al. Controversial issue: is it safe to employ mesenchymal stem cells in cell-based therapies? Exp Gerontol, 2008, 43:1018-1023.

[46] Bernardo ME, Zaffaroni N, Novara F, et al. Human bone marrow derived mesenchymal stem cells do not undergo transformation after long-term *in vitro* culture and do not exhibit telomere maintenance mechanisms. Cancer Res, 2007, 67:9142-9149.

[47] Meza-Zepeda LA, Noer A, Dahl JA, et al. High-resolution analysis of genetic stability of human adipose tissue stem cells cultured to senescence. J Cell Mol Med, 2008, 12:553-563.

[48] Sawada R, Ito T, Tsuchiya T. Changes in expression of genes related to cell proliferation in human mesenchymal stem cells during *in vitro* culture in comparison with cancer cells. J Artif Organs, 2006, 9:179-184.

[49] Rubio D, Garcia S, Paz MF, et al. Molecular characterization of spontaneous mesenchymal stem cell transformation. PLoS One, 2008, 3(1):e1398.

[50] Rubio D, Garcia-Castro J, Martín MC, et al. Spontaneous human adult stem cell transformation. Cancer Res, 2005, 65:3035-3039.

[51] Rubio D, Garcia S, De la Cueva T, et al. Human mesenchymal stem cell transformation is associated with a mesenchymal-epithelial transition. Exp Cell Res, 2008, 314:691-698.

[52] Guan L, Yu J, Zhong L, et al. Biological safety of human skin-derived stem cells after long-term *in vitro* culture. J Tissue Eng Regen Med, 2011, 5(2):97-103.

[53] Gou S, Wang C, Liu T, et al. Spontaneous differentiation of murine bone marrow-derived mesenchymal stem cells into adipocytes without malignant transformation after long-term culture. Cells Tissues Organs, 2010, 191(3):185-192.

[54] Rosland GV, Svendsen A, Torsvik A, et al. Long-term cultures of bone marrow-derived human mesenchymal stem cells frequently undergo spontaneous malignant transformation. Cancer Res, 2009, 69:5331-5339.

[55] Editorial. Identity crisis. Nature, 2009, 457:935-936.

[56] Leng L, Wang Y, He N, et al. Molecular imaging for assessment of mesenchymal stem cells mediated breast cancer therapy. Biomaterials, 2014, 35(19):5162-5170.

[57] Zhu Y, Sun Z, Han Q, et al. Human mesenchymal stem cells inhibit cancer cell proliferation by secreting DKK-1. Leukemia, 2009, 23(5):925-933.

[58] Bartholomew A, Sturgeon C, Siatskas M, et al. Mesenchymal stem cells suppress lymphocyte proliferation *in vitro* and prolong skin graft survival *in vivo*. Exp Hematol, 2002, 30:42-48.

[59] Devine SM, Cobbs C, Jennings M, et al. Mesenchymal stem cells distribute to a wide range of tissues following systemic infusion into non-human primates. Blood, 2003, 101:2999-3001.

[60] Saito T, Kuang JQ, Bittira B, et al. Xenotransplant cardiac chimera: immune tolerance of adult stem cells. Ann Thorac Surg, 2002, 74:19-24.

[61] Wang Y, Han ZB, Ma J, et al. A toxicity study of multiple-administration human umbilical cord mesenchymal stem cells in cynomolgus monkeys. Stem Cells Dev, 2012, 21(9):1401-1408.

[62] Li Y, McIntosh K, Chen J, et al. Allogeneic bone marrow stromal cells promote glial-axonal remodeling without immunologic sensitization after stroke in rats. Exp Neurol, 2006, 198:313-325.

[63] Liu L, Sun Z, Chen B, et al. *Ex vivo* expansion and *in vivo* infusion of bone marrow-derived Flk-1$^+$CD31$^-$CD34$^-$ mesenchymal stem cells: feasibility and safety from monkey to human. Stem Cells Dev, 2006, 15(3):349-357.

[64] Hofseth LJ, Hussain SP, Harris CC. p53: 25 years after its discovery, Trends Pharmacol Sci, 2004, 25:177-181.

[65] Armesilla-Diaz A, Elvira G, Silva A. p53 regulates the proliferation, differentiation and spontaneous transformation of mesenchymal stem cells. Exp Cell Res, 2009, 315(20):3598-3610.

[66] Thirabanjasak D, Tantiwongse K, Thorner PS. Angiomyeloproliferative lesions following autologous stem cell therapy. J Am Soc Nephrol, 2010, 21(7):1218-1222.

[67] Amariglio N, Hirshberg A, Scheithauer BW, et al. Donor-derived brain tumor following neural stem cell transplantation in an ataxia telangiectasia patient. PLoS Med, 2009, 6(2):e1000029.

[68] Poltavtseva RA, Marey MV, Aleksandrova MA, et al. Evaluation of progenitor cell cultures from human embryos for neurotransplantation. Brain Res Dev Brain Res, 2002, 134(1-2):149-154.

[69] Wakitani S, Okabe T, Horibe S, et al. Safety of autologous bone marrow-derived mesenchymal stem cell transplantation for cartilage repair in 41 patients with 45 joints followed for up to 11 years and 5 months. J Tissue Eng Regen Med, 2011, 5(2):146-150.

[70] Prasad VK, Lucas KG, Kleiner GI, et al. Efficacy and safety of *ex vivo* cultured adult human mesenchymal stem cells (Prochymal™) in pediatric patients with severe refractory acute graft-versus-host disease in a compassionate use study. Biol Blood Marrow Transplant, 2011, 17(4):534-541.

[71] Oh SK, Choi KH, Yoo JY, et al. A phase III clinical trial showing limited efficacy of autologous mesenchymal stem cell therapy for spinal cord injury. Neurosurgery, 2016, 78(3):436-447.

[72] Satti HS, Waheed A, Ahmed P, et al. Autologous mesenchymal stromal cell transplantation for spinal cord injury: a phase I pilot study. Cytotherapy, 2016, 18(4):518-522.

[73] Steinberg GK, Kondziolka D, Wechsler LR, et al. Clinical outcomes of transplanted modified bone marrow-derived mesenchymal stem cells in stroke: a phase 1/2a study. Stroke, 2016, 47(7):1817-1824.

[74] Fu Y, Yan Y, Qi Y, et al. Impact of autologous mesenchymal stem cell infusion on neuromyelitis optica spectrum disorder: a pilot, 2-year observational study. CNS Neurosci Ther, 2016, 22(8):677-685.

[75] Patel AN, Genovese J. Potential clinical applications of adult human mesenchymal stem cell (Prochymal®) therapy. Stem Cells Cloning, 2011, 4:61-72.

第五章　间充质干细胞的免疫调节机制

第一节　免疫系统的构成和功能

机体的免疫系统主要由免疫器官和组织、免疫细胞和免疫分子组成。免疫系统识别和清除抗原的过程通称为免疫应答，主要由固有免疫应答和适应性免疫应答组成。固有免疫是机体抵抗外界有害病原体入侵的第一道防线，主要由上皮细胞屏障、吞噬细胞（巨噬细胞和粒细胞）、树突细胞（DC）、NK细胞等固有免疫细胞及补体和细胞因子等免疫分子组成，固有免疫应答发生迅速，但特异性差，持续时间短，缺乏记忆性。适应性免疫由T淋巴细胞和B淋巴细胞（以下简称T细胞和B细胞）组成，分别介导细胞免疫应答和体液免疫应答，适应性免疫应答发生慢，但效应更为专一和强大，且具有免疫记忆。

骨髓是人体重要的免疫器官和造血器官，免疫系统的细胞来源于骨髓的多能造血干细胞（HSC）。HSC可分化为不同种系的祖细胞，包括髓样祖细胞和淋巴样祖细胞。髓样祖细胞最终分化为固有免疫系统的粒细胞（中性粒细胞、嗜酸性粒细胞和嗜碱性粒细胞）、巨噬细胞、DC和肥大细胞；而淋巴样祖细胞最终分化为适应性免疫系统的T细胞和B细胞。此外，固有免疫系统的NK细胞和浆细胞样DC（plasmacytoid DC，pDC）也来源于淋巴样祖细胞。因此，HSC分化成熟的过程也代表了免疫细胞的发育分化过程[1]。

胸腺和骨髓作为人体中枢免疫器官，分别是T细胞和B细胞发育、分化和成熟的场所。成熟淋巴细胞离开中枢免疫器官，随血液循环迁移到外周免疫器官（脾和淋巴结）或外周组织定居。部分淋巴细胞可在血液、淋巴液和淋巴组织间反复循环，有利于及时识别外来抗原，启动相应的免疫应答，并动员多种固有免疫细胞迁移到感染部位，发挥免疫效应[2]。T细胞是参与再循环的主要淋巴细胞。初始T细胞是指从未接触过抗原的成熟T细胞。在外周再循环过程中，初始T细胞识别DC提呈的主要组织相容性复合体（MHC）分子-抗原肽复合物，分化为不同类型的效应T细胞，发挥不同的效应机制。效应T细胞根据膜表面标志和生物功能分为$CD4^+$辅助性T细胞（helper T cell，Th细胞）和$CD8^+$细胞毒性T细胞（cytotoxic T lymphocyte，CTL）。Th细胞主要通过分泌细胞因子作用于其他细胞类型，包括巨噬细胞、B细胞、中性粒细胞等，影响它们的免疫功能。根据分泌的细胞因子和发挥不同的免疫效应，$CD4^+$ Th细胞可分为Th1、Th2、Th17和滤泡辅助性T细胞（follicular Th，Tfh）4类细胞亚群[3,4]。Th1细胞主要分泌干扰素-γ（IFN-γ），介导细胞免疫；Th2细胞主要分泌白介素-4（IL-4），调节B细胞的增殖和IgE抗体的类型转换，介导体液免疫；Th17细胞主要分泌IL-17，招募和活化中性粒细胞，介导组织炎症。$CD8^+$ CTL则主要通过释放颗粒酶或FasL/Fas通路诱导细胞凋亡，直接杀死病毒感染的细胞及肿瘤细胞。此外，CTL也能分泌细胞因子如IFN-γ等，促进细胞免疫[5]。$CD4^+CD25^+$调节性T细胞（Treg）是一类具有免疫抑制效应的$CD4^+$ T细胞亚群，抑制多种效应细胞介导的免疫应答。$CD4^+CD25^+$ Treg细胞直接从胸腺分化而来，天然存在于免疫系统。然而，研究发现体外低剂量 T 细胞受体（T cell receptor，TCR）刺激联合 IL-2 和 TGF-β 也可诱导 $CD4^+CD25^-$ T 细胞表达关键的 Treg 细胞转录因子 Foxp3，从而转变为具有免疫抑制活性的 $CD4^+CD25^+$ Treg 细胞[6]。

与固有免疫细胞相比，T 细胞和 B 细胞具有高度抗原受体多样性，能识别周围环境中数量众多的抗原，并针对某一特定抗原做出特异性应答，且具有记忆能力。当感染减退后，一部分抗原特异性淋巴细胞成为记忆性淋巴细胞长期存在于机体，当再次接触相同的抗原时，可迅速启动记忆免疫应答，保护机体免受危害[7]。机体产生有效免疫应答需要固有免疫和适应性免疫之间密切协作，适应性免疫应答依赖固有免疫细胞转导的信号，反之，适应性免疫应答能进一步提高固有免疫应答的效应。固有免疫细胞直接与外界环境中各类病原微生物接触，它们最初活化的方式决定了适应性免疫应答的强度和性质。适应性免疫应答的启动需要抗原提呈细胞捕获和提呈蛋白质抗原，DC 是启动 T 细胞免疫应答的关键抗原提呈细胞。DC 需要从不成熟阶段向成熟阶段发育才能获得活化 T 细胞的能力。不成熟的 DC 位于外周上皮组织和结缔组织，具有极强的抗原摄取、加工和处理能力，摄取和内吞从上皮屏障进入机体的微生物抗原，把蛋白质抗原处理为肽段使其能够与 MHC 分子结合为 MHC 分子-抗原肽复合物，同时 DC 分化成熟，表现为 MHC 分子、共刺激分子（CD80 和 CD86）和黏附分子的表达量增高，并高表达趋化因子受体 CCR7，携带 MHC 分子-抗原肽复合物向局部引流淋巴结迁移，提呈 MHC 分子-抗原肽复合物给 T 细胞，活化 T 细胞，启动 T 细胞免疫应答。相反，不成熟 DC 表达低水平的共刺激分子，提呈抗原后不能活化 T 细胞，而是诱导 T 细胞无能，参与外周免疫耐受[8]。此外，DC、巨噬细胞及上皮细胞等固有免疫细胞表达多种类型的模式识别受体（pattern recognition receptor，PRR），如 Toll 样受体（Toll like receptor，TLR）等。TLR 识别多种病原微生物及其产物共有的、分子结构高度保守的病原体相关分子模式（pathogen-associated molecular pattern，PAMP），如 TLR2 识别革兰氏阳性菌的肽聚糖、TLR4 识别革兰氏阴性菌的脂多糖（lipopolysaccharide，LPS）、TLR3 识别病毒双链 RNA 等[9]。PRR 的活化引起以 NF-κB 通路为主的信号转导通路的激活，活化的细胞分泌多种炎性细胞因子，参与机体的免疫应答，其中某些炎性细胞因子可作用于 DC 或直接作用于 T 细胞，对 Th 细胞亚群的分化起决定性作用，如活化的 DC 或巨噬细胞来源的 IL-12 促进 Th1 细胞的分化，而 IL-6 和 IL-23 促进 Th17 细胞的分化与存活[10]。

免疫系统的主要功能包括三个方面：免疫防御、免疫监视和免疫自身稳态的维持。免疫系统一方面识别和清除外界入侵的病原微生物，以及体内发生突变的肿瘤细胞和衰老凋亡的细胞；另一方面，免疫系统对自身组织抗原不起反应及维持免疫微环境的稳定。免疫自身稳态的维持主要依赖免疫耐受的诱导和免疫负反馈调节机制。免疫耐受包括中枢免疫耐受和外周免疫耐受。中枢免疫耐受在胸腺中诱导，主要通过与自身抗原相互作用，克隆清除表达与自身抗原具有高亲和力的自身反应性 T 细胞。但是中枢免疫耐受是不完全的，比如一些针对不在胸腺表达的外周组织抗原的自身反应性 T 细胞能逃避中枢免疫耐受。这样，外周免疫耐受的诱导对防止自身免疫应答和维持免疫稳态至关重要[11]。多种存在于外周的免疫细胞亚群具有调节免疫的功能，如 $CD4^+CD25^+$ Treg 细胞和耐受型 DC 等，均积极参与外周免疫耐受的诱导。此外，免疫系统可根据自身应答的强度实施正调节或负调节，通过负调节抑制过强的免疫应答，减轻对机体组织的损失，并在病原体清除之后，恢复自身内环境的稳态。一个重要的抑制 T 细胞应答的负反馈机制是活化的 T 细胞上调表达 CTLA-4 和 PD-1 等共抑制分子，竞争性抑制 B7-CD28 通路介导的 T 细胞活化，从而诱导免疫耐受或下调过强的 T 细胞免疫应答[12]。

人们很早就已经认识到免疫应答是把双刃剑，除了给机体带来免疫保护之外，当针对自身抗原或外界无害抗原发生异常免疫应答，或免疫调节功能紊乱时，会打破免疫稳态而发生过度的免疫应答，导致多种免疫相关疾病的发生。例如，Th1 细胞通过分泌多种效应细胞因子，动员和增强多种免疫细胞的效应功能，在机体抗感染免疫中发挥重要作用。然而，如果入侵的病原微生物有效地抵抗了 Th1 活化的巨噬细胞的杀菌作用后，持续活化的巨噬细胞过度释放各种杀菌成分、炎性细胞因子和生长因子，会导致宿主组织严重损伤及随后的发生纤维化，影响受感染器官的生理功能[13]。此外，机体针对外界环境中的一些无害蛋白质，如花粉、鸡蛋等，发生过度的 Th2 细胞应答，导致过敏性疾病的发生[14]；或针对自身抗原发生以 Th17 细胞应答为主的慢性炎症，导致自身免疫疾病的发生[15]。这些免疫相关疾病往往伴随着免疫抑制细胞如 $CD4^+CD25^+$ Treg 细胞数量或功能的下降，以及一种或多种免疫负调节机制的受损[16]。因此，治疗免疫相关疾病的主要原则是有效控制异常的或过度的免疫应答，恢复机体内免疫环境的稳态。

第二节　间充质干细胞的免疫学特性

目前普遍接受的 MSC 的主要免疫学特性是低免疫原性和免疫抑制能力。MSC 的低免疫原性表现为表达低水平的人白细胞抗原（HLA）/MHC Ⅰ 类分子，非组成性表达 MHC Ⅱ 类分子，不表达共刺激分子，以及不引起同种异源淋巴细胞的活化和增殖。然而，MSC 抑制 T 细胞增殖并不意味着 MSC 不具有启动免疫应答的能力。最近的研究表明 MSC 并不是绝对的免疫豁免细胞，而是在不同的环境下具有启动和抑制免疫应答的双重作用。在正常免疫活性小鼠体内，MSC 可致敏同种异源 T 细胞，表明了 MSC 具有免疫原性[17]。在适当的环境下，MSC 能作为抗原提呈细胞启动免疫应答，并促进记忆 T 细胞的扩增[18]。此外，最近的研究发现小鼠 MSC 可交叉提呈可溶性蛋白抗原，启动 $CD8^+$ T 细胞应答[19]。致炎性细胞因子能诱导 MSC 表达 MHC Ⅱ 类分子，其中 IFN-γ 的浓度对 MSC 表达 MHC Ⅱ 类分子的水平和 MSC 的免疫原性似乎尤为重要。MSC 自身分泌低水平的 IFN-γ 导致 MSC 表达 MHC Ⅱ 类分子，发挥抗原提呈细胞的功能，而随着 IFN-γ 浓度的上升，MSC 表达 MHC Ⅱ 类分子的水平和抗原提呈能力逐渐下降[20]。这样，在免疫应答的后期，伴随着活化的 T 细胞分泌的炎性细胞因子，尤其是 IFN-γ 水平的增高，MSC 丧失抗原提呈能力，转变成免疫抑制细胞。此外，IFN-γ 也调节 MSC 的负性共刺激分子 PD-L1 的表达，T 细胞产生的 IFN-γ 上调 MSC 表达 PD-L1 的水平，提高其对 T 细胞免疫应答的抑制能力[21]。因此，MSC 生理性的免疫学功能可能是病原体与其接触并将其活化后，MSC 上调 MHC Ⅱ 类分子的表达，作为抗原提呈细胞，启动保护性 T 细胞免疫应答，但随后微环境中细胞因子分泌谱及其表达水平的改变可能影响 MSC 的免疫调节功能，活化 T 细胞产生的高水平 IFN-γ 下调 MSC 表达 MHC Ⅱ 类分子的水平，关闭抗原提呈，同时上调 PD-L1 的表达，抑制 T 细胞过度活化,从而防止由过度的免疫应答可能造成的组织损伤。

第三节 间充质干细胞的免疫调节机制研究

虽然最初对 MSC 关注是因为其具有向多种组织细胞类型分化的可塑性和再生能力,但是目前更多的关注集中到 MSC 对免疫应答的调节能力。最早对 MSC 调节免疫应答的认识源于发现移植的同源异种 MSC 能逃避受体的免疫监视[22]。最能体现 MSC 的免疫抑制能力的例子是输入同种异源的 MSC 能够有效治疗对免疫抑制剂不敏感的重症急性移植物抗宿主病(aGVHD)[23]。体内外研究表明,MSC 能够直接抑制 T 细胞的增殖、分化及其效应机制,或者通过抑制 DC 的分化和成熟间接抑制 T 细胞的增殖,以及影响效应 T 细胞亚群的分化。除了 T 细胞和 DC,MSC 的靶细胞还包括众多其他类型的免疫细胞,如 NK 细胞、B 细胞、中性粒细胞、巨噬细胞等。MSC 通过调节这些细胞的功能,如细胞因子的分泌、细胞毒效应等,最终发挥抗炎效应和/或诱导免疫耐受状态(图 5-1)。目前对 MSC 的免疫调节机制还不完全清楚,但是普遍认为 MSC 主要通过细胞间接触和分泌可溶性因子发挥免疫抑制功能。

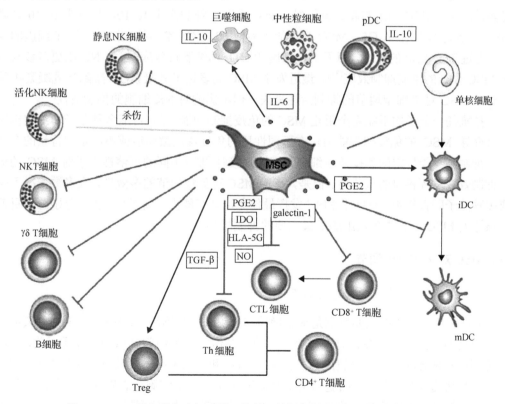

图 5-1 MSC 对各类免疫细胞的可能的调节机制(彩图请扫封底二维码)

IL-10. 白介素-10,IL-6. 白介素-6,PGE2. 前列腺素 E2,IDO. 吲哚胺 2,3-双加氧酶,HLA-G5. 人白细胞抗原-G5,NO. 一氧化氮,TGF-β. 转化生长因子-β,Treg. 调节性 T 细胞,Th 细胞. 辅助性 T 细胞,CTL 细胞. 细胞毒性 T 细胞,galectin-1. 半乳凝素-1,pDC. 浆细胞样树突细胞,iDC. 不成熟树突细胞,mDC. 成熟树突细胞多种可溶性因子参与 MSC 介导的免疫调节效应,它们可能直接来源于 MSC,或者通过 MSC 和靶细胞交互作用通过旁分泌产生。目前研究较多的可溶性因子是前列腺素 E2(PGE2)和吲哚胺 2,3-双加氧酶(IDO)。MSC 组成性表达环加氧酶 2(COX2),一种对 PGE2 合成起关键作用的酶。MSC 和外周血单个核细胞(PBMC)共培养可引起 PGE2 的分泌上调[24],炎性细胞因子 IFN-γ 和肿瘤坏死因子-α(TNF-α)能提高 MSC 释放 PGE2 的水平。从受体支气管肺泡灌洗液中分离的同种异源 MSC 具有合成 PGE2 的能力,体内外利用化学抑制剂阻断 PGE2 的生成能减轻 MSC 介导的免疫抑制功能[24-26],这些结果表明 PGE2 是参与 MSC 免疫调节效应的关键分子。MSC 不组成性表达 IDO,但是靶细胞来源的 IFN-γ 可诱导 MSC 表达 IDO[27]。IDO 是色氨酸降解为犬尿氨酸过程中的限速酶,而色氨酸是 T 细胞增殖所必需的氨基酸,其降解产物的堆积对 T 细胞增殖有抑制作用[28]。因此,在 IFN-γ 存在下,MSC 表达 IDO,加速色氨酸的代谢,抑制 T 细胞的增殖。其他参与 MSC 免疫调节的可溶性因子包括 TGF-β1、肝细胞生长因子(HGF)、IL-10、IL-6、可溶性 HLA-G5、一氧化氮(NO)和半乳凝素-1(galectin-1)等,它们可能通过诱导 Treg 细胞的生成、影响效应靶细胞的增殖、凋亡,以及效应因子的分泌等途径介导 MSC 的免疫抑制功能。其中,可溶性 HLA-G5 是介导人 MSC(hMSC)免疫调节功能的一个重要可溶性因子,MSC 通过与靶细胞接触,诱导活化 T 细胞产生 IL-10,反过来 IL-10 刺激 MSC 释放可溶性 HLA-G5。MSC 产生的可溶性 HLA-G5 参与抑制同种异源 T 细胞的增殖,促进 Treg 细胞的扩增,以及 NK 细胞和 CTL 的细胞毒效应[29,30]。NO 则更多参与小鼠 MSC 介导的免疫抑制效应[31]。最近发现 MSC 高表达半乳凝素-1,敲除半乳凝素-1 明显导致 MSC 对 T 细胞增殖的抑制活性丧失,但不影响对 NK 细胞的抑制活性[32]。

拮抗任一分子均不能完全阻止 MSC 的免疫抑制功能,很有可能多种分子和细胞机制联合介导 MSC 的免疫抑制能力,在不同性质的免疫炎症微环境或不同基因背景的个体中,抑制某一类特定的免疫效应细胞,细胞间接触和/或不同的可溶性因子对 MSC 的免疫抑制效应发挥各自的重要性。但是,基于 MSC 对免疫调节的多效性,大多数情况下需要多种分子联合作用,有助于 MSC 发挥最佳免疫调节能力。接下来,我们分别总结目前已报道的 MSC 对各类免疫细胞的最主要的抑制机制。

一、MSC 对 T 细胞的调节

(一)MSC 对效应 T 细胞的调节

T 细胞免疫应答的主要特点是细胞增殖和分泌效应细胞因子。MSC 对 T 细胞最显著的作用是抑制其增殖。MSC 对 T 细胞增殖的抑制作用与 T 细胞刺激物的性质无关,MSC 可抑制由同种异源抗原、丝裂原和多克隆抗 CD3 和 CD28 抗体刺激引起的 T 细胞增殖。早期研究认为 MSC 抑制 T 细胞增殖主要通过诱导活化 T 细胞分裂停滞[33],但随后的研究发现通过诱导 NO 依赖性的 T 细胞凋亡也可抑制 T 细胞增殖[34-36]。小鼠骨髓间充质干细胞(BM-MSC)与 T 细胞之间的直接接触可能活化程序性死亡-1(programmed death-1,PD-1)通路,抑制 T 细胞的活化,从而导致免疫抑制[37]。MSC 能够影响初始 T 细胞的活化,以及初始 T 细胞向不同 T 细胞亚群的分化和相应的细胞因子的分泌。在 Th1 存在的体外诱导分化条件下，MSC 降低 IFN-γ 的生成，但提高 IL-4 的生成[25]。在 Th17 存在的体外诱导分化条件下，人骨髓间充质干细胞（hBM-MSC）降低分化中的 Th17 细胞分泌 IL-17 和 IL-22 等炎性细胞因子，但提高免疫调节因子 IL-10 的分泌，而且 MSC 能够抑制从炎症病变组织分离的完全分化的 Th17 细胞克隆产生效应因子，并伴有 CD4$^+$CD25$^+$ Treg 细胞的关键转录因子 foxp3 的诱导表达，MSC 分泌的 PGE2 部分介导了该效应[38]。

CD8$^+$ CTL 的细胞毒效应是介导抗病毒免疫和同种异源移植排斥的主要效应机制。尽管 MSC 明显抑制由同种异源抗原刺激的特异性 CTL 的增殖和细胞毒效应，但对病毒特异性 CTL 的抑制作用有限[39]。体外 MSC 下调颗粒酶的表达，抑制 CTL 介导的细胞毒效应，而 MSC 自身能抵抗 CTL 的细胞毒效应[40,41]。

（二）MSC 对 CD4$^+$CD25$^+$ Treg 细胞的调节

作为抑制性 CD4$^+$ T 细胞亚群，CD4$^+$CD25$^+$ Treg 细胞对免疫稳态的维持及对自身免疫耐受的诱导起至关重要的作用。体外研究表明，MSC 可能通过多种机制影响 CD4$^+$CD25$^+$ Treg 细胞的免疫生物活性，如招募及促进 CD4$^+$CD25$^+$ Treg 细胞的分化，以及维持其免疫抑制功能[42]。同种异源 MSC 直接与 CD4$^+$ T 细胞接触，随后分泌 PGE2 和 TGF-β，促进 CD4$^+$CD25$^+$ Treg 细胞的分化和增殖[43-45]。MSC 以 IL-10 依赖方式分泌可溶性 HLA-G5，促进 CD4$^+$CD25$^+$ Treg 细胞的扩增[29]。体内研究发现，MSC 可能通过表达 IDO 促进 CD4$^+$CD25$^+$ Treg 细胞的产生，从而诱导多种器官移植的耐受[46,47]。多数 CD4$^+$ Th 细胞介导的免疫病理同时伴随着 CD4$^+$CD25$^+$ Treg 细胞的减少或功能紊乱。人脂肪间充质干细胞（hAD-MSC）治疗胶原诱导的关节炎小鼠模型的研究发现，MSC 抑制致炎性 Th1/Th17 细胞的扩增和炎性介质的释放，同时促进抗原特异性 CD4$^+$CD25$^+$ Treg 细胞的生成，进一步抑制自身反应活性的 T 细胞应答[48]。体内 MSC 治疗以 Th2 细胞介导为主的小鼠哮喘模型发现，MSC 通过其表达的 IL-4R 与 IL-4 和/或 IL-13 结合，激活信号转导及转录激活蛋白 6（signal transducer and activator of transcription 6，STAT6）通路，上调 TGF-β 的表达，反过来抑制以 Th2 应答为主的过敏性炎症环境中 IL-4 的产生，并伴随着肺部聚集更多的 CD4$^+$CD25$^+$ Treg 细胞[49]。因此，MSC 可能并不是单一地抑制某一类细胞亚群，而是通过准确感知它们所处炎症微环境的免疫失衡的性质来作出相应的调整[50]。

因此，MSC 抑制 T 细胞的增殖和 CTL 的细胞毒效应，影响不同效应 T 细胞亚群的分化，促进 CD4$^+$CD25$^+$ Treg 细胞的分化和扩增，最终调节 T 细胞免疫应答的强度和性质。此外，除了影响功能之外，研究发现小鼠 aGVHD 模型中输入 MSC 可使 T 细胞维持初始细胞的迁移特点，阻止活化的 T 细胞从周围淋巴组织迁移到外周组织发挥效应[51]。

二、MSC 对 DC 的调节

树突细胞（DC）是启动 T 细胞免疫应答的主要抗原提呈细胞，成熟 DC（mature DC，mDC）提呈从外周捕获的抗原给初始 T 细胞，并上调共刺激分子和炎性细胞因子，诱导 T 细胞的活化。相反，不成熟 DC（immture DC，iDC）提呈抗原与免疫耐受的诱导相关。目前认为 MSC 对 T 细胞增殖的抑制和对 Treg 细胞的诱导依赖单核细胞/DC。MSC 可能通过抑制 DC 的分化、迁移和成熟，影响 T 细胞免疫应答的启动或随后的 T 细胞亚群的分化，促进 CD4$^+$CD25$^+$ Treg 细胞的生成，从而间接地调节 T 细胞免疫应答的强度和性质。体外培养体系中有 MSC 存在可以阻止脐血 CD34$^+$细胞或外周血 CD14$^+$单核细胞向 DC 的分化[52]。小鼠 MSC 可以抑制 LPS 诱导的 DC 表面趋化因子受体 CCR7 的上调，以及明显降低 CCL19 诱导的体外 DC 迁移的能力[53]。MSC 也可以直接作用于 mDC，降低其表面共刺激分子 CD80（B7.1）和 CD86（B7.1）的表达水平与 IL-12 的分泌水平，导致 mDC 向 iDC 的表型转变[54]。此外，MSC 能改变 DC 的细胞因子分泌谱，如降低 TNF-α 的分泌水平，提高浆细胞样 DC（pDC）的免疫调节因子 IL-10 的分泌水平，进而影响随后的 T 细胞亚群的分化，导致致炎的 Th1 细胞向抗炎的 Th2 细胞或 Treg 细胞的转换[55]。MSC 产生几种可溶性因子，包括 IL-6、巨噬细胞集落刺激因子（M-CSF）和 PGE2 等，在 DC 分化和成熟过程中发挥重要作用。MSC 来源的高水平的 IL-6 可能部分参与抑制骨髓祖细胞向 DC 的分化和逆转 DC 的成熟[52,53]。随后的研究也表明，MSC 分泌的 PGE2 参与抑制 DC 的分化，其抑制作用主要发生在 DC 成熟的早期，即粒细胞-巨噬细胞集落刺激因子（GM-CSF）和 IL-4 诱导单核细胞向 iDC 分化阶段，但是对 LPS 所诱导的 DC 成熟没有影响[56]。此外，小鼠 MSC 可通过与 DC 的直接接触，影响 DC 的免疫原性，主要表现为提高 mDC 的抗凋亡能力，使其转变为一类具有调节功能的 DC 亚群，下调表达共刺激分子，而 CD11b 和 Notch 配体 Jagged-2 的表达上调，有效地抑制体内外 T 细胞的增殖[57]。不同组织来源的 MSC 对 DC 分化和成熟的抑制能力可能有差别，如脂肪组织来源的 MSC 可能比 BM-MSC 具有更强的抑制单核细胞向 DC 分化和表达共刺激分子的能力[58]。进一步的研究发现，体外低浓度的 MSC 抑制 DC 成熟可能通过细胞间接触，而不是分泌可溶性因子来实现的[59]。

总之，MSC 通过分泌可溶性因子或直接与 DC 接触，影响 DC 的各个阶段，如阻止 DC 的分化和成熟，甚至促使具有免疫原性的 mDC 转变为免疫耐受型 DC，最终抑制 T 细胞的免疫应答。

三、MSC 对 B 细胞的调节

B 细胞是适应性免疫应答的另外一个组成部分，通过分泌各种类型的免疫球蛋白介导体液免疫。B 细胞在骨髓中发育成熟，依赖于骨髓基质细胞与 B 细胞祖细胞的相互作用，基质细胞来源的细胞因子支持 B 细胞的存活和增殖。体外 MSC 对外周成熟 B 细胞的调节作用还有争议，部分研究表明 MSC 抑制 B 细胞的增殖和其向分泌免疫球蛋白的浆细胞分化[33,60]，小鼠 MSC 通过分泌 CCL2 直接抑制浆细胞分泌免疫球蛋白抗体[61]。但是，也有研究发现 B 细胞促进正常或者系统性红斑狼疮（SLE）患者来源的 B 细胞的存活、增殖和分化[62]。体内研究发现，注射 MSC 抑制小鼠实验性自身免疫脑脊髓炎（experimental autoimmune encephalomyelitis，EAE）模型中抗原特异性抗体的产生[63]。最近研究发现，hAD-MSC 可以抑制 B 细胞向浆细胞分化，同时促进调节性 B 细胞的产生[64]。不过，考虑到 B 细胞应答对 T 细胞的依赖性，如 T 细胞分泌的细胞因子 IL-4 或 IFN-γ 能促进 B 细胞增殖，影响 B 细胞抗体类型转化等，MSC 有可能通过抑制 T 细胞的免疫应答，减少 T 细胞相关因子的分泌水平，在体内间接影响 B 细胞的免疫应答。

四、MSC 对固有免疫 T 细胞的调节

γδ T 细胞和 NKT 细胞是主要的两类非常规 T 细胞，数量少，表达的 TCR 缺乏多样性，识别抗原谱窄，且不受 MHC 限制。γδ T 细胞主要分布于皮肤、小肠、肺等黏膜及皮下组织，识别多种病原体的共同抗原成分，活化的 γδ T 细胞分泌多种细胞因子。γδ T 细胞是产生 IL-17 和 IL-22 的主要固有免疫细胞，主要参与皮肤黏膜局部的早期抗感染免疫[64]。NKT 细胞主要识别 CD1d 分子提呈的糖脂类抗原，具有细胞毒效应和可分泌多种细胞因子，参与增强细胞免疫应答[65]。γδ T 细胞与 NKT 细胞在固有和适应性免疫之间起纽带作用，参与抗感染免疫、肿瘤排斥、自身免疫疾病的发病机制及移植耐受的维持。目前关于 MSC 对这两类 T 细胞的调节作用的研究非常有限，研究发现在体外 MSC 抑制由 anti-CD3 或非肽类磷酸抗原诱导的 γδ T 细胞增殖、细胞因子生成和细胞溶解效应，MSC 分泌的 PGE2 可能有助于促进该抑制效应[66]。MSC 对由 anti-CD3 刺激引起的 NKT 细胞的扩增和 IFN-γ 的分泌有抑制作用，但不影响 NKT 细胞的细胞毒效应。MSC 分泌的 PGE2，而不是 IDO 或 TGF-β，部分介导对 NKT 细胞增殖的抑制[67]。

五、MSC 对 NK 细胞的调节

NK 细胞是参与固有免疫应答的主要效应细胞，通过细胞毒效应和产生炎性细胞因子如 IFN-γ、TNF-α 等在早期抗病毒免疫和抗肿瘤免疫中发挥重要作用。NK 细胞表达的活化或抑制性受体，通过与靶细胞表达的特异性配体结合，严格调控其细胞毒效应，NK 细胞的膜表面活化性受体的表达水平与其细胞毒效应正相关。体外 MSC 和 NK 细胞共培养引起 NK 细胞活化受体 NKp30、NKp44 和 NKG2D 的表达水平下调，MSC 抑制由 IL-2 诱导的静息 NK 细胞的增殖、IFN-γ 的产生及其细胞毒效应等，阻断 IDO 和 PGE2 几乎完全恢复 NK 细胞的增殖，提示 MSC 表达的可溶性因子 PGE2 和 IDO 可能介导该抑制效应。但是，MSC 对预活化的 NK 细胞的增殖的抑制不完全，而且由 IL-2 活化的 NK 细胞能杀死自体或同源异体 MSC，可能是由于 MSC 表达低水平的 HLA-Ⅰ分子和识别活化 NK 受体的配体[68]。IFN-γ 至少可以部分地抑制 NK 细胞对 MSC 的细胞毒效应，可能主要通过上调 MSC 的 HLA-Ⅰ分子表达水平，增强 NK 细胞表面的抑制受体的活化，从而抑制 NK 细胞的细胞毒效应[69]。因此，NK 细胞活化的状态及是否存在富含 IFN-γ 的微环境决定体外 MSC 和 NK 细胞相互作用的最终结果。值得注意的是，最近的研究显示低危急性淋巴细胞白血病患者来源的 MSC 可以促进 NK 细胞的活化、细胞毒效应和细胞因子的分泌[70]。

六、MSC 对中性粒细胞的调节

中性粒细胞来源于骨髓，是数量最多的外周白细胞，也是最早进入感染局部，对入侵病原体发挥吞噬、杀伤和清除作用的重要固有免疫细胞，同时，中性粒细胞也是多种慢性炎症和自身免疫系统疾病的重要效应细胞，在慢性炎症病变部位大量聚集，参与组织损伤。在体外，MSC 分泌 IL-6，延迟静息和活化的中性粒细胞的凋亡，并伴随着活性氧（reactive oxygen species，ROS）生成的下降[71]。虽然中性粒细胞产生 ROS 对于清除

病原体是必需的，但过度或持续活化的中性粒细胞发生呼吸爆发，产生过度的 ROS，会导致严重的宿主组织损伤。中性粒细胞呼吸爆发常见于缺血灌注性损伤、急性肺损伤、慢性阻塞性肺病和类风湿关节炎（rheumatoid arthritis，RA）等疾病。因此，MSC 可能通过延迟中性粒细胞的凋亡来减轻呼吸爆发导致的组织损伤。从另外一方面来看，MSC 延迟静息中性粒细胞的凋亡也可能有利于中性粒细胞在骨髓和肺的储备，以便于感染发生的第一时间释放成熟的中性粒细胞清除病原体。最近的研究报道，在败血症中，MSC 可通过提高中性粒细胞的存活，增强其吞噬能力来促进细菌的清除[72]。

七、MSC 对巨噬细胞的调节

巨噬细胞是重要的固有免疫细胞，广泛分布于全身各组织，具有强大的吞噬、杀菌和清除凋亡细胞及其他异物的能力，同时也是参与细胞免疫的重要效应细胞。根据活化状态和功能，巨噬细胞能分为两类，一类是促炎的 M1 型巨噬细胞，同时具有杀菌活性；另一类是具有组织修复功能的 M2 型巨噬细胞。MSC 可诱导巨噬细胞从 M1 型向 M2 型转化，加速组织损伤的修复过程[73-75]。最近发现，脐血间充质干细胞（UCB-MSC）可通过分泌 IL-6 来诱导 M2 型巨噬细胞极化，改善 2 型糖尿病的胰岛素抵抗。此外，最近研究报道 MSC 可直接促进创伤脑组织中小胶质细胞向 M2 型巨噬细胞的极化[76]。体外研究显示，BM-MSC 通过细胞接触和分泌 PGE2、IL-6 和 IDO 促进单核细胞向 M2 型巨噬细胞成熟[50,77,78]。MSC 可影响巨噬细胞的细胞因子分泌谱，即降低致炎因子的分泌水平，提高抑炎因子的分泌水平，以减轻过度的炎症造成的组织损伤。在小鼠败血症模型中，LPS 或 TNF-α 活化的 MSC 分泌大量 PGE2，促进巨噬细胞分泌 IL-10，从而引起中性粒细胞滞留于外周血液中以清除细菌，同时防止中性粒细胞到达炎症局部引起氧化损伤，从而平衡了中性粒细胞的抗菌效应和组织损伤效应[79]。此外，MSC 也可提高巨噬细胞对凋亡细胞清除能力，缓解炎症[80]。值得注意的是，最近的研究表明，在多种小鼠肿瘤模型中，MSC 可通过促进肿瘤原位的 M2 型巨噬细胞的浸润，协助肿瘤的发生发展和药物抵抗[81-83]。

八、MSC 对肥大细胞的调节

肥大细胞是参与过敏反应的主要固有免疫细胞[84]，在宿主防御和自身免疫中也起重要的作用[85,86]。BM-MSC 在多种小鼠过敏模型中可抑制肥大细胞的脱颗粒和分泌炎性细胞因子的能力，主要是通过 MSC 分泌的 PGE2 及 TGF-β1 介导的[87-92]。有趣的是，肥大细胞反过来也可影响 MSC 的分化和功能[93]，如经 IgE 刺激的肥大细胞可活化 MSC，诱导其释放多种生长因子，调节 CD34$^+$ 祖细胞的增殖[94]。

第四节　间充质干细胞免疫调节能力的可塑性

MSC 的免疫调节能力不是与生俱来的，多种因素可影响 MSC 的免疫调节能力，其中微环境中的炎性细胞因子可能是最重要的影响因素。IFN-γ 是调节 MSC 免疫功能的最关键的细胞因子，对 MSC 免疫功能的调节作用依赖其浓度高低[95]。在生理状况下，低浓度的 IFN-γ 不能诱导对 MSC 免疫抑制能力起关键作用的诱导型一氧化氮合酶(inducible nitric oxide synthase, iNOS) 或 IDO 的大量产生,但足以上调趋化因子的分泌,招募效应 T 细胞在 MSC 周边聚集,从而加重炎症。此外,低浓度的 IFN-γ 可促进 MSC 的抗原提呈功能,活化抗原特异性 CTL[19]。相反,高浓度的 IFN-γ 可诱导 MSC 的免疫抑制能力。同时,多种致炎因子与 IFN-γ 协同作用,进一步提高 MSC 的免疫抑制能力。例如,TNF-α 和 IL-1β 可与 IFN-γ 发挥协同作用,促进 MSC 高表达 NO 生成的关键酶 iNOS,通过 NO 依赖的方式抑制 T 细胞免疫应答[34];TNF-α 与 IFN-γ 也可引起 MSC 的 COX2 和 PGE2 表达上调,通过 PGE2 依赖的方式抑制免疫应答。近期研究发现,同时加入 IL-17 可通过提高 RNA 结合蛋白 AUF1 介导的 iNOS 编码 RNA 的稳定性,增强 MSC 介导的免疫抑制[96]。TNF-α 和 IFN-γ 还促进 MSC 上调表达 CD54,诱导 MSC 与 Th17 细胞的黏附,以细胞接触依赖方式抑制 Th17 细胞的免疫应答效应[38]。此外,经炎性细胞因子作用的 MSC 上调表达血管细胞黏附分子-1 (VCAM-1) 和细胞间黏附分子-1 (intercellular adhesion molecule-1, ICAM-1),有助于 MSC 与 T 细胞的黏附,从而发挥其抑制 T 细胞增殖的功能[97]。免疫抑制细胞因子如 TGF-β 也经常存在于炎症环境中,发挥抗炎作用。当 TGF-β 与 IFN-γ 和 TNF-α 同时存在时,通过活化 TGF-β 下游的 Smad3,下调 iNOS 或 IDO 的表达,从而降低 MSC 的免疫抑制能力[98]。有意思的是,MSC 本身产生大量的 TGF-β,可能通过负反馈调节发挥部分维持炎症的作用。IL-10 经常和 TGF-β 共同作用,减弱 MSC 的免疫抑制能力[99]。因此,这些常见的免疫抑制细胞因子可通过作用于 MSC,减弱其免疫抑制能力,提高免疫应答效应。综上所述,MSC 的免疫调节能力依赖于所处微环境中炎性介质的性质和浓度,炎症状态从根本上决定了 MSC 的免疫调节命运及其疗效,提示 MSC 免疫调控具有可塑性。在免疫应答过程中,炎症状态随着时间、免疫系统活化物和许多其他因子的不同发生动态改变,因此,不同的炎症阶段 MSC 抑制免疫应答的水平存在显著差异。例如,在 GVHD 和 EAE 小鼠模型的急性炎症期进行 MSC 治疗可有效抑制炎症,但在炎症发生前或缓解期给予 MSC 则治疗效果明显减弱[100,101]。

免疫抑制剂经常用于治疗免疫性疾病,这样一个重要的问题出现了:免疫抑制剂是否也能影响 MSC 的免疫调节功能?研究发现,MSC 治疗小鼠 GVHD 的同时使用常见的免疫抑制剂环孢菌素 A 可消除 MSC 诱导的受体小鼠的免疫耐受[102]。同样,地塞米松也能在体内外消除 MSC 介导的免疫抑制。机制研究表明,地塞米松可阻止小鼠 MSC 表达 iNOS 或 hMSC 表达 IDO,导致炎症失控[103]。因此,免疫抑制剂和 TGF-β 一样具有减弱炎症环境中 MSC 的免疫抑制活性的能力。这个发现对临床治疗使用免疫抑制剂和 MSC 具有重要的提示作用。临床研究也支持该发现,如 MSC 对免疫抑制剂治疗不应答的急性炎症疾病的疗效最好[104,105],且临床报道 MSC 成功治疗了严重的、对药物无反应的 SLE 和克罗恩病 (CD)[104,106]。在这些对免疫抑制剂无反应的患者中,持续炎症可能赋予 MSC 免疫抑制功能。

MSC 具有固有免疫细胞的某些功能,且表达多种功能性的 TLR。最近研究表明,MSC 的 TLR 信号途径可能影响其分化、增殖、迁移和免疫抑制功能[107],但确切的作用还有待进一步的阐明。TLR 信号转导通路参与多种免疫相关疾病的发生和发展,可能通过启动异常免疫应答或维持慢性感染来加重机体免疫损伤,因此,理解 TLR 通路如何影响MSC免疫抑制能力对开发MSC有效治疗免疫相关疾病的策略非常重要。MSC表达的TLR的活化也能引起MSC免疫调节功能的转变。研究表明，致炎因子通过TLR4引起MSC向MSC1表型分化，主要表现为产生IL-6、IL-8和TGF-β；而抗炎信号通过TLR3引起MSC向MSC2表型分化[108,109]，主要表现为上调表达IDO、COX2、IL-4和IL-1RA。不同的MSC表型对炎症有相反的效应。提示不同的TLR通路活化后可能对MSC的免疫调节能力有截然不同的影响。理论上，MSC诱导的免疫抑制可能导致机体对病原微生物的易感性，然而，MSC表达功能性TLR，多种病原体及其产物可以活化MSC的TLR通路，一方面可能促进MSC的免疫抑制能力，从而抑制过度的免疫应答，限制组织损伤的程度和范围，有助于炎症的缓解；另一方面TLR活化的MSC可能分泌多种致炎细胞因子，逆转MSC对免疫应答的抑制效应，在炎症缓解后恢复机体的抗感染免疫。因此，阐明TLR通路活化影响MSC免疫调节能力的具体机制对MSC临床应用的有效性和安全性是非常重要的。

第五节　间充质干细胞治疗免疫相关疾病的临床应用研究

MSC用于治疗自身免疫系统疾病的研究源于应用MSC治疗小鼠EAE。注射自体的MSC能明显降低中枢神经组织中炎症细胞的浸润，抑制自身反应性T细胞的增殖，改善小鼠EAE的症状[110]。随后，大量研究利用多种小鼠免疫相关炎症疾病模型支持MSC治疗能够抑制不同类型免疫细胞介导的过度免疫应答导致的组织损伤，且具有明显的治疗效果，如MSC能缓解由胶原诱导的关节炎、1型糖尿病、自身免疫性心肌炎、败血症和急性肺损伤等临床症状，以及降低实验动物的死亡率[111-114]。不同于一般的免疫抑制治疗，研究显示MSC治疗不会增加条件致病菌的危险程度[115]。临床数据表明，MSC对CD具有较好的疗效，已开始临床Ⅲ期试验[115,116]。数十个MSC治疗多发性硬化（multiple sclerosis，MS）的临床前期试验也在世界范围内展开[117,118]。虽然MSC的低免疫原性有助于它们的异体应用，但是在某些情况下，MSC具有启动免疫应答的能力。因此针对某些疾病，如慢性器官性自身免疫系统疾病，使用自体MSC可能是最佳选择。研究表明来自自身免疫系统疾病患者的MSC表型正常，具有正常的造血支持和免疫抑制能力[119,120]，如果能确保体外扩增出足够的治疗用自体MSC，使用自体MSC治疗器官性自身免疫系统疾病是有可能的。鉴于MSC免疫调节能力的可塑性，国际细胞治疗协会（ISCT）建议临床应用前评价由炎性细胞因子诱导的MSC免疫调节能力[121]。此外，临床应用MSC治疗需要注意的是，MSC的免疫抑制能力和分泌生长因子的能力可能有助于肿瘤细胞逃避免疫监视与肿瘤细胞的增殖。目前MSC对肿瘤体内外生长的影响的研究存在不一致的结果[122,123]，但最近一篇研究显示肿瘤微环境中被炎性细胞因子活化的MSC可能通过分泌某些趋化因子，诱导具有免疫抑制能力的髓样细胞迁移到肿瘤局部，从而发挥促肿瘤效应[83]。因此，进一步阐明MSC影响肿瘤发生发展的可能机制，对临床安全有效地应用MSC治疗免疫相关疾病非常重要。

（何　睿）

参 考 文 献

[1] Arnold JN, Magiera L, Kraman M, et al. Tumoral immune suppression by macrophages expressing fibroblast activation protein-alpha and heme oxygenase-1. Cancer Immunology Research, 2014, 2(2):121-126.

[2] Brestoff JR, Artis D. Immune regulation of metabolic homeostasis in health and disease. Cell, 2015, 161(1):146-160.

[3] Murphy KM, Reiner SL. The lineage decisions of helper T cells. Nature Reviews Immunology, 2002, 2(12):933-944.

[4] Yamauchi K, Piao HM, Nakadate T, et al. Progress in allergy signal research on mast cells: the role of histamine in goblet cell hyperplasia in allergic airway inflammation-a study using the Hdc knockout mouse. Journal of Pharmacological Sciences, 2008, 106(3):354-360.

[5] Wong P, Pamer EG. CD8 T cell responses to infectious pathogens. Annual Review of Immunology, 2003, 21:29-70.

[6] Fontenot JD, Rudensky AY. A well adapted regulatory contrivance: regulatory T cell development and the forkhead family transcription factor Foxp3. Nature Immunology, 2005, 6(4):331-337.

[7] Sallusto F, Geginat J, Lanzavecchia A. Central memory and effector memory T cell subsets: function, generation, and maintenance. Annual Review of Immunology, 2004, 22:745-763.

[8] Steinman RM, Hemmi H. Dendritic cells: translating innate to adaptive immunity. Curr Top Microbiol, 2006, 311:17-58.

[9] van den Berg TK, Yoder JA, Litman GW. On the origins of adaptive immunity: innate immune receptors join the tale. Trends Immunol, 2004, 25(1):11-16.

[10] Shortman K, Liu YJ. Mouse and human dendritic cell subtypes. Nature Reviews Immunology, 2002, 2(3):151-161.

[11] Goodnow CC, Sprent J, Fazekas de St Groth B, et al. Cellular and genetic mechanisms of self tolerance and autoimmunity. Nature, 2005, 435(7042):590-597.

[12] Greenwald RJ, Freeman GJ, Sharpe AH. The B7 family revisited. Annual Review of Immunology, 2005, 23:515-548.

[13] Tanaka S, Fukuda A, Akiyama T, et al. Molecular mechanisms regulating life and death of the osteoclast. Bone, 2003, 32(5):S69.

[14] Paul WE, Zhu JF. How are T(h)2-type immune responses initiated and amplified? Nature Reviews Immunology, 2010, 10(4):225-235.

[15] Miossec P, Korn T, Kuchroo VK. Interleukin-17 and type 17 helper T cells. The New England Journal of Medicine, 2009, 361(9):888-898.

[16] Sakaguchi S, Yamaguchi T, Nomura T, et al. Regulatory T cells and immune tolerance. Cell, 2008, 133(5):775-787.

[17] Eliopoulos N, Stagg J, Lejeune L, et al. Allogeneic marrow stromal cells are immune rejected by MHC class I - and class II -mismatched recipient mice. Blood, 2005, 106(13):4057-4065.

[18] Nauta AJ, Westerhuis G, Kruisselbrink AB, et al. Donor-derived mesenchymal stem cells are immunogenic in an allogeneic host and stimulate donor graft rejection in a nonmyeloablative setting. Blood, 2006, 108(6):2114-2120.

[19] Francois M, Romieu-Mourez R, Stock-Martineau S, et al. Mesenchymal stromal cells cross-present soluble exogenous antigens as part of their antigen-presenting cell properties. Blood, 2009, 114(13):2632-2638.

[20] Chan JL, Tang KC, Patel AP, et al. Antigen-presenting property of mesenchymal stem cells occurs during a narrow window at low levels of interferon-gamma. Blood, 2006, 107(12):4817-4824.

[21] Stagg J, Pommey S, Eliopoulos N, et al. Interferon-gamma-stimulated marrow stromal cells: a new type of nonhematopoietic antigen-presenting cell. Blood, 2006, 107(6):2570-2577.

[22] Liechty KW, MacKenzie TC, Shaaban AF, et al. Human mesenchymal stem cells engraft and demonstrate site-specific differentiation after in utero transplantation in sheep. Nature Medicine, 2000, 6(11):1282-1286.

[23] Le Blanc K, Rasmusson I, Sundberg B, et al. Treatment of severe acute graft-versus-host disease with third party haploidentical mesenchymal stem cells. Lancet, 2004, 363(9419):1439-1441.

[24] Tse WT, Pendleton JD, Beyer WM, et al. Suppression of allogeneic T-cell proliferation by human marrow stromal cells: implications in transplantation. Transplantation, 2003, 75(3):389-397.

[25] Aggarwal S, Pittenger MF. Human mesenchymal stem cells modulate allogeneic immune cell responses. Blood, 2005, 105(4):1815-1822.

[26] Jarvinen L, Badri L, Wettlaufer S, et al. Lung resident mesenchymal stem cells isolated from human lung allografts inhibit T cell proliferation via a soluble mediator. J Immunol, 2008, 181(6):4389-4396.

[27] English K, Barry FP, Field-Corbett CP, et al. IFN-gamma and TNF-alpha differentially regulate immunomodulation by murine mesenchymal stem cells. Immunology Letters, 2007, 110(2):91-100.

[28] Meisel R, Zibert A, Laryea M, et al. Human bone marrow stromal cells inhibit allogeneic T-cell responses by indoleamine 2,3-dioxygenase-mediated tryptophan degradation. Blood, 2004, 103(12):4619-4621.

[29] Selmani Z, Naji A, Zidi I, et al. Human leukocyte antigen-G5 secretion by human mesenchymal stem cells is required to suppress T lymphocyte and natural killer function and to induce $CD4^+$ CD25 high $FOXP3^+$ regulatory T cells. Stem Cells, 2008, 26(1):212-222.

[30] Morandi F, Raffaghello L, Bianchi G, et al. Immunogenicity of human mesenchymal stem cells in HLA-class I -restricted T-cell responses against viral or tumor-associated antigens. Stem Cells, 2008, 26(5):1275-1287.

[31] Sato K, Ozaki K, Oh I, et al. Nitric oxide plays a critical role in suppression of T-cell proliferation by mesenchymal stem cells. Blood, 2007, 109(1):228-234.

[32] Giunta D, Fuentes N, Pazo V, et al. Creation of a hyponatremia registry supported by an industry-derived quality control methodology. Applied Clinical Informatics, 2011, 2(1):86-93.

[33] Glennie S, Soeiro I, Dyson PJ, et al. Bone marrow mesenchymal stem cells induce division arrest anergy of activated T cells. Blood, 2005, 105(7):2821-2827.

[34] Ren G, Zhang L, Zhao X, et al. Mesenchymal stem cell-mediated immunosuppression occurs via concerted action of chemokines and nitric oxide. Cell Stem Cell, 2008, 2(2):141-150.

[35] Lim JH, Kim JS, Yoon IH, et al. Immunomodulation of delayed-type hypersensitivity responses by mesenchymal stem cells is associated with bystander T cell apoptosis in the draining lymph node. J Immunol, 2010, 185(7):4022-4029.

[36] Cutler AJ, Limbani V, Girdlestone J, et al. Umbilical cord-derived mesenchymal stromal cells modulate monocyte function to suppress T cell proliferation. J Immunol, 2010, 185(11):6617-6623.

[37] Augello A, Tasso R, Negrini SM, et al. Bone marrow mesenchymal progenitor cells inhibit lymphocyte proliferation by activation of the programmed death 1 pathway. Eur J Immunol, 2005, 35(5):1482-1490.

[38] Ghannam S, Pene J, Moquet-Torcy G, et al. Mesenchymal stem cells inhibit human Th17 cell differentiation and function and induce a T regulatory cell phenotype. J Immunol, 2010, 185(1):302-312.

[39] Karlsson H, Samarasinghe S, Ball LM, et al. Mesenchymal stem cells exert differential effects on alloantigen and virus-specific T-cell responses. Blood, 2008, 112(3):532-541.

[40] Rasmusson I, Ringden O, Sundberg B, et al. Mesenchymal stem cells inhibit the formation of cytotoxic T lymphocytes, but not activated cytotoxic T lymphocytes or natural killer cells. Transplantation, 2003, 76(8):1208-1213.

[41] Rasmusson I, Uhlin M, Le Blanc K, et al. Mesenchymal stem cells fail to trigger effector functions of cytotoxic T lymphocytes. Journal of Leukocyte Biology, 2007, 82(4):887-893.

[42] Di Ianni M, Del Papa B, De Ioanni M, et al. Mesenchymal cells recruit and regulate T regulatory cells. Experimental Hematology, 2008, 36(3):309-318.

[43] Patel SA, Meyer JR, Greco SJ, et al. Mesenchymal stem cells protect breast cancer cells through regulatory T cells: role of mesenchymal stem cell-derived TGF-beta. J Immunol, 2010, 184(10):5885-5894.

[44] English K, Ryan JM, Tobin L, et al. Cell contact, prostaglandin E(2) and transforming growth factor beta 1 play non-redundant roles in human mesenchymal stem cell induction of $CD4^+$ CD25(high) forkhead box P3+ regulatory T cells. Clinical and Experimental Immunology, 2009, 156(1):149-160.

[45] Maccario R, Podesta M, Moretta A, et al. Interaction of human mesenchymal stem cells with cells involved in alloantigen-specific immune response favors the differentiation of CD4$^+$ T-cell subsets expressing a regulatory/suppressive phenotype. Haematologica, 2005, 90(4):516-525.

[46] Yagi H, Soto-Gutierrez A, Parekkadan B, et al. Mesenchymal stem cells: mechanisms of immunomodulation and homing. Cell Transplantation, 2010, 19(6):667-679.

[47] Casiraghi F, Azzollini N, Cassis P, et al. Pretransplant infusion of mesenchymal stem cells prolongs the survival of a semiallogeneic heart transplant through the generation of regulatory T cells. J Immunol, 2008, 181(6):3933-3946.

[48] Gonzalez MA, Gonzalez-Rey E, Rico L, et al. Treatment of experimental arthritis by inducing immune tolerance with human adipose-derived mesenchymal stem cells. Arthritis and Rheumatism, 2009, 60(4):1006-1019.

[49] Nemeth K, Keane-Myers A, Brown JM, et al. Bone marrow stromal cells use TGF-beta to suppress allergic responses in a mouse model of ragweed-induced asthma. Proceedings of the National Academy of Sciences of the United States of America, 2010, 107(12):5652-5657.

[50] 0Melief SM, Schrama E, Brugman MH, et al. Multipotent stromal cells induce human regulatory T cells through a novel pathway involving skewing of monocytes toward anti-inflammatory macrophages. Stem Cells, 2013, 31(9):1980-1991.

[51] Li H, Guo Z, Jiang X, et al. Mesenchymal stem cells alter migratory property of T and dendritic cells to delay the development of murine lethal acute graft-versus-host disease. Stem Cells, 2008, 26(10):2531-2541.

[52] Nauta AJ, Kruisselbrink AB, Lurvink E, et al. Mesenchymal stem cells inhibit generation and function of both CD34$^+$-derived and monocyte-derived dendritic cells. J Immunol, 2006, 177(4):2080-2087.

[53] English K, Barry FP, Mahon BP. Murine mesenchymal stem cells suppress dendritic cell migration, maturation and antigen presentation. Immunology Letters, 2008, 115(1):50-58.

[54] Jiang XX, Zhang Y, Liu B, et al. Human mesenchymal stem cells inhibit differentiation and function of monocyte-derived dendritic cells. Blood, 2005, 105(10):4120-4126.

[55] Djouad F, Charbonnier LM, Bouffi C, et al. Mesenchymal stem cells inhibit the differentiation of dendritic cells through an interleukin-6-dependent mechanism. Stem Cells, 2007, 25(8):2025-2032.

[56] Spaggiari GM, Abdelrazik H, Becchetti F, et al. MSCs inhibit monocyte-derived DC maturation and function by selectively interfering with the generation of immature DCs: central role of MSC-derived prostaglandin E2. Blood, 2009, 113(26):6576-6583.

[57] Zhang B, Liu R, Shi D, et al. Mesenchymal stem cells induce mature dendritic cells into a novel Jagged-2-dependent regulatory dendritic cell population. Blood, 2009, 113(1):46-57.

[58] Ivanova-Todorova E, Bochev I, Mourdjeva M, et al. Adipose tissue-derived mesenchymal stem cells are more potent suppressors of dendritic cells differentiation compared to bone marrow-derived mesenchymal stem cells. Immunology Letters, 2009, 126(1-2):37-42.

[59] Aldinucci A, Rizzetto L, Pieri L, et al. Inhibition of immune synapse by altered dendritic cell actin distribution: a new pathway of mesenchymal stem cell immune regulation. J Immunol, 2010, 185(9):5102-5110.

[60] Corcione A, Benvenuto F, Ferretti E, et al. Human mesenchymal stem cells modulate B-cell functions. Blood, 2006, 107(1):367-372.

[61] Rafei M, Hsieh J, Fortier S, et al. Mesenchymal stromal cell-derived CCL2 suppresses plasma cell immunoglobulin production via STAT3 inactivation and PAX5 induction. Blood, 2008, 112(13):4991-4998.

[62] Traggiai E, Volpi S, Schena F, et al. Bone marrow-derived mesenchymal stem cells induce both polyclonal expansion and differentiation of B cells isolated from healthy donors and systemic lupus erythematosus patients. Stem Cells, 2008, 26(2):562-569.

[63] Gerdoni E, Gallo B, Casazza S, et al. Mesenchymal stem cells effectively modulate pathogenic immune response in experimental autoimmune encephalomyelitis. Annals of Neurology, 2007, 61(3):219-227.

[64] Shin TH, Lee BC, Choi SW, et al. Human adipose tissue-derived mesenchymal stem cells alleviate atopic dermatitis via regulation of B lymphocyte maturation. Oncotarget, 2017, 8(1):512-522.

[65] Godfrey DI, Stankovic S, Baxter AG. Raising the NKT cell family. Nature Immunology, 2010, 11(3):197-206.

[66] Martinet L, Fleury-Cappellesso S, Gadelorge M, et al. A regulatory cross-talk between Vgamma9 Vdelta2 T lymphocytes and mesenchymal stem cells. Eur J Immunol, 2009, 39(3):752-762.

[67] Prigione I, Benvenuto F, Bocca P, et al. Reciprocal interactions between human mesenchymal stem cells and gamma delta T cells or invariant natural killer T cells. Stem Cells, 2009, 27(3):693-702.

[68] Spaggiari GM, Capobianco A, Abdelrazik H, et al. Mesenchymal stem cells inhibit natural killer-cell proliferation, cytotoxicity, and cytokine production: role of indoleamine 2,3-dioxygenase and prostaglandin E2. Blood, 2008, 111(3):1327-1333.

[69] Spaggiari GM, Capobianco A, Becchetti S, et al. Mesenchymal stem cell-natural killer cell interactions: evidence that activated NK cells are capable of killing MSCs, whereas MSCs can inhibit IL-2-induced NK-cell proliferation. Blood, 2006, 107(4):1484-1490.

[70] Entrena A, Varas A, Vazquez M, et al. Mesenchymal stem cells derived from low risk acute lymphoblastic leukemia patients promote NK cell antitumor activity. Cancer Letters, 2015, 363(2):156-165.

[71] Raffaghello L, Bianchi G, Bertolotto M, et al. Human mesenchymal stem cells inhibit neutrophil apoptosis: a model for neutrophil preservation in the bone marrow niche. Stem Cells, 2008, 26(1):151-162.

[72] Hall SR, Tsoyi K, Ith B, et al. Mesenchymal stromal cells improve survival during sepsis in the absence of heme oxygenase-1: the importance of neutrophils. Stem Cells, 2013, 31(2):397-407.

[73] Cho DI, Kim MR, Jeong HY, et al. Mesenchymal stem cells reciprocally regulate the M1/M2 balance in mouse bone marrow-derived macrophages. Experimental & Molecular Medicine, 2014, 46:e70.

[74] Frangogiannis NG. Regulation of the inflammatory response in cardiac repair. Circulation Research, 2012, 110(1):159-173.

[75] Zhang QZ, Su WR, Shi SH, et al. Human gingiva-derived mesenchymal stem cells elicit polarization of m2 macrophages and enhance cutaneous wound healing. Stem Cells, 2010, 28(10):1856-1868.

[76] Zanier ER, Pischiutta F, Riganti L, et al. Bone marrow mesenchymal stromal cells drive protective M2 microglia polarization after brain trauma. Neurotherapeutics: the Journal of the American Society for Experimental NeuroTherapeutics, 2014, 11(3):679-695.

[77] Xie Z, Hao H, Tong C, et al. Human umbilical cord-derived mesenchymal stem cells elicit macrophages into an anti-inflammatory phenotype to alleviate insulin resistance in type 2 diabetic rats. Stem Cells, 2016, 34(3):627-639.

[78] Francois M, Romieu-Mourez R, Li M, et al. Human MSC suppression correlates with cytokine induction of indoleamine 2,3-dioxygenase and bystander M2 macrophage differentiation. Molecular Therapy: the Journal of the American Society of Gene Therapy, 2012, 20(1):187-195.

[79] Nemeth K, Leelahavanichkul A, Yuen PS, et al. Bone marrow stromal cells attenuate sepsis via prostaglandin E(2)-dependent reprogramming of host macrophages to increase their interleukin-10 production. Nature Medicine, 2009, 15(1):42-49.

[80] Mei SH, Haitsma JJ, Dos Santos CC, et al. Mesenchymal stem cells reduce inflammation while enhancing bacterial clearance and improving survival in sepsis. American Journal of Respiratory and Critical Care Medicine, 2010, 182(8):1047-1057.

[81] Jia XH, Feng GW, Wang ZL, et al. Activation of mesenchymal stem cells by macrophages promotes tumor progression through immune suppressive effects. Oncotarget, 2016, 7(15):20934-20944.

[82] Jia XH, Du Y, Mao D, et al. Zoledronic acid prevents the tumor-promoting effects of mesenchymal stem cells via MCP-1 dependent recruitment of macrophages. Oncotarget, 2015, 6(28):26018-26028.

[83] Ren G, Zhao X, Wang Y, et al. CCR2-dependent recruitment of macrophages by tumor-educated mesenchymal stromal cells promotes tumor development and is mimicked by TNFalpha. Cell Stem Cell, 2012, 11(6):812-824.

[84] Amin K. The role of mast cells in allergic inflammation. Respiratory Medicine, 2012, 106(1):9-14.

[85] Theoharides TC, Alysandratos KD, Angelidou A, et al. Mast cells and inflammation. Biochimica et Biophysica Acta, 2012, 1822(1):21-33.

[86] Heybeli C. Mast cells, mastocytosis, and related disorders. The New England Journal of Medicine, 2015, 373(19):1885.

[87] Brown JM, Nemeth K, Kushnir-Sukhov NM, et al. Bone marrow stromal cells inhibit mast cell function via a COX2-dependent mechanism. Clinical and Experimental Allergy: Journal of the British Society for Allergy and Clinical Immunology, 2011, 41(4):526-534.

[88] Kim HS, Yun JW, Shin TH, et al. Human umbilical cord blood mesenchymal stem cell-derived PGE2 and TGF-beta1 alleviate atopic dermatitis by reducing mast cell degranulation. Stem Cells, 2015, 33(4):1254-1266.

[89] Liu J, Kuwabara A, Kamio Y, et al. Human mesenchymal stem cell-derived microvesicles prevent the rupture of intracranial aneurysm in part by suppression of mast cell activation via a PGE2-dependent mechanism. Stem Cells, 2016, 34(12):2943-2955.

[90] Su W, Wan Q, Huang J, et al. Culture medium from TNF-alpha-stimulated mesenchymal stem cells attenuates allergic conjunctivitis through multiple antiallergic mechanisms. The Journal of Allergy and Clinical Immunology, 2015, 136(2):423-428.

[91] Kim A, Yu HY, Heo J, et al. Mesenchymal stem cells protect against the tissue fibrosis of ketamine-induced cystitis in rat bladder. Scientific Reports, 2016, 6:30881.

[92] Su WR, Zhang QZ, Shi SH, et al. Human gingiva-derived mesenchymal stromal cells attenuate contact hypersensitivity via prostaglandin E2-dependent mechanisms. Stem Cells, 2011, 29(11):1849-1860.

[93] Nazari M, Ni NC, Ludke A, et al. Mast cells promote proliferation and migration and inhibit differentiation of mesenchymal stem cells through PDGF. Journal of Molecular and Cellular Cardiology, 2016, 94:32-42.

[94] Allakhverdi Z, Comeau MR, Armant M, et al. Mast cell-activated bone marrow mesenchymal stromal cells regulate proliferation and lineage commitment of CD34(+) progenitor cells. Frontiers in Immunology, 2013, 4:461.

[95] Li W, Ren G, Huang Y, et al. Mesenchymal stem cells: a double-edged sword in regulating immune responses. Cell Death and Differentiation, 2012, 19(9):1505-1513.

[96] Han X, Yang Q, Lin L, et al. Interleukin-17 enhances immunosuppression by mesenchymal stem cells. Cell Death and Differentiation, 2014, 21(11):1758-1768.

[97] Ren G, Zhao X, Zhang L, et al. Inflammatory cytokine-induced intercellular adhesion molecule-1 and vascular cell adhesion molecule-1 in mesenchymal stem cells are critical for immunosuppression. J Immunol, 2010, 184(5):2321-2328.

[98] Xu C, Yu P, Han X, et al. TGF-beta promotes immune responses in the presence of mesenchymal stem cells. J Immunol, 2014, 192(1):103-109.

[99] Li MO, Flavell RA. Contextual regulation of inflammation: a duet by transforming growth factor-beta and interleukin-10. Immunity, 2008, 28(4):468-476.

[100] Sudres M, Norol F, Trenado A, et al. Bone marrow mesenchymal stem cells suppress lymphocyte proliferation in vitro but fail to prevent graft-versus-host disease in mice. J Immunol, 2006, 176(12):7761-7767.

[101] Constantin G, Marconi S, Rossi B, et al. Adipose-derived mesenchymal stem cells ameliorate chronic experimental autoimmune encephalomyelitis. Stem Cells, 2009, 27(10):2624-2635.

[102] Inoue S, Popp FC, Koehl GE, et al. Immunomodulatory effects of mesenchymal stem cells in a rat organ transplant model. Transplantation, 2006, 81(11):1589-1595.

[103] Chen X, Gan Y, Li W, et al. The interaction between mesenchymal stem cells and steroids during inflammation. Cell Death & Disease, 2014, 5:e1009.

[104] Le Blanc K, Frassoni F, Ball L, et al. Mesenchymal stem cells for treatment of steroid-resistant, severe, acute graft-versus-host disease: a phase II study. Lancet, 2008, 371(9624):1579-1586.

[105] Sun L, Wang D, Liang J, et al. Umbilical cord mesenchymal stem cell transplantation in severe and refractory systemic lupus erythematosus. Arthritis and Rheumatism, 2010, 62(8):2467-2475.

[106] Dalal J, Gandy K, Domen J. Role of mesenchymal stem cell therapy in Crohn's disease. Pediatric Research, 2012, 71(4 Pt 2):445-451.

[107] Tomchuck SL, Zwezdaryk KJ, Coffelt SB, et al. Toll-like receptors on human mesenchymal stem cells drive their migration and immunomodulating responses. Stem Cells, 2008, 26(1):99-107.

[108] Bernardo ME, Fibbe WE. Mesenchymal stromal cells: sensors and switchers of inflammation. Cell Stem Cell, 2013, 13(4):392-402.

[109] Waterman RS, Tomchuck SL, Henkle SL, et al. A new mesenchymal stem cell (MSC) paradigm: polarization into a pro-inflammatory MSC1 or an immunosuppressive MSC2 phenotype. PLoS One, 2010, 5(4):e10088.

[110] Zappia E, Casazza S, Pedemonte E, et al. Mesenchymal stem cells ameliorate experimental autoimmune encephalomyelitis inducing T-cell anergy. Blood, 2005, 106(5):1755-1761.

[111] Chen FH, Tuan RS. Mesenchymal stem cells in arthritic diseases. Arthritis Research & Therapy, 2008, 10(5):223.

[112] Vija L, Farge D, Gautier JF, et al. Mesenchymal stem cells: stem cell therapy perspectives for type 1 diabetes. Diabetes & Metabolism, 2009, 35(2):85-93.

[113] Okada H, Suzuki J, Futamatsu H, et al. Attenuation of autoimmune myocarditis in rats by mesenchymal stem cell transplantation through enhanced expression of hepatocyte growth factor. International Heart Journal, 2007, 48(5):649-661.

[114] Gupta N, Su X, Popov B, et al. Intrapulmonary delivery of bone marrow-derived mesenchymal stem cells improves survival and attenuates endotoxin-induced acute lung injury in mice. J Immunol, 2007, 179(3):1855-1863.

[115] Gregoire C, Lechanteur C, Briquet A, et al. Review article: mesenchymal stromal cell therapy for inflammatory bowel diseases. Alimentary Pharmacology & Therapeutics, 2017, 45(2):205-221.

[116] Auletta JJ, Bartholomew AM, Maziarz RT, et al. The potential of mesenchymal stromal cells as a novel cellular therapy for multiple sclerosis. Immunotherapy, 2012, 4(5):529-547.

[117] Harris VK, Vyshkina T, Sadiq SA, Clinical safety of intrathecal administration of mesenchymal stromal cell-derived neural progenitors in multiple sclerosis. Cytotherapy, 2016, 18(12):1476-1482.

[118] Salem HK, Thiemermann C. Mesenchymal stromal cells: current understanding and clinical status. Stem Cells, 2010, 28(3):585-596.

[119] Roubelakis MG, Pappa KI, Bitsika V, et al. Molecular and proteomic characterization of human mesenchymal stem cells derived from amniotic fluid: comparison to bone marrow mesenchymal stem cells. Stem Cells and Development, 2007, 16(6):931-952.

[120] Papadaki HA, Tsagournisakis M, Mastorodemos V, et al. Normal bone marrow hematopoietic stem cell reserves and normal stromal cell function support the use of autologous stem cell transplantation in patients with multiple sclerosis. Bone Marrow Transplantation, 2005, 36(12):1053-1063.

[121] Krampera M, Galipeau J, Shi Y, et al. Immunological characterization of multipotent mesenchymal stromal cells-The International Society for Cellular Therapy (ISCT) working proposal. Cytotherapy, 2013, 15(9):1054-1061.

[122] Ramasamy R, Lam EW, Soeiro I, et al. Mesenchymal stem cells inhibit proliferation and apoptosis of tumor cells: impact on *in vivo* tumor growth. Leukemia, 2007, 21(2):304-310.

[123] Ame-Thomas P, Maby-El Hajjami H, Monvoisin C, et al. Human mesenchymal stem cells isolated from bone marrow and lymphoid organs support tumor B-cell growth: role of stromal cells in follicular lymphoma pathogenesis. Blood, 2007, 109(2):693-702.

第六章 间充质干细胞与治疗性血管新生

第一节 血管新生的概念

在人体中存在着复杂的血管网，其为组织细胞提供了生存、增殖、分化所必需的营养和氧气。这一庞大的血管网既包括稳定的、较大的动脉及静脉，也包括细小并存在动态变化的微血管结构，如毛细血管、小动脉及小静脉。

在胚胎时期，血管的生成依靠血管发生（vasculogenesis）机制，新生的脉管系统来源于内皮祖细胞（EPC），亦称为成血管细胞（angioblast）。随后，新生血管利用"出芽"的方式从已经存在的"亲代"血管上产生，原始的血管床随着血管新生（angiogenesis）这一过程不断扩大，逐渐形成我们成年后基本的血管网。当然，血管生成、发展的方式有多种，但是目前认为我们血管网的形成主要依靠血管新生过程。正常的生理性血管新生在胚胎发育、伤口愈合及骨折修复等多种生理过程中作为一种基本的过程存在。这种类型的血管形成方式与多种因子的相关作用紧密相关，如生长因子、趋化因子、黏附分子、血管新生抑制因子及各种酶等。

较小的毛细血管是由内皮细胞（endothelial cell，EC）围成的，空心，周围被周细胞（pericyte）、基底膜（basement membrane）及细胞外基质（ECM）包围支撑。在成年时期，大多数的血管及内皮细胞维持着一种静止的状态，很少被激活去形成新生的血管分支，包围在已形成血管周围的基底膜和周细胞可以阻止内皮细胞离开其原位置。只有在特殊的情况，如缺氧、炎症等状态的刺激下，血管新生机制才会被激活。目前，血管新生主要由以下过程组成。首先，周细胞分离，基底膜及 ECM 被蛋白酶分解，这一过程导致已存在的成熟血管不稳定，造成血管通透性增加。另外，蛋白质水解作用破坏了 ECM，造成各种促血管新生因子的释放，如血管内皮生长因子（VEGF）、成纤维细胞生长因子-2（FGF-2）及基质细胞衍生因子-1（SDF-1）等。这些生长因子释放后，会与内皮细胞相关受体结合，从而激活内皮细胞使之产生血管分支。随着内皮细胞的增殖、迁移，稳定的分支逐渐发育为成熟的毛细血管，新的 ECM 开始在血管外形成，周细胞包围在内皮细胞外，此时，成熟的、有功能的血管形成（图 6-1）。若无周细胞包围，那么我们认为此时的血管处于不稳定的状态，组成细胞可以快速分离回到初始状态。

血管新生作为体内血管形成的重要过程，对其进行调控至关重要。促进与抑制血管新生主要依靠微环境中相关细胞因子的调控。其中，促血管新生因子包括上文提到的 VEGF、FGF-2、SDF-1 及胎盘生长因子（PLGF）、表皮生长因子（epidermal growth factor，EGF）等。其中，VEGF 被认为是机体健康及疾病状态中最重要的血管新生相关细胞因子，能直接作用于血管内皮细胞促进其增殖，增加血管通透性。此外，血管新生抑制分子也是维持体内血管形成平衡的重要部分，包括血小板反应蛋白（thrombospondin，TSP）、内皮抑素（endostatin）、血管抑素（angiostatin）和血管生成素-2（angiopoietin-2，Ang-2）等都属于内源性血管生成抑制剂。这些重要的血管新生调控因子维持着体内血管生成的平衡状态，而一旦这种平衡被打破，则面临出现一系列疾病。一方面，广泛的、无调控的血管新生可能是导致类风湿关节炎（RA）、银屑病（psoriasis，PS）、致盲性眼疾乃至癌症生长和转移的原因。另一方面，治疗性血管新生可以增加患处血流，促进局部组织细胞存活。目前广受关注的组织工程与再生医学治疗过程中，血管新生不足是导致移植细胞死亡的重要原因之一。

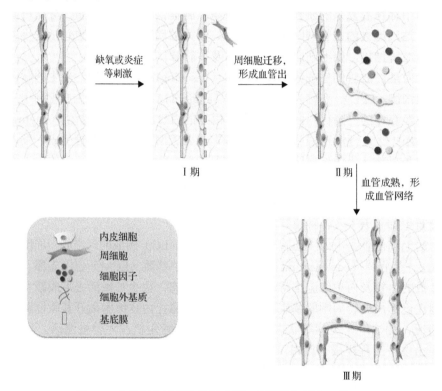

图 6-1　血管新生过程的关键步骤（彩图请扫封底二维码）

在组织损伤的情况下，我们需要快速、可控的治疗性血管新生，以促进受损组织再生与修复。例如，由血栓形成或血管痉挛导致组织供血不足而引起的心肌梗死（myocardial infarction，MI），目前的主流治疗方案如经皮冠脉介入术（percutaneous coronary intervention，PCI）及冠状动脉旁路移植术（coronary artery bypass grafting，CABG）已广泛施行，但许多慢性缺血性心脏病（chronic ischemic heart disease，CHD）患者并不能从中获益。此外，由糖尿病导致的缺血性慢性伤口，局部血管受损，导致组织供血不足，伤口难以愈合，大量患者面临截肢的风险。基于这些问题，我们利用间充质干细胞（MSC）促进血管新生的这一特性，可以为缺血性疾病，如心肌梗死、慢性伤口、脑卒中、外周动脉疾病（peripheral arterial disease，PAD）等提供新的治疗思路，利用血管新生增加患处血流，改善微循环，加速组织修复。

作为一把双刃剑，血管新生可以在机体受到损伤时产生新生血管，及时供应组织再生所需的氧气与营养，但同时不受控制的血管新生与肿瘤的生长、转移密切相关，且与多种免疫系统疾病有关。综上所述，基于血管新生在临床应用中的重要作用，实现对血管新生的人为调控将具有重要的意义。

第二节　间充质干细胞在血管新生中的角色

MSC 来源广泛，只表达低水平的人白细胞抗原（HLA）/主要组织相容性复合体 I（MHC I）类分子，具有低免疫原性和免疫抑制能力，成为干细胞治疗中的重要工具。MSC 具有多种细胞分化潜能，包括脂肪细胞、成骨细胞及内皮细胞等，此外，已经在 MSC 的分泌物中发现了许多促血管新生因子。所有这些证据都表明，MSC 在体内外均可以改变内皮细胞的行为并诱导血管新生（图6-2）。

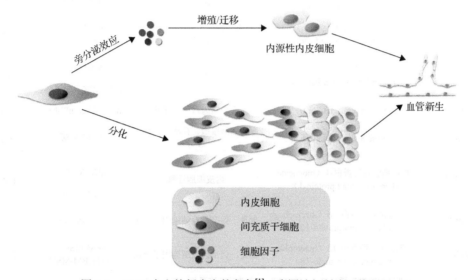

图 6-2　MSC 在血管新生中的角色[1]（彩图请扫封底二维码）

因此，MSC 被认为是血运障碍、由血管损伤引起的疾病（如慢性伤口、脑卒中和心肌梗死）的重要的治疗工具。我们将详细介绍从 MSC 分泌物中发现的与血管新生相关的细胞因子及其对内皮细胞行为的调节能力，以及 MSC 的内皮细胞分化能力。

一、MSC 的旁分泌作用促进血管新生

典型的 MSC 分泌谱包括生长因子、细胞因子、细胞外基质蛋白酶、激素和脂类介质（通常为低丰度）等。处于不同微环境中的细胞将显示出独特的表达谱。因此，分析 MSC 的分泌物必须考虑细胞培养中微环境的影响。目前，可用于研究细胞分泌物表达的方法包括基于细胞因子特定生物活性而设计的生物学检测法、基因检测，以及基于抗原-抗体特异性结合的免疫学检测法。

（一）细胞因子

作为促进血管新生最重要的机制，多种可溶性因子参与 MSC 介导的促血管新生过程，它们可能直接来源于 MSC，或者通过 MSC 和靶细胞交互作用经旁分泌产生（表 6-1）。

表 6-1　MSC 分泌的血管新生因子[11]

血管新生因子	全称	功能	MSC 来源	参考文献
Ang-1	血管生成素-1（angiopoietin-1）	增强血管稳定性，促进内皮细胞存活、周细胞招募	hUC-MSC；BM-MSC	[2-4]
Ang-2	血管生成素-2（angiopoietin-2）	促进内皮细胞迁移和出芽	BM-MSC	[3,4]
Cyr61	富半胱氨酸蛋白（cysteine-rich 61）	细胞黏附与内皮细胞迁移	BM-MSC	[5]
FGF-2	成纤维细胞生长因子-2（fibroblast growth factor-2）	内皮细胞增殖、迁移，细胞外基质的重塑	hBM-MSC	[6]
HGF	肝细胞生长因子（hepatocyte growth factor）	内皮细胞与平滑肌细胞的增殖、迁移	AD-MSC；小鼠 AD-MSC	[7,8]
IGF-1	胰岛素样生长因子-1（insulin-like-growth factor-1）	内皮细胞增殖、存活	AD-MSC；hBM-MSC；hP-MSC	[6,7]
IL-6	白介素-6（interleukin-6）	内皮细胞增殖、迁移	hBM-MSC；hPl-MSC	[6]
IL-8	白介素-8（interleukin-8）	内皮细胞增殖、存活、迁移与成管	hBM-MSC	[6]
MCP-1	单核细胞趋化蛋白-1（monocyte chemoattractant protein-1）	内皮细胞迁移	hUC-MSC；hBM-MSC	[2,6]
MIP-1α	巨噬细胞炎性蛋白-1α（macrophage inflammatory protein-1α）	促炎，调节炎性细胞因子分泌	小鼠 BM-MSC	[9]
MIP-1β	巨噬细胞炎性蛋白-1β（macrophage inflammatory protein-1β）	促炎，调节炎性细胞因子分泌	hBM-MSC；小鼠 BM-MSC	[6,9]
NAP-2（CXCL7）	中性粒细胞激活蛋白-2（neutrophil activating protein-2）	中性粒细胞募集，EC 迁移，VEGF 和 MMP 释放	hBM-MSC	[10]
PLGF	胎盘生长因子（placental growth factor）	促进血管生成	小鼠 AD-MSC	[8]
TGF-β	转化生长因子-β（transforming growth factor-β）	促进血管稳定、成管和细胞外基质合成	hUC-MSC；hAD-MSC	[2]
TIMP-1	金属蛋白酶组织抑制物-1（tissue inhibitor of metalloproteinase-1）	金属蛋白酶的抑制剂	hBM-MSC	[10]
TIMP-2	金属蛋白酶组织抑制物-2（tissue inhibitor of metalloproteinase-2）	金属蛋白酶的抑制剂	hBM-MSC	[10]
VEGF	血管内皮生长因子（vascular endothelial growth factor）	增加血管渗透性，促进 ECM 降解与 EC 细胞存活、增殖、迁移和成管	hUC-MSC；小鼠 BM-MSC；hBM-MSC；hAD-MSC	[2,3,6]

目前，通过分子生物学方式已经鉴定了多种细胞因子，包括 VEGF、FGF-2、Ang-1、单核细胞趋化蛋白-1（MCP-1）、白介素-6（IL-6）、IL-8 及 PLGF 等。多种实验表明，胎

盘间充质干细胞（P-MSC）可以产生 IL-6，而脂肪间充质干细胞（AD-MSC）可以分泌大量的 VEGF、肝细胞生长因子（HGF）和转化生长因子-β（TGF-β）[12]。

在体外和体内的实验中研究者发现，MSC 可以通过分泌 VEGF、MCP-1 和 IL-6 刺激血管新生，同时，在使用包含 VEGF、MCP-1 和 IL-6 的中和抗体预处理后可以显著抑制这一作用[13]。其中，MCP-1 和 IL-6 是 MSC 分泌的细胞因子，研究显示它们具有促进血管新生，促进内皮细胞存活、增殖和抗细胞凋亡的特性[9,14]。VEGF 可以调节内皮细胞的迁移和分化，并促进受损组织中血管生成和内皮细胞募集[15]，MSC 通过产生 VEGF 调节人脐静脉内皮细胞（human umbilical vein endothelial cell，HUVEC）的行为，促进新生血管的形成[16]。VEGF 的受体包括酪氨酸激酶受体 VEGFR1（Flt1）和 VEGFR2（KDR/Flk1）[17]。VEGFR2 可调节各种细胞信号通路，包括促分裂原活化的蛋白激酶（mitogen-activated protein kinase，MAPK）、磷酸肌醇-3-激酶（phosphoinositide-3-kinase）、Akt 及 Src 和 Rac 途径的激活。有研究显示，骨髓间充质干细胞（BM-MSC）缺乏一些促血管新生方面的受体，如 VEGF 受体，但表达 $VEGF_{165}$ 的共受体神经纤毛蛋白-1（neuropilin-1，NRP1），NRP1 与血小板源生长因子（PDGF）受体共定位于西胞膜上，这种串流（cross-talk）机制在 MSC 的增殖、迁移及成管行为中起着重要作用。

改变细胞的培养条件可以上调 MSC 分泌的一系列血管生成因子。研究证明，TGF-α 可以显著诱导多种生长因子如 VEGF、HGF、IL-6、IL-8、PDGF-BB 和 Ang-2 的表达，MEK/MAPK 和 PI3K/Akt 信号通路均参与这一过程，促进了 BM-MSC 对上述因子的分泌。所以，体内实验中，经 TGF-α 处理的 BM-MSC 的条件培养基（condition medium，CM）与未经 TGF-α 处理的相比，可以诱导更多的血管产生[18]。此外，还有研究表明，缺氧、肿瘤坏死因子-α（TNF-α）和脂多糖（LPS）可以通过诱导转录因子 NF-κB 的产生来刺激 VEGF、FGF-2、HGF 和胰岛素样生长因子-1（IGF-1）的产生。而在 AD-MSC 中，LPS 增加了包括 IL-6 和 TNF-α 在内的几种趋化因子的分泌，而 FGF-2 和 EGF 使得 HGF 的表达上调[19]。在缺氧条件下，与真皮成纤维细胞的 CM 相比，BM-MSC 的 CM 含有较高含量的 VEGF、IGF-1、SDF-1 和红细胞生成素（erythropoietin，EPO）[20]。

综上所述，我们可以看出 MSC 具有自分泌和旁分泌活性，其参与细胞存活、器官功能重建和组织血管新生等重要生理过程。上述 MSC 产生的生长因子如 VEGF、MCP-1、HGF、TGF-β 和 IL-6 等均属于重要的促血管生成因子，通过这些促血管生成因子的作用，MSC 可以通过增加微血管密度进一步增加组织灌注来治疗组织损伤并保持器官功能，在临床上，已有大量试验证明 MSC 可促进血管新生和组织修复，并用于 PAD 和心血管疾病（cardiovascular diseases，CVD）的治疗[21]。

（二）外泌体及 miRNA

血管新生相关细胞因子是血管生成的关键因素之一，其可以支持内皮细胞生长、增殖和迁移。然而，多项研究表明，MSC 和其他细胞之间的沟通不仅可以通过细胞因子，还可通过外泌体（exosome）发挥功能。外泌体是由脂质、蛋白质、细胞骨架元件、分子伴侣和信号分子动态组成的囊泡样小体[22]，由细胞主动向胞外分泌，大小均一。研究证明，外泌体中也含有遗传物质，包括信使 RNA（mRNA）和微 RNA（microRNA， miRNA）[23]。多种类型的细胞，包括心肌细胞、血管内皮细胞和 MSC，均可分泌含有 miRNA 及细胞因子的外泌体[24]，而 miRNA 介导内皮细胞与心血管细胞之间的通信也被证实[25]。有结果表明，通过转移遗传物质和促血管生成分子，外泌体在血管新生过程中有重要作用。在人来源的 MSC 的培养基中，科学家已经证实了外泌体的存在，将这些外泌体注射到缺血-再灌注模型大鼠中，其显示出卓越的心肌保护作用[26]。以上结果表明，MSC 分泌的外泌体可能具有促进血管新生的潜力，而进行具体临床应用还需要进一步研究。

二、MSC 调节内皮细胞的行为

在体内，血管新生的过程是一个复杂的过程，涉及内皮细胞的增殖、迁移、成管及成熟，要找出 MSC 在这一过程中扮演的角色，需要对这一过程进行体外模拟，并对血管新生的不同阶段施以相应干预措施，从而确定 MSC 对内皮细胞行为进行调节的机制。

首先，大量研究证明，MSC 可以促进内皮细胞增殖[27-30]，在将 MSC 与 HUVEC 共培养后，人 MSC（hMSC）可以通过旁分泌机制激活相邻 HUVEC 中的 NF-κB 信号转导通路，增加 IL-6 与 IL-8 的表达，从而促进新生血管的形成。此外，已证明各种不同来源的 MSC 可诱导体内、外内皮细胞迁移，包括脐带华通氏胶来源的间充质干细胞（WJ-MSC）、羊膜液来源的间充质干细胞（amniotic fluid-derived mesenchymal stem cell，AF-MSC）、AD-MSC 和 BM-MSC[31]。同时，在侵袭实验中，MSC 被证实具有上调不同的蛋白酶表达，促进内皮细胞降解周围 ECM 的能力，这一行为可进一步诱导内皮细胞迁移[32]。此外，在体外 3D 培养中，MSC 表现出更加显著的促进内皮细胞增殖和迁移的能力，这一过程可能是由单位体积内较高浓度的 VEGF、FGF-2、Ang 和 IL-11 导致的。另外，相关研究显示，人骨髓间充质干细胞（hBM-MSC）与内皮细胞直接共培养在基质胶（matrigel）上后，增强了先前存在的血管的持久性，这表明 BM-MSC 不仅促进血管形成，亦增强了 3D 基质胶共培养系统中这些血管的复杂程度，而且增强了新形成血管的稳定性及其在机体中的功能性[33]，在直接接触培养中，这一效应更加明显。另一项研究描述了 BM-MSC 与管状内皮细胞以周细胞样方式（pericyte-like manner）结合。在体内实验中，hBM-MSC 以血管祖细胞（perivascular precursor cell）的形式存在，有效地稳定了体内新生血管。这一特性提醒我们，MSC 可以促进组织工程化血管移植后的存活。另外，由于活性氧（ROS）的产生，加入大量的 BM-MSC 可以抑制血管的形成。观察到的抗血管生成作用（anti-angiogenic effect）被认为是由大量使用 MSC 导致的细胞毒效应引起的，因为当添加的 BM-MSC 数量减少时，内皮细胞的毒性显著降低。这些结果表明，基于 MSC 的治疗应该避免使用高浓度的移植细胞，因为它可以潜在地诱导新形成血管的内皮细胞凋亡。

在体外实验中，MSC 也具有保护内皮细胞、减少细胞凋亡的作用。P-MSC 通过 STAT3 信号转导通路显著减少氧化应激相关凋亡。缺氧培养 AD-MSC 后的条件培养基使内皮细胞凋亡显著减少。此外，BM-MSC 的分泌物抑制由缺氧诱导的内皮细胞凋亡，而这一行为可能通过激活 PI3K/Akt 而不是胞外信号调节激酶（extracellular signal-regulated kinase，ERK）信号通路。

总之，MSC 参与体内血管生成的所有阶段，不仅包括早期步骤如内皮细胞的募集、迁移，还包括血管成熟、维持等后期阶段，这一结论更加支持了我们的假设，即 MSC

可以作为治疗缺血性疾病的重要工具。

第三节 间充质干细胞的内皮分化

在 MSC 通过旁分泌效应调节内皮细胞行为，促进血管新生的同时，其作为干细胞，具有多向分化潜能，那么 MSC 是否可以直接分化为内皮细胞，参与到血管新生的过程中呢？如前所述，关于 MSC 移植后向内皮细胞的分化，一直是一个有争议的话题，部分研究认为在 MSC 移植到损伤位置后，主要由其分泌的细胞因子参与血管新生过程，而其直接的内皮分化过程在治疗中所起的效果并不明显。但是，MSC 作为多能干细胞，存在巨大的可塑性，已经有大量研究成功实现了 MSC 的内皮分化（表 6-2）。

表 6-2 体外诱导 MSC 分化实验[11]

MSC 来源	培养时间	分化培养基	内皮标志表达情况	MSC 内皮分化的功能性表现	参考文献
大鼠 BM-MSC	7 天	VEGF 培养	CD31、factor Ⅶ、VEGFR2 和 tPA	未检测	[34]
大鼠 BM-MSC	8 天	施加剪切力（24h）+ VEGF 培养（7 天）	CD31、factor Ⅶ、VEGFR2 和 tPA	未检测	[34]
小鼠 AD-MSC	12 天	内皮细胞培养基（无 FGF-2）+ ITS + 10ng/mL FGF-2	CD34、Tie-2、VEGFR2 和 VE-cadherin	基质胶上成管，低密度脂蛋白（LDL）摄取增加	[35]
猪 BM-MSC	10 天	内皮细胞培养基 + 50ng/mL VEGF	vWF、VE-cadherin、CD31	基质胶上成管，LDL 摄取增加	[36]
hBM-MSC	11 天	内皮细胞培养基+施加剪切力(10 天) + 基质胶（1 天）	VEGFR1、vWF、CD31	LDL 摄取增加	[37]
hBM-MSC	14 天	基础培养基 +100ng/mL VEGF	CD34、VEGFR1 和 VEGFR2，无 Tie-2 和 vWF 表达	基质胶上成管失败	[38]
AD-MSC	7 天	内皮细胞培养基 +5%胎牛血清	VEGFR1、Ang-1 和 VE-cadherin，无 VEGFR2 和 CD31 表达	未检测	[39]
hBM-MSC	14 或 28 天	基础培养基	VE-cadherin、PECAM-1、vWF 和 VEGFR1	基质胶上成管，LDL 摄取增加	[40]

MSC 向内皮细胞分化的过程主要通过典型的内皮细胞表面分子如 CD31、CD34、VEGFR1、VEGFR2 和 vWF 的上调来检测。考虑到细胞表面标志物的变化不足以充分证明 MSC 的内皮分化，还可以通过功能测试，包括在基质胶上的成管实验和对乙酰化低密度脂蛋白（acetylated-low density lipoprotein, ac-LDL）的摄取等，对分化后的 MSC 进行鉴定。尽管如此，基于这些常用实验直接将这些分化的 MSC 定义为完全成熟的和功能性的内皮细胞可能是不准确的，而将这些细胞定义为内皮细胞样细胞（endothelial cells-like cell）可能更为准确。此外，在不同研究中发现，MSC 在相同实验条件下，其促进血管新生的能力并不完全相同，而影响 MSC 促血管新生能力的因素包括细胞来源、细胞培养时间及培养条件，移植途径和移植细胞数量，以及 MSC 本身的异质性等。

一、体外分化

目前，最常用的促进 MSC 内皮分化过程是通过 VEGF 进行诱导。最早的促 MSC 内皮分化实验中，直接将 VEGF 加入到 BM-MSC 的培养基中进行，结果显示细胞表面内皮细胞标志物表达显著增加，其中包括 VEGFR1、VEGFR2、VE-钙黏着蛋白（VE-cadherin）、血管细胞黏附分子-1（VCAM-1）和血管性假血友病因子（von Willebrand factor，vWF）。电子显微镜分析显示，MSC 存在典型的内皮形态特征，包括 W-P 小体（Weibel-Palade body）、紧密的胞间连接等。在此基础上，利用含有 VEGF、EGF、FGF-2 及 IGF-1 等生长因子培养基培养的 hBM-MSC，显示 CD31、CD144、VEGFR2、CD105 和 CD34 的表达显著升高，并且能够在基质胶上成管[41]。最近研究者发现，衍生自乳牙的 MSC，可以分化成 VEGFR2/CD31 阳性的内皮细胞样细胞，而通过使用干扰小 RNA（small interfering RNA，siRNA）对 ERK 信号转导通路中 MEK1 进行抑制后则完全消除了这种分化，揭示了 VEGF/MEK1/ERK 信号转导通路可能是 MSC 内皮分化过程的重要一环[42]。也有研究认为，在 MSC 向内皮细胞分化的过程中，Notch 信号通路起关键作用。使用辛伐他汀诱导大鼠 BM-MSC 后，显示 CD31、vWF、VE-cadherin、VEGFR1 和 VEGFR2 的表达增加，同时可以在基质胶上成管，而对 Notch 信号转导通路 Notch 1 分子应用其 siRNA 靶向抑制可以抑制这种内皮分化。此外，Whyte 等[40]的研究也验证了 Notch 信号在内皮分化中的作用。

通常，干细胞的命运主要依靠细胞因子进行调节，如生长因子、趋化因子、激素等。最新数据表明，除可溶性因子外，物理作用也可以成功诱导 MSC 的内皮分化。这种调节工具包括机械刺激、剪切应力施加，以及将细胞接种在弹性纳米纤维水凝胶建立三维培养体系等。例如，hP-MSC 经受剪切应力 24h 后显示出有 vWF 和血小板内皮细胞黏附分子-1（platelet endothelial cell adhesion molecule-1，PECAM-1）的表达、LDL 摄取和基质胶上成管能力显著增加[43]。目前，已经确定了一些对 MSC 内皮分化过程起关键作用的影响因素，如 Notch 信号通路、细胞因子的施加等。但对 MSC 内皮分化过程的了解并不完全，我们需要进一步研究，以充分了解这一过程潜在的分子机制，从而对促进功能性血管的发育及 MSC 的临床应用有关键作用。

二、体内分化

如上所述，在大多数研究中，MSC 的移植成功率仍然较低。因此，部分研究者考虑 MSC 移植后促血管新生的能力主要依靠其旁分泌作用而不是内皮分化作用。在大多数实验中，MSC 移植后，只有很少部分细胞发生了内皮分化，而这部分细胞的治疗作用相对较低。尽管体内 MSC 的完全内皮分化存在争议，但通过预处理的 MSC 依然可能通过内皮分化促进体内血管新生，因为这种预处理可能会促进 MSC 向更"促血管生成"的细胞类型转化。例如，使用 VEGF 孵育 hP-MSC 后，与对照组相比有更高的体内内皮分化程度[44]。还有 Whyte 等[40]证明了高密度移植的 BM-MSC 后（其分化为内皮细胞样细胞），在体内测定中显示出更高的血管密度。另外，AD-MSC 与 EGM-2 孵育 7 天后有更高的 Ang-1 分泌水平，在大鼠股动脉损伤模型中，与对照组相比，预处理后的细胞具有更好的促血管内膜新生的能力,证明了预处理后的 MSC 的体内内皮分化能力增加。

第四节 间充质干细胞作为载体搭载血管新生因子

自然状态下的 MSC 具有巨大的治疗潜力,应用转基因技术可以改善其疗效甚至扩大 MSC 的临床应用范围。为获得稳定表达的基因工程细胞,可以利用病毒载体将目的基因整合到目的细胞基因组中。MSC 可以很容易地接受病毒载体,包括基于逆转录病毒的病毒载体系统,如慢病毒、腺病毒和腺相关病毒等。经病毒转染的 MSC 可以相对容易、高效地产生可以与细胞质、膜结合的蛋白质产物。目前,已有许多动物研究报道了 MSC 在体内各种病理过程中可作为基因递送载体的成功应用。尽管如此,也有许多使用病毒载体的临床试验出现了与这些载体相关的不良反应,包括毒性、免疫排斥和致癌性。而一些非病毒载体转染方式,如应用磷酸钙、脂质体、类脂质体(niosome)和纳米颗粒等进行细胞转染,这种非病毒系统的基因表达短暂,且转染效率相对较低,目前这些基因递送方法仍然限于体外实验。

通过基因工程改造的 MSC 已被广泛应用于各种动物模型。例如,使用慢病毒载体,研究者设计 BM-MSC 过表达 Ang-1,从而成功诱导内皮细胞存活,增加了新生血管的稳定性。在 LPS 诱导肺损伤小鼠模型中,与单独进行 BM-MSC 或 Ang-1 注射相比,携带 Ang-1 的 BM-MSC 组肺损伤症状明显减轻,肺功能得到一定程度的恢复[45]。此外,HGF 和 VEGF 过表达 BM-MSC 治疗小鼠 MI 模型中,与空载体转染的 MSC 组相比,治疗组取得了更好的治疗效果,包括血管新生增加、瘢痕减小和左心室功能提升等[46]。

尽管还处于实验阶段,但 MSC 作为载体搭载血管新生因子进行治疗是一种有效的手段,可以增加治疗效果,减少由移植细胞死亡带来的治疗失败。

第五节 间充质干细胞促进血管新生的进展与展望

一、MSC 治疗相关疾病的进展

(一)皮肤损伤

伤口愈合是一个非常复杂的、有多因素涉及的过程,包括炎症反应与上皮再生、血管新生之间的平衡,以及 ECM 的沉积和重塑。作为世界范围内流行的慢性病,糖尿病可以危及多个组织器官,导致血管损伤,伤口愈合不良成为糖尿病患者最严重的并发症之一。目前已有的治疗方法只能达到局部和暂时的愈合,因此找到一种有效的治疗方法迫在眉睫。

目前,已经有关于 MSC 在皮肤损伤方面应用的实验。例如,应用 BM-MSC 治疗糖尿病小鼠的皮肤损伤,与对照组相比,可以看到伤口加速愈合,创面再上皮化(re-epithelialization)增加,伤口血管密度明显提高。此外,使用浓缩的 MSC 条件培养基或缺氧条件下的培养基亦可以显著加速伤口愈合,在加入针对 VEGF 和 FGF-2 的中和抗体后,可以发现这一过程被抑制,说明这一过程主要依靠移植的 MSC 释放的 FGF-2

和VEGF介导[4]。MSC还可以通过募集巨噬细胞和EPC到损伤部位进一步促进血管新生，减少瘢痕形成[20]。

（二）外周动脉疾病

缺血是一种限制器官或组织正常血液供应后引起的血管系统病理变化，例如血管收缩或血栓形成（阻塞血管）可导致缺血。常见的缺血性疾病包括PAD，指由动脉粥样硬化和炎症等原因引起的后肢大动脉阻塞。PAD可引起各种症状，包括组织的颜色和温度变化，慢性伤口和不愈合的溃疡，肌肉无力和绞痛，以及肌肉萎缩、坏疽等。由于目前不存在有效的药物治疗，1/3的患者需要通过截肢解决肢体坏疽的问题。

由于MSC促进血管新生的潜力，其被视为是有效的治疗方案，可以通过促进血管新生恢复肢体血流，最终修复受损肢体功能。在基础实验中，一般通过对动物重要的后肢动脉（如股动脉）进行结扎或切断，阻塞血流，导致肢体缺血来模仿PAD的病例过程。有研究显示，在结扎股动脉后，局部注射MSC可以显著改善肢体功能，后肢血管密度增加，肢体坏死发生率降低，肌肉萎缩减轻。但并没有观察到明显的MSC整合入血管的现象，因此可能主要由MSC的旁分泌作用引起VEGF和FGF-2水平提升，产生治疗效果。此外，有II期临床试验表明，动脉内注入自体骨髓一年后可以显著增加周边血液灌注，降低截肢风险[47]。除了直接的细胞治疗外，MSC来源的外泌体局部注射后，同样可以显著改善局部血流，减少肢体坏死。

（三）心肌梗死

相关数据显示，我国由缺血性心脏病导致的死亡占城乡居民死亡原因的首位。即使患者存活下来，由心肌梗死造成的心脏扩张或心力衰竭通常会导致患者生活质量严重下降乃至死亡。目前，心肌梗死的治疗方法的基本原理是通过促进心肌细胞分化和提供生长因子来诱导血管新生，或刺激内源性心脏干细胞迁移，从而修复损伤的心脏组织。在动物模型中，一般通过结扎冠状动脉，如左前降支和左回旋支引起心肌梗死。

目前，有大量研究证明，在各种小鼠或大鼠模型中，不同来源的MSC，包括骨髓、脂肪、胎盘及羊膜液，均对缺血性心脏病有治疗作用，在这些研究中，细胞移植方式包括心肌注射、心内膜注射及静脉注射等。通过细胞移植可以增加血管新生，显著改善心室功能。在这种治疗效果的背后，可以发现移植后HGF、VEGF及Ang-1的表达显著上升[48]。此外，最近有实验对大型动物模型进行了研究（包括猪、狗和羊），心肌梗死后进行MSC的移植可以显著改善左心室射血分数，缓解心肌损伤。此外，与MSC治疗相关的安全性评价也正在进行，对不同注射剂量及注射方式的安全性进行了深入研究。例如，有随机、双盲临床试验显示，静脉注射同种异体BM-MSC进行治疗是安全的，与安慰剂组相比，经细胞治疗的患者的肺功能及左心室功能有明显提高[49]。

（四）脑缺血/脑卒中

脑卒中，也称为脑血管意外，是由脑供血不足引起的局灶性神经功能受损。其中，由缺血引起的脑卒中是最易复发的类型，这种缺血可由血栓形成导致血液循环受阻，从而导致氧气供应降低引起。脑的暂时或永久性损伤与否取决于受影响的脑区、缺血时间与治疗及时性。脑卒中也是造成全世界成人残疾及死亡的最主要原因。

因此，新型治疗方式的主要目标是尽快恢复受影响脑区的功能，改善局部血流供应，促进受损神经细胞再生。MSC 治疗可以提供一种替代方法来诱导损伤区域功能的恢复，促进血运重建，诱导细胞再生。这些年来，已经开发出多种动物脑卒中模型，以便模拟人脑卒中后的状态。在最初的设想中，希望 MSC 会转分化为神经元和神经胶质细胞，促进脑损伤的恢复，现在研究发现 MSC 主要通过抑制神经细胞凋亡，增强内源性神经细胞再生和促血管新生作用来改善缺血后损伤。事实上，多项研究集中描述了 MSC 的促血管新生作用，以及这一作用在改善卒中动物恢复中所起的关键作用。有研究组报道显示，BM-MSC 移植后脑组织中 Ang-1 和 Ang-2 表达明显上调，区域血流量上升，功能明显得到改善[50]。此外，AD-MSC 同样有关于促进脑功能恢复，减少细胞死亡及促进细胞增殖、神经发生和并表达血管生成标志物（如 VEGF）方面的报道[51]。

二、挑战与展望

综上所述，由于 MSC 具有多向分化潜能及促血管新生的特性，其有望成为重要的治疗工具。近几年来，大量的研究已经鉴定了许多从 MSC 的条件培养基中分离出的血管生成因子，如 VEGF、FGF-2 和 MCP-1。此外，有新的证据表明 MSC 在免疫调节方面的作用也有助于增强其治疗效果。

然而，一些未解决的问题仍然阻碍了这些细胞广泛应用于临床治疗中。首先，MSC 并没有被广泛认可的细胞标志物，导致原代细胞的异质性，这使得在不同的研究背景下，治疗效果会出现差异性，同时也为 MSC 的大规模临床应用增加了障碍。此外，在大多数研究中，MSC 的移植成活率极低，在体内的停留时间不够理想也成为其临床应用中的一个障碍。除有效性之外，细胞治疗的安全性也是我们关注的重点，在目前的研究中，尽管有大量实验结果证明了 MSC 治疗的安全性，但长期的随访观察依然是必需的。最关键的一点是，有研究认为 MSC 有促进肿瘤生长的风险[52]。如前所述，MSC 可以促进体内血管新生，那么其对肿瘤的作用就成为关注的重点。虽然部分研究表明，MSC 具有肿瘤抑制功能[53]，但也有其他研究结果指出 MSC 可以在各种癌症模型如腺癌、黑素瘤及乳腺癌中促进肿瘤生长[54]。虽然 MSC 促进肿瘤生长背后的分子机制尚未完全了解，但有几篇文章确实表明 MSC 通过促进血管新生刺激肿瘤生长。

总而言之，随着研究的深入，我们对 MSC 在血管新生过程中作用的研究已经取得了巨大进步。同时大量研究证明，MSC 可以成功应用于心肌梗死、脑卒中、局部缺血和慢性伤口愈合的临床治疗中，这种基础研究的成功转化也让我们对 MSC 的临床应用越来越乐观。虽然细胞治疗成为日常治疗的道路依旧漫长而艰辛，但我们相信通过技术革新，MSC 可以成功应用于缺血性疾病的临床治疗中。

（陶红燕　李宗金）

参 考 文 献

[1] Tao H, Han Z, Han ZC, et al. Proangiogenic features of mesenchymal stem cells and their therapeutic applications. Stem Cells International, 2016, 5:1314709.

[2] Han KH, Kim AK, Kim MH, et al. Enhancement of angiogenic effects by hypoxia-preconditioned human umbilical cord-derived mesenchymal stem cells in a mouse model of hindlimb ischemia. Cell Biology International, 2016, 40:27-35.

[3] Rahbarghazi R, Nassiri SM, Ahmadi SH, et al. Dynamic induction of pro-angiogenic milieu after transplantation of marrow-derived mesenchymal stem cells in experimental myocardial infarction. International Journal of Cardiology, 2014, 173:453-466.

[4] Wu Y, Chen L, Scott PG, et al. Mesenchymal stem cells enhance wound healing through differentiation and angiogenesis. Stem Cells, 2007, 25:2648-2659.

[5] Estrada R, Na LI, Sarojini H, et al. Secretome from mesenchymal stem cells induces angiogenesis via Cyr61. Journal of Cellular Physiology, 2009, 219:563-571.

[6] Ostanin AA, Petrovskii YL, Shevela EY, et al. Multiplex analysis of cytokines, chemokines, growth factors, MMP-9 and TIMP-1 produced by human bone marrow, adipose tissue, and placental mesenchymal stromal cells. Bulletin of Experimental Biology and Medicine, 2011, 151:133-141.

[7] Gómez-Mauricio G, Moscoso I, Martín-Cancho MF, et al. Combined administration of mesenchymal stem cells overexpressing IGF-1 and HGF enhances neovascularization but moderately improves cardiac regeneration in a porcine model. Stem Cell Research & Therapy, 2016, 7:94.

[8] Efimenko A, Starostina E, Kalinina N, et al. Angiogenic properties of aged adipose derived mesenchymal stem cells after hypoxic conditioning. Journal of Translational Medicine, 2010, 9:10.

[9] Boomsma RA, Geenen DL. Mesenchymal stem cells secrete multiple cytokines that promote angiogenesis and have contrasting effects on chemotaxis and apoptosis. PLoS One, 2012, 7:e35685.

[10] Hung SC, Pochampally RR, Chen SC, et al. Angiogenic effects of human multipotent stromal cell conditioned medium activate the PI3K-Akt pathway in hypoxic endothelial cells to inhibit apoptosis, increase survival, and stimulate angiogenesis. Stem Cells, 2007, 25:2363-2370.

[11] Bronckaers A, Hilkens P, Martens W, et al. Mesenchymal stem/stromal cells as a pharmacological and therapeutic approach to accelerate angiogenesis. Pharmacology & Therapeutics, 2014, 143:181-196.

[12] Rehman J. Secretion of angiogenic and antiapoptotic factors by human adipose stromal cells. Circulation, 2004, 109:1292-1298.

[13] Kwon HM, Hur SM, Park KY, et al. Multiple paracrine factors secreted by mesenchymal stem cells contribute to angiogenesis. Vascular Pharmacology, 2014, 63:19-28.

[14] Botto S, Streblow DN, Defilippis V, et al. IL-6 in human cytomegalovirus secretome promotes angiogenesis and survival of endothelial cells through the stimulation of survivin. Blood, 2011, 117:352-361.

[15] Shimamura M, Nakagami H, Koriyama H, et al. Gene therapy and cell-based therapies for therapeutic angiogenesis in peripheral artery disease. BioMed Research International, 2013, (12):186215.

[16] Beckermann BM, Kallifatidis G, Groth A, et al. VEGF expression by mesenchymal stem cells contributes to angiogenesis in pancreatic carcinoma. British Journal of Cancer, 2008, 99:622-631.

[17] Berendsen AD, Olsen BR. How vascular endothelial growth factor-A (VEGF) regulates differentiation of mesenchymal stem cells. The Journal of Histochemistry and Cytochemistry : Official Journal of the Histochemistry Society, 2014, 62:103-108.

[18] Luca AD, Gallo M, Aldinucci D, et al. Role of the EGFR ligand/receptor system in the secretion of angiogenic factors in mesenchymal stem cells. Journal of Cellular Physiology, 2011, 226:2131-2138.

[19] Kilroy GE, Foster SJ, Wu X, et al. Cytokine profile of human adipose-derived stem cells: expression of angiogenic, hematopoietic, and pro-inflammatory factors. Journal of Cellular Physiology, 2007, 212:702-709.

[20] Chen L, Tredget EE, Wu PY, et al. Paracrine factors of mesenchymal stem cells recruit macrophages and endothelial lineage cells and enhance wound healing. PLoS One, 2008, 3:e1886.

[21] Timmers L, Lim SK, Hoefer IE, et al. Human mesenchymal stem cell-conditioned medium improves cardiac function following myocardial infarction. Stem Cell Research, 2011, 6:206-214.

[22] Yeh YY, Ozer HG, Lehman AM, et al. Characterization of CLL exosomes reveals a distinct microRNA signature and enhanced secretion by activation of BCR signaling. Blood, 2015, 125:3297-3305.

[23] Valadi H, Ekström K, Bossios A, et al. Exosome-mediated transfer of mRNAs and microRNAs is a novel mechanism of genetic exchange between cells. Nature Cell Biology, 2007, 9:654-659.

[24] Waldenstrom A, Ronquist G. Role of exosomes in myocardial remodeling. Circulation Research, 2014, 114: 315-324.

[25] Hergenreider E, Heydt S, Tréguer K, et al. Atheroprotective communication between endothelial cells and smooth muscle cells through miRNAs. Nature Cell Biology, 2012, 14:249-256.

[26] Lai RC, Arslan F, Lee MM, et al. Exosome secreted by MSC reduces myocardial ischemia/reperfusion injury. Stem Cell Research, 2010, 4:214-222.

[27] Lee WY, Tsai HW, Chiang JH, et al. Core-shell cell bodies composed of human cbMSCs and HUVECs for functional vasculogenesis. Biomaterials, 2011, 32:8446-8455.

[28] Baiguera S, Ribatti D. Endothelialization approaches for viable engineered tissues. Angiogenesis, 2013, 16:1-14.

[29] Hofmann A, Ritz U, Verrier S, et al. The effect of human osteoblasts on proliferation and neo-vessel formation of human umbilical vein endothelial cells in a long-term 3D co-culture on polyurethane scaffolds. Biomaterials, 2008, 29:4217-4226.

[30] Li J, Ma Y, Teng R, et al. Transcriptional profiling reveals crosstalk between mesenchymal stem cells and endothelial cells promoting prevascularization by reciprocal mechanisms. Stem Cells and Development, 2015, 24:610-623.

[31] Burlacu A, Grigorescu G, Rosca AM, et al. Factors secreted by mesenchymal stem cells and endothelial progenitor cells have complementary effects on angiogenesis *in vitro*. Stem Cells and Development, 2013, 22:643-653.

[32] Kachgal S, Putnam AJ. Mesenchymal stem cells from adipose and bone marrow promote angiogenesis via distinct cytokine and protease expression mechanisms. Angiogenesis, 2011, 14:47-59.

[33] Duffy GP, Ahsan T, O'Brien T, et al. Bone marrow-derived mesenchymal stem cells promote angiogenic processes in a time- and dose-dependent manner *in vitro*. Tissue Engineering. Part A, 2009, 15:2459-2470.

[34] Bai K, Huang Y, Jia X, et al. Endothelium oriented differentiation of bone marrow mesenchymal stem cells under chemical and mechanical stimulations. Journal of Biomechanics, 2010, 43:1176-1181.

[35] Konno M1, Hamazaki TS, Fukuda S, et al. Efficiently differentiating vascular endothelial cells from adipose tissue-derived mesenchymal stem cells in serum-free culture. Biochemical and Biophysical Research Communications, 2010, 400:461-465.

[36] Pankajakshan D, Kansal V, Agrawal DK. *In vitro* differentiation of bone marrow derived porcine mesenchymal stem cells to endothelial cells. Journal of Tissue Engineering and Regenerative Medicine, 2013, 7:911-920.

[37] Portalska KJ, Groen N, Krenning G, et al. The effect of donor variation and senescence on endothelial differentiation of human mesenchymal stromal cells. Tissue Engineering Part A, 2013, 19:2318-2329.

[38] Roobrouck VD, Clavel C, Jacobs SA, et al. Differentiation potential of human postnatal mesenchymal stem cells, mesoangioblasts, and multipotent adult progenitor cells reflected in their transcriptome and partially influenced by the culture conditions. Stem Cells, 2011, 29:871-882.

[39] Takahashi M, Suzuki E, Oba S, et al. Adipose tissue-derived stem cells inhibit neointimal formation in a paracrine fashion in rat femoral artery. American Journal of Physiology Heart and Circulatory Physiology, 2010, 298:H415-H423.

[40] Whyte JL, Ball SG, Shuttleworth CA, et al. Density of human bone marrow stromal cells regulates commitment to vascular lineages. Stem Cell Research, 2011, 6:238-250.

[41] Liu JW, Dunoyergeindre S, Serrebeinier V, et al. Characterization of endothelial-like cells derived from human mesenchymal stem cells. Journal of Thrombosis and Haemostasis: JTH, 2007, 5:826-834.

[42] Bento LW, Zhang Z, Imai A, et al. Endothelial differentiation of SHED requires MEK1/ERK signaling. Journal of Dental Research, 2013, 92:51-57.

[43] Wu CC, Chao YC, Chen CN, et al. Synergism of biochemical and mechanical stimuli in the differentiation of human placenta-derived multipotent cells into endothelial cells. Journal of Biomechanics, 2008, 41:813-821.

[44] Lee MY, Huang JP, Chen YY, et al. Angiogenesis in differentiated placental multipotent mesenchymal stromal cells is dependent on integrin alpha5beta1. PLoS One, 2009, 4:e6913.

[45] Xu J, Qu J, Cao L, et al. Mesenchymal stem cell-based angiopoietin-1 gene therapy for acute lung injury induced by lipopolysaccharide in mice. The Journal of Pathology, 2008, 214:472-481.

[46] Deuse T, Peter C, Fedak PWM, et al. Hepatocyte growth factor or vascular endothelial growth factor gene transfer maximizes mesenchymal stem cell-based myocardial salvage after acute myocardial infarction. Circulation, 2009, 120:S247-S254.

[47] Alessandro S, Ciro M, Chiara B, et al. A phase II trial of autologous transplantation of bone marrow stem cells for critical limb ischemia: results of the Naples and Pietra ligure evaluation of stem cells study. Stem Cells Translational Medicine, 2012, 1:572-578.

[48] Kim SW, Zhang HZ, Kim CE, et al. Amniotic mesenchymal stem cells with robust chemotactic properties are effective in the treatment of a myocardial infarction model. International Journal of Cardiology, 168:1062-1069.

[49] Hare JM, Traverse JH, Henry TD, et al. A randomized, double-blind, placebo-controlled, dose-escalation study of intravenous adult human mesenchymal stem cells (prochymal) after acute myocardial infarction. Journal of the American College of Cardiology, 2009, 54:2277-2286.

[50] Li HL, Jiang L, Jiang XM, et al. Human mesenchymal stem cells increases expression of alpha-tubulin and angiopoietin 1 and 2 in focal cerebral ischemia and reperfusion. Current Neurovascular Research, 2013, 10:103-111.

[51] Gutiérrez-Fernández M, Rodríguez-Frutos B, Ramos-Cejudo J, et al. Effects of intravenous administration of allogenic bone marrow- and adipose tissue-derived mesenchymal stem cells on functional recovery and brain repair markers in experimental ischemic stroke. Stem Cell Research & Therapy, 2013, 4:11.

[52] Wang J, Wang Y, Wang S, et al. Bone marrow-derived mesenchymal stem cell-secreted IL-8 promotes the angiogenesis and growth of colorectal cancer. Oncotarget, 2015, 6:42825-42837.

[53] Khakoo AY, Pati S, Anderson SA, et al. Human mesenchymal stem cells exert potent antitumorigenic effects in a model of Kaposi's sarcoma. The Journal of Experimental Medicine, 2016, 203:1235-1247.

[54] Jeon ES, Lee IH, Heo SC, et al. Mesenchymal stem cells stimulate angiogenesis in a murine xenograft model of A549 human adenocarcinoma through an LPA1 receptor-dependent mechanism. Biochimica et Biophysica Acta, 2010, 1801:1205-1213.

第七章 间充质干细胞的生物学特性及在造血干细胞移植中的应用

间充质干细胞（MSC）是一类具有自我更新和多向分化潜能的干细胞，由 Caplan[1] 于 1991 年首次提出，2005 年国际细胞治疗协会（ISCT）提议将具有多向分化潜能的间充质基质细胞（mesenchymal stromal cell）也归为 MSC[2]。MSC 首先从骨髓中成功分离，随后其来源不断扩展，可以从脐血（umbilical cord blood, UCB）、脂肪组织（adipose tissue, AT）、肌肉、牙髓（dental pulp）等组织中分离获得[3-5]。随着对 MSC 的深入研究，发现在特定的微环境影响下，MSC 可定向分化为内胚层、中胚层及外胚层三个胚层来源组织的细胞，如骨、脂肪、肝、肾、皮肤、肌肉、神经甚至胰腺等组织的成熟细胞，具有免疫抑制、造血支持及组织修复等功能[6-12]。目前多项动物及临床前期试验将 MSC 用于治疗自身免疫系统疾病及炎症修复，如 1 型糖尿病、克罗恩病（CD）、多发性硬化（MS）、自身结缔组织病等[13-21]，在造血干细胞移植（HSCT）方面，目前已有多项临床研究将 MSC 用于急性（aGVHD）、慢性移植物抗宿主病（chronic GVHD, cGVHD）的防治，以及促进造血干细胞（HSC）的植入。

第一节 间充质干细胞的生物学特性

一、MSC 的来源

MSC 来源于中胚层，具有自我更新能力和多向分化潜能。MSC 首先在骨髓中发现，但骨髓中 MSC 含量有限，仅占骨髓细胞的 0.01%，从骨髓中分离的 MSC 还可能合并病毒感染。同时随着年龄的增长，骨髓间充质干细胞（BM-MSC）的增殖和分化能力逐渐降低[22,23]。此外，正常骨髓供者来源有限，因此 BM-MSC 在临床上的应用较为局限。除骨髓外，MSC 还存在于脂肪、脐血、胎盘、胎肝等组织。与 BM-MSC 相比，脐带间充质干细胞（UC-MSC）表达多种胚胎干细胞（ESC）的特有分子标记，具有分化潜力大、增殖能力强、免疫原性低、取材方便、易于工业化制备等特征，且无道德伦理问题限制，使其在临床应用中具备显著的优势。人脐带大约重 40g，包括 2 条脐动脉和 1 条静脉，研究者已成功从完整脐带组织及脐带某些特殊部位，如华通氏胶、脐带静脉上皮组织、羊膜中分离出 MSC[24-26]。从脐带不同部位分离的 MSC 具有各自的特点，其中来自脐带静脉的 MSC 更倾向于分化为上皮细胞[27]。而华通氏胶富含透明质酸、胶原蛋白、黏多糖，因此华通氏胶来源的 MSC 分化为结缔组织相关细胞的可能性更大[28,29]。但 Xu 等[30] 的研究表明，脐带不同部位来源的 MSC 具有相同的免疫调节、造血支持特性。MSC 在体外培养时对塑料具有黏附性，在塑料培养瓶中呈贴壁生长并且早于其他细胞，可应用此特性对其分离、培养。Gottipamula 等[31]研究发现，无血清培养基中培养的 MSC 形态学及表型均未改变，传代增殖能力及免疫调节能力强，纯度较高。冷冻保存环境下，MSC 的特性，包括免疫原性、免疫调节能力、造血支持作用均不会发生改变[32]。

二、MSC 的表面标志

2006 年 ISCT 提出了 MSC 的最低鉴定标准：①体外培养时对塑料具有黏附性；②表达表面抗原 CD105、CD73、CD90，不表达造血细胞表面标志：CD45、CD34、CD14、CD11b、CD79、CD19、HLA-DR；③具有多向分化潜能，在不同的培养条件下，可诱导分化成成骨细胞、脂肪细胞和软骨细胞等[6]。随着流式细胞术水平的提高，以及对 MSC 认识逐步加深，发现 MSC 不表达内皮细胞表面抗原 CD31，以及活化 T 细胞作用的共刺激分子，如 CD80、CD86、CD40、CD40L 等。此外，MSC 低表达主要组织相容性复合体 I（MHC I），不表达 MHC II 抗原。作为免疫调节及造血支持细胞，CD44、CD73、CD90、CD105、CD166 等被认为是其特异性表面抗原[33]。近年研究发现了新的 MSC 表面标记，如基质前体抗原-1（stromal precursor antigen-1，STRO-1）、阶段特异性胚胎抗原-4（stage specific embryonic antigen-4，SSEA-4）、CD49a、CD271、CD146、瘦素受体等[34,35]。Te Boome 等[36]研究发现，糖皮质激素耐药的 aGVHD 患者输注 MSC 后，一些特异性表面标志物表达情况发生改变是提示预后良好的指标。

三、MSC 的低免疫原性和免疫抑制性

MSC 的低免疫原性主要表现为不表达 MHC II，低表达 MHC I 类分子，不表达共刺激分子 CD40、CD40L、CD80、CD86 等。因此当 MSC 输注到异基因受者体内后，很难被机体免疫系统识别，可能逃脱受体的免疫监视。MSC 对 T 细胞具有抑制作用，具体的作用机制目前尚不清楚，有研究认为 MSC 是通过直接接触作用抑制淋巴细胞增殖和分化，或者是通过分泌调节因子间接发挥作用，如白介素-6（IL-6）、IL-10 和转化生长因子-β（TGF-β）[37]。Ren 等[38]研究认为，MSC 只有在干扰素-γ（IFN-γ）和其他致炎因子包括肿瘤坏死因子-α（TNF-α）、IL-1α、IL-1β 等存在的情况下才能发挥免疫抑制作用。这些细胞因子作用于 MSC，通过促进化学因子及 2,3-去氧吲哚胺的释放来抑制 T 细胞的增殖，并通过下调色氨酸水平诱导 T 细胞凋亡[39]。树突细胞（DC）是启动 T 细胞免疫应答的主要抗原提呈细胞，成熟 DC（mDC）通过提呈从外周捕获的抗原给初始 T 细胞来上调共刺激分子的表达水平，促进炎性细胞因子分泌，从而诱导 T 细胞活化。在体外培养过程中，发现 MSC 的存在可以阻止脐血 $CD34^+$ 细胞或外周血 $CD14^+$ 单核细胞分化为 DC[40]。MSC 还可以通过直接作用于 mDC，降低其表面共刺激分子 CD80 和 CD86 的表达水平，从而降低 mDC 的水平[41]。此外，MSC 还可以通过抑制 B 细胞与 NK 细胞的功能来发挥免疫抑制功能[42,43]。

第二节　间充质干细胞在造血干细胞移植中的应用

造血干细胞移植（HSCT）是治疗血液系统恶性疾病的重要手段，但 HSC 移植后出

现的造血恢复延迟、免疫重建缓慢及移植物抗宿主病（GVHD）等均会影响移植的成败。作为造血微环境的重要组成成分，MSC 来源广泛、取材方便、可行体外扩增，这使其在 HSCT 中应用具有巨大潜力。目前，世界范围内多家临床研究中心已经尝试应用 MSC 防治 aGVHD、cGVHD，以及利用 MSC 与 HSC 联合输注促进移植成功等。

一、MSC 促进造血重建及 HSC 植入

经 HSCT 后患者的造血微环境被大剂量化疗、放疗及造血系统肿瘤所破坏[44,45]，而 MSC 作为骨髓造血微环境的重要组成成分，是成纤维细胞、成骨细胞、脂肪细胞及其他骨髓基质细胞的前体，通过细胞-细胞之间相互接触，可以提供造血所需的生长因子、黏附分子、基质蛋白等，如 IL-6、IL-7、IL-8、IL-11 和干细胞因子（stem cell factor，SCF），促进造血微环境的恢复，加快 HSC 的增殖和分化，从而促进 HSC 的植活及造血细胞的恢复[46,47]。此外，MSC 还可以通过免疫调节作用，改善免疫环境和 T 细胞亚群，诱导免疫耐受，减少移植排斥现象的发生[48]。

王萍等[49]在动物实验中发现 HSCT 联合 MSC 输注后，外周血及骨髓中有核细胞数量显著提高，且回输的 MSC 可长期存在于造血微环境中。2000 年经大剂量化疗后的乳腺癌患者接受 MSC 和 HSC 联合输注是第一项证明 MSC 可以促进造血恢复的临床试验[50]。Fang 等[51]于 2006 年报道了 1 例第 1 次植入失败的重度再生障碍性贫血（severe aplastic anemia，SAA）患者，行相同供者第 2 次异基因造血干细胞移植（allo-HSCT）时，联合输注了 MSC，最终 HSC 植入成功。2011 年的一项随机临床对照研究中，恶性血液肿瘤患者接受单倍体异基因造血干细胞移植（haploidentical HSCT，halpo-HSCT）时，联合输注 MSC，可提高体内血小板生成素（Thrombopoietin，TPO）水平，从而促进血小板恢复[52]。Kassis 等[53]于 2011 年提出 MSC 输注后可以增加 T 细胞的数量。这可能与促进造血恢复的机制有关，但遗憾的是目前缺乏调节 T 细胞的历史对照数据。2013 年 Lee 等[54]报道了 16 例急性白血病患儿行全相合无关供者脐血造血干细胞移植（UCB-HSCT），7 例试验组患者联合输注 MSC，9 例对照组患者未输注。结果试验组患者粒系及血小板恢复时间显著缩短，存在的差异有统计学意义。同时与对照组相比，试验组移植后的中位生存时间延长，aGVHD、cGVHD 发生率及发生程度均较低，无巨细胞病毒（CMV）血症发生。2013 年 Wang 等[55]报道的再生障碍性贫血（AA）患者单中心研究资料显示，halpo-HSCT 联合 MSC 输注可以降低 HSCT 的失败率。2014 年的一项研究中，22 例患者随机分配到 MSC+UCB 或单纯 UCB 组，结果显示 MSC+UCB 组患者中性粒细胞恢复时间明显短于 UCB 组（$P=0.635$）。尽管多数研究认为 MSC 输注是安全、有效的，但仍然存在争议。2011 年 Bernardo 等[56]研究表明，13 例患血液系统疾病的儿童[其中急性淋巴细胞白血病（acute lymphoblastic leukemia，ALL）6 例，急性髓性白血病（acute myeloid leukemia，AML）1 例，骨髓增生异常综合征/幼年型粒-单核细胞白血病 2 例，噬血性淋巴组织细胞增生症 4 例]行 UCB-HSCT 联合 MSC 输注，其造血恢复时间与 39 例匹配的对照组患者无明显差异（$P=0.05$）。

二、MSC 在急性 GVHD 中的作用

对于人白细胞抗原（HLA）相合的 HSCT，急性 GVHD（aGVHD）的发生率为 30%~50%；而 HLA 不全相合的 HSCT，其发生率可高达 60%~80%。目前免疫抑制药物是 GVHD 的主要治疗药物，其中类固醇激素仍是 GVHD 的一线治疗药物，但是部分 aGVHD 患者对类固醇激素耐药，此外还有一部分 aGVHD 患者可能发展为 cGVHD，此类患者由于感染、GVHD 相关全血细胞减少及多器官功能衰竭等，预后往往很差，5 年生存率常不超过 30%[57,58]。由于 MSC 具有免疫调节及造血支持作用，因此近年来被认为可以作为 GVHD 的替代治疗方案。

2004 年 Le Blanc 等[59]最先报道了 1 例 9 岁的 ALL 患儿接受 MSC 输注成功治疗 aGVHD 的研究，该患儿在接受了 HLA 相合的无关供者外周血干细胞移植后，于移植后第 11 天开始出现 aGVHD 症状，以皮肤、胃肠道及肝脏为主，对糖皮质激素、英利昔单抗、霉酚酸酯、甲氨蝶呤等免疫抑制治疗均无反应，于移植后第 73 天予以母亲 BM-MSC 输注，输注过程中未见明显的不良反应，且于输注后 2 周 aGVHD 得到完全控制。2008 年 Le Blanc 等[60]在一项规模较最大的 II 期多中心临床研究中，于 2001~2007 年共纳入 55 例类固醇激素耐药的 aGVHD 患者，其中 25 例为儿童，II、III、IV 级 aGVHD 分别为 5 例、25 例、25 例，aGVHD 发生部位以胃肠道和皮肤、肝脏为主，其中 45 例患者有不止 1 个器官受累。所有患者均接受了 BM-MSC 治疗，MSC 中有 5 份来自 HLA 相合供者，18 份为亲缘单倍体供者，69 份为 HLA 不相合无关供者。55 例患者中，27 例患者接受 1 次输注，22 例接受 2 次输注，6 例接受 3~5 次输注，每次输注 MSC 的平均剂量为 $1.4×10^6$cells/kg。该项研究中 39 例患者得到缓解，总体治疗反应率为 71%，其中接受 1 次治疗的反应率为 52.7%（29/55）。该项研究还表明，MSC 的供者年龄、HLA 相合程度对疗效无明显影响，但儿童患者的总体反应率显著高于成年（21/25 比 18/30，$P=0.07$）。所有患者均未出现近期或远期 MSC 输注相关不良反应。2013 年以色列一项由 5 家医疗中心进行的临床研究[61]，共纳入 50 例类固醇激素抵抗的 aGVHD 患者，其中儿童 25 例，成人 25 例，中位年龄 19 岁，IV 级 GVHD 有 42 例。患者输注的 MSC 有 62 份来自 HLA 全相合第三方供者，5 份来自 HLA 不全相合第三方供者，5 份为与 HSC 来源相同的 HLA 全相合供者，2 份为与 HSC 来源相同的 HLA 不全相合供者，MSC 共输注 1 次者 9 例，输注 2 次者 18 例，3 次及以上者 21 例，每次 MSC 的中位输注剂量为 $1.0×10^6$cells/kg。结果显示 33 例患者（33/50，66%）初步治疗有效，但仅有 17 例患者症状完全缓解（17/50，34%），所有患者均未观察到 MSC 输注相关早期及迟发不良反应。上述研究均证实 MSC 治疗 aGVHD 安全、有效，但遗憾的是均缺乏对照组。2015 年 Zhao 等[62]将 47 例难治性 aGVHD 患者基于自愿原则分为 MSC 治疗组（$n=28$）和对照组（$n=19$）。结果 MSC 组患者整体反应率显著高于对照组（75% 比 42.1%，$P=0.023$），且两组患者 aGVHD 治疗期间巨细胞病毒（CMV）、EB 病毒（EBV）感染风险及肿瘤复发风险无显著性差异。

2002 年 Lee 等[63]首先报道了 1 例异体 MSC 联合 HSC 共移植治疗高危 AML 的研究，其 HSC 来源于父亲的外周血，于第 2 次输注 HSC 后 1h，予以父亲 BM-MSC $1.5×10^6$cells/kg。患者移植后未接受预防性的免疫抑制治疗。移植 12 天后，患者的中性

粒细胞计数大于 $0.5×10^6$ cells/L，血小板计数大于 $20×10^9$ cells/L。移植 18 天后，聚合酶链反应（PCR）检测结果显示为完全供者植入。患者移植后曾出现 CMV 肺炎，应用更昔洛韦和免疫球蛋白治疗后好转，共随访 31 个月未见急、cGVHD 发生，亦未观察到 AML 复发。因此 MSC 联合 HSC 共移植可能可以预防 GVHD 的发生。2005 年 Lazarus 等[64]报道了 7 家医疗中心的一项 I 期临床研究，该研究共纳入 46 例血液系统疾病患者，所有患者均接受 MSC 联合 HSC 共移植，移植后常规接受免疫抑制剂治疗预防 GVHD，结果显示 46 例患者中有 23 例出现 aGVHD，其中仅 5 例为III级 aGVHD，2 例为IV级 aGVHD，重度 aGVHD 的发生率明显降低，而且未观察到明显的 MSC 输注相关不良反应。上述研究虽初步证实 MSC 联合 HSC 输注预防 aGVHD 安全、有效，但是缺乏对照组是其共同的缺陷。2008 年 Ning 等[65]在一项开放性随机对照研究中发现，MSC+HSC 组中II~IV级 aGVHD 发生率仅为 11%，HSCT 组则高达 53%，但是两组数据的差异并不具有统计学意义。2010 年 Baron 等[66]报道的一项历史对照研究表明，移植后 100 天，MSC 联合 HSC 组 aGVHD 累积发生率为 45%，而仅接受 HSCT 的对照组为 56%，但两组差异无统计学意义。2012 年 Kuzmina 等[67]报道的一项临床随机对照试验中，37 例患者随机分配到 HSCT、HSC+MSC 两组，结果发现 MSC+HSC 组 aGVHD 发生率显著低于 HSCT 组，且差异有统计学意义（对照组II~IV级 aGVHD 发生率为 38.9%，MSC 与 HSC 共移植组为 5.3%，$P=0.002$），而且未观察到 MSC 输注相关不良反应，不过该研究中 MSC 并非是与 HSC 同期输注，而是在 HSC 输注后的血象恢复期予以输注 MSC。

三、MSC 在慢性 GVHD 中的作用

HSCT 后生存期超过 100 天的患者中，一半以上患者会发生慢性 GVHD（cGVHD），是限制 HSCT 在临床应用的重要原因之一[68]。然而超过 1/3 的 cGVHD 患者一线治疗（主要包括糖皮质激素）无效[69]。MSC 在 aGVHD 的预防及糖皮质激素抵抗的 aGVHD 治疗方面已经取得一定疗效，近年来研究者尝试研究 MSC 在 cGVHD 治疗中的作用。

2006 年 Ringdén 等[70]首先报道了 MSC 在一例肝 cGVHD 患者中发挥作用的研究。2009 年张乐施等[71]应用 MSC 治疗 12 例糖皮质激素耐药的 cGVHD 患者，中位年龄 28.5 岁，其中 1 例输注 MSC 1 次，10 例输注 2 次，1 例输注 3 次，受累器官包括全身皮肤、肺、肝、口腔、眼睛等，结果显示 3 例获得完全缓解并完全停用免疫抑制剂，6 例获得部分缓解且免疫抑制剂成功减量，总体有效率为 75%（9/12）。2010 年 Weng 等[72]纳入 19 例难治性 cGVHD 患者，所有患者均给予健康志愿者 BM-MSC 治疗，中位输注剂量为 $0.6×10^6$ cells/kg，结果 19 例患者中 4 例获得完全缓解，10 例为部分缓解，总体有效率为 73.7%（14/19）。存活的 14 例患者中有 5 例可停用免疫抑制剂，5 例减量超过 50%。中位随访时间为 697 天，2 年生存率达 77.7%，所有患者均未观察到 MSC 输注相关不良反应。此外，MSC 对口腔黏膜、消化道、肝、皮肤部位的 cGVHD 效果更好。2010 年 Zhou 等[73]应用 MSC 治疗 4 例 cGVHD 相关难治性硬皮病样改变，中位年龄 41 岁，2 例患者输注 MSC 4 次，1 例输注 8 次，另一例输注 7 次，中位随访时间为 14.1 个月，4 例患者临床症状均得到改善，所有患者均未观察到 MSC 输注相关不良反应，且均未复发。2012 年 Weng 等[74]应用 MSC 治疗 22 例 cGVHD 相关性干眼症，其中 12 例患者的临床症状及

Schirmer 泪液分泌试验均有明显改善,该研究中 MSC 输注无效组和有效组在年龄、性别、MSC 供者来源等方面均无显著差异。由于缺乏较好的 cGVHD 动物模型,因此目前 MSC 在 cGVHD 中的作用机制尚不明确[75]。此外,MSC 在 cGVHD 中的作用仍不肯定,尚需要较大规模的随机对照试验进一步证实。

四、MSC 在 HSCT 中与感染的关系

造血干细胞移植(HSCT)后,患者由于移植前大剂量化疗,以及移植后使用免疫抑制剂防治 GVHD,受到各种病原体感染的机会增加,尤其是不全相合 HSCT,由于移植后免疫重建及造血恢复延迟,GVHD 发生率高,感染机会进一步增加。由于 MSC 具有免疫抑制作用,因此其在 HSCT 中应用可能增加患者感染的风险,但目前尚无统一意见。

2012 年,Forslöw 等[76]在一项历史回顾性研究中提出 MSC 输注可能会增加肺炎相关死亡的发生率。2014 年 Moermants 等[77]报道的一项回顾性研究中,无关供者 HSC 联合 MSC 输注组共 30 例,单纯无关供者 HSCT 组共 28 例,结果显示 HSC+MSC 组 1 年累计肺部感染率为 48%,对照组则仅为 15%,差异有统计学意义($P<0.01$),其中 HSC+MSC 组真菌感染率增加,而 CMV 感染率两组无明显差异。2015 年张晓婷等[78]统计 22 例 HSC+MSC 组和 27 例单纯 HSCT 患者的感染发生情况,其中 MSC+HSC 组肺部感染发生率为 68.2%,对照组为 44.4%,但差异无统计学意义($P>0.05$),不过 MSC+HSC 组 CMV 感染率显著高于对照组(81.8%比 51.9%,$P=0.028$)。

相反,2014 年韩冬梅等[79]对 83 例单倍体相合 HSCT 患者的术后感染情况进行了监测,其中 HSCT 组共 42 例,HSC 联合 MSC 组共 41 例。结果显示,HSCT 组共有 21 例合并肺部感染,发生率为 50%;HSC 联合 MSC 组有 15 例合并肺部感染,发生率为 36.6%,HSC+MSC 组肺部感染率低于 HSCT 组,不过两者之间的差异无统计学意义($P>0.05$);HSCT 组合并 CMV 血症共 31 例,发生率为 73.8%,HSC+MSC 组合并 CMV 血症共 32 例,约占 78%,两者无统计学差异($P>0.05$)。两组感染率无差异可能与单倍体相合 HSCT 本身预处理强度大,免疫抑制药物作用强,感染发生率较高有关。此外,MSC 可以促进植活,缩短粒系植入时间,感染时间窗可缩短。2015 年 Zhao[62]等在 MSC 输注治疗 aGVHD 的研究中发现,MSC 输注组并未增加 CMV、EBV 血症的发生率分别为 39.3%、25%,对照组分别为 36.8%、31.5%,两组差异无统计学意义(P 分别为 0.621、0.325)。因此 MSC 输注是否增加感染发生率仍需进一步扩大样本量进行大规模随机对照研究验证。

五、MSC 在 HSCT 中与肿瘤发生的关系

由于 MSC 具有免疫抑制和分泌生长因子的作用,可能有助于肿瘤细胞逃避免疫监视,并促进肿瘤细胞的增殖,因此输注 MSC 最大的风险是可能减弱移植物抗肿瘤(graft versus tumor,GVT)作用,增加原发疾病复发率,或者是促进新的恶性肿瘤发生。但是目前各研究结果并不一致。

Ning 等[65]通过小样本非盲临床对照试验发现,MSC 联合 HSC 共移植组复发率为 60%(6/10),显著高于对照组的 20%(3/15),差异有统计学意义($P=0.02$)。此外,HSC+MSC 组中位复发时间早于对照组(63 天比 177 天)。2011 年,Liu 等[52]进行的Ⅱ期临床非盲

随机对照试验中，HSC 联合 MSC 共移植组疾病复发率为 12.8%，对照组复发率为 9.3%；两组 2 年生存率分别为 69.7%、64.3%，差异无统计学意义（P=0.737）。

相反，Baron 等[66]研究表明，MSC 联合 HSC 共移植组（n=20）和对照组（n=16）1 年累计复发率分别为 30%、25%，差异无统计学意义。但 1 年整体生存率试验组显著优于对照组（80%比 44%，P=0.02）。因此，HSC+MSC 共移植可能减少了致死性 GVHD 的发生，但 GVT 作用并无减弱。2015 年 Zhao 等[62]在 MSC 输注治疗 aGVHD 的研究中发现，MSC 组 28 例患者中位随访时间 322.5 天，共有 2 例患者复发，3 年累积肿瘤复发率为 17.5%，对照组 19 例患者中位随访时间 256.5 天，只有 1 例患者复发，3 年累积肿瘤复发率为 7.7%，两组的差异无统计学意义（P=0.725）。目前 MSC 输注与肿瘤发生的关系仍有争议，仍需进一步证实。

第三节 小　　结

MSC 来源丰富、制备简单、具有多向分化潜能和低免疫原性，虽然各项临床研究结果不尽相同，但是大部分结果证明其在 HSCT 中应用安全、可行，是一项很有潜力的细胞治疗方案。目前较为统一的认识是，MSC 可促进移植后造血恢复，预防 aGVHD 的发生，并对难治性 aGVHD 有较好的疗效，但 MSC 对 cGVHD 的作用目前仍存在较多争议。除此以外，还存在很多问题。一方面，MSC 的输注是否会增加肿瘤复发率，是否会增加移植后机会性感染，以及 HSC 联合 MSC 共移植后 GVHD 的预防方案如何制定，仍需要进一步的大规模随机对照试验证实。另一方面，MSC 的来源、制备流程、给药时机、给药途径、给药剂量、给药间隔和疗程等问题亟待进一步规范化。

（竺晓凡）

参 考 文 献

[1] Caplan AI. Mesenchymal stem cells. J Orthop Res, 1991, 9(5):641-650.

[2] Horwitz EM, Le Blanc K, Dominici M, et al. Clarification of the nomenclature for MSC: the international society for cellular therapy position statement. Cytotherapy, 2005, 7(5):393-395.

[3] Haynesworth SE, Baber MA, Caplan AI. Cell surface antigens on human marrow-derived mesenchymal cells are detected by monoclonal antibodies. Bone, 1992, 13(1):69-80.

[4] da Silva Meirelles L, Chagastelles PC, Nardi NB. Mesenchymal stem cells reside in virtually all post-natal organs and tissues. J Cell Sci, 2006, 119(11):2204-2213.

[5] Kern S, Eichler H, Stoeve J, et al. Comparative analysis of mesenchymal stem cells from bone marrow, umbilical cord blood, or adipose tissue. Stem Cells, 2006, 24(5):1294-1301.

[6] Dominici M, Le Blanc K, Mueller I, et al. Minimal criteria for defining multipotent mesenchymal stromal cells. The International Society for Cellular Therapy position statement. Cytotherapy, 2006, 8(4):315-317.

[7] Choi YS, Dusting GJ, Stubbs S, et al. Differentiation of human adipose-derived stem cells into beating cardiomyocytes. J Cell Mol Med, 2010, 14(4):878-889.

[8] Safford KM, Hicok KC, Safford SD, et al. Neurogenic differentiation of murine and human adipose-derived stromal cells. Biochem Biophys Res Commun, 2002, 294(2):371-379.

[9] Seo BM, Miura M, Gronthos S, et al. Investigation of multipotent postnatal stem cells from human periodontal ligament. Lancet, 2004, 364(9429):149-155.

[10] Kuroda Y, Kitada M, Wakao S, et al. Unique multipotent cells in adult human mesenchymal cell populations. Proc Natl Acad Sci USA, 2010, 107(19):8639-8643.

[11] Guo F, Parker Kerrigan BC, Yang D, et al. Post-transcriptional regulatory network of epithelial-to-mesenchymal and mesenchymal-to-epithelial transitions. J Hematol Oncol, 2014, 7:19.

[12] Yin X, Zhang BH, Zheng SS, et al. Coexpression of gene Oct4 and Nanog initiates stem cell characteristics in hepatocellular carcinoma and promotes epithelial-mesenchymal transition through activation of Stat3/Snail signaling. J Hematol Oncol, 2015, 8:23.

[13] Zappia E, Casazza S, Pedemonte E, et al. Mesenchymal stem cells ameliorate experimental autoimmune encephalomyelitis inducing T-cell anergy. Blood, 2005, 106(5):1755-1761.

[14] Chen FH, Tuan RS. Mesenchymal stem cells in arthritic diseases. Arthritis Res, 2008, 10(5):223.

[15] Vija L, Farge D, Gautier JF, et al. Mesenchymal stem cells: stem cell therapy perspectives for type 1 diabetes. Diabetes Metab, 2009, 35(2):85-93.

[16] Okada H, Suzuki J, Futamatsu H, et al. Attenuation of autoimmune myocarditis in rats by mesenchymal stem cell transplantation through enhanced expression of hepatocyte growth factor. Int Heart J, 2007, 48(5):649-661.

[17] Gupta N, Su X, Popov B, et al. Intrapulmonary delivery of bone marrow-derived mesenchymal stem cells improves survival and attenuates endotoxin-induced acute lung injury in mice. J Immunol, 2007, 179: 1855-1863.

[18] Mazzini L, Mareschi K, Ferrero I, et al. Mesenchymal stromal cell transplantation in amyotrophic lateral sclerosis: a long-term safety study. Cytotherapy, 2012, 14(1):56-60.

[19] Trachtenberg B, Velazquez DL, Williams AR, et al. Rationale and design of the transendocardial injection of autologous human cells (bone marrow or mesenchymal) in chronic ischemic left ventricular dysfunction and heart failure secondary to myocardial infarction (TAC-HFT) trial: a randomized, double-blind, placebo-controlled study of safety and efficacy. Am Heart J, 2011, 161(3):487-493.

[20] Jiang R, Han Z, Zhuo G, et al. Transplantation of placenta-derived mesenchymal stem cells in type 2 diabetes: a pilot study. Front Med, 2011, 5(1):94-100.

[21] Zhang Z, Lin H, Shi M, et al. Human umbilical cord mesenchymal stem cells improve liver function and ascites in decompensated liver cirrhosis patients. J Gastroenterol Hepatol, 2012, 27(Suppl 2):112-120.

[22] Mueller SM, Glowacki J. Age-related decline in the osteogenic potential of human bone marrow cells cultured in three-dimensional collagen sponges. J Cell Biochem, 2001, 82:583-590.

[23] Kolf CM, Cho E, Tuan RS. Mesenchymal stromal cells. Biology of adult mesenchymal stem cells: regulation of niche, self-renewal and differentiation. Arthritis Res Ther, 2007, 9(1):204.

[24] Lazarus HM, Haynesworth SE, Gerson SL, et al. *Ex vivo* expansion and subsequent infusion of human bone marrow-derived stromal progenitor cells (mesenchymal progenitor cells): implications for therapeutic use. Bone Marrow Transplant, 1995, 16:557-564.

[25] Romanov YA, Svintsitskaya VA, Smirnov VN. Searching for alternative sources of postnatal human mesenchymal stem cells: candidate MSC-like cells from umbilical cord. Stem Cells, 2003, 21:105-110.

[26] In't Anker PS, Scherjon SA, Kleijburg-van der Keur C, et al. Isolation of mesenchymal stem cells of fetal or maternal origin from human placenta. Stem Cells, 2004, 22(7):1338-1345.

[27] Perry TE, Kaushal S, Sutherland FW, et al. Thoracic surgery directors association award. Bone marrow as a cell source for tissue engineering heart valves. Ann Thorac Surg, 2003, 75:761-767.

[28] Sobolewski K, Bankowski E, Chyczewski L. Collagen and glycosaminoglycans of Wharton's jelly. Biol Neonate, 1997, 71:11-21.

[29] Takechi K, Kuwabara Y, Mizuno M. Ultrastructural and immunohistochemical studies of Wharton's jelly umbilical cord cells. Placenta, 1993, 14:235-245.

[30] Xu M, Zhang B, Liu Y. et al. The immunologic and hematopoietic profiles of mesenchymal stem cells derived from different sections of human umbilical cord. Acta Biochim Biophys Sin (Shanghai), 2014, 46(12):1056-1065.

[31] Gottipamula S, Ashiwin KM, Muttigi MS, et al. Isolation,expansion and characterization of bone marrow-derived mesenchymal stromal cells in serum-free conditions. Cell Tissue Res, 2014, 356(1):123-135.

[32] Luetzkendorf J, Nerger K, Hering J, et al. Cryopreservation does not alter main characteristics of good manufacturing process-grade human multipotent mesenchymal stromal cells including immunomodulating potential and lack of malignant transformation. Cytotherapy, 2015, 17(2):186-198.

[33] Fong CY, Richards M, Manasi N, et al. Comparative growth behaviour and characterization of stem cells from human Wharton's jelly. Reprod Biomed Online, 2007, 15:708-718.

[34] Zhou BO, Yue R, Murphy MM, et al. Leptin-receptor-expressing mesenchymal stromal cells represent the main source of bone formed by adult bone marrow. Cell Stem Cell, 2014, 15(2):154-168.

[35] Tormin A, Li O, Brune JC, et al. CD146 expression on primary nonhematopoietic bone marrow stem cells is correlated with in situ localization. Blood, 2011, 117(19):5067-5077.

[36] Te Boome LC, Mansilla C, van der Wagen LE, et al. Biomarker profiling of steroid-resistant acute GVHD in patients after infusion of mesenchymal stromal cells. Leukemia, 2015, 29(9):1839-1846.

[37] Tu Z, Li Q, Bu H, et al. Mesenchymal stem cells inhibit complement activation by secreting factor H. Stem Cells Dev, 2010, 19:1803-1809.

[38] Ren G, Zhang L, Zhao X, et al. Mesenchymal stem cell-mediated immunosuppression occurs via concerted action of chemokines and nitric oxide. Cell Stem Cell, 2008, 2:141-150.

[39] Meisel R, Zibert A, Laryea M, et al. Human bone marrow stromal cells inhibit allogeneic T-cell responses by indoleamine 2,3-dioxygenase-mediated tryptophan degradation. Blood, 2004, 103:4619-4621.

[40] Prigione I, Benvenuto F, Bocca P, et al. Reciprocal interactions between human mesenchymal stem cells and gamma delta T cells or invariant natural killer T cells. Stem Cell, 2009, 27(3):693-702.

[41] English K, Barry FP, Mahon BP. Murine mesenchymal stem cells suppress dendritic cell migration, maturation and antigen presentation. Immunol Lett, 2008, 115(1):50-58.

[42] Spaggiari GM, Capobianco A, Becchetti S, et al. Mesenchymal stem cell-natural killer cell interactions: evidence that activated NK cells are capable of killing MSCs，whereas MSCs can inhibit IL-2-induced NK-cell proliferation. Blood, 2006, 107(4):1484-1490.

[43] Corcione A, Benvenuto F, Ferretti E, et al. Human mesenchymal stem cells modulate B-cell functions. Blood, 2006, 107(1):367-372.

[44] Arai Y, Aoki K, Takeda J, et al. Clinical significance of high-dose cytarabine added to cyclophosphamide/total-body irradiation in bone marrow or peripheral blood stem cell transplantation for myeloid malignancy. J Hematol Oncol, 2015, 8:102.

[45] Chang YJ, Zhao XY, Xu LP, et al. Donor-specific anti-human leukocyte antigen antibodies were associated with primary graft failure after unmanipulated haploidentical blood and marrow transplantation: a prospective study with randomly assigned training and validation sets. J Hematol Oncol, 2015, 8:84.

[46] Meuleman N, Tondreau T, Ahmad I, et al. Infusion of mesenchymal stromal cells can aid hematopoietic recovery following allogeneic hematopoietic stem cell myeloablative transplant: a pilot study. Stem Cells Dev, 2009, 18(9):1247-1252.

[47] Dazzi F, Ramasamy R, Glennie S, et al. The role of mesenchymal stem cells in haemopoiesis. Blood Rev, 2006, 20(3):161-171.

[48] Ball LM, Bernardo ME, Roelofs H, et al. Cotransplantation of *ex vivo* expanded mesenchymal stem cells accelerates lymphocyte recovery and may reduce the risk of graft failure in haploidentical hematopoietic stem-cell transplantation. Blood, 2007, 110(7):2764-2767.

[49] 王萍, 房佰俊, 宋永平, 等. 小鼠脂肪源间充质干细胞在重型再生障碍性贫血中的应用. 中国组织工程研究与临床康复, 2008, 12(43):8426-8430.

[50] Koc ON, Gerson SL, Cooper BW, et al. Rapid hematopoietic recovery after coinfusion of autologous-blood stem cells and culture-expanded marrow mesenchymal stem cells in advanced breast cancer patients receiving high-dose chemotherapy. J Clin Oncol, 2000, 18(2):307-316.

[51] Fang B, Li N, Song Y, et al. Cotransplantation of haploidentical mesenchymal stem cells to enhance engraftment of hematopoietic stem cells and to reduce the risk of graft failure in two children with severe aplastic anemia. Pediatr Transplant, 2009, 13(4):499-502.

[52] Liu K, Chen Y, Zeng Y, et al. Coinfusion of mesenchymal stomal cells facilitates platelet recovery without increasing leukemia recurrence in haploidentical hematopoietic stem cell transplantation: a randomized, controlled clinical study. Stem Cells Dev, 2011, 20(10):1679-1685.

[53] Kassis I, Vaknin-Dembinsky A, Karusis D. Bone marrow mesenchymal stem cell infusion on hematopoiesis in mice with aplastic anemia. Zhongguo Shi Yan Xue Ye Xue Za Zhi, 2007, 15:1005-1008.

[54] Lee SH, Lee MW, Yoo KH, et al. Co-transplantation of third-party umbilical cord blood-derived MSCs promotes engraftment in children undergoing unrelated umbilical cord blood transplantation. Bone Marrow Transplant, 2013, 48(8):1040-1045.

[55] Wang H, Wang Z, Zheng X, et al. Hematopoietic stem cell transplantation with umbilical cord multipotent stromal cell infusion for the treatment of aplastic anemia-a single-center experience. Cytotherapy, 2013, 15(9):1118-1125.

[56] Bernardo ME, Ball LM, Cometa AM, et al. Co-infusion of *ex vivo*-expanded, parental MSCs prevents life-threatening acute GVHD, but does not reduce the risk of graft failure in pediatric patients undergoing allogeneic umbilical cord blood transplantation. Bone Marrow Transplant, 2011, 46(2):200-207.

[57] Ferrara JL, Levine JE, Reddy P, et al. Graft-versus-host disease. Lancet, 2009, 373(9674):1550-1561.

[58] Ringdén O, Nilsson B. Death by graft-versus-host disease associated with HLA mismatch, high recipient age, low marrow cell dose, and splenectomy. Transplantation, 1985, 40(1):39-44.

[59] Le Blanc K, Rasmusson I, Sundberg B, et al. Treatment of severe acute graft-versus-host disease with third party haploidentical mesenchymal stem cells. Lancet, 2004, 363(9419):1439-1441.

[60] Le Blanc K, Frassoni F, Ball L, et al. Mesenchymal stem cells for treatment of steroid-resistant, severe, acute graft-versus-host disease: a phase Ⅱ study. Lancet, 2008, 371:1579-1586.

[61] Resnick IB, Barkats C, Shapira MY, et al. Treatment of severe steroid resistant acute GVHD with mesenchymal stromal cells (MSC). Am J Blood Res, 2013, 3(3):225-238.

[62] Zhao K, Lou R, Huang F, et al. Immunomodulation effects of mesenchymal stromal cells on acute graft-versus-host disease after hematopoietic stem cell transplantation. Biol Blood Marrow Transplant, 2015, 21(1):97-104.

[63] Lee ST, Jang JH, Cheong JW, et al. Treatment of high risk acute myelogenous leukaemia by myeloablative chemoradiotherapy followed by co-infusion of T cell-depleted haematopoietic stem cells and culture-expanded marrow mesenchymal stem cells from a related donor with one fully mismatched human leucocyte antigen haplotype. Br J Haematol, 2002, 118(4):1128-1131.

[64] Lazarus HM, Koc ON, Devine SM, et al. Cotransplantation of HLA-identical sibling culture-expanded mesenchymal stem cells and hematopoietic stem cells in hematologic malignancy patients. Biol Blood and Marrow Transplant, 2005, 11(5):389-398.

[65] Ning H, Yang F, Jiang M, et al. The correlation between cotransplantation of mesenchymal stem cells and higher recurrence rate in hematologic malignancy patients: outcome of a pilot clinical study. Leukemia, 2008, 22(3):593-599.

[66] Baron F, Lechanteur C, Willems E, et al. Cotransplantation of mesenchymal stem cells might prevent death from graft-versus-host disease (GVHD) without abrogating graft-versus-tumor effects after HLA-mismatched allogeneic transplantation following nonmyeloablative conditioning. Biol Blood Marrow Transplant, 2010, 16(6):838-847

[67] Kuzmina LA, Petinati NA, Parovichnikova EN, et al. Multipotent mesenchymal stromal cells for the prophylaxis of acute graft-versus-host disease-a phase Ⅱ study. Stem Cells Int, 2012, (1):968213.

[68] Stewart BL, Storer B, Storek J, et al. Duration of immunosuppressive treatment for chronic graft-versus-host disease. Blood, 2004, 104:3501-3506.

[69] Lee SJ, Vogelsang G, Flowers ME. Chronic graft-versus-host disease. Biol Blood Marrow Transplant, 2003, 9:215-233.

[70] Ringdén O, Uzunel M, Rasmusson I, et al. Mesenchymal stem cells for treatment of therapy-resistant graft-versus-host disease. Transplantation, 2006, 81:1390-1397.

[71] 张乐施, 刘启发, 黄科, 等. 间充质干细胞治疗糖皮质激素耐药性慢性移植物抗宿主病临床疗效观察. 中华内科杂志, 2009, 48(7):542-546.

[72] Weng JY, Du X, Geng SX, et al. Mesenchymal stem cell as salvage treatment for refractory chronic GVHD. Bone Marrow Transplantation, 2010, 45:1732-1740.

[73] Zhou H, Guo M, Bian C, et al. Efficacy of bone marrow derived mesenchymal stem cells in the treatment of sclerodermatous chronic graft-versus-host disease: clinical report. Biol Blood Marrow Transplant, 2010, 16(3):403-412.

[74] Weng J, He C, Lai P, et al. Mesenchymal stromal cells treatment attenuates dry eye in patients with chronic graft-versus-host disease. Mol Ther, 2012, 20(12):2347-2354.

[75] Chu YW, Gress RE. Murine models of chronic graft-versus-host disease: insights and unresolved issues. Biol Blood Marrow Transplant, 2008, 14(4):365-378.

[76] Forslöw U, Blennow O, LeBlanc K, et al. Treatment with mesenchymal stromal cells is a risk factor for pneumoniarelated death after allogeneic hematopoietic stem cell transplantation. Eur J Haematol, 2012, 89(3):220-227.

[77] Moermans C, Lechanteur C, Baudoux E, et al. Impact of cotransplantation of mesenchymal stem cells on lung function after unrelated allogeneic hematopoietic stem cell transplantation following non-myeloablative conditioning. Transplantation, 2014, 98(3):348-353.

[78] 张晓婷, 段连宁, 丁丽, 等. 非血缘供者异基因外周血造血干细胞联合脐带间充质干细胞移植治疗恶性血液病临床研究. 中国实验血液学杂志, 2015, 23(5):1445-1450.

[79] 韩冬梅, 王志东, 丁丽, 等. 脐带间充质干细胞在单倍体相合造血干细胞移植中对肺部感染的影响. 中国实验血液学杂志, 2014, 22(4):1084-1088.

第八章 间充质干细胞在心血管疾病治疗中的应用

第一节 心血管疾病的概述

在心血管疾病（CVD）研究领域，由于心肌细胞再生能力有限，心肌梗死及由其导致的心力衰竭和先天性心脏病均已成为危害人类健康的头号杀手。在美国诊断为心力衰竭的成人中，50%会在5年内死亡，且大约每死亡9人就有1人是由于心力衰竭[1]。虽然外科冠状动脉旁路移植术（CABG）和内科经皮冠脉介入术（PCI）研究取得了很大进展，但是在以心力衰竭为代表的终末期心脏病治疗方面目前还没有大的突破。另外，约5%的先天性心脏病患儿为单心室、左心发育不良等无法完成双心室矫治的患儿，他们在2岁前需要接受2次甚至更多次手术，使他们的生活质量降低，但不可避免地会发展为终末期心力衰竭[2]。有收缩功能的健康心肌细胞是保证正常心脏拥有收缩和舒张功能的前提与基础。如果少量心肌细胞坏死，剩余的健康心肌细胞能够通过心肌纤维增生来增加心肌收缩力，并通过降低室壁张力和心肌氧耗量来维持正常心脏的功能。如果大量心肌细胞坏死、凋亡或功能异常而出现严重心力衰竭，只能通过心脏移植或安装左心辅助装置来改善患者的心脏功能，但是潜在的感染、血栓形成限制了左心辅助装置应用的发展。由于供体缺乏、免疫排斥和巨额经济费用等，又限制了心脏移植作为常规治疗方案在终末期心力衰竭中的临床应用[3,4]。目前对于一些复杂先天性心脏病，如法洛四联症、永存动脉干、合并室间隔缺损和肺动脉狭窄的大动脉错位、肺动脉狭窄的右室双出口、左心发育不良综合征等，大部分需要应用同种或异种外管道重建右室-肺动脉的连续性[如拉斯泰利手术（Rastelli operation）、达穆斯-凯-斯坦塞尔动脉转位术（Damus-Kaye-Stansel arterial switch procedure）等]，或用带单瓣补片加宽右室流出道。这类管道或补片优良的血流动力学特性、良好的生物相容性使复杂先天性心脏病的外科治疗疗效和预后得到明显提高。然而，这类管道或补片存在的主要问题在于内层蛋白沉着、钙化和梗阻，远期衰败率较高，耐久性差，而且无收缩能力，无法辅助右室泵功能。重要的是，这类补片无生物活性，不能随着患儿的生长而生长，这也是远期衰败率较高的原因[5,6]。

鉴于此，基于自体干细胞移植的再生医学研究有可能会在某些领域取得突破。干细胞移植成功的关键是能够再生出足够数量的有正常功能的心肌细胞，从而提高心脏的收缩功能[7]。Hendrikxv等[8]报道了第一个自体骨髓干细胞移植的前瞻性随机对照研究，发现移植组的局部心肌收缩功能较对照组有明显改善。骨髓干细胞移植后能够通过促进心肌细胞分化、血管新生，抑制心肌细胞凋亡及旁分泌作用促进心脏功能的恢复。但是，单纯干细胞移植，不可避免地存在细胞存活率低的问题，导致其对大面积心肌梗死（室壁瘤形成）和先天性心脏缺损无能为力。由此引出心脏组织工程学（cardiac tissue engineering），随着细胞分化、材料工程、组织构建、再血管化等领域的发展，心脏组织工程学研究得到快速发展，本章就间充质干细胞（MSC）移植治疗CVD领域的一些基本问题做一探讨,并介绍 MSC 在 CVD 中的应用及基础研究。

第二节　干细胞移植的方法

一、静脉注射干细胞

心肌梗死后心肌组织能表达并释放某些生物活性因子促进干细胞向梗死心脏归巢,但干细胞靶向迁移并聚集到心脏损伤区域的机制尚不完全明了[9,10]。研究结果认为,趋化作用是细胞迁移的主要机制之一。MSC 定向迁移与其细胞表面及细胞质内的多种趋化因子受体(chemokine receptor)有关。趋化因子(chemokine)是一类可引起趋化反应的小分子细胞因子超家族,可分为 4 个家族,以 CXC 和 C-C 亚族成员为主。其特有的分泌性糖蛋白可激活细胞表面表达的相应趋化因子受体,并吸引这些细胞沿趋化因子梯度从低到高迁移。其中,淋巴系统特异性趋化因子(lymphatic chemokine)可吸引成熟树突细胞(mDC)由外周进入淋巴结,由此使幼稚 T 细胞进入淋巴结并成为成熟 T 细胞,并参与炎症反应等一系列重要的生理病理活动。

静脉注射(intravenous infusion)干细胞操作简单、创伤小、并发症发生率低,可以反复多次注射是其最大的优点。Hare 等[11]报道了第一个静脉注射 MSC 治疗心肌梗死的临床研究,认为静脉注射 MSC 安全、可操作性强,患者能从中获益,有一定的临床价值。该方法的主要缺点是大量移植的 MSC 分布于肺、脾、肝及网状内皮系统,可能会导致治疗效价降低并对其他器官造成影响,目前尚缺少相关数据。

二、外科心肌内注射

外科心肌内注射(surgical myocardial injection)主要用于需要同时行冠状动脉搭桥术的慢性心肌梗死患者,从 1999 年开展以来[12],临床研究报道较多,结果较为满意。该方法主要优点是直接进行心肌内注射,局部细胞数量和细胞迁移率明显提高;缺点是,可能导致心律失常,以及其为创伤性操作,临床应用限定为特定病例。对于心肌菲薄的心肌梗死患者,操作风险,包括心肌穿透伤等明显增加,目前临床上对于此类患者,要求超声下心肌厚度至少应大于 4mm。

三、经冠状动脉注射干细胞

冠状动脉注射(intracoronary artery infusion)干细胞是临床研究中最常用的方法之一,该方法一般与经皮冠脉介入术(PCI)同时进行,细胞通过特殊的球囊导管输送到再通的冠状动脉。细胞直接输送到冠状动脉,避免了静脉注射干细胞时的细胞损耗和再分布[13],大部分临床试验,包括新英格兰医学期刊(The New England Journal of Medicine, NEJM)的 3 篇文章[14-16]都认为该方法安全、有效,对急性心肌梗死患者心脏功能的恢复是有利的。但争议依然存在,有研究认为该方法增加了患者其他并发症,包括心律失常等的发生率,从而对其安全性提出质疑[17]。Surder 等[18]研究发现在急性心梗后的 5~7 天或者 4 周,经冠状动脉注射骨髓干细胞并没有促进心脏功能的恢复。目前仍需要进一步的研究以评估冠状动脉注射干细胞对患者心脏功能及其他临床相关问题的影响。

四、经冠状静脉注射干细胞

经冠状静脉注射（retrograde coronary venous infusion）心肌保护液已经广泛用于心脏外科手术，能获得冠状动脉注射干细胞相同的心肌保护效果。Vicario 等[19]的研究证明，冠状静脉注射干细胞是可行的。进而他们开始临床试验，对 15 位慢性心绞痛患者进行了注册研究，结果表明该方法切实可行，患者的心功能得到改善[20]。

该方法可以单独操作，其操作方法和冠状动脉注射一样容易完成。由于冠状静脉系统是完全开放的，而冠状动脉存在严重狭窄，这样冠状静脉注射理论上可以获得更好的治疗效果，这对那些冠状动脉严重狭窄、侧支循环差的患者尤为适用。此外，这些患者因为心功能较差往往无法接受 CABG 或 PCI [21,22]。然后，患者能否从冠状静脉注射获得更好的治疗效果还需要更多的临床试验来证明，特别是和冠状动脉注射的对比研究。

五、心内膜注射干细胞

心内膜注射（transendocardial injection）干细胞可以作为一种独立的移植方法，这种干细胞移植方法是在电机械标测（electromechanical mapping，EMM）的引导下完成的。Fuchs 等[23]首先报道了在 EMM 导引下往心内膜注射干细胞的研究，证明该方法具有一定的安全性和有效性，并可促进局部侧支血管的形成和心脏功能的恢复。最近，Chan 等[24]报道了心内膜注射干细胞治疗严重冠状动脉疾病的随机对照研究，结果表明，心内膜注射干细胞能够减少局部心肌细胞的凋亡、促进心肌灌注的恢复，进而促进心脏功能的恢复。但 Perin 等及 Santoso 等的研究却没有得到相似的数据[25,26]。

EMM 可以根据病变细胞坏死透过器官壁的情况来精确地区分梗死心肌和存活心肌[27,28]，这对临床上进行细胞移植非常重要，通过该方法，我们能够精确地将干细胞输送到需要的心肌部位。理论上，在 EMM 导引下往心内膜注射骨髓间充质干细胞（BM-MSC）与冠状动脉注射和外科直接心肌内注射相比，能够使患者获得更好的治疗效果，但这还需要更多的临床数据来证明。

除此之外，还有注射自体干细胞到损伤心肌部位，干细胞移植与细胞因子结合等治疗方法[29]。

第三节 干细胞移植的作用机制

一、心肌细胞再生

心肌梗死后，梗死区心肌细胞数量的减少导致局部心肌收缩无力，影响心脏射血功能。然而，由 Orlic 等[30]提出在心肌微环境下，移植的 MSC 能够分化为心肌细胞，表达心肌特异性基因，增加心肌细胞的数量，从而提高心脏的射血功能。但是这种假设目前还存在争议，不同的学术群体间还没有达成共识[31]。

二、细胞融合

细胞融合（cell fusion）是在自发或人工诱导下，两个不同基因型的细胞或原生质体融合形成一个杂种细胞。基本过程包括细胞融合形成异核体（heterokaryon），异核体通过细胞有丝分裂进行核融合，最终形成单核的杂种细胞。研究发现，MSC 移植到局部心肌梗死区后，在梗死区周围发现形成新的融合心肌细胞，而不是 MSC 分化成的心肌细胞[32]。有研究认为，细胞移植后，细胞融合和细胞分化是相辅相成的，共同促进移植的干细胞向心肌细胞表型分化，并促进原始的心脏干细胞分化为成熟的心肌细胞[33]。目前还没有研究融合或分化细胞在机体内的转归及对心脏功能恢复贡献的报道。

三、旁分泌

虽然干细胞在心肌微环境下能够通过细胞分化或细胞融合形成新的心肌细胞，但是由于定位于心肌内的移植细胞数量少，分化的细胞更少，很难解释细胞移植对心脏功能的改善作用，因此干细胞移植修复心脏损害还存在其他的机制。与此同时，有实验研究发现移植的 MSC 能够分泌促血管新生因子，如血管内皮生长因子（VEGF）、碱性成纤维细胞生长因子（bFGF）、干细胞因子（SCF）等，通过组织间隙作用于周围细胞，发挥重要的旁分泌（paracrine）作用。具体表现为促进血管新生、抑制心肌细胞凋亡、促进心肌祖细胞增生分化[34,35]。另外，移植的 MSC 还能够调控血管内皮细胞及平滑肌细胞增殖，且通过抑制心肌纤维化来改善心室重构，提高心脏功能[36]。

四、激活内源性修复

在 MSC 心肌移植机制研究中，我们过多地认为其治疗机制为移植干细胞通过相应作用促进心脏功能的恢复。有研究认为，MSC 移植启动内源性修复（endogenous repair），激活体内干细胞龛（stem cell niche），干细胞龛是成体干细胞的集中存储部位，通过特定的细胞外基质（ECM）和龛细胞提供特殊的微环境，维持干细胞的自我更新能力和多向分化潜能[37]。移植的 MSC 促进内源性干细胞向心肌梗死部位定向趋化移动，分化为心脏所需的心肌细胞、内皮细胞或成纤维细胞，从而修复心脏损伤[38]。

五、神经芽生

Pak 等[39]研究认为，MSC 移植后能够促进局部神经再生，增加交感神经密度、心脏收缩力，从而改善心脏功能。目前，有研究证据证明神经芽生（nerve sprouting）在 MSC 心脏移植重构中发挥作用，但是可能导致室性心律失常，目前仍需要进一步的研究[40,41]。

第四节 干细胞移植存在的问题

一、细胞存活

MSC 移植后必须存活，然后与宿主心肌发生新的整合，才能促进心脏功能的恢复。MSC 移植后面临的主要问题就是细胞存活，研究发现经冠状动脉注射骨髓干细胞 75min

后，只有1%~2%的注射细胞能够在心肌内存活[7,42]。Blocklet等[43]发现只有约5.5%的移植细胞能够在心肌注射周围存活。Suzuki等[44]的研究直接将干细胞注射至局部心肌内，经动态分析发现：注射10min后约有44%细胞存活，24h后约有15%，而3天后这一数字下降至8%。细胞存活率下降可能是由注射时的机械损伤及心脏跳动导致的压迫及血流的冲击导致的，另一重要原因是死亡的细胞释放免疫调节物质产生炎症反应，而新移植的细胞还没有与宿主细胞建立有效的联系，从而导致细胞死亡[45]。目前有研究采用细胞预处理[46,47]、微粒包装后移植[48]、基因修饰[49]等方法来提高移植细胞的存活率。

二、室性心律失常

细胞移植后导致的心律失常也是值得关注的问题。Menasche等[50]在患者行CABG时进行骨骼肌成肌细胞移植，有位患者在移植后的第11天、12天、13天、22天因出现不同程度的室性心动过速而安装体内除颤仪。Siminiak等[51]报道了2例患者在移植早期出现室性心动过速，2例在2周后出现。导致心律失常的原因包括与缝隙连接分布不均、骨骼肌和心肌中同型离子通道存在差异、穿刺点有炎症反应等，由于干细胞不纯，掺杂的其他细胞也可能会加剧这种反应。

三、再狭窄

Kang等[52]研究发现，干细胞移植后低灌注区减少、冠脉血流储备增加、功能容量和心功能得到改善，但干细胞组10人中有5人出现了支架植入处的再狭窄。他们认为冠状动脉注射干细胞可能会促进冠状动脉支架局部内皮细胞增生，从而加快再狭窄的发生。Vulliet等[53]研究发现在急性心肌梗死模型中，BM-MSC移植后出现ST段抬高、T波改变，7天后检查心肌组织发现了多个梗死病变区，表明冠脉注射干细胞后出现急性心肌缺血和亚急性心肌微梗死。采用药物涂层支架可能可以预防支架局部内皮细胞增生导致的再狭窄。

四、钙化

移植区心肌钙化是另一问题。Yoon等[54]研究了骨髓干细胞移植是否会导致心肌钙化的出现，他们选择7周龄雌性Fisher-344大鼠，随机分为3组：全骨髓细胞组，骨髓干细胞组，对照组。全骨髓干细胞移植2周后，超声显示14只存活大鼠中有4只出现了不同程度的心肌钙化，发生率为28.5%，钙化区碱性磷酸酶（ALP）染色呈阳性，荧光显微镜下观察，钙化区被染色的全骨髓细胞包围，表明移植的骨髓细胞与钙化密切相关。相反，纯化的骨髓干细胞组和对照组没有出现明显的心肌钙化，认为直接移植未经筛选的骨髓干细胞可引起明显的心肌钙化，而移植纯化的干细胞能避免出现心肌钙化。

由于干细胞的分化调控机制还未完全阐明，移植后的干细胞会不会分化为不期望的细胞类型甚至肿瘤也是需要解决的问题[55]。

第五节 干细胞构建组织工程心肌

对于大面积心肌梗死（室壁瘤形成）和先天性心脏病患者，如果通过组织工程方法构建具有活性的组织工程心肌，将其移植后若能与自体心肌发生电机械和功能整合，进而改善心脏的收缩和舒张功能，无疑会对心脏外科的发展起到重要作用。理想的组织工程心肌应具备以下特点：①可收缩性；②电生理的稳定性；③良好的机械强度和柔韧性；④移植后快速血管化或已血管化；⑤无免疫原性[56]。目前已经有采用胎儿或新生儿心肌细胞成功构建组织工程心肌的报道[57,58]。MSC 有自体来源，具有旺盛的增殖能力，在体外条件下可以分化为心肌细胞，是非常有前途的种子细胞。

一、支架材料

在组织工程研究中支架材料起到 ECM 作用，是细胞附着的基本框架和代谢场所，决定新生组织的形状和大小。理想的支架材料应具有以下特点：良好的表面活性和生物相容性，材料本身或其降解产物在体内不引起炎症反应；可塑性和有一定的机械强度；生物可降解性及降解的速度应与细胞的生长、组织的形成相一致；移植后能够与自体组织发生整合并快速血管化[59]。目前还没有开发出具备所有这些特性的支架材料。胶原是一种天然材料，具有良好的生物相容性、可降解性和可塑性，能为细胞的生长提供一定的空间，刺激细胞合成新的胶原蛋白。近年来，多种天然聚合物[如基质胶（matrigel）胶原、壳聚糖、明胶等]和合成聚合物（如聚己内酯、聚羟乙酸、聚乳酸、聚羟基丁酸戊酯，以及聚羟乙酸和聚乳酸的共聚物）已应用于心肌组织工程研究中[60]。

近年，仿生支架的研究取得很大进展，影响仿生支架设计的重要因素包括生物材料的选择和支架的制备工艺。生物材料的生物相容性受许多因素的影响，包括材料表面的亲疏水性、电荷性质、表面能、化学官能团、生物活性因子等。氧化石墨烯是一种性能优异的新型碳材料，表面带有大量含氧活性基团，具有良好的生物相容性，并可对其表面进行化学功能化修饰[61]。还原型氧化石墨烯表面含有丰富的官能团，具有一定的导电性，并且能够促细胞黏附、细胞因子聚集及介导干细胞定向分化。提高其导电性能的好处是能够在体外构建心肌过程中模拟心肌电信号，促进心肌细胞的发育、分化和成熟，从而构建有生理活性、有一定韧性的组织工程心肌组织补片[62,63]。

如何促进细胞在支架材料上黏附、生长及控制支架材料的降解与细胞生长、组织形成相一致也是一个非常有意思的课题。到目前，我们对细胞与支架材料之间的信号通路包括整合素、细胞因子及激素之间的关系和相互影响还知之甚少，更为重要的是，细胞黏附率低及支架降解造成的炎症反应还是阻碍研究进展的一个重要问题[64]。研究者正在研究一些更新型的空隙率更好的支架材料和促进细胞生长的技术方法，如降解时间、生物相容性、易于外科裁剪都是要考虑的因素。

二、细胞种植方法和技术

细胞种植是体外构建心肌块的第一步，这是组织块构建成功非常关键的一步。研究表明，高密度细胞种植有利于细胞的黏附和三维组织块的形成[65]。Birla 等[66]研究认为，细胞种植密度与组织工程心肌形成、力学变化、再血管化和移植后细胞迁移、存活等密切相关。但是低密度种植组具有较好的韧性和力学表现。这说明细胞种植密度影响细胞在组织块中的分布和组织的生物学特性，或提示协调一致的细胞种植对形成性能良好的组织工程心肌块至关重要[67]。组织在形成过程中还需要给种植细胞提供一定的生长空间，利于细胞的增殖和迁移。但是合适的种植密度和种植时间还有待进一步研究阐明。

促进细胞在支架材料上快速生长，特别是向深层生长形成组织是另一个需要研究的课题。研究者研究了很多种植技术，包括静态种植、在旋转力作用下的动态种植。动态种植是目前研究最多、应用最广泛的种植方法，也取得了很好的结果，但是这种方法在低密度种植时效果很差。平衡密度梯度离心下种植优于静态种植和旋转瓶中的动态种植，该方法种植效率高，细胞分布均匀，生长状态良好，重要的是细胞能够在三维支架上均匀分布，与其他种植方法只在表面分布是显著不同的，特别是在细胞数较少时[68,69]。

三、生物反应器

近几年，在体外构建心肌过程中模拟体内的生长环境方面取得了很大的研究进展。应用传统的细胞培养方法和培养系统很难满足细胞生长所需的高标准要求。自然机体细胞都是在机体提供的动力微环境中生长的，常规的体外单层培养方法不能提供组织正常生长发育所需的环境条件。通过模拟体内组织生长所处的动力微环境，构建组织培养系统，可为体外细胞的生长提供较为理想的环境。最主要的就是利用生物反应器模拟体内环境[70]。近年来，研究者已设计出了多种能提供动力环境的培养系统，如机械搅拌式生物反应器、灌注培养生物反应器、模拟微重力旋转生物反应器、固体旋转生物反应器等[71]。

有研究比较了机械搅拌式生物反应器及模拟微重力旋转生物反应器对体外心肌组织的培养效果，证实模拟微重力旋转生物反应器培养的心肌组织有更高的有氧代谢活力。由于这种生物反应器提供给心肌细胞或组织的剪切力非常低，在体外培养过程中细胞能保持与其他细胞的三维接触，所以培养的细胞或组织更接近于自体心肌[72]。Hollweck 等[73]设计了一种新型的生物反应器体外构建心肌组织，他们将分化的 UC-MSC 种植在钛包被的聚四氟乙烯支架上，然后在这种生物反应器中培养，在培养过程中提供波动式灌注，结果表明所形成的心肌组织有高的细胞活性和代谢率。可是，目前大多数生物反应器不能为较厚的组织提供足够的养分和氧气，研究设计可调节的能感受搏动或机械刺激变化的适合培养临床可使用的组织块的生物反应器是一大挑战[74]。Tandon 等[75]研究开发的便携式、可植入或可降解的生物反应器是未来的方向。

四、再血管化

目前阻碍组织工程进一步发展的一个问题是如何使构建的组织工程移植物在体外培养和重塑过程中存活以至于移植后能够快速血管化来保证移植物的活性，使其能够与自体组织发生整合，成为自体组织的一部分。因为自体的血管很难在短时间内长入构建的组织工程移植物，所以这成为目前研究的一个难点。随着研究的深入，目前可以在体外构建有一定厚度的组织工程移植物,并通过添加促血管生成的因子如血管内皮生长因子(VEGF)、胰岛素样生长因子(IGF)、bFGF 等促进血管的生成和组织工程心肌的存活与体外生长[76,77]。

Sekine 等[78]将内皮细胞与体外构建的心肌组织片共培养,发现共培养的内皮细胞在心肌片上形成血管网状结构,将体外构建的富含内皮细胞的心肌组织片移植到梗死心肌周围,发现心脏功能得到恢复,且相对于不含内皮细胞的心肌片移植,局部形成了更多的微血管。He 等[79]将由胚胎干细胞(ESC)分化来的心肌细胞、脐带静脉内皮细胞、胚胎成纤维细胞种植在液态胶原支架上,在生物反应器内进行混合培养,研究发现脐带静脉内皮细胞、胚胎成纤维细胞能明显促进组织工程心肌的再血管化。但是,新生血管的生成依然缓慢,同时新生的血管还没有达到携带血液的功能,这方面的研究进展将为最终研制出临床可用的组织工程移植物起到至关重要的作用。

五、细胞片心肌组织工程

为了避免支架材料降解速度与细胞生长速度不匹配和炎症反应等不良反应,Shimizu 等[80]开发了温度反应性细胞片层(cell sheet engineering)培养技术,不采用任何支架材料在体外构建组织工程移植物。该技术利用 N-异丙基丙烯酰胺聚合物(PIPAAm)在不同温度下与细胞黏附能力不同的特点收集细胞。PIPAAm 在 37℃具有高疏水性,细胞能够在 PIPAAm 上快速黏附、增殖、分化;当温度下降至 32℃以下时,PIPAAm 具有高亲水性,能够通过水化作用让细胞从 PIPAAm 表面解离出来,而不需要经过酶的消化,达到收集细胞的目的。动物实验研究证明,将由干细胞构建的细胞片层移植到心肌梗死区能够促进局部血管新生及心脏收缩和舒张功能的恢复[81-83]。有研究发现,将体外构建的组织工程心肌片移植到受损的心肌部位,这些细胞片表达血管新生相关基因,形成内皮网络样结构,并且与宿主自体微血管相连接,从而改善心脏的功能[84]。Furuta 等[85]进一步验证了细胞片移植后能否与宿主心肌建立信号连接,形成电机械偶联,他们发现移植后心肌组织片能够形成双向的动作电位,没有出现自发的或者诱发的心律失常。这说明移植的心肌组织片能够与宿主心肌发生整合而不会出现严重的心律失常。

六、动物实验研究

Kenar 等[86]以 UC-MSC 为种子细胞,体外构建三维组织工程心肌,构建的组织工程心肌块发生协调一致的收缩并对物理或药物刺激产生反应。组织学和透射电镜检查发现,构建的心肌块在结构和功能上都初步具备了新生儿心肌组织的特点。Krupnick 等[87]将 BM-MSC 种植在三维基质支架上,体外培养后移植到鼠的左心室梗死部位,发现体外构建的组织工程块移植后没有出现明显的炎症反应,阻止了室壁瘤的进一步扩张,并且没有发生室性心律失常。超声随访显示,大鼠的心室舒张和收缩期容积都较对照组明显增加。更为重要的是,MSC 在心肌微环境下能够分化为具有心肌细胞表型的细胞。

Lee 等[88]采用脐血间充质干细胞(UCB-MSC)和脐带内皮细胞,体外构建血管化的组织工程心肌组织,并将构建的心肌组织移植到大鼠的心肌梗死部位,体外研究发现这些细胞能够协同一致的生长，构建的组织韧性强，VEGF 和 IGF-1 表达增加，并且观察到 UCB-MSC 分化为具有平滑肌细胞和心肌细胞表型的细胞。体内移植的实验组织在心脏表面生长良好，具有更高的血管新生能力，与对照相比，组织工程心肌移植后有效地避免了心脏的进一步扩大，使梗死心肌局部的收缩功能增强。因此认为，可以在体外成功构建具有收缩功能、有一定韧性、可以外科裁剪使用的组织工程心肌块，而且构建的心肌组织移植后能够存活并与宿主心肌发生整合，改善心脏的收缩功能。

Matsumura 等[89]将 BM-MSC 种植到支架材料上，体外构建组织工程血管补片，并将构建的血管补片移植到狗的下腔静脉处，没有出现血管狭窄、血栓形成等并发症。Mettler 等[90]提取羊的血管内皮细胞和 BM-MSC 种植到支架材料上，构建组织工程血管补片并移植到右室流出道至肺动脉部位，术后未出现狭窄、血栓或动脉瘤形成等并发症，心脏超声显示流出道通畅，无跨瓣压差。Gottlieb 等[91]成功采用 BM-MSC 和可降解支架材料构建了组织工程带瓣管道，并将构建的自体带瓣管道移植到羊的肺动脉瓣部位，重建右室流出道。术后 6 周心脏超声提示最大跨瓣压差为 17mmHg①，肺动脉瓣基本为微量反流，但随着时间的延长，反流渐加重；核磁共振成像（magnetic resonance imaging，MRI）发现组织工程带瓣管道没有出现皱缩、扩张、钙化等并发症；组织学检查发现随着支架材料的降解，干细胞在瓣膜支架上的分布和表型与自体瓣膜非常相似。

还有研究以由 ESC [92]及诱导多能干细胞（iPSC）[93]分化来的心肌细胞为种子细胞，体外构建组织工程心肌。这些研究结果显示，在外科治疗之前，提取患者的干细胞体外构建组织工程血管补片或带瓣管道，将具有非常诱人的前景。

七、临床研究

在动物实验研究的基础上，Shin'oka 等[94]进一步将由 BM-MSC 构建的组织工程血管应用到临床先天性心脏缺损修复重建中，入选的患儿在术前抽取骨髓，分离 BM-MSC 并将其种植在可降解的支架材料上，构建组织工程血管。共有 42 名患儿接受了组织工程血管的移植重建手术，其中 23 例为功能性单心室需行心外管道全腔静脉-肺动脉连接术（extracardiac total cavopulmonary connection，TCPC）的患儿，19 例为先天性心脏缺损需修补的患儿；随访过程中，所有患儿没有出现补片相关的血栓形成、狭窄及阻塞等并发症。最近，Hibino 等[95]报道了 25 例采用由 BM-MSC 构建的组织工程血管作为心外管道的行 TCPC 患儿的长期随访结果：平均随访时间为 5.8 年，没有出现组织工程血管相关的术后死亡、血管破裂、感染等并发症，1 例患儿出现附壁血栓，经服用华法林治愈；4 例出现血管管腔狭窄，采用经皮血管腔内成形术治愈；血管造影没有发现瘤状突出和钙化，随时间的推移，血管内径增加，具有生长潜能，但管腔狭窄的原因需要进一步的随访和研究。这些令人鼓舞的研究结果显示，组织工程血管或心肌补片将会在先天性心脏病的外科治疗领域发挥重要的作用，成为临床重建材料的重要补充，并将会进一步取代这些材料，从而提高先天性心脏病外科治疗的整体水平和远期结果。

---

① 1mmHg=1.333 22×10^{2}Pa

第六节 问题与展望

总之，心血管疾病，尤其是冠状动脉硬化性心脏病，发病率高，严重威胁着人类身体健康，细胞治疗可能会为这类心血管疾病带来新的治疗选择。MSC 移植及组织工程研究在动物实验中取得良好结果，并已成功应用于临床。BM-MSC 在细胞移植方面已经取得非常好的结果，但是仍然需要进一步的长期研究来评价其远期结果，UC-MSC 具有自体来源、可以冻存、更强的增殖分化潜能等特点，使其成为细胞移植领域非常有竞争力的细胞来源。除直接移植外，细胞移植合并基因治疗及动员自体干细胞修复受损的心肌也是非常有前途的治疗选择，但我们仍然需要进一步的研究来评估心肌再生对心功能的影响及可能的机制。

对于严重冠状动脉狭窄引起的大面积心肌梗死，如室壁瘤形成及先天性心脏病患者，可以将 MSC 与组织工程结合起来，体外构建有活性、植入后能与机体发生电机械融合的组织工程心肌来代替梗死心肌，修复心脏缺损，这将是干细胞治疗的又一选择。但是仍然需要进一步的长期研究来评价其远期结果，并和未种植细胞或种植其他细胞的血管补片进行对比研究来证明其有效性。今后进一步的研究方向包括：寻找合适的支架材料、最佳的细胞来源、种植方法及体外构建条件，生物反应器的应用，各种细胞活性因子的释放和控制，移植物再血管化等方面。相信在不久的将来，随着上述问题的解决，最终能够构建出具有生物活性、可以外科裁剪使用的组织工程补片材料。

（武开宏）

参考文献

[1] Go AS, Mozaffarian D, Roger VL, et al. Heart disease and stroke statistics-2013 update: a report from the American Heart Association. Circulation, 2013, 127(1):e236-e245.

[2] Ohuchi H, Kagisaki K, Miyazaki A, et al. Impact of the evolution of the Fontan operation on early and late mortality: a single-center experience of 405 patients over 3 decades. Ann Thorac Surg, 2011, 92(4):1457-1466.

[3] Almond CS, Singh TP, Gauvreau K, et al. Extracorporeal membrane oxygenation for bridge to heart transplantation among children in the United States: analysis of data from the Organ Procurement and Transplant Network and Extracorporeal Life Support Organization Registry. Circulation, 2011, 123(25):2975-2984.

[4] Moreno SG, Novielli N, Cooper NJ. Cost-effectiveness of the implantable Heart Mate II left ventricular assist device for patients awaiting heart transplantation. J Heart Lung Transplant, 2012, 31(5):450-458.

[5] Ruzmetov M, Geiss DM, Fortuna RS. Outcomes of pericardial bovine xenografts for right ventricular outflow tract reconstruction in children and young adults. J Heart Valve Dis, 2013, 22(2):209-214.

[6] Flameng W, De Visscher G, Mesure L, et al. Coating with fibronectin and stromal cell-derived factor-1alpha of decellularized homografts used for right ventricular outflow tract reconstruction eliminates immune response-related degeneration. J Thorac Cardiovasc Surg, 2013, 147(4):1398-1404.

[7] Wu KH, Mo XM, Han ZC, et al. Stem cell engraftment and survival in the ischemic heart. Ann Thorac Surg, 2011, 92(5):1917-1925.

[8] Hendrikx M, Hensen K, Clijsters C, et al. Recovery of regional but not global contractile function by the direct intramyocardial autologous bone marrow transplantation: results from a randomized controlled clinical trial. Circulation, 2006, 114(1 Suppl):I101-I107.

[9] Min JY, Huang X, Xiang M, et al. Homing of intravenously infused embryonic stem cell-derived cells to injured hearts after myocardial infarction. J Thorac Cardiovasc Surg, 2006, 131(4):889-897.

[10] Boomsma RA, Swaminathan PD, Geenen DL. Intravenously injected mesenchymal stem cells home to viable myocardium after coronary occlusion and preserve systolic function without altering infarct size. Int J Cardiol, 2007, 122(1):17-28.

[11] Hare JM, Traverse JH, Henry TD, et al. A randomized, double-blind, placebo-controlled, dose-escalation study of intravenous adult human mesenchymal stem cells (prochymal) after acute myocardial infarction. J Am Coll Cardiol, 2009, 54(24):2277-2286.

[12] Hamano K, Nishida M, Hirata K, et al. Local implantation of autologous bone marrow cells for therapeutic angiogenesis in patients with ischemic heart disease: clinical trial and preliminary results. Jpn Circ J, 2001, 65(9):845-847.

[13] Strauer BE, Brehm M, Zeus T, et al. Repair of infarcted myocardium by autologous intracoronary mononuclear bone marrow cell transplantation in humans. Circulation, 2002, 106(15):1913-1918.

[14] Assmus B, Honold J, Schachinger V, et al. Transcoronary transplantation of progenitor cells after myocardial infarction. N Engl J Med, 2006, 355(12):1222-1232.

[15] Schachinger V, Erbs S, Elsasser A, et al. Intracoronary bone marrow-derived progenitor cells in acute myocardial infarction. N Engl J Med, 2006, 355(12):1210-1221.

[16] Lunde K, Solheim S, Aakhus S, et al. Intracoronary injection of mononuclear bone marrow cells in acute myocardial infarction.N Engl J Med, 2006, 355(12):1199-1209.

[17] Penicka M, Horak J, Kobylka P, et al. Intracoronary injection of autologous bone marrow-derived mononuclear cells in patients with large anterior acute myocardial infarction: a prematurely terminated randomized study. J Am Coll Cardiol, 2007, 49(24):2373-1374.

[18] Surder D, Manka R, Moccetti T, et al. Effect of bone marrow-derived mononuclear cell treatment, early or late after acute myocardial infarction: twelve months CMR and long-term clinical results. Circ Res, 2016, 119(3):481-490.

[19] Vicario J, Piva J, Pierini A, et al. Transcoronary sinus delivery of autologous bone marrow and angiogenesis in pig models with myocardial injury. Cardiovasc Radiat Med, 2002, 3(2):91-94.

[20] Vicario J, Campo C, Piva J, et al. One-year follow-up of transcoronary sinus administration of autologous bone marrow in patients with chronic refractory angina. Cardiovasc Revasc Med, 2005, 6(3):99-107.

[21] Yokoyama S, Fukuda N, Li Y, et al. A strategy of retrograde injection of bone marrow mononuclear cells into the myocardium for the treatment of ischemic heart disease. J Mol Cell Cardiol, 2006, 40(1):24-34.

[22] Wu K, Mo X, Lu S, et al. Retrograde delivery of stem cells: promising delivery strategy for myocardial regenerative therapy. Clin Transplant, 2011, 25(6):830-833.

[23] Fuchs S, Baffour R, Zhou YF, et al. Transendocardial delivery of autologous bone marrow enhances collateral perfusion and regional function in pigs with chronic experimental myocardial ischemia. J Am Coll Cardiol, 2001, 37(6):1726-1732.

[24] Chan CW, Kwong YL, Kwong RY, et al. Improvement of myocardial perfusion reserve detected by cardiovascular magnetic resonance after direct endomyocardial implantation of autologous bone marrow cells in patients with severe coronary artery disease. J Cardiovasc Magn Reson, 2010, 12:6.

[25] Santoso T, Siu CW, Irawan C, et al. Endomyocardial implantation of autologous bone marrow mononuclear cells in advanced ischemic heart failure: a randomized placebo-controlled trial (END-HF). J Cardiovasc Transl Res, 2014, 7(6):545-552.

[26] Perin EC, Willerson JT, Pepine CJ, et al. Effect of transendocardial delivery of autologous bone marrow mononuclear cells on functional capacity, left ventricular function, and perfusion in chronic heart failure: the FOCUS-CCTRN trial. JAMA, 2012, 307(16):1717-1726.

[27] Kornowski R, Hong MK, Gepstein L, et al. Preliminary animal and clinical experiences using an electromechanical endocardial mapping procedure to distinguish infarcted from healthy myocardium. Circulation, 1998, 98(11):1116-1124.

[28] Perin EC, Silva GV, Sarmento-Leite R, et al. Assessing myocardial viability and infarct transmurality with left ventricular electromechanical mapping in patients with stable coronary artery disease: validation by delayed-enhancement magnetic resonance imaging. Circulation, 2002, 106(8):957-961.

[29] Li N, Wang C, Jia L, et al. Heart regeneration, stem cells, and cytokines. Regen Med Res, 2014, 2(1):6.

[30] Orlic D. Adult bone marrow stem cells regenerate myocardium in ischemic heart disease. Ann N Y Acad Sci, 2003, 996:152-157.

[31] Lin ZB, Qian B, Yang YZ, et al. Isolation, characterization and cardiac differentiation of human thymus tissue derived mesenchymal stromal cells. J Cell Biochem, 2015, 116(7):1205-1212.

[32] Nygren JM, Jovinge S, Breitbach M, et al. Bone marrow-derived hematopoietic cells generate cardiomyocytes at a low frequency through cell fusion, but not transdifferentiation. Nat Med, 2004, 10(5):494-501.

[33] Oh H, Bradfute SB, Gallardo TD, et al. Cardiac progenitor cells from adult myocardium: homing, differentiation, and fusion after infarction. Proc Natl Acad Sci USA, 2003, 100(21):12313-12318.

[34] Uemura R, Xu M, Ahmad N, et al. Bone marrow stem cells prevent left ventricular remodeling of ischemic heart through paracrine signaling. Circ Res, 2006, 98(11):1414-1421.

[35] Cai M, Shen R, Song L, et al. Bone marrow mesenchymal stem cells (BM-MSCs) improve heart function in swine myocardial infarction model through paracrine effects. Sci Rep, 2016, 6:28250.

[36] Kinnaird T, Stabile E, Burnett MS, et al. Local delivery of marrow-derived stromal cells augments collateral perfusion through paracrine mechanisms. Circulation, 2004, 109(12):1543-1549.

[37] Kfoury Y, Scadden DT. Mesenchymal cell contributions to the stem cell niche. Cell Stem Cell, 2015, 16(3):239-253.

[38] Beltrami AP, Barlucchi L, Torella D, et al. Adult cardiac stem cells are multipotent and support myocardial regeneration. Cell, 2003, 114(6):763-776.

[39] Pak HN, Qayyum M, Kim DT, et al. Mesenchymal stem cell injection induces cardiac nerve sprouting and increased tenascin expression in a Swine model of myocardial infarction. J Cardiovasc Electrophysiol, 2003, 14(8):841-848.

[40] Chen J, Zheng S, Huang H, et al. Mesenchymal stem cells enhanced cardiac nerve sprouting via nerve growth factor in a rat model of myocardial infarction. Curr Pharm Des, 2014, 20(12):2023-2029.

[41] Gao H, Yin J, Shi Y, et al. Targeted $P2X_7$ R shRNA delivery attenuates sympathetic nerve sprouting and ameliorates cardiac dysfunction in rats with myocardial infarction. Cardiovasc Ther, 2017, 35(2).

[42] Hofmann M, Wollert KC, Meyer GP, et al. Monitoring of bone marrow cell homing into the infarcted human myocardium. Circulation, 2005, 111(17):2198-2202.

[43] Blocklet D, Toungouz M, Berkenboom G, et al. Myocardial homing of nonmobilized peripheral-blood $CD34^+$ cells after intracoronary injection. Stem Cells, 2006, 24(2):333-336.

[44] Suzuki K, Murtuza B, Beauchamp JR, et al. Role of interleukin-1beta in acute inflammation and graft death after cell transplantation to the heart. Circulation, 2004, 110(11 Suppl 1):II219-II224.

[45] Karpov AA, Udalova DV, Pliss MG, et al. Can the outcomes of mesenchymal stem cell-based therapy for myocardial infarction be improved? Providing weapons and armour to cells. Cell Prolif, 2017, 50(2).

[46] Wu KH, Mo XM, Han ZC, et al. Cardiac cell therapy: pre-conditioning effects in cell-delivery strategies. Cytotherapy, 2012, 14(3):260-266.

[47] Khan I, Ali A, Akhter MA, et al. Preconditioning of mesenchymal stem cells with 2,4-dinitrophenol improves cardiac function in infarcted rats. Life Sci, 2016, 162:60-69.

[48] Mayfield AE, Tilokee EL, Latham N, et al. The effect of encapsulation of cardiac stem cells within matrix-enriched hydrogel capsules on cell survival, post-ischemic cell retention and cardiac function. Biomaterials, 2014, 35(1):133-142.

[49] Huang J, Zhang Z, Guo J, et al. Genetic modification of mesenchymal stem cells overexpressing CCR1 increases cell viability, migration, engraftment, and capillary density in the injured myocardium. Circ Res, 2010, 106(11):1753-1762.

[50] Menasche P, Hagege AA, Vilquin JT, et al. Autologous skeletal myoblast transplantation for severe postinfarction left ventricular dysfunction. J Am Coll Cardiol, 2003, 41(7):1078-1083.

[51] Siminiak T, Kalawski R, Fiszer D, et al. Autologous skeletal myoblast transplantation for the treatment of postinfarction myocardial injury: phase I clinical study with 12 months of follow-up. Am Heart J, 2004, 148(3):531-537.

[52] Kang HJ, Kim HS, Zhang SY, et al. Effects of intracoronary infusion of peripheral blood stem-cells mobilised with granulocyte-colony stimulating factor on left ventricular systolic function and restenosis after coronary stenting in myocardial infarction: the MAGIC cell randomised clinical trial. Lancet, 2004, 363(9411):751-756.

[53] Vulliet PR, Greeley M, Halloran SM, et al. Intra-coronary arterial injection of mesenchymal stromal cells and microinfarction in dogs. Lancet, 2004, 363(9411):783-784.

[54] Yoon YS, Park JS, Tkebuchava T, et al. Unexpected severe calcification after transplantation of bone marrow cells in acute myocardial infarction. Circulation, 2004, 109(25):3154-3157.

[55] Morrison SJ, Kimble J. Asymmetric and symmetric stem-cell divisions in development and cancer. Nature, 2006, 441(7097):1068-1074.

[56] 武开宏, 孙剑, 莫绪明. 体外构建组织工程心肌块的研究进展. 中华实验外科杂志, 2015, 32(5):1215-1516.

[57] Li RK, Jia ZQ, Weisel RD, et al. Survival and function of bioengineered cardiac grafts. Circulation, 1999, 100(19 Suppl):II63-II69.

[58] Leor J, Aboulafia-Etzion S, Dar A, et al. Bioengineered cardiac grafts: a new approach to repair the infarcted myocardium? Circulation, 2000, 102(19 Suppl 3):III56-III61.

[59] Wu K, Liu YL, Cui B, et al. Application of stem cells for cardiovascular grafts tissue engineering. Transpl Immunol, 2006, 16(1):1-7.

[60] Cutts J, Nikkhah M, Brafman DA. Biomaterial approaches for stem cell-based myocardial tissue engineering. Biomark Insights, 2015, 10(Suppl 1):77-90.

[61] Gurunathan S, Kim JH. Synthesis, toxicity, biocompatibility, and biomedical applications of graphene and graphene-related materials. Int J Nanomedicine, 2016, 11:1927-1945.

[62] Jeong JT, Choi MK, Sim Y, et al. Effect of graphene oxide ratio on the cell adhesion and growth behavior on a graphene oxide-coated silicon substrate. Sci Rep, 2016, 6:33835.

[63] Zhang XR, Hu XQ, Jia XL, et al. Cell studies of hybridized carbon nanofibers containing bioactive glass nanoparticles using bone mesenchymal stromal cells. Sci Rep, 2016, 6:38685.

[64] Zimmermann WH, Melnychenko I, Eschenhagen T. Engineered heart tissue for regeneration of diseased hearts. Biomaterials, 2004, 25(9):1639-1647.

[65] Carrier RL, Papadaki M, Rupnick M, et al. Cardiac tissue engineering: cell seeding, cultivation parameters, and tissue construct characterization. Biotechnol Bioeng, 1999, 64(5):580-589.

[66] Birla R, Dhawan V, Huang YC, et al. Force characteristics of in vivo tissue-engineered myocardial constructs using varying cell seeding densities. Artif Organs, 2008, 32(9):684-691.

[67] Dar A, Shachar M, Leor J, et al. Optimization of cardiac cell seeding and distribution in 3D porous alginate scaffolds. Biotechnol Bioeng, 2002, 80(3):305-312.

[68] Godbey WT, Hindy SB, Sherman ME, et al. A novel use of centrifugal force for cell seeding into porous scaffolds. Biomaterials, 2004, 25(14):2799-2805.

[69] Mohebbi-Kalhori D, Rukhlova M, Ajji A, et al. A novel automated cell-seeding device for tissue engineering of tubular scaffolds: design and functional validation. J Tissue Eng Regen Med, 2011, 6(9):710-720.

[70] Kensah G, Gruh I, Viering J, et al. A novel miniaturized multimodal bioreactor for continuous *in situ* assessment of bioartificial cardiac tissue during stimulation and maturation. Tissue Eng Part C Methods, 2011, 17(4):463-473.

[71] Mertsching H, Hansmann J. Bioreactor technology in cardiovascular tissue engineering. Adv Biochem Eng Biotechnol, 2009, 112:29-37.

[72] Kofidis T, Akhyari P, Boublik J, et al. *In vitro* engineering of heart muscle: artificial myocardial tissue. J Thorac Cardiovasc Surg, 2002, 124(1):63-69.

[73] Hollweck T, Akra B, Haussler S, et al. A novel pulsatile bioreactor for mechanical stimulation of tissue engineered cardiac constructs. J Funct Biomater, 2011, 2(3):107-118.

[74] Massai D, Cerino G, Gallo D, et al. Bioreactors as engineering support to treat cardiac muscle and vascular disease. J Healthc Eng, 2013, 4(3):329-370.

[75] Tandon N, Taubman A, Cimetta E, et al. Portable bioreactor for perfusion and electrical stimulation of engineered cardiac tissue. Conf Proc IEEE Eng Med Biol Soc, 2013, 2013:6219-6223.

[76] Perets A, Baruch Y, Weisbuch F, et al. Enhancing the vascularization of three-dimensional porous alginate scaffolds by incorporating controlled release basic fibroblast growth factor microspheres. J Biomed Mater Res A, 2003, 65(4):489-497.

[77] Cittadini A, Monti MG, Petrillo V, et al. Complementary therapeutic effects of dual delivery of insulin-like growth factor-1 and vascular endothelial growth factor by gelatin microspheres in experimental heart failure. Eur J Heart Fail, 2011, 13(12):1264-1274.

[78] Sekine H, Shimizu T, Hobo K, et al. Endothelial cell coculture within tissue-engineered cardiomyocyte sheets enhances neovascularization and improves cardiac function of ischemic hearts. Circulation, 2008, 118(14 Suppl):S145-S152.

[79] He W, Ye L, Li S, et al. Construction of vascularized cardiac tissue from genetically modified mouse embryonic stem cells. J Heart Lung Transplant, 2012, 31(2):204-212.

[80] Shimizu T, Yamato M, Kikuchi A, et al. Cell sheet engineering for myocardial tissue reconstruction. Biomaterials, 2003, 24(13):2309-2316.

[81] Miyahara Y, Nagaya N, Kataoka M, et al. Monolayered mesenchymal stem cells repair scarred myocardium after myocardial infarction. Nat Med, 2006, 12(4):459-565.

[82] Bel A, Planat-Bernard V, Saito A, et al. Composite cell sheets: a further step toward safe and effective myocardial regeneration by cardiac progenitors derived from embryonic stem cells. Circulation, 2010, 122(11 Suppl):S118-S123.

[83] Miyagawa S, Saito A, Sakaguchi T, et al. Impaired myocardium regeneration with skeletal cell sheets-a preclinical trial for tissue-engineered regeneration therapy. Transplantation, 2010, 90(4):364-372.

[84] Sekiya S, Shimizu T, Yamato M, et al. Bioengineered cardiac cell sheet grafts have intrinsic angiogenic potential. Biochem Biophys Res Commun, 2006, 341(2):573-582.

[85] Furuta A, Miyoshi S, Itabashi Y, et al. Pulsatile cardiac tissue grafts using a novel three-dimensional cell sheet manipulation technique functionally integrates with the host heart, *in vivo*. Circ Res, 2006, 98(5):705-712.

[86] Kenar H, Kose GT, Toner M, et al. A 3D aligned microfibrous myocardial tissue construct cultured under transient perfusion. Biomaterials, 2011, 32(23):5320-5329.

[87] Krupnick AS, Kreisel D, Engels FH, et al. A novel small animal model of left ventricular tissue engineering. J Heart Lung Transplant, 2002, 21(2):233-243.

[88] Lee WY, Wei HJ, Wang JJ, et al. Vascularization and restoration of heart function in rat myocardial infarction using transplantation of human cbMSC/HUVEC core-shell bodies. Biomaterials, 2012, 33(7):2127-2136.

[89] Matsumura G, Miyagawa-Tomita S, Shin'oka T, et al. First evidence that bone marrow cells contribute to the construction of tissue-engineered vascular autografts *in vivo*. Circulation, 2003, 108(14):1729-1734.

[90] Mettler BA, Sales VL, Stucken CL, et al. Stem cell-derived, tissue-engineered pulmonary artery augmentation patches *in vivo*. Ann Thorac Surg, 2008, 86(1):132-140.

[91] Gottlieb D, Kunal T, Emani S, et al. *In vivo* monitoring of function of autologous engineered pulmonary valve. J Thorac Cardiovasc Surg, 2010, 139(3):723-731.

[92] Mihic A, Li J, Miyagi Y, et al. The effect of cyclic stretch on maturation and 3D tissue formation of human embryonic stem cell-derived cardiomyocytes. Biomaterials, 2014, 35(9):2798-2808.

[93] Kawamura M, Miyagawa S, Miki K, et al. Feasibility, safety, and therapeutic efficacy of human induced pluripotent stem cell-derived cardiomyocyte sheets in a porcine ischemic cardiomyopathy model. Circulation, 2012, 126(11 Suppl 1):S29-S37.

[94] Shin'oka T, Matsumura G, Hibino N, et al. Midterm clinical result of tissue-engineered vascular autografts seeded with autologous bone marrow cells. J Thorac Cardiovasc Surg, 2005, 129(6):1330-1338.

[95] Hibino N, McGillicuddy E, Matsumura G, et al. Late-term results of tissue-engineered vascular grafts in humans. J Thorac Cardiovasc Surg, 2010, 139(2):431-436.

第九章 间充质干细胞在慢性肺疾病治疗中的应用

慢性肺疾病（如慢性阻塞性肺疾病和肺纤维化）的转归通常是肺心病，传统治疗不能有效抑制病情进展。间充质干细胞（MSC）是来源于发育早期中胚层的一类多能干细胞，具有增殖能力强、多向分化潜能的特性，属于成体干细胞的一种。MSC 能够被诱导分化为 II 型肺泡上皮细胞，同时 MSC 具有调节免疫、促进血管新生等功能，在慢性肺疾病的治疗中显示出应用潜力。在本章中，我们将详细描述 MSC 治疗慢性肺疾病的临床前和临床研究，同时列举在线注册的 MSC 治疗慢性肺疾病的临床试验。

第一节 慢性肺疾病治疗现状

慢性肺疾病主要包括慢性阻塞性肺疾病、肺纤维化和肺结节病。由于 MSC 治疗肺结节病的研究论文极少，在此不予陈述。

慢性阻塞性肺疾病（chronic obstructive pulmonary disease，COPD）是一种具有气流阻塞特征的慢性支气管炎和/或肺气肿，可进一步发展为肺心病和呼吸衰竭的常见慢性疾病，不仅累及肺，还可累及全身多个器官系统，并发症包括心力衰竭、抑郁症、肌营养不良等[1]，严重影响患者的生活质量，给患者和社会造成沉重的经济负担[2]。2014 年 12 月至 2015 年 12 月，一项针对 66 752 位成年人的流行病学调查显示，中国 COPD 的总患病率为 13.6%，其中男性患病率 19.0%，女性患病率 8.1%，意味着我国约有 1.69 亿 COPD 患者[3]。COPD 的治疗药物主要包括支气管扩张剂（如 β2 受体激动剂、胆碱能受体阻断剂和甲基黄嘌呤）、糖皮质激素、祛痰剂、抗氧化剂（如 N-乙酰半胱氨酸、羧甲司坦等），但目前尚没有一种药物能够阻止疾病进展并有效地抑制气道炎症反应[4]。

肺纤维化是以成纤维细胞增殖及大量细胞外基质（ECM）聚集并伴炎症损伤、组织结构破坏为特征的一大类肺疾病的终末期改变，也就是正常肺泡组织被损坏后经过异常修复导致结构异常（瘢痕形成）。绝大部分肺纤维化患者病因不明（特发性），这类疾病称为特发性肺纤维化（idiopathic pulmonary fibrosis，IPF），是间质性肺疾病中一大类。流行病学资料显示，IPF 发病率呈不断上升趋势，由于发病机制不清，病死率高。虽然抗纤维化药物吡非尼酮和表皮生长因子受体（epidermal growth factor receptor，EGFR）抑制剂乙磺酸尼达尼布显示有一定的治疗作用，但并不能修复受损的肺组织[5]。因此，积极寻找新的治疗方法来改善 IPF 患者预后，提高其生存率已迫在眉睫。

MSC 能够释放血管内皮生长因子（VEGF）、前列腺素 E2（PGE2）、白介素-10（IL-10）、肿瘤坏死因子刺激基因-6（TSG-6）、一氧化氮（NO）和转化生长因子-β1（TGF-β1），调控 Th1/Th2 比例，抑制 Th17 分化，具有抗炎、免疫调节、抗凋亡、抗纤维化、促进血管新生等作用，同时能够分化为 II 型肺泡上皮细胞，其在慢性肺疾病的治疗中的应用价值得到越来越多临床前和临床研究的验证，并显示出优于传统治疗的潜力[6]。

第二节　间充质干细胞治疗慢性肺疾病的临床前研究

一、MSC 治疗 COPD 的临床前研究

目前已有大量不同来源 MSC 治疗慢性阻塞性肺疾病（COPD）的临床前研究论文，所应用 MSC 主要包括人骨髓间充质干细胞（hBM-MSC）、人脂肪间充质干细胞（hAD-MSC）、人脐血间充质干细胞（hUCB-MSC）、大鼠和小鼠 AD-MSC、大鼠和小鼠 BM-MSC 等。

日本学者 Shigemura 等[7]于 2006 年发表了第一篇 MSC 治疗 COPD 的临床前研究论文，应用大鼠自体 AD-MSC 治疗肺气肿大鼠模型（由弹性蛋白酶诱导）。AD-MSC 通过静脉输注移植入大鼠体内后，能够抑制肺泡细胞凋亡、增强肺上皮细胞增殖和促进肺血管生成，从而促进肺气肿大鼠模型肺功能的恢复。同时，肺组织中肝细胞生长因子（HGF）的表达显著增加，HGF 对肺组织的修复能发挥多重促进作用，因此可能是 MSC 发挥治疗作用的机制之一。

2006 年日本另一个研究小组通过左右主支气管将自体骨髓单个核细胞（BMMC，含 MSC）移植入由弹性蛋白酶诱导的兔肺气肿模型，治疗后肺功能[一秒用力呼吸容积（forced expiratory volume in first second，FEV1）、用力肺活量（forced vital capacity，FVC）、呼气流量峰值（peak expiratory flow，PEF）]均显著改善，肺泡空间明显减小。同时，支气管肺泡灌洗液内凋亡细胞数量和基质金属蛋白酶-2（MMP-2）的水平均显著降低，而具增殖活性的细胞数量增多。BM-MSC 可能通过减轻炎症、降低 MMP-2 表达、抑制细胞凋亡和促进肺泡细胞增殖来发挥对兔 COPD 的治疗作用[8]。华中科技大学同济医学院 Zhen 研究小组报道异体输注大鼠 BM-MSC 对由木瓜蛋白酶诱导的 COPD 具有显著的治疗作用。研究小组发现 BM-MSC 能够迁移到病变肺泡组织并能够分化为 II 型肺泡上皮细胞，同时还能够通过调节 *Bcl-2* 和 *Bax* 基因表达抑制肺泡细胞凋亡[9]。该研究小组进一步深入研究后发现，当与由木瓜蛋白酶处理的肺组织体外共培养时，BM-MSC 分泌的 VEGF-A（主要功能是促进血管新生）显著增多，同时发生凋亡的肺泡细胞数量减少。当加入肿瘤坏死因子-α（TNF-α）抗体时，VEGF-A 的生成被抑制。这提示受损的肺泡组织通过释放 TNF-α 诱导 BM-MSC 分泌 VEGF-A，促进 COPD 患者肺泡组织血管新生，从而发挥治疗作用[10]。除 Zhen 研究小组外，华中科技大学同济医学院 Zhang 研究小组、上海交通大学医学院附属新华医院 Xu 研究小组、郑州大学第一附属医院 Ren 研究小组各发表研究论文，应用大鼠 BM-MSC 治疗由香烟烟雾和/或脂多糖（LPS）诱导的 COPD 大鼠模型，验证了 MSC 的治疗作用，并从不同侧面阐明了 MSC 发挥治疗作用的机制，如促进 MSC 归巢、抑制肺泡巨噬细胞合成环加氧酶 2（COX2）/PGE2、上调 TGF-β1 表达等[11-14]。

前面总结了自体或同种异体 MSC 对 COPD 发挥治疗作用的研究，下面陈述一下人源 MSC 对 COPD 动物模型发挥治疗作用的报道。

美国 Schweitzer 等[15]研究发现，静脉输注 hAD-MSC 对小鼠 COPD 模型同样具有治

疗作用，表现为由香烟烟雾刺激造成的炎性细胞浸润被抑制，凋亡肺细胞的数量减少，肺泡腔体积也显著减少。hAD-MSC 条件培养基的上清对肺血管内皮细胞损伤具有保护作用，提示 hAD-MSC 所分泌的细胞因子是发挥治疗作用的关键。韩国学者 Kim 等[16]通过静脉注射 hUCB-MSC 治疗由弹性蛋白酶诱导的小鼠肺气肿模型，发现 $5×10^4$cells/mL MSC 是发挥治疗作用的最佳剂量。作用机制与抑制 MMP-9 活性、促进 VEGF 表达有关。中国学者 Li 等[17]分别应用由人诱导多能干细胞分化的 MSC（iPSC-MSC）和 hBM-MSC 通过静脉移植治疗香烟烟雾刺激制备的 COPD 大鼠模型，发现两种不同人源的 MSC 均能减轻肺泡损伤程度，但 iPSC-MSC 向支气管上皮细胞转移线粒体的能力优于 BM-MSC。巴西学者 Peron 等[18]应用从人输卵管组织分离培养的 MSC 结合低剂量激光放射治疗由香烟烟雾刺激制备的 COPD 小鼠模型，发现 MSC 能够减轻肺的炎症程度、减少炎性细胞浸润和抑制促炎因子分泌。

中国学者 Liu 等[19]在一篇综述和 Meta 分析中总结了 2016 年之前在国际杂志上发表的 MSC 治疗 COPD 的临床研究，并根据第一作者、发表论文国家、发表年份、MSC 来源、模型动物、COPD 诱导剂、给药方式、给药剂量和时间、评估参数等信息进行了归类总结，如表 9-1 所示。

表 9-1 国际杂志上发表的 MSC 治疗慢性阻塞性肺疾病（COPD）的临床研究[19]

第一作者	国家	发表年份	MSC 来源	模型动物	COPD 诱导剂	给药方式	给药剂量和时间	评估参数
Shigemura N	日本	2006	大鼠脂肪 MSC（AD-MSC）	大鼠	猪胰弹性蛋白酶	静脉	$5×10^7$cells/0.2mL；注射猪胰弹性蛋白酶后 1 周	细胞凋亡和增殖、血氧分压、肝细胞生长因子（HGF）、运动能力
Yuhgetsu H	日本	2006	兔骨髓 MSC（BM-MSC）	兔	猪胰弹性蛋白酶	支气管内	$1×10^8$cells/2mL；注射猪胰弹性蛋白酶后 24h	细胞凋亡和增殖、肺泡体积、支气管肺泡灌洗液中总细胞和巨噬细胞数
Zhen GH	中国	2008	大鼠 BM-MSC	大鼠	木瓜蛋白酶	静脉	$4×10^6$cells/0.4mL；注射木瓜蛋白酶的同时	肺平均内衬间隔（mean linear intercept，MLI）、TUNEL 检测（terminal deoxynucleotidyl transferase-mediated dUTP-biotin nick end labeling assay）和半胱氨酸蛋白酶-3（Caspase-3）
Zhen GH	中国	2010	大鼠 BM-MSC	大鼠	木瓜蛋白酶	静脉	$4×10^6$cells/0.4mL；注射木瓜蛋白酶后 2h	MLI、TUNEL 检测和 Caspase-3、血管内皮生长因子-A（VEGF-A）
Huh JW	韩国	2011	大鼠 BM-MSC	大鼠	香烟烟雾	静脉	$6×10^5$cells/0.3mL；首次烟雾刺激后 6 个月	肺巨噬细胞、中性粒细胞、Caspase-3、肺容量
Katsha AM	日本	2011	C57BL/6 BM-MSC	C57BL/6 小鼠	猪胰弹性蛋白酶	气管内	$5×10^5$cells/0.2mL；注射猪胰弹性蛋白酶后 14 天	肺破坏指数；*IL-1*、*IL-1β* mRNA
Schweitzer KS	美国	2011	人 AD-MSC	大鼠	香烟烟雾	静脉	$3×10^5$cells；首次烟雾刺激后 2 个月	肺巨噬细胞、中性粒细胞、Caspase-3、肺容量、肺泡表面积

续表

第一作者	国家	发表年份	MSC 来源	模型动物	COPD 诱导剂	给药方式	给药剂量和时间	评估参数
Furuya N	日本	2012	大鼠 AD-MSC	大鼠	猪胰弹性蛋白酶	静脉	$2.5×10^6$cells/0.5mL；注射猪胰弹性蛋白酶后 7 天	动脉血氧分压（PaO_2）、HGF、细胞因子诱导中性粒细胞趋化因子 1（cytokine-induced neutrophil chemoattractant-1, CINC-1）、白介素 IL-1β（IL-1β）
Guan XJ	中国	2013	大鼠 BM-MSC	大鼠	香烟烟雾	气管内	$6×10^6$cells；首次烟雾刺激后 7 周	MLI、TUNEL 检测和 Caspase-3、肺活量、一秒用力呼吸容积（FEV1）、基质金属蛋白酶（MMP-9、MMP-12）、转化生长因子-β（TGF-β）、VEGF
Antunes MA	巴西	2014	C57BL/6 BM-MSC、AD-MSC 和肺 MSC	C57BL/6 小鼠	猪胰弹性蛋白酶	气管内和静脉	$1×10^5$cells；注射猪胰弹性蛋白酶后 3 周	正常肺容积（%）、肺过度充气（%）、TUNEL 检测、中性粒细胞
Feizpour A	伊朗	2014	豚鼠 AD-MSC	豚鼠	香烟烟雾	气管内和静脉	$1×10^6$cells/0.3mL；烟雾刺激后第 1 天和 14 天	组胺引起最大反应的 50% 有效浓度（50% effective concentration, EC50）、血清或肺泡灌洗液中 IL-8 和白细胞水平
Ghorbani A	伊朗	2014	豚鼠 AD-MSC	豚鼠	香烟烟雾	气管内和静脉	$1×10^6$cells/0.3mL；烟雾刺激后第 1 天和 14 天	肺气肿评分、肺泡灌洗液中巯基含量、血清丙二醛（MDA）、中性粒细胞
Li X	中国	2014	人 iPSC 分化的 MSC、人 BM-MSC	大鼠	香烟烟雾	静脉	$3×10^6$cells/0.3mL；烟雾刺激后第 29 天和 43 天	肺毛状体（trichome）
Li YQ	中国	2014	大鼠 AD-MSC	大鼠	香烟烟雾+脂多糖（LPS）	气管	$4×10^5$cells/0.2mL；烟雾刺激和注射 LPS 后第 4 周和 8 周	MLI、TUNEL 检测、平均肺泡面积
Tibboel J	加拿大	2014	C57BL/6 BM-MSC	C57BL/6 小鼠	猪胰弹性蛋白酶	气管内和静脉	$5×10^5$cells/0.1mL；$1×10^5$cells/0.1mL；注射猪胰弹性蛋白酶前和后	MLI、肺动态顺应性、平均用力呼气量
Zhang WG	中国	2014	大鼠 BM-MSC	大鼠	香烟烟雾	静脉	$4×10^6$cells/0.2mL；首次烟雾刺激后第 20 天和 62 天	MLI、TUNEL 检测、IL-6
Zhao YM	中国	2014	大鼠 BM-MSC	大鼠	香烟烟雾+脂多糖（LPS）	静脉	$5×10^6$cells；首次烟雾刺激后第 36 天	平均肺泡数量、肺泡面积
Gu W	中国	2015	大鼠 BM-MSC	大鼠	香烟烟雾	气管	$5×10^6$cells/0.2mL；烟雾刺激后 7 周	MLI、环加氧酶 2（COX2）、前列腺素 E2（PGE2）、IL-6、IL-10
Kim YS	韩国	2015	人脐带 MSC（UC-MSC）	C57BL/6 小鼠	猪胰弹性蛋白酶	静脉	多个不同剂量；注射猪胰弹性蛋白酶后 7 天	MLI、VEGF
Peron JP	巴西	2015	人输卵管 MSC	C57BL/6 小鼠	香烟烟雾+放射	腹腔或鼻腔	$1×10^6$cells；第 60 天和 67 天	细胞凋亡和增殖、中性粒细胞、气道黏膜、胶原

该论文 Meta 分析的统计结果表明，MSC 对 COPD 动物模型的肺结构和功能均有显著改善作用。在减少肺泡过度通气和胶原纤维含量方面，气管或支气管内灌注 MSC 的效果似乎优于静脉注射，提示气管或支气管内灌注可能是 MSC 治疗呼吸道疾病更加安全、有效的方式[19]。

总之，无论是骨髓、脂肪、脐血还是由 iPSC 诱导分化来的 MSC，在治疗 COPD 动物模型的过程中均显示出有良好的安全性和有效性。发挥治疗作用的机制涉及 MSC 抑制促炎因子（如 IL-1β、TNF-α、IL-6 和 PGE2）释放，刺激抗炎因子（如 IL-10）释放，并上调促进组织修复的因子（VEGF、HGF、EGF 和 TGF-β1）表达等。

二、MSC 治疗肺纤维化的临床前研究

2003 年，Ortiz 等[20]通过颈静脉为由博来霉素（bleomycin）诱导的肺纤维化小鼠注射小鼠 BM-MSC（5×10^5cells/只，悬浮于 200μL PBS 中），发现 MSC 能够定位在肺损伤部位，并显示出上皮细胞样形态。与对照组相比，MSC 治疗组小鼠 II 型肺泡上皮细胞的数量显著增多，博来霉素诱导的炎症反应和胶原沉积明显减轻，肺纤维化的进展被显著抑制。在一个由二氧化硅（SiO_2）诱导的肺纤维化小鼠模型中，人 MSC（hMSC）移植后发现正常肺细胞取代了纤维化细胞，并且肺纤维化的表现（如胶原沉积和炎症反应）显著减轻。研究者进一步探究了 hMSC 分泌的微泡对肺纤维化的治疗作用，发现虽然治疗作用比 MSC 弱，但也能显著抑制肺纤维化过程中的胶原沉积和炎症进展。因此，分化为肺泡上皮细胞并替换受损肺泡上皮细胞，以及分泌微泡发挥肺泡细胞保护作用，可能是 hMSC 治疗 COPD 的关键机制[21]。澳大利亚学者 Moodley 等[22]通过静脉为由博来霉素诱导的肺纤维化小鼠注射人脐带间充质干细胞（hUC-MSC），发现 2 周后 hUC-MSC 只定位于肺发生炎症和纤维化的部位，在健康肺组织中检测不到。hUC-MSC 能抑制炎症反应，并减少 TGF-β、干扰素-γ（IFN-γ）、巨噬细胞移动抑制因子（macrophage migration inhibition factor，MIF）和 TNF-α 等炎性细胞因子的表达。hUC-MSC 治疗后胶原沉积显著减少，可能是 Smad2 蛋白磷酸化被抑制的结果（TGF-β 活性）。同时，hUC-MSC 能增加 MMP-2 的水平，并且抑制 MMP-2 的内源性抑制分子，从而有利于沉积胶原的降解。这些研究结果提示，hUC-MSC 具有抗纤维化作用，并促进肺损伤组织修复。

MSC 除本身具抗肺纤维化的作用外，还可以作为治疗性外源基因的表达载体，增强其对肺纤维化的治疗作用。上海交通大学 Min 等[23]将血管紧张素转换酶 2 基因（angiotensin converting enzyme 2，*ACE2*）转染 hUC-MSC，制备高效表达 ACE2 的 hUC-MSC，即 ACE2-hUC-MSC。ACE2-hUC-MSC 对肺纤维化小鼠模型的治疗效果明显优于单独 ACE2 或 hUC-MSC 治疗。其治疗机制是 ACE2-hUC-MSC 能显著下调丙二醛（malondialdehyde，MDA）、氧化型谷胱甘肽（glutathione oxidized，GSSG）、TNF-α、IFN-γ、TGF-β、IL-1、IL-2、IL-6、1 型胶原、MMP、金属蛋白酶组织抑制物（tissue inhibitor of metalloproteinase，TIMP）和羟脯氨酸等促肺纤维化因子的表达，同时上调超氧化物歧化酶（superoxide dismutase，SOD）、谷胱甘肽（glutathione，GSH）、ACE2 和 IL-10 等抑制肺纤维化因子的表达。因此，通过基因转染可使 ACE2 和 hUC-MSC 对肺纤维化发挥协同治疗作用。另外，MSC 还可以与传统的免疫抑制剂共用来对肺纤维化发挥治疗效果。

Xu 等[24]的研究发现,免疫抑制剂环磷酰胺对由博来霉素诱导的肺纤维化小鼠没有治疗作用,甚至对小鼠有害,但与 BM-MSC 联合应用对肺纤维化有明显治疗作用。

博来霉素是目前制备肺纤维化模型应用较多的诱导剂,Srour 和 Thébaud[25]将 2015 年之前发表的 MSC 治疗由博来霉素诱导的肺纤维化动物模型的研究进行了总结。在此将其中有明确 MSC 来源和注射剂量信息的研究列为表 9-2。

表 9-2 MSC 治疗由博来霉素诱导的肺纤维化动物模型的研究[25]

第一作者	模型动物	细胞来源	剂量($\times 10^6$cells)	给药时间(造模后)和途径	观察结果时间点
Cargnon	C57BL/6 小鼠	人和小鼠羊膜/绒毛膜 MSC	1 和 4	15min;气管内或腹腔内	第 3 天、7 天、9 天和 14 天
Gao	SD 大鼠	人脐带 MSC(UC-MSC)	0.25	3 天;静脉	第 0 天、3 天、7 天和 14 天
Garcia	C57BL/6 小鼠	人羊水 MSC	1	2h 或 14 天;静脉	第 3 天、14 天和 28 天
Gazdhar	Fisher F344 大鼠	Fisher F344 大鼠骨髓 MSC(BM-MSC)	3	7 天;气管内	第 14 天、21 天
Huang	Wistar 大鼠	Wistar 大鼠 BM-MSC	2.5	0 天和 7 天;静脉	第 7 天、14 天和 28 天
Jun	C57BL/6 小鼠	C57BL/6 小鼠肺 MSC	0.15~0.25	0 天;静脉	第 14 天
Kumamoto	C57BL/6 小鼠	C57BL/6 小鼠 BM-MSC	0.5	3 天;静脉	第 10 天
Lee	SD 大鼠	SD 大鼠 BM-MSC	1	4 天;静脉	第 0 天、7 天、14 天、21 天和 28 天
Lee	C57BL/6 小鼠	人脂肪 MSC(AD-MSC)	0.3	8 周、10 周、12 周和 14 周;静脉	第 16 周
Moodley	SCID(重度联合免疫缺陷)小鼠	人 UC-MSC	1	24h;静脉	第 7 天、14 天和 28 天
Moodley	C57BL/6 小鼠	人 BM-MSC;人羊膜 MSC	1	10 天;静脉	第 1 周和 3 周
Ono	C57BL/6 小鼠	人 BM-MSC	0.5	24h;静脉	第 14 天
Ortiz	C57BL/6 小鼠	Balb/C 小鼠 BM-MSC	0.5	0 或 7 天;静脉	第 14 天
Ortiz	C57BL/6 小鼠	Balb/C 小鼠 BM-MSC	0.5	0h;静脉	第 3 天、7 天和 14 天
Zhao	SD 大鼠	SD 大鼠 BM-MSC	0.5	12h;静脉	第 2 周

第三节 间充质干细胞治疗慢性肺疾病的临床研究

一、MSC 治疗 COPD 的临床研究

Ribeiro-Paes 等[26]首次进行了干细胞治疗慢性阻塞性肺疾病(COPD)的临床研究,评估患者自体骨髓单个核细胞(BMMC)治疗晚期肺气肿(Ⅳ级呼吸困难)的安全性和有效性。在抽取自身骨髓(10mL)并分离 BMMC 前,对肺气肿患者进行临床检查并注射粒细胞集落刺激因子(G-CSF)。将分离的 BMMC(每位患者 50×10^8cells)通过外周静脉为患者输注。在 12 个月的随访期间,COPD 患者的生活质量得到显著改善,临床状况稳定。肺活量测定结果显示,所有患者在细胞治疗后最初的 30 天功能指标有所改善。

然而经过 30~90 天，肺功能指标趋于降低，但没有低至基线值。在该研究中，BMMC 治疗晚期 COPD 患者的安全性得到验证，并且对抑制或减缓疾病的进展具有一定程度的作用。Stessuk 等[27]在 2013 年开展了一项与上述类似的临床研究，同样是应用 BMMC，研究方案也相同，所不同的是随访时间延长至 3 年。结果显示，BMMC 治疗是安全的，没有发生明显的不良反应，所有 4 位患者的情绪状态和体能都得到显著的改善，其中 2 位的肺功能改善显著，FVC 分别从 21%增至 36.5%和从 34%增至 58%。这些晚期肺气肿患者肺功能得到改善的机制可以用 BMMC 的旁分泌效应及输注后血浆中炎症相关蛋白水平的降低来解释。

Weiss 等[28]开展了一项利用 BM-MSC 治疗中至重度 COPD 患者的随机双盲安慰剂对照临床试验。MSC 产品的名称是 Prochymal，由美国生物制药公司 Osiris 研发和生产，来自于健康青年捐献者的骨髓。该研究共纳入 62 位患者，随机分为 MSC 治疗组（$n=30$）和安慰剂对照组（$n=32$）。患者每月通过静脉输注 1 次 MSC，共输注 4 次，每次输注的剂量为 $100×10^6$cells，随访期为 2 年。评估的终点事件是安全性、肺功能、生活质量和全身炎症状态。结果显示，没有与 MSC 输注相关的严重不良反应发生，表明 MSC 安全性良好。在肺功能和生活质量方面，MSC 治疗组与安慰剂组之间没有显著差异。然而，在治疗早期，MSC 治疗组外周血 C 反应蛋白（C-reactive protein，CRP）水平的下降程度显著高于对照组。因此，BM-MSC 治疗中至重度 COPD 是安全的，但其长期效果还有待优化治疗方案进行更进一步的验证。

目前，对于严重肺气肿，唯一的治疗方法是肺减容术（lung volume reduction surgery，LVRS），切除肺损伤最严重的部分。Stolk 等[29]于 2016 年在荷兰进行了一项 I 期前瞻性临床试验，评估重度肺气肿患者 LVRS 前后应用自体 BM-MSC 的安全性和可行性。9 位患者进行了骨髓抽吸，平均体积为 158mL±64mL，三个扩增周期后 8 位患者的 MSC 达到目标数量。患者分别于第二次 LVRS 手术前 4 周和 3 周各输注 1 次 BM-MSC，共输注 2 次。所有患者在 BM-MSC 输注后 48h 内生命状态稳定，在第二次输注后 48h 和 3 周（第二次 LVRS 的前一天）时均未观察到与 MSC 注射有关的不良反应。临床随访 12 个月，与基线值相比，FEV1 增加了 390mL±240mL。所有患者的体重均显著增加，平均增加 4.6kg（1~10kg）。CD31 免疫组化显示，肺泡隔内皮细胞标记分子 CD31 表达量增加三倍。CD31 表达增加提示 BM-MSC 治疗可刺激肺病变部位的微血管内皮细胞增殖。基因表达分析显示，LVRS 和 MSC 治疗后活检组织中 IL-10 和肿瘤坏死因子刺激基因-6（TSG-6）的 mRNA 表达水平增高。这些结果表明，自体 BM-MSC 治疗重度肺气肿是可行和安全的。

二、MSC 治疗肺纤维化的临床研究

虽然在肺纤维化的临床前研究中 MSC 显示出具有良好的安全性和显著的治疗作用，但目前已发表的临床试验结果只显示了 MSC 临床应用的安全性，却没有显示出有显著的治疗作用。这可能与 MSC 来源、剂量、方案，以及患者的入组条件都有一定的关系，未来需要在这些方面进一步进行优化。下面对已发表的 MSC 治疗肺纤维化的几篇报道进行总结。

一项 I b 期非随机临床试验研究了 3 次支气管内注入自体 AD-MSC（$50×10^4$cells/kg 体重）治疗轻至重度 IPF（$n=14$）患者的安全性。主要终点事件是 12 个月内治疗相关不

良事件的发生率，次要终点是在连续时间点（基线时间点、首次输注后6个和12个月）评估患者肺功能、运动能力和生活质量的改变程度。结果显示，所有患者均未发生严重或临床有意义的不良事件，包括短期注射毒性和长期异位组织形成。3个时间点的监测表明，细胞治疗后患者的肺功能参数和生活质量没有恶化。该临床试验证明了通过支气管内注入SDSC治疗肺纤维化的安全性[30]。

在另一项剂量递增Ⅰb临床试验中，8位中至重度IPF患者接受无亲缘关系供者的胎盘间充质干细胞（P-MSC）治疗，外周静脉输注，剂量为$1×10^6$cells/kg体重（$n=4$）或$2×10^6$cells/kg体重（$n=4$），随访6个月。结果显示，患者对两种不同剂量耐受性良好，只有轻微的和短暂的急性不良反应。MSC输注15min后动脉血氧饱和度（arterial oxygen saturation，SaO_2）一过性下降，平均下降幅度1%（0~2%），但血流动力学无变化。治疗后6个月时肺功能、6min步行距离和胸部CT与基线相比无明显变化，同时没有发现纤维化继续恶化的证据。因此，中至重度IPF患者静脉输注MSC是可行的，具有良好的短期安全性[31]。

2017年发表的一项临床试验中，9位轻至中度特发性纤维化患者被分为3个剂量组，分别通过静脉单次输注$20×10^6$个、$100×10^6$个和$200×10^6$个异体hMSC。所输注的hMSC来源于与患者无血缘关系的年轻人骨髓。临床试验的主要终点是治疗后4周内出现严重不良事件的发生率。临床试验期间未发现显著的与治疗相关的不良反应。有两位患者死亡，但与治疗无关，是由IPF进展（疾病恶化和/或急性加重）造成的。输注60周后，患者FVC平均下降3%，一氧化碳弥散功能平均下降5.4%。该临床研究说明，单次注射hMSC治疗中至重度IPF是安全的，但并未显示显著的治疗作用[32]。

第四节　间充质干细胞治疗慢性肺疾病的注册临床试验

一、MSC治疗COPD的注册临床试验

美国国立卫生研究院（National Institutes of Health，NIH）临床试验网站（ClinicalTrials.gov）上注册的MSC治疗慢性阻塞性肺疾病（COPD）的临床试验（进行中或完成但结果尚未发表论文的临床试验，撤销或状态不明的临床试验未列出）如表9-3所示。

表9-3　在ClinicalTrials.gov上注册的MSC治疗慢性阻塞性肺疾病（COPD）的临床试验

注册题目	注册编号	干细胞种类	给药方式	国家	完成情况	临床分期
Safety, Tolerability and Preliminary Efficacy of Adipose Derive Stem Cells for Patients With COPD	NCT02161744	自体脂肪间充质干细胞	静脉输注	美国	进行中	Ⅰ期
Safety and Efficacy of Adipose Derived Stem Cells for Chronic Obstructive Pulmonary Disease	NCT02216630	自体脂肪间充质干细胞	静脉输注	美国	完成	Ⅰ期+Ⅱ期
Outcomes Data of Adipose Stem Cells to Treat Chronic Obstructive Pulmonary Disease	NCT02348060	自体脂肪血管基质片段（SVF）细胞	静脉输注	美国	进行中	

注：在"中国临床试验注册中心"网站上没有检索到MSC治疗COPD的临床试验

二、MSC 治疗肺纤维化的注册临床试验

ClinicalTrial.gov 网站上注册的 MSC 治疗肺纤维化的临床试验（进行中或完成但结果尚未发表论文的临床试验，撤销或状态不明的临床试验未列出），如表 9-4 所示。

表 9-4 在 ClinicalTrials.gov 上注册的 MSC 治疗肺纤维化的临床试验

注册题目	注册编号	干细胞种类	给药方式	国家	完成情况	临床分期
Study of Autologous Mesenchymal Stem Cells to Treat Idiopathic Pulmonary Fibrosis	NCT01919827	自体脂肪间充质干细胞	支气管内灌注	西班牙	完成	Ⅰ期
A Study on Radiation-induced Pulmonary Fibrosis Treated With Clinical Grade Umbilical Cord Mesenchymal Stem Cells	NCT02277145	异体脐带间充质干细胞	纤维支气管镜灌注	中国	进行中	Ⅰ期
Safety and Efficacy of Allogeneic Mesenchymal Stem Cells in Patients With Rapidly Progressive Interstitial Lung Disease	NCT02594839	异体骨髓间充质干细胞	静脉输注	俄罗斯	完成	Ⅰ期+Ⅱ期
Allogeneic Human Cells (hMSC) in Patients With Idiopathic Pulmonary Fibrosis Via Intravenous Delivery (AETHER)	NCT02013700	异体骨髓间充质干细胞	静脉输注	美国	进行中	Ⅰ期
Clinical Efficacy and Safety of Autologous Lung Stem Cell Transplantation in Patients With Idiopathic Pulmonary Fibrosis	NCT02745184	自体肺干细胞	纤维支气管镜灌注	中国	进行中	Ⅰ期+Ⅱ期

注：在"中国临床试验注册中心"网站上没有检索到 MSC 治疗肺纤维化的临床试验

（王伟强 孙 军）

参 考 文 献

[1] Vestbo J, Hurd SS, Agustí AG, et al. Global strategy for the diagnosis, management, and prevention of chronic obstructive pulmonary disease: GOLD executive summary. Am J Respir Crit Care Med, 2013, 187(4):347-365.

[2] Halbert RJ, Natoli JL, Gano A, et al. Global burden of COPD: systematic review and meta-analysis. Eur Respir J, 2006, 28(3): 523-532.

[3] Fang L, Gao P, Bao H, et al. Chronic obstructive pulmonary disease in China: a nationwide prevalence study. Lancet Respir Med, 2018, 6(6):421-430.

[4] 沈宁. 慢性阻塞性肺病诊治进展. 临床药物治疗杂志, 2012, 10(05):29-33.

[5] Hagmeyer L, Treml M, Priegnitz C, et al. Successful concomitant therapy with pirfenidone and nintedanib in idiopathic pulmonary fibrosis: a case report. Respiration, 2016, 91(4):327-332.

[6] Antoniou KM, Karagiannis K, Tsitoura E, et al. Clinical applications of mesenchymal stem cells in chronic lung diseases. Biomed Rep, 2018, 8(4):314-318.

[7] Shigemura N, Okumura M, Mizuno S, et al. Autologous transplantation of adipose tissue-derived stromal cells ameliorates pulmonary emphysema. Am J Transplant, 2006, 6(11):2592-2600.

[8] Yuhgetsu H, Ohno Y, Funaguchi N, et al. Beneficial effects of autologous bone marrow mononuclear cell transplantation against elastase-induced emphysema in rabbits. Exp Lung Res, 2006, 32(9):413-426.

[9] Zhen G, Liu H, Gu N, et al. Mesenchymal stem cells transplantation protects against rat pulmonary emphysema. Front Biosci, 2008, 13:3415-3422.

[10] Zhen G, Xue Z, Zhao J, et al. Mesenchymal stem cell transplantation increases expression of vascular endothelial growth factor in papain-induced emphysematous lungs and inhibits apoptosis of lung cells. Cytotherapy, 2010, 12(5):605-614.

[11] Gu W, Song L, Li XM, et al. Mesenchymal stem cells alleviate airway inflammation and emphysema in COPD through down-regulation of cyclooxygenase-2 via p38 and ERK MAPK pathways. Sci Rep, 2015, 5:8733.

[12] Zhang WG, He L, Shi XM, et al. Regulation of transplanted mesenchymal stem cells by the lung progenitor niche in rats with chronic obstructive pulmonary disease. Respir Res, 2014, 15:33.

[13] Guan XJ, Song L, Han FF, et al. Mesenchymal stem cells protect cigarette smoke-damaged lung and pulmonary function partly via VEGF-VEGF receptors. J Cell Biochem, 2013, 114(2):323-335.

[14] Zhao Y, Xu A, Xu Q, et al. Bone marrow mesenchymal stem cell transplantation for treatment of emphysemic rats. Int J Clin Exp Med, 2014, 7(4):968-972.

[15] Schweitzer KS, Johnstone BH, Garrison J, et al. Adipose stem cell treatment in mice attenuates lung and systemic injury induced by cigarette smoking. Am J Respir Crit Care Med, 2011, 183(2):215-225.

[16] Kim YS, Kim JY, Huh JW, et al. The therapeutic effects of optimal dose of mesenchymal stem cells in a murine model of an elastase induced-emphysema. Tuberc Respir Dis (Seoul), 2015, 78(3):239-245.

[17] Li X, Zhang Y, Yeung SC, et al. Mitochondrial transfer of induced pluripotent stem cell-derived mesenchymal stem cells to airway epithelial cells attenuates cigarette smoke-induced damage. Am J Respir Cell Mol Biol, 2014, 51(3):455-565.

[18] Peron JP, de Brito AA, Pelatti M, et al. Human tubal-derived mesenchymal stromal cells associated with Low level laser therapy significantly reduces cigarette smoke-induced COPD in C57BL/6 mice. PLoS One, 2015, 10(8):e0136942.

[19] Liu X, Fang Q, Kim H. Preclinical studies of mesenchymal stem cell (MSC) administration in chronic obstructive pulmonary disease (COPD): a systematic review and meta-analysis. PLoS One, 2016, 11(6):e0157099.

[20] Ortiz LA, Gambelli F, McBride C, et al. Mesenchymal stem cell engraftment in lung is enhanced in response to bleomycin exposure and ameliorates its fibrotic effects. Proc Natl Acad Sci USA, 2003, 100(14):8407-8411.

[21] Choi M, Ban T, Rhim T. Therapeutic use of stem cell transplantation for cell replacement or cytoprotective effect of microvesicle released from mesenchymal stem cell. Mol Cells, 2014, 37(2):133-139.

[22] Moodley Y, Vaghjiani V, Chan J, et al. Anti-inflammatory effects of adult stem cells in sustained lung injury: a comparative study. PLoS One, 2013, 8(8):e69299.

[23] Min F, Gao F, Li Q, et al. Therapeutic effect of human umbilical cord mesenchymal stem cells modified by angiotensin-converting enzyme 2 gene on bleomycin-induced lung fibrosis injury. Mol Med Rep, 2015, 11(4):2387-2396.

[24] Xu J, Li L, Xiong J, et al. Cyclophosphamide combined with bone marrow mesenchymal stromal cells protects against bleomycin-induced lung fibrosis in mice. Ann Clin Lab Sci, 2015, 45(3):292-300.

[25] Srour N, Thébaud B. Mesenchymal stromal cells in animal bleomycin pulmonary fibrosis models: a systematic review. Stem Cells Transl Med, 2015, 4(12):1500-1510.

[26] Ribeiro-Paes JT, Bilaqui A, Greco OT, et al. Unicentric study of cell therapy in chronic obstructive pulmonary disease/pulmonary emphysema. Int J Chron Obstruct Pulmon Dis, 2011, 6:63-71.

[27] Stessuk T, Ruiz MA, Greco OT, et al. Phase I clinical trial of cell therapy in patients with advanced chronic obstructive pulmonary disease: follow-up of up to 3 years. Rev Bras Hematol Hemoter, 2013, 35(5):352-357.

[28] Weiss DJ, Casaburi R, Flannery R, et al. A placebo-controlled, randomized trial of mesenchymal stem cells in COPD. Chest, 2013, 143(6):1590-1598.

[29] Stolk J, Broekman W, Mauad T, et al. A phase Ⅰ study for intravenous autologous mesenchymal stromal cell administration to patients with severe emphysema. QJM, 2016, 109(5):331-336.

[30] Tzouvelekis A, Paspaliaris V, Koliakos G, et al. A prospective, non-randomized, no placebo-controlled, phase Ⅰb clinical trial to study the safety of the adipose derived stromal cells-stromal vascular fraction in idiopathic pulmonary fibrosis. J Transl Med, 2013, 11:171.

[31] Chambers DC, Enever D, Ilic N, et al. A phase 1b study of placenta-derived mesenchymal stromal cells in patients with idiopathic pulmonary fibrosis. Respirology, 2014, 19(7):1013-1018.

[32] Glassberg MK, Minkiewicz J, Toonkel RL, et al. Allogeneic human mesenchymal stem cells in patients with idiopathic pulmonary fibrosis via intravenous delivery (AETHER): a phase Ⅰ safety clinical trial. Chest, 2017, 151(5):971-981.

第十章 间充质干细胞在自身免疫系统疾病治疗中的应用

第一节 自身免疫系统疾病

自身免疫系统疾病是由机体免疫系统对自身组织和器官发生了免疫应答并造成组织损伤和功能障碍的一类疾病。目前患有各种自身免疫系统疾病的人群约占全世界人口的5%，而且近80%患者的临床病症与免疫系统功能异常相关。在自身免疫系统疾病中，类风湿关节炎（RA）、系统性红斑狼疮（SLE）、移植物抗宿主病（GVHD）和银屑病（PS）等是常见的自身免疫系统疾病[1,2]。

自身免疫系统疾病之所以成为临床难治性疾病，其主要原因在于：①自身免疫调控机制复杂、临床表现多样，为复杂多易感基因性疾病；②至今尚未发现可导致疾病发生的确切的自身抗原，因此临床上无法开展特异性治疗；③目前临床仅能采用的免疫抑制治疗常导致严重感染，甚至肿瘤发生，直接导致60%的患者死亡。

自身免疫系统疾病的发病机制十分复杂，在遗传和环境因素影响下，固有免疫和适应性免疫功能紊乱参与其发病，各种异常活化的免疫细胞、各种受体及各种细胞因子间相互调控，构成复杂多变的免疫失衡网[3]。导致自身免疫系统疾病发病的最基本因素是自身免疫耐受机制被破坏。在正常生理状况下，高亲和力的自身反应性T、B细胞在中枢（胸腺）选择中被去掉。在中枢选择中未被清除的低亲和力的自身反应性T、B细胞被释放到外周血，在外周循环中遭遇缺乏细胞表面共刺激分子的不成熟树突细胞（iDC），经凋亡机制或调节性T细胞（Treg）机制导致自身反应性T细胞无能或被抑制，产生外周免疫耐受[2]。因此，导致自身免疫耐受被破坏的关键机制是DC异常活化成熟，而Toll样受体（TLR）等模式识别受体（PRR）异常表达则是连接抗原（外源性/内源性配体）和DC（或肥大细胞）成熟的桥梁。

自从有人提出自身免疫系统疾病是一种造血干细胞（HSC）疾病后，应用自体和异体造血干细胞移植（HSCT）治疗自身免疫系统疾病便成为研究热点，并在SLE中取得了良好的效果[4]。但由于移植物中及患者体内存在病态的HSC及T、B细胞，移植后复发是自体HSCT的致命缺陷，而移植物抗宿主反应又是同种异体HSCT最严重的并发症。间充质干细胞（MSC）具有低免疫原性的特点和免疫抑制的功能，使其在治疗自身免疫系统疾病方面具有独特优势[5]。

第二节 间充质干细胞驱动的免疫调节

MSC是一群具有多向分化潜能的、广泛存在于造血系统及结缔组织的成体干细胞。骨髓中存在大量的MSC，是MSC研究的主要来源[6,7]。MSC可以自我更新并分化为成熟后代，在一定条件下可分化为中胚层及神经外胚层来源的多种间质细胞，如成骨细胞、软骨细胞、脂肪细胞、肌肉细胞、成纤维细胞、内皮细胞、神经元和胶质细胞等[8]。包裹内皮微血管的周细胞，可能是组织 MSC 的代表，因为它们在体内广泛分布，但这个问题仍有争议[9]。用生长因子如基础成纤维细胞生长因子（FGF）、血小板裂解物或专用培养基培养，可以在体外培养骨髓、脐带或脂肪组织来源的 MSC [10,11]。MSC 的主要任务是在骨、软骨、脂肪和组织间质中产生具有支持功能的细胞类型。此外，MSC 可以分化为内皮细胞，从而有利于组织血管形成。

近年来的研究发现，MSC 还介导广泛的免疫调节功能[12,13]。有报道称 MSC 与同种异体 T 细胞共同培养不刺激 T 细胞增殖；能延长同种异体皮肤移植后的存活；与 HSC 共移植能降低 GVHD 的发生等。MSC 的另一个研究领域涉及再生医学和组织工程。MSC 调节免疫反应和促组织再生的能力通常是互补的，这可能有助于增强 MSC 在自身免疫/炎性疾病的临床前模型中的治疗活性。这些研究为 MSC 移植治疗自身免疫系统疾病及炎症相关性疾病定了基础。

大量关于先天免疫细胞和 MSC 之间相互作用的研究已揭示了 MSC 广泛的免疫调节功能：①抑制人中性粒细胞的促免疫活性，MSC 抑制呼吸爆发并通过白介素-6（IL-6）依赖和 STAT3 依赖机制延长中性粒细胞存活[14,15]；②抑制单核细胞或 $CD34^+$ 造血祖细胞分化为成熟 DC（mDC），导致不成熟 DC（iDC）的积累[16]；③促进产生 IL-10 的浆细胞样 DC（pDC）的分化[17]；④促进巨噬细胞向 M2 样表型分化[18]；⑤尽管 MSC 可能被由 IL-2 激活的 NK 细胞杀死，但是可抑制 NK 细胞增殖、细胞因子产生和细胞毒效应[19]。

MSC 促进 T 细胞静止状态下的存活，并诱导活化 T 细胞的分裂停滞，使其积聚于 G_0/G_1 期[20]。MSC 还抑制 T 细胞增殖，同时通过诱导 pDC 产生 IL-10 促进 Treg 细胞的发育[21,22]。MSC 抑制 $CD4^+$ T 细胞活化，抑制了 T 细胞依赖性的 B 细胞向浆细胞的分化。此外，MSC 直接作用于 B 细胞，抑制其增殖、向浆细胞分化和趋化性[23]。最后，MSC 抑制表达 Vδ2 的 γδ T 细胞的增殖和扩散，并且维持 NKT 细胞恒定，调节两种 T 细胞群体在适应性和固有免疫之间保持平衡。此外，MSC 抑制 Vδ2 T 细胞产生干扰素-γ（IFN-γ），但被后者通过 T 细胞受体（TCR）依赖性机制杀死[24]。

以上这些研究已经在以不同比例和不同时间间隔孵育分离的免疫细胞群体与 MSC 的实验中进行，但仍有一些技术问题值得考虑。首先，MSC 在实验室进行大量的体外扩增必须具备以下方面的条件：①MSC 的来源（如骨髓来源、脂肪组织来源、脐带来源）；②用于促进 MSC 扩增的培养基和生长因子的组成；③收集和检测培养 MSC 的方法。关于 MSC 来源，人们对可以轻易分离和扩增的脂肪间充质干细胞（AD-MSC）的兴趣越来越大，此种来源的 MSC 可产生比骨髓来源更高比例的 MSC 祖细胞，并且和骨髓间充质干细胞（BM-MSC）具有相同的免疫抑制作用[25]。体外扩增的 MSC 必须经过 2 或 3 次传代，因为长时间的培养会导致衰老 MSC 的比例增高[26]。分离的免疫细胞群体与 MSC 的共培养是以不同比例进行的，这可能无法实现模拟体内特定的组织微环境。

MSC 介导的免疫抑制活性大部分由 MSC 暴露于促炎细胞因子如 IFN-γ、肿瘤坏死因子-α（TNF-α）和 IL-1β 中引发。IFN-γ 的"许可"是 MSC 能够执行各种免疫调节功能的关键步骤[27]。此外，MSC 通过表达各种 TLR，识别微环境中内源性和外源性危险信号。外源性配体的实例包括病毒、细菌、真菌和原生动物组分，而内源性配体包括热休克蛋白和 RNA。在与不同信号相互作用时，MSC 获得更高的免疫抑制或更高的免疫促进活性。例如，通过 TLR3 活化的 MSC 显示出增强的免疫抑制活性，并被定义为 MSC1，而在 TLR4 活化时，MSC 表现为与 T 细胞活化相关的促免疫激活表型（MSC2）[28,29]（图 10-1）。因此，微环境的组成发生变化，MSC 引发或调控其旁分泌作用增强 MSC 的可塑性，促使 MSC 从免疫抑制表型向免疫促进表型转换，反之亦然。

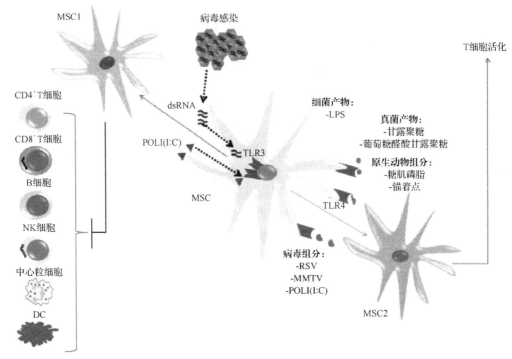

图 10-1　MSC 的可塑性[29]（彩图请扫封底二维码）
MSC. 间充质干细胞，NK 细胞. 自然杀伤细胞，DC. 树突细胞，dsDNA. 双链 DNA，TLR. Toll 样受体，
LPS. 脂多糖，GPI. 糖基磷脂酰肌醇，RSV. 呼吸道合胞病毒，MMTV. 鼠乳腺瘤病毒

第三节　间充质干细胞治疗自身免疫系统疾病的临床前模型

本节我们将主要介绍 MSC 在典型自身免疫系统疾病模型中获得的实验结果，即系统性红斑狼疮（SLE）、类风湿关节炎（RA）、多发性硬化（MS）、克罗恩病（CD）和移植物抗宿主病（GVHD）。

系统性红斑狼疮（SLE）的典型动物模型由淋巴细胞增殖（lpr）自发性纯合突变的 MRL 小鼠（MRL/lpr）构建，其表征为肾小球性肾炎、血管炎和自身抗体产生。其他的 SLE 模型包括 B6-lpr、NZB/WF1 和 BXSB 小鼠[30]。在不同模型中开展的关于 BM-MSC 治疗潜力的研究结果存在很大差异。在 MRL/lpr 模型中，注射同种异体 MSC 会使肾功能得到改善，肾功能和肾小球结构完全正常化，并且与 C3 和肾小球 IgG 沉积减少有关。而在 NZB/WF1 小鼠中进行的其他研究，全身 MSC 给药未能发挥任何有益作用[31]。最近，有研究报道了脐带间充质干细胞（UC-MSC）以剂量依赖的方式改善 NZB/WF1 小鼠的狼疮性肾炎，主要作用机制是抑制肾单核细胞趋化蛋白-1（MCP-1）和高迁移率族 1 蛋白（high-mobility group box 1 protein，HMGB1）并增加 Foxp3$^+$ Treg 细胞数量[32]。给 NZB/WF1 小鼠注射人脂肪 MSC（hAD-MSC），可以延长治疗动物的存活时间，减少组织学和免疫学异常及蛋白尿发生率。动物体内抗双链 DNA（dsDNA）自身抗体和血尿素氮水平显著降低，而血清粒细胞-巨噬细胞集落刺激因子（GM-CSF）、IL-4 和 IL-10 水平增高，且 CD4$^+$Foxp3$^+$ Treg 细胞比例增加[33]。

在类风湿关节炎（RA）临床前模型中有关同源基因或同种异体 BM-MSC 潜在治疗活性的研究结果一直是相冲突的。在由胶原诱导的关节炎模型中，系统地施用 MSC 有时能改善关节炎[34]，有时则对这种疾病无效甚至导致恶化[35]。仅在 RA 的早期施用 MSC 才能检测到疾病得到改善。最近，在两种不同的 RA 模型中研究了 MSC 的治疗活性，即由弗氏佐剂诱导的关节炎和出生后大约 1 个月自发形成严重侵蚀性关节炎的 K/BxN 小鼠。在两种模型中，MSC 强烈抑制脾细胞和成纤维样滑膜细胞的体外增殖，并降低后者的侵袭能力。相比之下，体内 MSC 的全身给药完全无效。然而，当小鼠在关节炎发作之前用蛋白酶体抑制剂硼替佐米进行短暂治疗，然后在疾病的活动期接受 MSC 的全身输注，硼替佐米加 MSC 的组合疗法显示出显著的治疗活性，病症从以下四方面得到改善：①关节炎评分降低；②关节结构保持正常，炎症反应明显减轻，无血管翳形成，骨破坏减轻或滑膜增厚；③血液中的 TLR 表达和 Treg 细胞比例得到恢复；④成纤维样滑膜细胞和脾细胞增殖正常，细胞因子分泌恢复正常[36]。因此，实验性关节炎的炎症环境导致的 MSC 免疫调节特性完全丧失这一现象，能被硼替佐米预处理所逆转。

接种髓鞘肽疫苗诱导实验性自身免疫脑脊髓炎（EAE）可以产生不同的临床疾病过程（单相、复发缓解、慢性），并且可以代表 MS 研究的模型。Th17 和 Th1 细胞被认为是神经炎症的主要影响因子，但其他细胞群体如 B 细胞、DC 和巨噬细胞也起着重要作用。患有进行性或复发性 EAE 的小鼠在全身施用同基因或同种异体 MSC 后，其疾病症状显示出得到改善，此过程与髓鞘特异性和非特异性 T 细胞与 B 细胞的反应相关。输注的 MSC 抑制 Th17 和 Th1 细胞反应，并减少中枢神经系统（CNS）免疫细胞的膨胀[37]。同时有研究显示，MSC 治疗增加了 Th2 细胞因子 IL-4 和 IL-5 的积累及体内 Treg 细胞的产生，这两者都有助于减少 EAE 症状。在 EAE 模型中，MSC 极化 CD4$^+$ T 细胞的分子机制包括吲哚胺 2,3-双加氧酶（IDO）和 MCP-1/CC 趋化因子配体 2（MCP-1/CCL2）[38]。尽管 MSC 在 EAE 模型中具有一定的效果，但临床上即使是大脑内给药，这些细胞的移植效果也很差。此外，没有证据表明体内 MSC 可以反向分化成神经细胞。因此，最近有报道说，MSC 在 EAE 模型中的治疗效果由 MSC 分泌的可溶性因子发挥作用，特别是肝细胞生长因子（HGF）与其受体（Met）的结合，在免疫和神经细胞上都有表达[39]。

两种最常使用的克罗恩病（CD）模型是由葡聚糖硫酸钠（dextra sulfate sodium，DSS）诱导的结肠炎大鼠和由 2,4,6-三硝基苯磺酸（trinitro-benzene-sulfonic acid，TNBS）诱导的结肠炎 SJL/J 小鼠。给这些模型静脉注射 MSC，能减轻炎症反应，抑制 Th1 细胞因子反应，并诱导 Treg 细胞的反应。经 MSC 处理的动物显示有更好的存活率和体重减轻快速逆转[40]。组织学分析表明，巨噬细胞和中性粒细胞的透壁膨胀减少，固有层中 CD4$^+$ T 细胞的炎症反应减少，产生黏蛋白的上皮和杯状细胞及隐窝得以保存。在小鼠的结肠中局部植入MSC能加速结肠炎的痊愈[41]。在所研究的CD模型中以上这些效果都可以通过注射来自不同来源的MSC（包括同基因、同种异体和异种）实现。

同种异体造血干细胞移植（HSCT）与供体淋巴细胞破坏宿主器官导致的移植物抗宿主病（GVHD）的发展过程始终相关。GVHD可能遵循急性（aGVHD）或慢性（cGVHD）的病程发展，并可能是致命的。已经建立了许多GVHD模型并用于检测MSC在预防或治疗这种疾病方面的疗效[42]。然而，人与小鼠的MSC至少存在四个方面的不同：①小鼠MSC比人MSC(hMSC)的免疫抑制能力要弱；②小鼠MSC的体外扩增能力要比hMSC更弱，效果更差；③小鼠MSC更倾向于自发地进行永生化和转化；④IDO是hMSC强大的免疫抑制介质，不参与小鼠MSC介导的免疫调节[43]。

不同小鼠GVHD模型的研究结果，不论是主要组织相容性复合体（MHC）错配或者是单倍体，都可以总结为两点：①移植时单次输注MSC不能有效阻止GVHD，而多次移植（至少是时间点上的多次注射）可能是有益的，虽然这种效果未在所有的模型中都表现，还取决于MSC的来源、剂量和输注时间；②GVHD位点处移植MSC表达水平低或不存在，这再次指向旁分泌作用机制。为了实现预防和治疗GVHD，MSC正确的输注方式、重复输注的时间点及适当的免疫环境激活MSC，这些都影响其执行免疫抑制活动。

第四节　间充质干细胞在自身免疫系统疾病治疗中的临床应用

鉴于MSC在治疗人类疾病方面的作用，研究人员彻底检测了患者来源MSC在体外扩增和免疫调节活性方面的质量。尽管存在一些例外[44]，但许多研究已经证明，来自患有自身免疫系统疾病/感染患者的MSC在体外增殖能力存在障碍，但与正常对照MSC表现出相似的T细胞增殖抑制作用[45]。

一、MSC与血液系统自身免疫系统疾病

目前应用干细胞治疗血液疾病最有效的便是造血干细胞（HSC）。临床上已经开始利用各种组织来源的HSC治疗造血恶性肿瘤和遗传性疾病，包括遗传性贫血和免疫缺陷。同时根据临床情况，可以对患者进行自体或相匹配的同种异体/第三方HSCT。进行同种异体/第三方HSCT时，尽管同时采取了免疫抑制剂进行治疗，但GVHD形式的免疫排斥仍然是引起患者发病率和死亡率增加的主要原因，30%~40%的同种异体HSCT患者会发生GVHD[46]。部分学者已在动物身上发现，预防性应用MSC可减少HSCT后cGVHD的发生率。

MSC在GVHD的临床应用比在其他类型的免疫/炎症介导的疾病中的应用发展迅速，这在很大程度上归因于一例患有严重GVHD的儿科患者，注射单倍体BM-MSC具有显著疗效的病例报告[47]。这种情况的科学依据主要依赖于部分体外研究显示同种异体BM-MSC对淋巴细胞的增殖和功能起抑制作用，并且有临床试验数据支持了MSC和HSC共移植的安全性[48]。基于这一个成功案例的报告，随后在不同的研究中观察了MSC与HSC联合移植对GVHD的疗效，获得的最终临床结论为：MSC既不能促进移植物的植入，也不能阻止T细胞充足的移植物的排斥反应，但可能对T细胞缺乏的移植发挥有益作用[42]。对于

皮质类固醇难治性的Ⅱ、Ⅳ级aGVHD，也已经通过输注不同来源[人白细胞抗原（HLA）配型相同、HLA-单倍体配型相同、第三方来源]的BM-MSC和采取不同的输注方案进行治疗[49-51]。欧洲有一项试验为MSC的疗效提供了明确的证据（55例中有50例完全有效，9例病情得到改善）[51]。最近，在采用HLA单倍体同源HSC移植预防cGVHD后，又展开了有关UC-MSC移植有效性和安全性的多中心、随机双盲对照试验的Ⅱ期研究。研究结果显示重复输注MSC可能会抑制cGVHD的症状，并且使Th1/Th2细胞比例失衡现象发生逆转，接受MSC治疗的患者耐受性良好，复发率也没有增加[52]。其他研究发现，MSC能减少cGVHD患者的难治性血细胞减少[53]和干眼症等并发症的发生[54]。

目前MSC治疗GVHD作用机制的研究仅限于对外周血淋巴细胞的免疫表型进行分析，其研究结果支持MSC发挥体内免疫调节作用，抑制致病性T细胞亚群Th1和Th17细胞的增殖，促进Treg细胞增殖。涉及的MSC细胞因子包括前列腺素E2（PGE2）和一氧化氮（NO），并且可以通过IFN-γ预处理MSC来增强其作用[55,56]。综上所述，移植物内MSC缺失可能是HSCT后cGVHD发生率高的原因，机制如下：①MSC可加快骨髓造血微环境的重建，HSC与MSC共输注后，造血功能快速重建，可减少感染的发生率及严重程度，这对GVHD有一定的预防作用，既往动物和临床移植结果也表明在保护性环境下MSC可以减少GVHD的发生率。②MSC可诱导Th0向Th2转化，使与GVHD发生有关的$CD4^+$ Th1细胞相对减少，即便在Th1辅助下细胞毒性T细胞（CTL）激活，MSC对CTL也有增殖抑制效应。③$CD4^+CD25^+$ TR/TS细胞的数量增加，MSC借此类细胞减少HSCT后GVHD的发生率。④在移植后继续使用粒细胞集落刺激因子（G-CSF）可能有利于减少GVHD的发生率，已知MSC可分泌G-CSF[57]。

近年来，MSC在再生障碍性贫血（AA）和免疫性血小板减少性紫癜（immune thrombocytopenic purpura, ITP）治疗中应用的研究也日趋完善。AA是一组由多种原因引起的HSC数量下降和功能异常，导致全血细胞减少的HSC免疫疾病。目前在临床上MSC可以单独输注或者联合HSC共同移植治疗AA[58]。治疗重型AA的有效方法为免疫抑制治疗和HSCT。如果可找到全相合的供体，HSCT可作为重型AA的首选疗法。已有大量临床试验证实MSC对于促进HSC的造血重建具有肯定的作用。联合移植MSC和HSC可明显增加HSC的植入水平，提高HSC移植物的成活率[59]。

ITP也称特发性血小板减少性紫癜，是一种由自身免疫介导的血小板减少综合征，是临床最常见的出血性疾病，严重威胁人的生命健康。ITP主要以血小板过度破坏和/或血小板生成减少为特点，抗原特异性自身抗体介导的血小板破坏和CTL对血小板的直接溶解是血小板破坏增多的重要原因，而免疫介导的巨核细胞成熟障碍和凋亡异常则导致ITP患者血小板生成减少[60]。研究发现，ITP患者MSC存在增殖缺陷和抑制功能减弱，输注正常MSC或用药物（沙利度胺）纠正异常MSC，恢复患者MSC功能可以诱导mDC耐受及抑制T细胞增殖[61,62]。

MSC是体内骨髓局部免疫微环境的重要组成部分，可以构成免疫抑制位点，调节正常造血及修复造血微环境，促进骨髓造血。众多研究证明，MSC具有造血支持、促进造血重建等作用。在体外培养中发现MSC能分泌造血因子，并能表达多种表面黏附分子，在造血调控中具有一定的作用。同时，MSC在一定条件下具有向基质细胞分化的能力，可弥补由基质细胞缺乏导致的造血功能障碍。

二、MSC 与皮肤系统自身免疫系统疾病

基于 MSC 的动态免疫抑制特点，研究发现具有皮肤表征的自身免疫系统疾病，如系统性红斑狼疮（SLE）、皮肌炎（dermatomyositis，DM）和系统性硬化症（systemic scleredema，SSc）已经成为 MSC 疗法的有效靶标[63]。

自 2010 年以来，300 余例难治性系统性红斑狼疮（SLE）患者已经接受 MSC 治疗，患者基本都是静脉给予同种异体 BM-MSC 或 UC-MSC，但缺乏适当的对照组进行直接比较[64]。总的来说，采用 MSC 治疗的患者耐受性良好，没有明显的不良反应，肾功能有明显改善，疾病活动评分提高，蛋白尿减少并且抗 dsDNA 抗体水平降低。在 MSC 输注后，SLE 患者体内循环的 Treg 细胞增加，Th1 和 Th2 细胞因子之间的平衡得到归一化。在肺泡出血或血细胞减少的 SLE 患者中也观察到使用 MSC 治疗后有积极效果[65]。另一项研究中，有 10 例耐药性多发性肌炎或 DM 患者，在经过同种异体 BM-MSC 或 UC-MSC 治疗后，所有患者的病情总体上有改善，患者的慢性非愈合性皮肤溃疡得到缓解，并且通常耐受性良好[66]。而对于 SSc，Scuderi 等[66]发现，6 例患者在接受一次皮下注射自体 AD-MSC 治疗后，皮肤收紧性得到显著改善，疾病进展得到控制。尽管这些临床试验的设计存在局限性，可以在未来的对照试验中，有针对性地选择适当的对照患者和范围。就目前而言，这些 MSC 应用于皮肤自身免疫系统疾病的结果是令人鼓舞的。

此外，MSC 在治疗银屑病（PS）、特应性皮炎（atopic dermatitis，AD）等皮肤炎性疾病方面也表现出有很大的潜在价值。PS 俗称"牛皮癣"，是皮肤免疫性疾病，被世界卫生组织（World Health Organization，WHO）列为世界十大顽症之一。MSC 与 Th-17 细胞相互作用，抑制促炎细胞因子的产生并诱导 Treg 细胞表型和功能改变来有效治疗 PS。AD 是另一种涉及异常 T 细胞活性的慢性炎性皮肤病。Ra 等[63]报道了 AD 患者经自体 AD-MSC 治疗后临床症状明显改善，无任何不良反应。这些进一步证明 MSC 治疗皮肤自身免疫系统疾病和耐药性皮炎的潜力。

三、MSC 与关节自身免疫系统疾病

MSC 是治疗关节疾病的重要选择之一，因为软骨不能再生，而 MSC 是软骨这些组织的内源性祖细胞。目前 MSC 治疗主要针对两大实体关节疾病：骨关节炎（OA）和类风湿关节炎（RA）。

骨关节炎（OA）是最常见的关节疾病，是由"磨损和撕裂"引起关节软骨逐渐恶化，随后引起免疫反应，进一步导致关节损伤。由于软骨不能再生，OA 是一种进行性和不可逆转的疾病，其发病率随年龄和体重增加而增加。虽然免疫损伤不是导致 OA 发病的主要原因，但是炎症反应始终伴随患者的疼痛和关节僵硬症状。此外，为了发挥联合修复作用，减轻软骨破坏和免疫攻击的恶性循环也是十分必要的。因此，MSC 治疗特别适用于 OA，因为可以同时实现软骨再生和免疫抑制效应。实际上，小动物和大动物研究都证实，MSC 可以减轻 OA 的炎症，并允许软骨修复[68,69]。目前，有 33 项注册临床试验，其中 9 项在 I 期，16 项为联合 I/II 期，8 项为 II 期。超过 18%的研究已经公布了

MSC 治疗 OA 的安全性和有效性[69]。总体而言，这些研究中 MSC 治疗在症状改善方面效果突出，包括软骨再生、疼痛缓解和关节修复。

虽然临床资料显示 MSC 治疗 OA 通常是有效的，但令人惊讶的是，对于类风湿关节炎（RA）不是这种情况。在 RA 临床前模型中获得了一些延迟于临床研究的相反结果。即使在动物模型中，关于 MSC 治疗效果也有不同的结果。到目前为止，只有 5 项临床试验用 MSC 治疗 RA，Ⅰ期试验 1 项，Ⅰ/Ⅱ期 3 项，Ⅱ/Ⅲ期 1 项。迄今为止最大的一项研究是一项非随机对照试验，共纳入了 172 例传统药物无疗效的 RA 患者。其中 136 例患者静脉注射 UC-MSC（$4×10^7$cells），其余 36 例作为对照组仅接受细胞的培养基注射。随后所有患者均接受改良型抗风湿药（disease-modifying anti-rheumatic drug，DMARD）的治疗。多数患者在输注 MSC 后没有出现任何不良反应，仅 4% 的患者出现发冷或发烧。另外接受 DMARD 和 MSC 治疗的患者显示出疾病得到明显缓解，这与外周血中 Treg 细胞数量增加关系密切[70]。显然，RA 和 OA 的发病机制之间存在着较细微的差异，解决这一问题有可能能解释 MSC 治疗两种关节疾病的疗效差异。

四、MSC 与神经系统自身免疫系统疾病

MSC 作为一种多能干细胞，对由中枢神经系统（CNS）缺血损伤或外伤等造成的功能缺失的治疗作用显著，与其他细胞相比具有明显的优点：①获取方便，可从患者本人骨髓取材；②可在体外快速培养扩增，并定向诱导分化；③在宿主脑组织中可长期生存并进行整合；④避开了免疫排斥的难题等。大量动物实验已证实，MSC 经各种途径移植后均可到达病损部位，并分化成神经干细胞（NSC）、成熟神经元和胶质细胞，能安全和有效地促进中枢神经损伤后的功能恢复[71,72]。

多发性硬化（MS）是 CNS 最常见的自身免疫系统疾病，属于 CNS 炎性和脱髓鞘疾病。通常为复发性、偶发性疾病，称为复发缓解性 MS，并且经常演变为以进行性神经功能障碍为特征的慢性疾病，称为继发性进行性 MS。目前的研究已经发现 Th1、IL-17A 和分泌型 CD4（Th17）细胞参与了这种自身免疫系统疾病的发病机制[73]。长期以来，人们已知 MS 是由 $CD4^+$ T 细胞介导的自身免疫系统疾病，其靶向作用于髓鞘碱性蛋白（myelin basic protein，MBP），产生的脱髓鞘导致神经元损伤和传导障碍。MS 的临床症状是进行性的，包括视力模糊、失明、部分或全部麻痹、记忆和认知缺陷[74]。迄今为止，已有 23 项使用 MSC 治疗 MS 的临床试验，不同来源（自体，脐带或胎盘）的 MSC 已经按照不同浓度和时间表，通过静脉或鞘内注射，施用于少数继发性 MS 患者。所有研究都显示了 MSC 给药的可行性，几乎没有任何的不良反应，并且发现与免疫调节现象有关的临床疗效的原理，如体内参与循环的 Treg 细胞明显增多[75]。完善 MSC 治疗 MS 研究的详细试验设计和临床疗效，将对解决难治性 CNS 疾病的治疗问题有重要作用。成人 MSC 为神经系统疾病的细胞和基因治疗提供了新的思路与广阔前景。

五、MSC 与消化系统自身免疫系统疾病

人消化系统自身免疫系统疾病主要为炎性肠病（inflammatory bowel disease，IBD）。IBD 主要包括克罗恩病（CD）和溃疡性结肠炎（ulcerative colitis，UC），IBD 的病因和

疾病进展是多因素的，但IBD的关键是对肠微生物产生不受控制的免疫应答引起的。CD和UC都是逐渐致命的，没有有效的治疗方法，这使得MSC成为治疗这些慢性炎性疾病极具吸引力的选择。

目前，已有19项有关MSC治疗IBD的注册临床试验。总体而言，MSC用于治疗IBD，特别是治疗CD形成瘘似乎是安全且高度可行的[76]。CD的主要治疗挑战之一是复杂的肛周结石。最近，已发表了第一个随机的双盲的采用安慰剂作对照组的多中心Ⅲ期临床研究，评估了给患者单次腔内注射同种异体MSC制剂的安全性和有效性。这项研究证实了以前在小型患者群中获得的成果的可靠性。患者分为两组，每组105例，随机分配接受MSC或安慰剂。接受MSC的患者组疾病有明显缓解。不良反应在安慰剂组中更多，包括肛门脓肿和垂体痛[77]。这项研究是一个里程碑式的研究，明确地证明了MSC治疗CD复合肛门周围软组织中的病灶的有效性。研究发现其作用机制可能与MSC抑制致病性Th1和Th17反应的能力，以及促炎因子IL-1β、IL-6、IL-17、TNF-α、IFN-γ等的血清水平，同时增强Treg细胞和脾髓系抑制性细胞（myeloid-derived suppressor cell，MDSC）的数量[78]有关。在TNBS诱导的动物模型中，注射MSC导致免疫细胞浸润和TNF-α表达水平降低，但损伤部位转化生长因子-β（TGF-β）水平升高[79]。为了提高MSC治疗IBD的疗效，研究者用黏膜分泌物细胞黏附分子-1（MAdCAM-1）和血管细胞黏附分子-1（VCAM-1）的抗体包被这些干细胞，可以增加MSC向发炎肠道区域的传递。用IL-12p40或IL-37β修饰MSC，其免疫抑制作用也会增强[80]。

六、MSC与内分泌系统自身免疫系统疾病

1型糖尿病是一种慢性由多因素诱导的胰腺特发性或遗传性自身免疫系统疾病，导致胰岛素分泌β细胞发生进行性破坏。属于儿童和青少年性疾病，通常首次出现在5~7岁。典型的症状是多饮、多尿和多食，血糖明显增高。普通人群1型糖尿病患病率为0.1%~0.5%，发病率为30~50/100 000人。发病率每年以3%的惊人速度增长。最终导致多器官功能障碍，如糖尿病性视网膜病变、酮症酸中毒、肾病、神经病变、心血管疾病甚至终末期器官功能衰竭。胰岛素替代疗法是目前唯一公认和被接受的治疗方法。

近年来，使用MSC治疗1型糖尿病的报道有很多，主要通过由MSC向胰岛素分泌细胞分化来发挥作用。另外已经有实验证明，转化为胰岛细胞的外源性MSC同时具有在肝脏及胰腺内定位存活及表达胰岛素的功能，但是对其临床治疗效果还是存在争议的。随后多种MSC联合胰岛细胞移植物治疗成为研究的新热点，并且临床效果良好[81]。Vanikar等[82]对1型糖尿病患者输注分泌胰岛素的AD-MSC（AD-MSC-ISC）和骨髓造血干细胞（BM-HSC），患者平均外源性胰岛素需求量下降，Hb1Ac水平下降，血清c-肽水平增加，并且都没有糖尿病酮症酸中毒事件发生。2015年，Thakkar等[83]在20例1型糖尿病患者中进行随机试验，将自体和异体AD-MSC-ISC与BM-HSC通过门内途径、胸腺循环途径和皮下脂肪层途径输注到肝中。

皮下组织是免疫特异部位，部分细胞皮下脂肪层输注，作为胰岛素供应的"备用库"。研究发现，自体MSC对高血糖症的长期控制效果优于同种异体MSC。1型糖尿病患者可以储存自己的脂肪组织用来治疗自己的疾病。其主要作用机制可能是MSC具有向损伤

部位迁移并进行修复的特性，通过分泌细胞因子、增生及转化来完成相应的损伤修复。同时，MSC 具有较低的免疫原性及较强的免疫调节功能，可以产生一些细胞因子和生长因子，通过旁分泌机制来提高周围细胞的存活率。新血管的产生也依赖于干细胞释放血管源性细胞因子。

七、MSC 与炎症性气道和肺部免疫系统疾病

在肺部的许多疾病中发现的急性和慢性肺损伤总是涉及异常的免疫活化与纤维化，包括阻塞性疾病[如慢性阻塞性肺疾病（COPD）和哮喘]和限制性疾病[如特发性肺纤维化（IPF）和急性呼吸窘迫综合征（acute respiratory distress syndrome，ARDS）][84]。而 MSC 特别适合用于治疗肺部疾病，因为研究表明 MSC 可能通过静脉将绝大多数（通常为 80%~90%）的细胞产物迅速运输到肺部。在肺部有炎症的情况下，运输比例增加。最近的一项研究还表明，肺部可能是 MSC 独特的龛。

特定的炎症/免疫过程即使在同类肺部疾病中也是不同的，如 COPD 与哮喘。在 COPD 中，由肺泡巨噬细胞、CTL 和中性粒细胞介导的炎症反应导致气流的进行性限制，发生小气道纤维化和肺泡破坏[85]。在哮喘中，肥大细胞、嗜酸性粒细胞和 Th2 淋巴细胞使气道高反应性和支气管收缩加重[86]。在由弹性蛋白酶诱导的肺气肿和由香烟烟雾诱导的 COPD 啮齿动物模型中，MSC 输注可减轻肺部破坏和异常炎症[87]。MSC 分泌的表皮生长因子（EGF）诱导分泌型白细胞蛋白酶抑制剂（SLPI）产生，保护上皮组织免于被丝氨酸蛋白酶降解[88]。由吸烟引起的肺损伤大鼠模型中输注 MSC，也导致促炎细胞因子，如 TNF-α、IL-1β、IL-6 和 MCP-1/CCL2 的下调，血管内皮生长因子（VEGF）和 TGF-β 的上调[89]。此外，MSC 治疗可以抑制肺泡巨噬细胞中环加氧酶 2（COX2）和 COX2 介导的 PGE2 产生以减少炎症。对于哮喘，在吸入甲苯二异氰酸酯、卵清蛋白或蟑螂提取物的啮齿动物疾病模型中，MSC 治疗通过产生 Treg 细胞并抑制 Th2 应答来调节免疫环境，疾病症状得到逆转，Th2 细胞因子（包括 IL-4、IL-5 和 IL-13）、免疫球蛋白 E（IgE）水平、基质金属蛋白酶（MMP）沉积和黏液生成都发生降低[90]。

MSC 治疗纤维化的肺部疾病似乎也是有效的。在由博莱霉素诱导的 IPF 小鼠模型中，MSC 分泌的 IL-1 受体拮抗剂（IL-1RA）发挥抗炎和抗纤维化的作用。并且 UC-MSC 的输注对 IPF 模型也具有治疗效果。除了抗炎作用外，MSC 通过增加常驻肺支气管肺泡干细胞群数量来减轻纤维化，从而修复和再生健康的肺实体[91]。

迄今为止，已经注册了使用 MSC 治疗肺功能障碍的 29 项临床研究。靶向疾病包括哮喘、COPD、ARDS、支气管肺发育不良（bronchopulmonary dysplasia，BPD）和纤维化。同时有几份关于 MSC 治疗各种肺部疾病的实验报告已发表，其中最大一项是评估采用同种异体骨髓干细胞治疗 62 名 COPD 患者的实验[92]。结果显示安全性良好，但并没有表现出有很好的疗效。其他已发表的研究是有关各种组织来源的同种异体 MSC 的 I 期临床试验，其中治疗 ARDS 有两项试验，一项使用 AD-MSC，另一项使用 BM-MSC[93]。此外，还有一项使用脐血间充质干细胞（UCB-MSC，气管内递送）治疗早产 BPD 和一项使用胎盘间充质干细胞（P-MSC）治疗 IPF[69,94]。所有报告都显示了 MSC 输注的安全性，但功效不佳。MSC 在临床前动物研究中显示出的强有力的疗效目前似乎无法在临床

研究中被复制,这可能是靶向肺部疾病存在多样性的结果。另外,MSC 组织来源的差异是否影响其归巢能力也是一个关键问题。因此,仔细选择患者群体和开展组织特异性 MSC 是否疗效不同的更多研究是有必要的。

第五节 间充质干细胞在自身免疫系统疾病治疗中面临的问题

本章中,我们介绍了 MSC 在自身免疫系统疾病临床前模型中的研究成果和已经完成的少数临床试验的结果。在这些研究中 MSC 发挥作用的机制仍是未知的,但 MSC 可能主要通过可溶性介质直接影响多种免疫细胞群体的命运和功能。研究发现,MSC 可以抑制 B 细胞和 T 细胞及树突细胞(DC)的增殖与分化;它们还可以通过抑制活性氧(ROS)的产生来抑制中性粒细胞凋亡和呼吸爆发。此外,MSC 促进单核细胞向 M2 表型极化,并改变内皮细胞对促炎细胞因子的应答。虽然 MSC 已经存在这些效应器功能,但它们在自身免疫系统疾病中的确切作用模式尚未可知。由于许多免疫细胞和内皮功能已经涉及慢性炎性疾病与自身免疫系统疾病,我们推测 MSC 可以通过抑制这些效应细胞功能来发挥有益作用(图 10-2)。

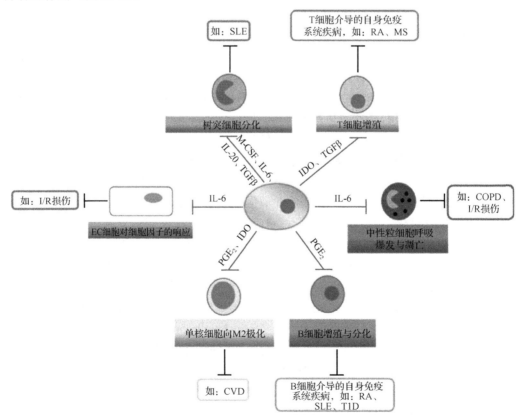

图 10-2 MSC 的免疫调节作用和在自身免疫系统疾病中的潜在治疗作用(彩图请扫封底二维码)
I/R 损伤. 缺血/再灌注损伤,SLE. 系统性红斑狼疮,RA. 类风湿关节炎,MS. 多发性硬化,M-CSF. 巨噬细胞集落刺激因子,IL-6. 白介素-6,IL-10. 白介素-10,IDO. 吲哚胺 2,3-双加氧酶,TGF-β. 转化生长因子-β,PGE2. 前列腺素 E2,COPD. 慢性阻塞性肺疾病,CVD. 心血管疾病,T1D. 1 型糖尿病

对最近出版数据进行统计，约有 500 项临床试验进行了登记，其中很大一部分针对自身免疫系统/炎性疾病[95]。迄今为止发表的大量研究结果证明了 MSC 治疗的安全性，患者没有出现或伴随轻微的不良反应，并且输注细胞未发生恶性转化现象。MSC 输注的潜在风险是通过其免疫抑制活性诱导或加速肿瘤的生长，但这需要对治疗患者进行更长时间的随访才能对此问题做出明确的判断。大多数公布的临床试验都是不受人为控制的，通常属于招募少数患者的非随机研究，在许多目前的 I/II 期研究中没有设置适当的安慰剂对照。另外，自身免疫系统疾病有波动的症状恶化和周期性改善。因此，难以确定 MSC 施用后疾病的任何改善是否直接来自于 MSC 治疗或病症的自然病史。不同试验中使用的 MSC 扩增方法不同，MSC 的来源也是不同的。在 MSC 系统地作为治疗自身免疫/炎症疾病的有效手段之前，仍有许多工作要做。

另外，自身免疫的动物模型，特别是本章中讨论的动物模型，不能概括人类自身免疫系统疾病的所有特征。MSC 的临床前疗效，尽管为开展临床试验提供了坚实的科学背景，但并不能预期任何临床结果。由 MSC 处理的自身免疫系统疾病动物模型展现出不同的或经常相反的治疗效果，进一步加剧了这一顾虑。目前我们对输注体内的 MSC 的作用机制了解甚少。MSC 表达一组严格受限的趋化因子受体，并在体外迁移到一些趋化因子上，如 CXCL12。事实上到达目标组织的 MSC 比例非常低，并且大多数效应似乎由旁分泌作用发挥。这些发现开辟了新的治疗观点，因为 MSC 衍生细胞制剂的临床应用能绕过管理人体细胞治疗的严格规则。另一个问题是，如果 MSC 在接受者中迅速消除，如何才能发挥持久的治疗作用。最近的研究表明，MSC 预处理后的肺单核细胞/巨噬细胞可以对眼睛内的同种异体免疫和自身免疫产生耐受性[96]。最后，我们能否提高 MSC 的治疗效果？在临床前模型中至少有三种策略可能有希望：①MSC 的转基因工程，过表达与特异性疾病模型相关的蛋白质（如 EAE 中的睫状神经营养因子）；②在输注之前体外预处理 MSC（如缺血性疾病的缺氧预处理）；③使用靶向组织特异性的 MSC（如 CD44 表面受体的酶修饰，使用生物化学技术将归巢分子偶联到细胞表面）[91]。这些方法虽然十分有前景，但将这些方法应用于临床还任重而道远。

（张 颢）

参 考 文 献

[1] 李成荣. 自身免疫性疾病免疫发病机制概况. 中国实用儿科杂志, 2010, 25(5):411-414.

[2] 沈浩, 王颖, 王佳琦, 等. 重要自身免疫性疾病的发病机制及其免疫干预策略. 上海交通大学学报, 2012, 32(9): 1128-1134.

[3] Crozat K, Vivier E, Dalod M. Crosstalk between components of the innate immune system:promoting anti-microbial defenses and avoiding immunopathologie. Immunol Rev, 2009, 227(1):129-149.

[4] 孙凌云. 间质干细胞治疗系统性红斑狼疮的机制和应用. 内科理论与实践, 2008, 3:158-161.

[5] 顾伟杰, 赵广. 间充质干细胞在自身免疫病及免疫相关性皮肤病治疗中的应用. 北京医学, 2011, 33(11):922-924.

[6] Pittenger MF, Mackay AM, Beck SC, et al. Multilineage potential of adult human mesenchymal stem cells. Science, 1999, 284:143.

[7] Caplan AI. Mesenchymal stem cells. J Orthop Res, 1991, 9:641.

[8] Sacchetti B, Funari A, Michienzi S, et al. Self-renewing osteoprogenitors in bone marrow sinusoids can organize a hematopoietic microenvironment. Cell, 2007, 131:324.

[9] Sacchetti B, Funari A, Remoli C, et al. No identical "mesenchymal stem cells" at different times and sites: human committed progenitors of distinct origin and differentiation potential are incorporated as adventitial cells in microvessels. Stem Cell Reports, 2016, 6:897.

[10] Mosna F, Sensebé L, Krampera M. Human bone marrow and adipose tissue mesenchymal stem cells: a user's guide. Stem Cells Dev, 2010, 19:1449.

[11] Bieback K, Kinzebach S, Karagianni M. Translating research into clinical scale manufacturing of mesenchymal stromal cells. Stem Cells Int, 2011, 2010:193519.

[12] Keating A. Mesenchymal stromal cells: new directions. Cell Stem Cell, 2012, 10:709.

[13] Bernardo ME, Fibbe WE. Mesenchymal stromal cells: sensors and switchers of inflammation. Cell Stem Cell, 2013, 13:392.

[14] Raffaghello L, Bianchi G, Bertolotto M, et al. Human mesenchymal stem cells inhibit neutrophil apoptosis: a model for neutrophil preservation in the bone marrow niche. Stem Cells, 2008, 26:151.

[15] Cassatella MA, Mosna F, Micheletti A. et al. Toll-like receptor-3-activated human mesenchymal stromal cells signifcantly prolong the survival and function of neutrophils. Stem Cells, 2011, 29:1001.

[16] Nauta AJ, Kruisselbrink AB, Lurvink E, et al. Mesenchymal stem cells inhibit generation and function of both $CD34^{+}$-derived and monocyte-derived dendritic cells. J Immunol, 2006, 177:2080.

[17] Aggarwal S, Pittenger MF. Human mesenchymal stem cells modulate alloogeneic immune cell responses. Blood, 2005, 105:1815.

[18] François M, Romieu-Mourez R, Li M, et al. Human MSC suppression correlates with cytokine induction of indoleamine 2,3-dioxygenase and bystander M2 macrophage differentiation. Mol Ther, 2012, 20:187.

[19] Spaggiari GM, Capobianco A, Becchetti S, et al. Mesenchymal stem cell-natural killer cell interactions: evidence that activated NK cells are capable of killing MSCs, whereas MSCs can inhibit IL-2-induced NK-cell proliferation. Blood, 2006, 107:1484.

[20] Benvenuto F, Ferrari S, Gerdoni E, et al. Human mesenchymal stem cells promote survival of T cells in a quiescent state. Stem Cells, 2007, 25:1753.

[21] Bartholomew A, Sturgeon C, Siatskas M, et al. Mesenchymal stem cells suppress lymphocyte proliferation *in vitro* and prolong skin graft survival *in vivo*. Exp Hematol, 2002, 30:42.

[22] Maccario R, Podestà M, Moretta A, et al. Interaction of human mesenchymal stem cells with cells involved in alloantigen-specifc immune response favors the differentiation of $CD4^{+}$ T-cell subsets expressing a regulatory/suppressive phenotype. Haematologica, 2005, 90:516.

[23] Corcione A, Benvenuto F, Ferretti E, et al. Human mesenchymal stem cells modulate B-cell functions. Blood, 2006, 107:367.

[24] Prigione I, Benvenuto F, Bocca P, et al. Reciprocal interactions between human mesenchymal stem cells and gammadelta T cells or invariant natural killer T cells. Stem Cells, 2009, 27:693.

[25] Bourin P, Bunnell BA, Casteilla L, et al. Stromal cells from the adipose tissue-derived stromal vascular fraction and culture expanded adipose tissue-derived stromal/stem cells: a joint statement of the International Federation for Adipose Therapeutics and Science (IFATS) and the International Society for Cellular Therapy (ISCT). Cytotherapy, 2013, 15:641.

[26] Schellenberg A, Stiehl T, Horn P, et al. Population dynamics of mesenchymal stromal cells during culture expansion. Cytotherapy, 2012, 14:401.

[27] Krampera M. Mesenchymal stromal cell 'licensing': a multistep process. Leukemia, 2011, 9:1408.

[28] Zhao X, Liu D, Gong W, et al. The toll-like receptor 3 ligand, poly(I:C), improves immunosuppressive function and therapeutic effect of mesenchymal stem cells on sepsis via inhibiting MiR-143. Stem Cells, 2014, 32:521.

[29] Waterman RS, Tomchuck SL, Henkle SL, et al. A new mesenchymal stem cell (MSC) paradigm: polarization into a pro-inflammatory MSC1 or an immunosuppressive MSC2 phenotype. PLoS One, 2010, 5:e10088.

[30] Pistoia V, Raffaghello L. Mesenchymal stromal cells and autoimmunity. International Immunology, 2017, 29(2):49-58.

[31] Youd M, Blickarz C, Woodworth L, et al. Allogeneic mesenchymal stem cells do not protect NZBxNZW F1 mice from developing lupus disease. Clin Exp Immunol, 2010, 161:176

[32] Gu Z, Akiyama K, Ma X, et al. Transplantation of umbilical cord mesenchymal stem cells alleviates lupus nephritis in MRL/lpr mice. Lupus, 2010, 19:1502.

[33] Choi EW, Shin IS, Park SY, et al. Reversal of serologic, immunologic, and histologic dysfunction in mice with systemic lupus erythematosus by long-term serial adipose tissue-derived mesenchymal stem cell transplantation. Arthritis Rheum, 2012, 64:243.

[34] González MA, Gonzalez-Rey E, Rico L, et al. Treatment of experimental arthritis by inducing immune tolerance with human adipose-derived mesenchymal stem cells. Arthritis Rheum, 2009, 60:1006.

[35] Schurgers E, Kelchtermans H, Mitera T, et al. Discrepancy between the *in vitro* and *in vivo* effects of murine mesenchymal stem cells on T-cell proliferation and collagen-induced arthritis. Arthritis Res. Ther, 2010, 12:R31.

[36] Papadopoulou A, Yiangou M, Athanasiou E, et al. Mesenchymal stem cells are conditionally therapeutic in preclinical models of rheumatoid arthritis. Ann Rheum Dis, 2012, 71:1733.

[37] Luz-Crawford P, Kurte M, Bravo-Alegría J, et al. Mesenchymal stem cells generate a $CD4^+CD25^+Foxp3^+$ regulatory T cell population during the differentiation process of Th1 and Th17 cells. Stem Cell Res Ther, 2013, 4:65.

[38] Rafei M, Campeau PM, Aguilar-Mahecha A, et al. Mesenchymal stromal cells ameliorate experimental autoimmune encephalomyelitis by inhibiting CD4 Th17 T cells in a CC chemokine ligand 2-dependent manner. J Immunol, 2009, 182: 5994.

[39] Bai L, Lennon DP, Caplan AI, et al. Hepatocyte growth factor mediates mesenchymal stem cell-induced recovery in multiple sclerosis models. Nat Neurosci, 2012, 15:862.

[40] González MA, Gonzalez-Rey E, Rico L, et al. Adipose-derived mesenchymal stem cells alleviate experimental colitis by inhibiting inflammatory and autoimmune responses. Gastroenterology, 2009, 136:978.

[41] Hayashi Y, Tsuji S, Tsujii M, et al. Topical implantation of mesenchymal stem cells has beneficial effects on healing of experimental colitis in rats. J Pharmacol. Exp Ther, 2008, 326:523.

[42] Baron F, Storb R. Mesenchymal stromal cells: a new tool against graft-versus-host disease? Biol Blood Marrow Transplant, 2012, 18:822.

[43] Meisel R, Brockers S, Heseler K, et al. Human but not murine multipotent mesenchymal stromal cells exhibit broadspectrum antimicrobial effector function mediated by indoleamine 2,3-dioxygenase. Leukemia, 2011, 25:648.

[44] Zhang D, Li H, Ma L, et al. The defective bone marrowderived mesenchymal stem cells in patients with chronic immune thrombocytopenia. Autoimmunity, 2014, 47:519.

[45] Calkoen FG, Brinkman DM, Vervat C, et al, Mesenchymal stromal cells isolated from children with systemic juvenile idiopathic arthritis suppress innate and adaptive immune responses. Cytotherapy, 2013, 15:280.

[46] Flowers ME, Martin PJ. How we treat chronic graft-versus-host disease. Blood, 2015, 125(4):606-615.

[47] Le Blanc K, Rasmusson I, Sundberg B, et al. Treatment of severe acute graft-versus-host disease with third party haploidentical mesenchymal stem cells. Lancet, 2004, 363(9419):1439-1441.

[48] Bartholomew A, Sturgeon C, Siatskas M, et al. Mesenchymal stem cells suppress lymphocyte proliferation *in vitro* and prolong skin graft survival *in vivo*. Exp Hematol, 2002, 30(1):42-48.

[49] Sánchez-Guijo F, Caballero-Velázquez T, López-Villar O, et al. Sequential third-party mesenchymal stromal cell therapy for refractory acute graft-versus-host disease. Biol Blood Marrow Transplant, 2014, 20:1580.

[50] Muroi K, Miyamura K, Okada M, et al. Bone marrowderived mesenchymal stem cells (JR-031) for steroid-refractory grade III or IV acute graft-versus-host disease: a phase II/III study. Int J Hematol, 2016, 103:243.

[51] Le Blanc K, Frassoni F, Ball L, et al. Mesenchymal stem cells for treatment of steroid-resistant, severe, acute graft-versushost disease: a phase II study. Lancet, 2008, 371:1579.

[52] Gao L, Zhang Y, Hu B, et al. Phase II multicenter, randomized, double-blind controlled study of efficacy and safety of umbilical cord-derived mesenchymal stromal cells in the prophylaxis of chronic graft-versus-host disease after HLA-haploidentical stem-cell transplantation. J Clin Oncol, 2016, 34:2843.

[53] Sanchez-Guijo FM, Lopez-Villar O, Lopez-Anglada L, et al. Allogeneic mesenchymal stem cell therapy for refractory cytopenias after hematopoietic stem cell transplantation. Transfusion, 2012, 52(5):1086-1091.

[54] Weng J, He C, Lai P, et al. Mesenchymal stromal cells treatment attenuates dry eye in patients with chronic graft-versus-host disease. Mol Ther, 2012, 20(12):2347-2354.

[55] Tobin LM, Healy ME, English K, et al. Human mesenchymal stem cells suppress donor CD4(+) T cell proliferation and reduce pathology in a humanized mouse model of acute graft-versus-host disease. Clin Exp Immunol, 2013, 172(2):333-348

[56] Auletta JJ, Eid SK, Wuttisarnwattana P, et al. Human mesenchymal stromal cells attenuate graft-versus-host disease and maintain graft-versus-leukemia activity following experimental allogeneic bone marrow transplantation. Stem Cells, 2015, 33(2):601-614.

[57] 黄金棋, 林艳娟. 间充质干细胞治疗自身免疫性疾病及其细胞生物治疗的致瘤性. 中国组织工程研究与临床康复, 2010, 14(45):8478-8482.

[58] 管英华, 谢扬虎, 魏晓巍, 等. 骨髓间充质干细胞与再生障碍性贫血的研究进展. 中华全科医学, 2012, 10(7):1131-1133.

[59] 成柯君. 骨髓间质干细胞治疗再生障碍性贫血的研究进展. 世界最新医学信息文摘, 2015, 15(59):27.

[60] Pérez-Simón JA, Tabera S, Sarasquete ME, et al. Mesenchymal stem cells are functionally abnormal in patients with immune thrombocytopenicpurpura. Cytotherapy, 2009, 11(6):698-705.

[61] Zhang JM, Feng FE, Wang QM, et al. Platelet-derived growth factor-BB protects mesenchymal stem cells (MSCs) derived from immune thrombocytopenia patients against apoptosis and senescence and maintains MSC-mediated immunosuppression. Stem Cells Transl Med, 2016, 5(12):1631-1643.

[62] Ma L, Zhou Z, Zhang D, et al. Immunosuppressive function of mesenchymal stem cells from human umbilical cord matrix in immune thrombocytopenia patients. Thromb Haemost, 2012, 107(5):937-950.

[63] Ra JC, Kang SK, Shin IS, et al. Stem cell treatment for patients with autoimmune disease by systemic infusion of culture-expanded autologous adipose tissue derived mesenchymal stem cells. J Transl Med, 2011, 9:181.

[64] Liang J, Zhang H, Hua B, et al. Allogenic mesenchymal stem cells transplantation in refractory systemic lupus erythematosus: a pilot clinical study. Ann Rheum Dis, 2010, 69:1423.

[65] Wang D, Li J, Zhang Y, et al. Umbilical cord mesenchymal stem cell transplantation in active and refractory systemic lupus erythematosus: a multicenter clinical study. Arthritis Res Ther, 2014, 16:R79.

[66] Scuderi N, Ceccarelli S, Onesti MG, et al. Human adipose-derived stromal cells for cell-based therapies in the treatment of systemic sclerosis. Cell Transplant, 2013, 22(5):779-795.

[67] Liu R, Wang Y, Zhao X, et al. Lymphocyte inhibition is compromised in mesenchymal stem cells from psoriatic skin. Eur J Dermatol EJD, 2014, 24(5):560-567.

[68] Horie M, Choi H, Lee RH, et al. Intra-articular injection of human mesenchymal stem cells (MSCs) promote rat meniscal regeneration by being activated to express Indian hedgehog that enhances expression of type II collagen. Osteoarthritis Cartilage, 2012, 20(10):1197-1207.

[69] Wang LT, Ting CH, Yen ML. Human mesenchymal stem cells (MSCs) for treatment towards immune-and inflammation-mediated diseases: review of current clinical trials. J Biomed Sci, 2016, 23(1):76.

[70] Wang L, Wang L, Cong X, et al. Human umbilical cord mesenchymal stem cell therapy for patients with active rheumatoid arthritis: safety and efficacy. Stem Cells Dev, 2013, 22:3192.

[71] Dezawa M, Kanno H, Hoshino M, et al. Specific induction of neu-ronal cells from bone marrow stromal cells and application for autologous transplantation. J Clin Invest, 2004, 113(12):1701-1710.

[72] Satake K, Lou J, Lenke LG. Migration of mesenchymal stem cells through cerebrospinal fluid into injured spinal cord tissue. Spine, 2004, 29(18):1971-1979.

[73] Hedegaard CJ, Krakauer M, Bendtzen K, et al. T helper cell type 1 (Th1), Th2 and Th17 responses to myelin basic protein and disease activity in multiple sclerosis. Immunology, 2008, 125(2):161-169.

[74] Feinstein A, Freeman J, Lo AC. Treatment of progressive multiple sclerosis: what works, what does not, and what is needed. Lancet Neurol, 2015, 14(2):194-207.

[75] Li JF, Zhang DJ, Geng T, et al. The potential of human umbilical cord-derived mesenchymal stem cells as a novel cellular therapy for multiple sclerosis. Cell Transplant, 2014, (Suppl 1):S113.

[76] Molendijk I, Bonsing BA, Roelofs H, et al. Allogeneic bone marrow-derived mesenchymal stromal cells promote healing of refractory perianal fistulas in patients with Crohn's disease. Gastroenterology, 2015, 149(4):918-927.

[77] Panés J, García-Olmo D, Van Assche G, et al. Expanded allogeneic adipose-derived mesenchymal stem cells (Cx601) for complex perianal fistulas in Crohn's disease: a phase 3 randomised, double-blind controlled trial. Lancet, 2016, 388:1281.

[78] Sala E, Genua M, Petti L, et al. Mesenchymal stem cells reduce colitis in mice via release of TSG6, independently of their localization to the intestine. Gastroenterology, 2015, 149(1):163-176.

[79] Zuo D, Tang Q, Fan H, et al. Modulation of nuclear factor-kappaB-mediated pro-inflammatory response is associated with exogenous administration of bone marrow-derived mesenchymal stem cells for treatment of experimental colitis. Mol Med Rep, 2015, 11(4):2741-2748.

[80] Wang WQ, Dong K, Zhou L, et al. IL-37b gene transfer enhances the therapeutic efficacy of mesenchumal stromal cells in DSS-induced colitis mice. Acta Pharmacol Sin, 2015, 36(11):1377-1187.

[81] Vanikar AV, Trivedi HL, Thakkar UG. Stem cell therapy emerging as the key player in treating type 1 diabetes mellitus. Cytotherapy, 2016, 18(9):1077-1086.

[82] Thakkar UG, Trivedi HL, Vanikar AV, et al. Insulin secreting adipose-derived mesenchymal stromal cells with bone marrow-derived hematopoietic stem cells from autologous and allogenic sources for type 1 diabetes mellitus. Cytotherapy, 2015, 17:940-947.

[83] Staeva TP, Chatenoud L, Insel R, et al. Recent lessons learned from prevention and recent-onset type 1 diabetes immunotherapy trials. Diabetes, 2013, 62:9-17.

[84] Inamdar AC, Inamdar AA. Mesenchymal stem cell therapy in lung disorders: pathogenesis of lung diseases and mechanism of action of mesenchymal stem cell. Exp Lung Res, 2013, 39(8):315-327.

[85] Vlahos R, Bozinovski S. Role of alveolar macrophages in chronic obstructive pulmonary disease. Front Immunol, 2014, 5: 435.

[86] Postma DS, Rabe KF. The asthma-COPD overlap syndrome. N Engl J Med, 2015, 373(13):1241-1249.

[87] Kim YS, Kim JY, Huh JW, et al. The therapeutic effects of optimal dose of mesenchymal stem cells in a murine model of an elastase induced-emphysema. Tuberc Respir Dis (Seoul), 2015, 78(3):239-245.

[88] Broekman W, Amatngalim GD, De Mooij-Eijk Y, et al. TNF-alpha and IL-1beta-activated human mesenchymal stromal cells increase airway epithelial wound healing *in vitro* via activation of the epidermal growth factor receptor. Respir Res, 2016, 17(1):3.

[89] Guan XJ, Song L, Han FF, et al. Mesenchymal stem cells protect cigarette smoke-damaged lung and pulmonary function partly via VEGF-VEGF receptors. J Cell Biochem, 2013, 114(2):323-335.

[90] Gao P, Zhou Y, Xian L, et al. Functional effects of TGF-beta1 on mesenchymal stem cell mobilization in cockroach allergen-induced asthma. J Immunol, 2014, 192(10):4560-4570.

[91] Tropea KA, Leder E, Aslam M, et al. Bronchioalveolar stem cells increase after mesenchymal stromal cell treatment in a mouse model of bronchopulmonary dysplasia. Am J Physiol Lung Cell Mol Physiol, 2012, 302(9):L829-L837.

[92] Weiss DJ, Casaburi R, Flannery R, et al. A placebo-controlled, randomized trial of mesenchymal stem cells in COPD. Chest, 2013, 143(6):1590-1598.

[93] Wilson JG, Liu KD, Zhuo H, et al. Mesenchymal stem (stromal) cells for treatment of ARDS: a phase 1 clinical trial. Lancet Respir Med, 2015, 3(1):24-32.

[94] Chang YS, Ahn SY, Yoo HS, et al. Mesenchymal stem cells for bronchopulmonary dysplasia: phase 1 dose-escalation clinical trial. J Pediatr, 2014, 164(5):966-972.

[95] Squillaro T, Peluso G, Galderisi U. Clinical trials with mesenchymal stem cells: an update. Cell Transplant, 2016, 23:829.

[96] Ko JH, Lee HJ, Jeong HJ, et al. Mesenchymal stem/stromal cells precondition lung monocytes/macrophages to produce tolerance against allo- and auto-immunity in the eye. Proc Natl Acad Sci USA, 2016, 113:158.

第十一章　间充质干细胞在下肢缺血性疾病治疗中的应用

第一节　下肢缺血性疾病的概述

一、下肢缺血性疾病的概念及特点

下肢缺血性疾病（limb ischemia，LI）是包括累及四肢动脉的一组急慢性疾病，可急性发病也可慢性发病。急性 LI 多由动脉栓塞或下肢动脉急性血栓所致，慢性疾病包括下肢闭塞性动脉硬化（arteriosclerosis obliterans，ASO）、血栓闭塞性脉管炎（Buerger 病）、主动脉型大动脉炎（Takayasu 病）、糖尿病足（diabetic foot）、雷诺氏病（Reynolds disease）等。重症肢体缺血（critical limb ischemia，CLI）是 LI 的终末阶段，无药物治疗可供选择。下肢缺血是一种常见症状，临床表现多伴有反复发作的静息痛、间歇性跛行，或溃疡、坏疽，并具有发病率高、致残率高、医疗费高、死亡率高和缺乏有效的治疗手段的特点。国内外临床常用的分期方法有 Fontaine 法和 Rutherford 法。

随着人们生活水平的提高及人口老龄化，糖尿病等引起的 LI 在我国的发病率呈逐年上升的趋势，预后较差。糖尿病患者中 20%~40%发生 LI，导致的下肢动脉狭窄、闭塞使得糖尿病患者下肢缺血、溃疡甚至坏死，也是糖尿病患者心血管事件及死亡率增加的重要原因。但由于糖尿病患者的 LI 存在血管病变弥漫、多阶段病变、远端流出道闭塞等特点，约 40%下肢缺血严重的患者经过一般治疗仍不能得到改善，截肢成为他们的必然选择，截肢后死亡率高达 20%~50%。探索缺血肢体血循环重建新的治疗策略，对减少患者截肢率，提高其生活质量具有非常重要的临床意义[1]。

二、下肢缺血性疾病的治疗现状

下肢缺血的治疗目的是增加患者行走距离，缓解疼痛，促进溃疡愈合和避免截肢，治疗机制是促进血管再生、增加侧支循环。传统的治疗方法主要包括戒烟、降脂、控制血糖、控制血压、保暖及适当的锻炼。常用的药物主要是抗凝和抗血小板治疗药物，还有一大类药物是各种促进血管扩张和侧支循环形成的药物，如前列环素类药物凯时、保达新等。传统手术治疗包括动脉切取栓术、动脉内膜剥脱术、动脉旁路移植术、间接改善下肢血运的手术（包括大网膜铺植术和腰交感神经节切除术）。血管腔内介入治疗因微创、可重复进行，部分过去需行动脉旁路搭桥的手术逐渐已被腔内介入治疗所替代。但是目前传统的药物治疗、介入手术、局部护理等治疗方式对于严重病变的患者往往是不够的，部分患者应用上述治疗方式往往无效，不能耐受外科手术治疗，或者术后复发，最后只能截肢治疗，甚至危及生命。尤其是糖尿病足和 Buerger 病患者，病变多累及下肢远端小动脉，过去这部分患者难以避免截肢，给患者带来巨大的经济负担和生活压力。因此，亟待探索新的治疗方式来降低糖尿病足患者截肢率，改善糖尿病足患者的生存质量。一些新的治疗方式如生长因子、自体血小板富集血浆凝胶、臭氧治疗、胎盘膜的使用不断涌现，但这些治疗方式还需要进一步评估和改进。目前兴起的干细胞治疗作为促进血管再生的新技术，给一些无法进行介入治疗或外科手术或术后复发者，尤其是糖尿病下肢缺血和Buerger病患者带来了新的希望。

第二节　间充质干细胞治疗下肢缺血性疾病的研究进展

一、MSC的组织再生与修复特性

我国的黄平平等[2]在国际上首次报告了动员后的自体外周血干细胞移植治疗肢体缺血性疾病的临床研究。许多国家相继在临床上开展了应用骨髓和外周血单个核细胞（BMMC和PBMC）治疗糖尿病足的项目，并且取得了一系列成果。目前应用于LI的干细胞的来源有$CD34^+$的单个核细胞、内皮祖细胞（EPC）和间充质干细胞（MSC），其中MSC研究得较为广泛。近年来的动物实验表明，骨髓干细胞移植能促使缺血的后肢生成新生血管。骨髓中最重要的干细胞之一是造血干细胞（HSC），将经骨髓穿刺抽出的血细胞置于平皿中培养，部分细胞呈圆形并能够分化增殖成各类型血细胞的是HSC。还有一部分细胞呈梭状并能贴壁生长，且可以增殖传代，这类细胞称为MSC。MSC在不同的条件下还可以向不同的组织分化，如向骨组织、软骨组织、肌肉组织、神经组织、肝组织、肺组织等分化。实验还发现，这些细胞可以定向分化成心肌细胞、血管内皮细胞和平滑肌细胞。

目前，MSC对下肢缺血的治疗性血管生成作用已被实验证实。作为成血管细胞来源之一，自体MSC具有来源于自体、取材方便、扩增迅速、受缺血组织趋化、可在缺氧条件下向内皮细胞、血管平滑肌细胞分化等优点。目前认为干细胞分化成血管涉及血管发生（vasculogenesis）、血管新生（angiogenesis）、动脉生成（arteriogenesis）和旁分泌（paracrine）4种机制。这4种机制具有协同作用，旁分泌机制可促进其他三种机制的形成、发展，共同恢复缺血后组织的供血和供氧。目前很多研究表明，MSC的旁分泌机制在治疗下肢缺血的过程中发挥了更重要的作用。干细胞主要通过分泌一些分子，如生长因子、细胞因子、趋化因子和细胞外微泡，通过旁分泌机制进入周围环境来起到促进血管新生的作用。动物和临床研究均证实，MSC在心肌梗死治疗中有促血管新生的作用。MSC还能改善大鼠缺血下肢的血流灌注，增加血管内皮生长因子（VEGF）和成纤维细胞生长因子（FGF）等血管新生因子的含量和血管内皮细胞数量。多数研究者认为，MSC的分化与它所处的微环境密切相关。在损伤或缺血的条件下，可能是病变组织的微环境中含有各种不同的因子，促进MSC向其所在环境需要的细胞或组织方向分化，其机制可能是当它在处于向内皮细胞分化的微环境下时，MSC的某些特征性的内皮细胞基因开放，通过促进血管新生相关蛋白的表达，使MSC直接分化为血管内皮细胞[3]。

以糖尿病下肢缺血为例，MSC的组织再生与修复特性表现在以下几方面：①MSC具有向不同损伤部位趋化、归巢的能力，这是其组织修复功能的基础和前提。②防止胰岛细胞损伤、促进损伤胰岛细胞再生。③MSC 能够分泌多种促血管新生的细胞因子，包括 VEGF、FGF 等，提示 MSC 可能在促进胰岛血管化这一过程中具有一定作用。

二、MSC 治疗下肢缺血的机制与现状

近年来，干细胞移植治疗下肢缺血这一领域取得了较快的发展。MSC 第一次由 Friedenstein[4]提出。最近的研究发现，MSC 在免疫调节中可以起到抗炎的作用，因此它们可以抑制免疫细胞的增殖[5]。骨髓间充质干细胞（BM-MSC）和脂肪间充质干细胞（AD-MSC）或局部使用种植在胶原支架上的同种异体非糖尿病 BM-MSC 在动物模型中已显示出可以促进伤口愈合及加速血管生成[6,7]。不断有报道显示，BM-MSC 的应用有效地实现了伤口愈合[8,9]。这些研究为干细胞移植在心血管内外科中临床应用提供了宝贵信息。

糖尿病足是糖尿病的严重并发症之一。Kuo 等[7]使用 MSC 治疗的糖尿病大鼠与对照组大鼠相比伤口范围明显减少，伤口完全愈合时间更短，MSC 使得局部炎症反应和 CD45 表达抑制显著减少。实验也表明，MSC 增强糖尿病患者伤口愈合可能与组织再生中生物标志物的增加有关。适当的措施可以提高 BM-MSC 移植的效果。Kim 等[10]使用含有 MSC 的三维（3D）胶原凝胶支架治疗 Sprague-Dawley 大鼠的全层皮肤缺损的实验结果表明，MSC 联合 3D 胶原可以促进新生血管形成，MSC 上调基质金属蛋白酶-9（MMP-9）的早期表达和 VEGF 的早期活化，明显加速伤口愈合。有研究通过移植 BM-MSC 促进糖尿病大鼠延迟的伤口愈合的疗效差别来比较采用皮下注射与肌肉注射干细胞的优劣，结果表明两种方式均对伤口愈合有积极作用，但肌内注射疗效优于皮下注射，BM-MSC 可以优先移动到缺血及损伤组织参与伤口愈合过程，并且肌肉注射的方式在伤口组织恢复后期可以表现出疗效的持久性和更高的 VEGF 表达水平[6,11]。

MSC 改善 2 型糖尿病患者高血糖的作用可能涉及多种作用和机制，包括增强 β 细胞功能和改善胰岛素抵抗。Si 等[12]的研究表明：在糖尿病早期阶段，MSC 的输注不仅可以增强 β 细胞功能，而且可以改善胰岛素抵抗，而晚期的 MSC 输注则只能改善胰岛素抵抗。其他临床试验证明了自体 BM-MSC 在人糖尿病足溃疡治疗中的安全性与有效性[13]。

值得考虑的是，大多数糖尿病足患者是老年人，因干细胞数目较低，所以增殖减弱、黏附和血管发生潜力相应减退。因此，人羊膜 MSC 和 UC-MSC 可以用作同种异体细胞移植的细胞来源。有证据表明，来自成体体细胞的胚胎干细胞（ESC）和诱导多能干细胞（iPSC）有益于伤口再生[14]。其中，人脐血是 MSC 和 HSC 的丰富来源，与从骨髓或自体外周血获取的干细胞相比有更大的分化潜力和扩增能力，具有低免疫原性，并且容易大量获得，还可参与新的毛细血管产生，因此它们适合治疗患有糖尿病下肢血管疾病的患者。

Pereira 等[15]发现肌肉注射人脐带间充质干细胞（hUC-MSC）可改善小鼠缺血肢体的血流灌注，其可能机制是 hUC-MSC 诱导内皮细胞同时上调转化生长因子-β2（TGF-β2）、血管生成素-2（Ang-2）、FGF 和肝细胞生长因子（HGF）的表达。此外，MSC 具有免疫抑制作用，可以在体内外抑制多种 T 细胞和 NK 细胞活性，下调促炎细胞因子如肿瘤坏死因子-α（TNF-α）等的表达[16]。这些性质使得 MSC 较其他干细胞更加适合应用于下肢

缺血患者的慢性伤口中。黄平平等（未发表数据）目前进行的临床研究，UC-MSC 移植组 13 例与自体动员后的外周血干细胞移植组 23 例初步证明了 UC-MSC 治疗缺血性疾病是安全和有效的方法；比较 HSC 与 UC-MSC 在临床上治疗糖尿病继发周围血管病变的疗效，结果显示均能显著改善临床指标，提高患者的生存质量和降低截肢率；发现 UC-MSC 存在使用范围更广、简便易行的优势，UC-MSC 的疗效有一定优势，但无显著性差异，可能与样本量偏少有关；探索出了 UC-MSC 治疗的细胞数量，每平方米下肢体表面积细胞数约为 $6.29×10^7$cells 和 $9.43×10^7$cells 均有效，高剂量组有较好的疗效。实验表明，hUC-MSC 治疗下肢缺血患者是一种安全的方法，并取得了一定的疗效；hUC-MSC 通过上调血清促血管新生因子水平，抑制高炎症状态和调节淋巴细胞亚群的不平衡来促进血管新生、伤口愈合。Qin 等[5]评估了 UC-MSC 治疗进行过血管成形术后的糖尿病足的效果。在 3 个月的随访期间，相对于对照组的患者，实验组中的患者在皮肤温度、踝肱压力指数、经皮氧压力和跛行距离等方面得到了更大和更稳定的改善，截肢的面积显著减小。

刘瑶等[17]的研究表明，骨髓和脐血单个核细胞采用小腿肌间注射的方法治疗 2 型糖尿病下肢血管病变均安全、有效，脐血单个核细胞的疗效不劣于自体骨髓单个核细胞（BMMC）。羊膜 MSC 也被认为是有潜力的干细胞资源。羊水中含有胎儿生长发育过程中胚胎组织来源的多种细胞，是胎儿 MSC 的丰富来源，但其促进伤口愈合的治疗潜力尚未得到广泛的研究。Kim 等[18]进行的动物实验表明，羊膜 MSC 通过分泌血管生成因子和定植分化能力发挥对慢性伤口的治疗潜力，与 AD-MSC 和真皮成纤维细胞相比，羊膜 MSC 中血管生成因子、胰岛素样生长因子-1（IGF-1）、表皮生长因子（EGF）和白介素-8（IL-8）显著上调，还表现出较高的植入率，并且在伤口区域表达角质形成细胞特异性蛋白和细胞角蛋白，表明对皮肤闭合有直接的促进作用。

最近有两项 I/II 期临床试验评估了 MSC 治疗下肢缺血的安全性及有效性。一项来自印度的随机双盲对照 I/II 期临床试验研究了同种异体 BM-MSC 治疗严重肢体缺血的安全性与有效性[19]，该试验设定了严格的纳入排除标准（表 11-1）。

表 11-1 纳入排除标准[19]

准入标准	排除标准
年龄在 18~60 岁印度男性和没有生育计划的印度女性确诊 CLI，且临床或血流动力学证实 Rutherford 分级为 II-4、III-5 或 III-6；伴有静息痛和/或坏疽的腹股沟以下的动脉闭塞疾病，不适合接受传统的血管重建治疗，或接受传统的血管重建治疗失败 肱踝指数（ABPI）≤0.6，或者踝部压力≤70mmHg，或者足部 $TcPO_2$≤60mmHg 有 2 型糖尿病的患者需要接受药物治疗并控制糖化血红蛋白 ≤8%，不伴有并发症	重症肢体缺血（CLI）患者可以接受外科手术或介入治疗；伴有急性或慢性感染 需接受临近趾骨水平的截肢 排除了 CLI 本身以外的原因干扰步态 1 型糖尿病 患者患有呼吸系统并发症，左室射血分数<25%，或 3 个月内发生过心梗或脑梗 患者有 MRA 检查禁忌证

如图 11-1 所示：在该试验中，共接收 28 名患者，其中 8 名不符合条件而被剔除，剩下的 20 名患者被随机平均分为 2 组，其中一组接受 MSC 治疗，另一组为安慰剂对照组。试验组中有 2 名失访，1 名因单次随访数据缺失被排除，最终实验组共有 9 人，对照组有 10 人纳入最后的统计分析。

第十一章 间充质干细胞在下肢缺血性疾病治疗中的应用

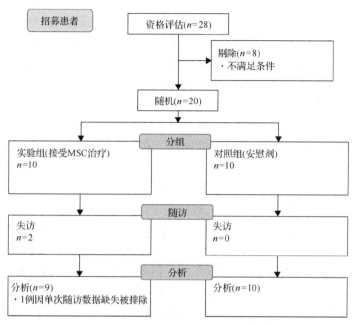

图 11-1 临床试验方案（印度）[19]

该试验的细胞产品要求：形态上在良好培养条件下为纤维细胞状或纺锤形，经过胰蛋白酶作用后为圆形未受损细胞；细胞计数每份在 $18\times10^7 \sim 22\times10^7$ cells，活性在 75% 以上，细胞表型 CD73、CD105、CD90、CD166>80%，CD34、CD45、CD133、CD14、CD19、HLA-DR<5%。

入组患者在接受治疗的 1 天前，治疗后 4 周、12 周、24 周、1 年、2 年接受随访。本研究在 6 个月的随访后揭盲。安全性评估包括：不良反应的发生、心电图评估、血液学及生化检查、定期生命体征的测量及体格检查。有效性评估包括：静息痛的缓解情况，坏疽和溃疡的愈合，踝压增加及肱踝指数（ankle-brachial pressure index，ABPI）增加，预防肢体截肢发生，静息痛评级 0~4 分（0 分为疼痛完全缓解、4 分为经对乙酰氨基酚等非甾体抗炎药治疗后仍不能缓解的疼痛）。由独立的医师对所有目标肢体的坏死及溃疡进行评估，并在每次随访时进行摄像记录。根据方案用激光多普勒来测量静息 ABPI。在研究的最后阶段记录截肢数量，并重点关注接受 BM-MSC 后在随访期内出现的任何不良反应。除了常规的实验室检查，该研究还测定了相关炎症细胞因子来评估细胞治疗引起的免疫改变，包括 IL-2、TNF-α、INF-γ，根据流式结果检查 CD4、CD8 及 CD25。20 名患者被随机分在 4 个研究中心。18 名患者完成了该试验，有 2 名患者因为严重的不良事件退出试验，而导致他们退出试验的原因均是自身疾病进展。在该试验过程中，没有感染、出血或其他有关病原微生物等并发症，也没有因为局部注射而导致的过敏反应及水肿的发生。该研究表明，在 CLI 患者进行异体 BM-MSC 注射是安全的。在注射研究用医学产品（investigational medicinal product，IMP）后的 1 个月、3 个月及 6 个月检测淋巴细胞表面分子（CD4/CD8/CD25），结果表明对照组和试验组在淋巴细胞表型及细胞因子水平上的差异没有统计学意义。这也表明这些异体细胞并没有导致 T 细胞增殖的作用。总之，这些数据表明 CLI 患者接受异体 BM-MSC 注射治疗并没有发生不利的免疫学作用。试验临床疗效

（表 11-2）：①ABPI 及踝压：在接受治疗 24 周后，ABPI 及踝压显著上升，在从基线至 6 个月的随访期内，在 BM-MSC 组中观察到 ABPI 平均改变了 0.22，而安慰剂组中 ABPI 没有观察到改变（$P=0.0018$）；在 BM-MSC 组中观察到踝压平均改变了 18.96mmHg，而安慰剂组相对于基线改变了 3.92mmHg（$P=0.047$）。②静息痛：在随访期内试验组与对照组相对于基线都显示出静息痛分数得到的改善，在 BM-MSC 组和安慰组分别是 2.00 和 3.00（$P=0.1099$）。③溃疡：入组的 13 名患者每名均有 1 处或多处溃疡，7 名患者分到 BM-MSC 治疗组，6 名分到对照组。在两年的随访中，仅 BM-MSC 组的 1 名患者溃疡未愈合，但面积有所减少，两组其他患者溃疡均愈合。④截肢率：4 名患者（每组 2 名）在筛查时即明确需要拇趾截断术的患者实施了拇趾截断术。在试验中，另外的 5 名（BM-MSC 组 3 名，对照组 2 名）采取了截肢手术。其中包括 4 例膝部以上的截肢，每组 2 例，在 BM-MSC 组中 1 例行脚趾截断术。两个组截肢的数量和水平相似。该临床研究支持了使用同种异体 BM-MSC 治疗 CLI 患者的安全性和在改善以 ABPI 及踝压等指标的积极作用，但是一些迫切需要截肢治疗的患者并没有从接受 BM-MSC 治疗中获益，这些患者由于疾病进展及已有的临床数据表明其预后不良。此外，与他们的团队先前做的研究相结合，表明肌肉注射的方式不劣于经动脉注射。

表 11-2　疗效数据[19]

	BM-MSC 组		对照组		P 值
	基线	6 个月	基线	6 个月	
静息痛（中位数）	3	1	3	0	0.1099*
ABPI 平均数（SD）	0.554（0.26）	0.768（0.15）	0.592（0.23）	0.596（0.14）	0.0018

* Kruskal-Wallis 检验

尽管 BM-MSC 在治疗下肢缺血方面有广阔潜力，然而使用自体全骨髓作为 MSC 的来源有其局限性：需要侵入性手术收获这些干细胞，很难大量生产作为干细胞商业产品。因此，人脐血间充质干细胞（hUCB-MSC）成为一个很好的选择。韩国的血管外科中心进行了单中心 hUCB-MSC 治疗外周动脉闭塞性疾病患者的Ⅰ期临床试验[20]，研究纳入了 8 名由 ASO 和血栓闭塞症（thromboangiitis obliterans，TAO）引起 CLI 的患者（表 11-3）。同时通过严格纳入排除标准，使用单剂量 $1×10^7$ cells hUCB-MSC 局部肌内注射单侧缺血肢体，膝盖下方缺血性小腿肌肉上沿胫骨和腓动脉分布的 20 个点注射，并在注射后第 7 天、1 个月、3 个月、6 个月进行随访。

表 11-3　患者基本特征（$n=8$）[20]

病例	年龄/性别	诊断	危险因素和并发症	临床症状	Rutherford 分级	既往治疗史
1	31/男	TAO	既往吸烟史	溃疡（G2）	5	1 次血栓切除术
2	46/男	TAO	既往吸烟史	跛行	3	无
3	72/男	ASO	既往吸烟史 HTN DM CAD	跛行	3	2 次旁路手术
4	56/男	ASO	既往吸烟史	静息痛	4	1 次旁路手术 1 次截肢

续表

病例	年龄/性别	诊断	危险因素和并发症	临床症状	Rutherford 分级	既往治疗史
5	56/男	ASO	吸烟	溃疡（G2）	5	3 次血栓切除术 2 次 PTA 3 次旁路手术 1 次截肢
6	77/男	TAO	既往吸烟史	溃疡（G2）	6	无
7	48/男	TAO	既往吸烟史	跛行	3	1 次血栓切除术
8	44/男	TAO	吸烟	溃疡（G2）	5	无

注：TAO. 血栓闭塞症，ASO. 闭塞性动脉硬化，HTN. 高血压，DM. 糖尿病，CAD. 冠状动脉疾病，PTA. 经皮腔内血管成形术，G2. 皮下组织缺损

以安全性作为Ⅰ期临床试验的主要终点，评估指标包括一系列不良事件的发生，如对 hUCB-MSC 的急性或过敏反应、慢性移植物抗宿主病（GVHD）和心血管事件（急性冠状动脉综合征）或脑血管事件（短暂性脑缺血发作和/或脑卒中）等，并根据临床严重程度分为轻、中、重度。次要终点：1 周、1 个月、3 个月与 6 个月后，肢体状况、足部溃疡愈合程度、踝臂指数、无痛步行距离，用常规血管造影术和独立评审员评分来评估血管生成等参数并与基线相比较。在 6 个月的研究中观察到不良事件 3 例。治疗当天，一名患者口腔轻度溃疡和腹泻。第二个患者经历了一次中度全身荨麻疹，接受抗组胺药一天后，荨麻疹消失。一名患者血清肌酐含量水平升高（基线时为 1.28mg/dL，但在一个具有病史的患者中检测到 1.75mg/dL），这名患者被怀疑有慢性肾脏疾病。最终，通过专家审查和数据审核，所有这些不良事件均被认为与干细胞治疗无关。

在接受肌肉注射后，患者平均踝肱指数（ankle brachial index，ABI）没有改善（基线时 0.51mg/dL，6 个月时 0.57mg/dL，$P>0.05$）。尽管进行跑步机测试时患者的无痛步行距离平均值有所增加（$n=5$，基线时 76.3m，6 个月时 189.4m），但没有统计学意义（$P>0.05$）（图 11-2）。溃疡愈合和截

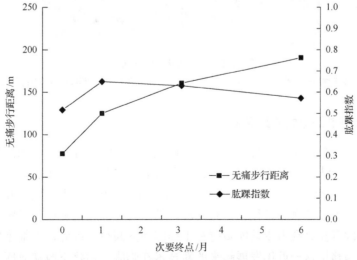

图 11-2 无痛步行距离（PFWD）和肱踝指数（ABPI）[20]

8 名患者只有 4 名测量了无痛步行距离

肢参与者中，有 4 人进入该研究时脚趾有难愈性溃疡，6 个月随访期间内有 3 名患者溃疡完全愈合且没有患者接受截肢（图 11-3）。注射 6 个月后，血管造影结果与基线相比，发现血管生成和动脉生成水平较基线有所增加，其中 8 名患者中的 3 名径流血管再通（图 11-4）。

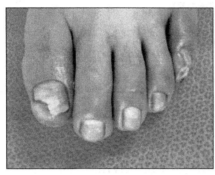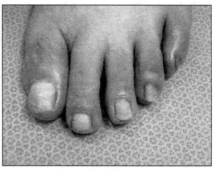

图 11-3　一名患者在 6 个月的随访期内溃疡完全愈合（彩图请扫封底二维码）
左. 未经治疗时第 5 脚趾的难愈溃疡；右. 在接受干细胞治疗 6 个月后第 5 脚趾完全愈合

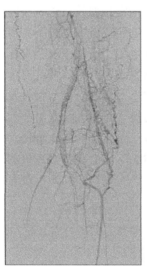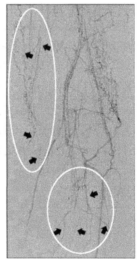

图 11-4　一名患者的血管造影
在接受干细胞治疗术后 6 个月显示与基线血管造影相比侧支血管的形成增加（黑色箭头）

三、MSC 治疗下肢缺血的未来展望

细胞治疗已被全世界公认为是 21 世纪新的临床医疗治疗技术，临床治疗探索正在逐步推进。MSC 治疗与基因治疗一起，已经应用于治疗心脏、脑、下肢缺血等疾病。然而有临床研究表明局部组织中 MSC 数量较少，且在骨髓移植后出现分化，并不能达到很好的疗效，尤其是下肢缺血患者往往伴有高龄、糖尿病、高脂血症或高半胱氨酸血症等症状，这些危险因素会导致内皮功能障碍和干细胞数量减少。因此，干细胞数量稀少和功能受损是干细胞移植这一近年发展起来的新技术在临床上治疗下肢缺血应用中亟待解决的问题。目前国内外开展了相关研究，旨在提高 MSC 的移植率和存活率，增加治疗下肢

缺血的疗效，这些方法包括联合基因、药物、细胞因子进行治疗等方式。

Xiao 等[21]的研究发现，TNF-α 可以提高缺血组织的干细胞移植率。这项研究是将干细胞暴露在不同浓度的 TNF-α 中测量其黏附因子及细胞表面特异标志物，并与动脉内皮细胞一起培养，通过测量细胞黏附率找出最佳的 TNF-α 浓度，将经该最佳浓度 TNF-α 预处理的细胞后注射到小鼠的缺血后肢，结果显示经过合适浓度 TNF-α 处理后的干细胞在缺血组织的聚集比对照组显著性增多，且干细胞表型在处理后并没有发生变化。

目前有各种转基因技术应用于治疗性血管生成，包括利用病毒转染相关基因或直接注入 DNA 质粒等，目前中国报道了经慢病毒转染表达人碱性成纤维细胞生长因子（hbFGF）基因的 MSC 移植存活率增加，移植细胞随后分化为血管内皮细胞，促进体内后肢血管的生成[22]。有报道称 MSC 联合神经轴突导向因子-1（netrin-1）蛋白可以显著提高下肢缺血的血管造影评分及血管密度，改进缺血肢体的功能，并且提高 VEGF 在血浆及受损组织中的水平。进一步进行实验分析指出，MSC 联合 netrin-1 可促进干细胞的迁移并且提高其形成管腔的能力[23]。利用仙台病毒（Sendai viral vector，SeV）作为载体，使用人 Ang-1 修饰的 MSC（SeV-hAng-1-modified MSC）强烈表达 p-Akt，可以增强移植到缺血部位的 MSC 存活能力[24]。

对于合并其他慢性疾病如慢性肾病（chronic kidney disease，CKD）患者，相关的病理生理条件进一步降低了移植 MSC 的存活能力和增殖能力。有学者的研究表明，利用岩藻依聚糖处理脂肪组织来源的 MSC 来改善其移植后增殖能力，并通过黏着斑激酶和磷脂酰肌醇-4,5-二磷酸 3-Akt-激酶上调细胞周期相关蛋白如 cyclin E、CDK2、cyclin D1 和 CDK4 的表达；此外，岩藻依聚糖通过 ERK-IDO-1 信号通路可以增强 MSC 的免疫调节活性[25]。也有研究表明，岩藻依聚糖可以对抗氧化应激作用，通过调节 MnSOD、Akt 调节凋亡相关蛋白和细胞活性氧（ROS）水平来保护 MSC 免受缺血诱导的细胞死亡的影响[26]，以上研究均表明岩藻依聚糖可以作为干细胞治疗下肢缺血的有效辅助剂。

有实验表明，PEP-1-CAT 转导的 MSC 通过 PEP-1-CAT 阻断 H_2O_2 诱导的 PI3K/Akt 的活性，这是调节 MSC 凋亡必需的信号通路，并且可以上调细胞的 SOD 活性，产生抑制 MSC 凋亡的一致作用。在下肢缺血大鼠模型体内，将 PEP-1-CAT 转导 MSC 植入缺血区域的生存能力增加 4 倍，并且 PEP-1-CAT 转导的 MSC 显著上调 VEGF 的表达，成功诱导了缺血部位的血管发生[27]。一些研究表明，VEGF 可以促进血管生成，因此可采用 VEGF 基因修饰的 MSC 作为提高递送 VEGF 的治疗手段。通过遗传修饰过表达 VEGF 的同种异体 MSC 用于缺血部位可以使 VEGF 表现出稳定性下降，其中一小部分 MSC/VEGF 可保留长达 4.5 个月，在这期间安全性得到证明，而有效性还需要更进一步的研究[28]。更令人惊喜的是，最近有研究表明，通过脉冲聚焦超声（pulsed focused ultrasound，pFUS）可以提高 MSC 治疗严重肢体缺血的效果。其机制可能是 pFUS 利用分子反应来促进输注 MCS 在下肢的定位，归巢到经过 pFUS 处理过的肌肉的 MSC 可以表达更多的 VEGF 和 IL-10。实验表明，pFUS 和 MSC 联合应用比单独应用可更为显著地增加缺血下肢的足部血液灌流量和血管密度[28,29]。

如上文所提及，旁分泌机制在干细胞治疗下肢缺血方面起到了重要的作用。在干细胞分泌的产物里，外泌体作为一个重要的旁分泌分子可以重新编程受损细胞。其为由细胞内溶酶体微粒内陷形成的多囊泡体，其外膜与细胞膜融合后释放到胞外基质中。在一项试验中，研究人员使用 iPSC 诱导分化来的 MSC（iMSC），并且经过超滤和纯化手段来得到 iMSC-Exo，经肌肉注射的方式进入小鼠下肢缺血部位可显著增加血管密度及血液灌流量。同时结果显示 iMSC-Exo 可以激活相关血管生成分子的表达，促进脐静脉内皮细胞的迁移、增殖和形成[30]。

此外，血管新生涉及多种血管生长因子和细胞的参与，是一个非常复杂的过程。探索各种血管生长因子在血管新生中的作用机制及和其干细胞之间的相互作用，结合细胞治疗和基因修饰（即对干细胞进行基因修饰），是目前解决从患者血液中分离的祖细胞数量较少和功能存在缺陷问题的重要技术[31]。

尽管干细胞移植已应用在很多领域，尤其是 MSC 有广阔的前景，并为 CLI 患者带来了新的希望，但是目前已有的研究肯定了 MSC 治疗下肢缺血的安全性，其有效性还需要更进一步深入研究。目前临床研究大多数为动物实验，实际应用 MSC 在人体上的试验较少且多为单臂试验，样本数偏小。因此开展多中心、大样本和系统规范的临床研究来进一步确定 MSC 治疗下肢缺血的作用是很有必要的。

（高弘烨　黄平平）

参 考 文 献

[1] Norgren L, Hiatt WR, Dormandy JA, et al. Inter-society consensus for the management of peripheral arterial disease (TASC II). J Vasc Surg, 2007, 45(Suppl S):S5-S67.

[2] 黄平平, 李尚珠, 韩明哲, 等. 自体外周血干细胞移植治疗下肢动脉硬化性闭塞症. 中华血液学杂志, 2003, 24:308-311.

[3] 王彤. 骨髓间充质干细胞临床研究进展. 北京: 人民卫生出版社, 2010.

[4] Tateishi-Yuyama E, Matsubara H, Murohara T, et al. Therapeutic angiogenesis for patients with limb ischaemi by autologous transplantation of bone-marrow cells: a pilot study and a randomised controlled trial. Lancet, 2002, 360:427-435.

[5] Qin HL, Zhu XH, Zhang B, et al. Clinical evaluation of human umbilical cord mesenchymal stem cell transplantation after angioplasty for diabetic foot. Exp Clin Endocrinol Diabetes, 2016, 124:497-503.

[6] Wan J, Xia L, Liang W, et al. Transplantation of bone marrow-derived mesenchymal stem cells promotes delayed wound healing in diabetic rats. J Diabetes Res, 2013, 2013:647107.

[7] Kuo YR, Wang CT, Cheng JT, et al. Bone marrow-derived mesenchymal stem cells enhanced diabetic wound healing through recruitment of tissue regeneration in a rat model of streptozotocin-induced diabetes. Plast Reconstr Surg, 2011, 128:872-880.

[8] Falanga V, Iwamoto S, Chartier M, et al. Autologous bone marrow-derived cultured mesenchymal stem cells delivered in a fibrin spray accelerate healing in murine and human cutaneous wounds. Tissue Eng, 2007, 13:1299-1312.

[9] Han SK, Yoon TH, Lee DG, et al. Potential of human bone marrow stromal cells to accelerate wound healing *in vitro*. Ann Plast Surg, 2005, 55:414-419.

[10] Kim CH, Lee JH, Won JH, et al. Mesenchymal stem cells improve wound healing *in vivo* via early activation of matrix metalloproteinase-9 and vascular endothelial growth factor. J Korean Med Sci, 2011, 26:726-733.

[11] Phinney DG, Prockop DJ. Concise review: mesenchymal stem/multipotent stromal cells: the state of transdifferentiation and modes of tissue repair-current views. Stem Cells, 2007, 25:2896-2902.

[12] Si Y, Zhao Y, Hao H, et al. Infusion of mesenchymal stem cells ameliorates hyperglycemia in type 2 diabetic rats: identification of a novel role in improving insulin sensitivity. Diabetes, 2012, 61:1616-1625.

[13] Dash NR, Dash SN, Routray P, et al. Targeting nonhealing ulcers of lower extremity in human through autologous bone marrow-derived mesenchymal stem cells. Rejuvenation Res, 2009, 12:359-366.

[14] Makrantonaki E, Wlaschek M, Scharffetter-Kochanek K. Pathogenesis of wound healing disorders in the elderly. J Dtsch Dermatol Ges, 2017, 15:255-275.

[15] Pereira AR, Mendes TF, Ministro A, et al. Therapeutic angiogenesis induced by human umbilical cord tissue-derived mesenchymal stromal cells in a murine model of hindlimb ischemia. Stem Cell Res Ther, 2016, 7:145.

[16] Ribeiro A, Laranjeira P, Mendes S, et al. Mesenchymal stem cells from umbilical cord matrix, adipose tissue and bone marrow exhibit different capability to suppress peripheral blood B, natural killer and T cells. Stem Cell Res Ther, 2013, 4:125.

[17] 刘璠, 周慧敏, 杨爱格, 等. 不同来源单个核细胞移植对 2 型糖尿病下肢血管病变的疗效比较. 中华细胞与干细胞: 电子版, 2016, 6:351-355.

[18] Kim CH, Lee JH, Won JH, et al. Mesenchymal stem cells improve wound healing *in vivo* via early activation of matrix metalloproteinase-9 and vascular endothelial growth factor. J Korean Med Sci, 2011, 26:726-733.

[19] Gupta PK, Chullikana A, Parakh R, et al. A double blind randomized placebo controlled phase Ⅰ/Ⅱ study assessing the safety and efficacy of allogeneic bone marrow derived mesenchymal stem cell in critical limb ischemia. J Transl Med, 2013, 11:143.

[20] Yang SS, Kim NR, Park KB, et al. A phase Ⅰ study of human cord blood-derived mesenchymal stem cell therapy in patients with peripheral arterial occlusive disease. Int J Stem Cells, 2013, 6:37-44.

[21] Xiao Q, Wang SK, Tian H, et al. TNF-alpha increases bone marrow mesenchymal stem cell migration to ischemic tissues. Cell Biochem Biophys, 2012, 62:409-414.

[22] Zhang JC, Zheng GF, Wu L, et al. Bone marrow mesenchymal stem cells overexpressing human basic fibroblast growth factor increase vasculogenesis in ischemic rats. Braz J Med Biol Res, 2014, 47:886-894.

[23] Li Q, Yao D, Ma J, et al. Transplantation of MSCs in combination with netrin-1 improves neoangiogenesis in a rat model of hind limb ischemia. J Surg Res, 2011, 166:162-169.

[24] Piao W, Wang H, Inoue M, et al. Transplantation of Sendai viral angiopoietin-1-modified mesenchymal stem cells for ischemic limb disease. Angiogenesis, 2010, 13:203-210.

[25] Lee JH, Ryu JM, Han YS, et al. Fucoidan improves bioactivity and vasculogenic potential of mesenchymal stem cells in murine hind limb ischemia associated with chronic kidney disease. J Mol Cell Cardiol, 2016, 97:169-179.

[26] Han YS, Lee JH, Jung JS, et al. Fucoidan protects mesenchymal stem cells against oxidative stress and enhances vascular regeneration in a murine hindlimb ischemia model. Int J Cardiol, 2015, 198:187-195.

[27] Zhang L, Dong XW, Wang JN, et al. PEP-1-CAT-transduced mesenchymal stem cells acquire an enhanced viability and promote ischemia-induced angiogenesis. PLoS One, 2012, 7:e52537.

[28] Beegle JR, Magner NL, Kalomoiris S, et al. Preclinical evaluation of mesenchymal stem cells overexpressing VEGF to treat critical limb ischemia. Mol Ther Methods Clin Dev, 2016, 3:16053.

[29] Tebebi PA, Kim SJ, Williams RA, et al. Improving the therapeutic efficacy of mesenchymal stromal cells to restore perfusion in critical limb ischemia through pulsed focused ultrasound. Sci Rep, 2017, 7:41550.

[30] Hu GW, Li Q, Niu X, et al. Exosomes secreted by human-induced pluripotent stem cell-derived mesenchymal stem cells attenuate limb ischemia by promoting angiogenesis in mice. Stem Cell Res Ther, 2015, 6:10.

[31] 谷涌泉, 郭连瑞. 下肢血管外科. 北京: 人民卫生出版社, 2010.

第十二章 间充质干细胞在神经损伤修复中的应用

第一节 间充质干细胞与神经损伤修复的概述

神经创伤及神经退行性等疾病导致的神经元损伤后的修复与再生问题，是困扰医学界的难题之一。长期以来，神经细胞被认为缺乏再生能力。20 世纪 90 年代初，Reynolds 等从成年小鼠纹状体中分离出能在体外不断分化增殖、具有多种分化潜能的细胞群，并提出了神经干细胞（NSC）的概念[1,2]。这一发现打破了神经元不能再生的传统观念，为许多难以治疗的神经系统疾病[如帕金森病、缺血性脑病、脊髓损伤（spinal cord injury）等]提供了新的治疗途径。已有大量研究证实成人脑内存在 NSC，在较低程度或少量神经元损伤时，这些脑内现存的 NSC 可以起到自我修复的作用。但当遇到大量神经元死亡时，仅靠自身的 NSC 将无法完成修复，因此就要移植新的干细胞来代替因神经损伤或退行性病变而丧失的神经元[3-5]。

间充质干细胞（MSC）已应用于治疗全身多系统疾病，其对多种系统损伤组织的修复研究已经取得重大突破，如骨损伤、软骨损伤、关节损伤、心脏损伤、肝脏损伤、颅脑损伤（trauma brain injury，TBI）、脊髓损伤等方面，尤其是 MSC 在神经系统损伤修复方面具有长远的发展前景[5]。MSC 具有分化成许多不同类型组织的能力且分化效率不同。目前，研究相对比较多的是骨髓间充质干细胞（BM-MSC），其可以分化成平滑肌细胞、成骨细胞、软骨细胞、心肌细胞和肝细胞，并可一定程度上分化为神经元[6-8]。此外，从脐血和脂肪等组织中分离的 MSC 也具有相似的特征，经诱导可以表达神经元表型，这为 MSC 修复神经损伤提供了理论基础[9-11]。

目前，干细胞与神经损伤修复是基础研究和临床研究的热点。其中 MSC 与神经损伤修复相关的基础研究更为广泛，取得的成果更为丰富，主要包括 MSC 在中枢神经系统（CNS）损伤修复中的应用及 MSC 在周围神经损伤（peripheral nerve injury，PNI）修复中的应用。本章针对 MSC 在 CNS 损伤修复中应用的研究，包括 MSC 移入缺血损伤脑区，MSC 替代损伤的神经细胞，MSC 分泌神经营养因子，MSC 促进内源性神经祖细胞的增殖分化[12-14]等进行介绍。在 PNI 领域，神经再生修复目前主要采用组织工程学方法[10,15-16]。随着对干细胞、生物材料及局部微环境机制等因素研究的深入，干细胞神经组织工程修复 PNI 的效果将大大提高[17-20]。

已有充分临床证据证明，采用 MSC 治疗神经损伤取得了良好的效果，MSC 的临床研究已经在许多国家相继展开，我国自 2016 年起也启动了第一批以干细胞为发展导向的国家科技重点研发计划，随着 MSC 及其相关技术的日益成熟，干细胞将逐渐走进临床应用[28-31]。

第二节 间充质干细胞与神经损伤修复的基础研究

一、MSC 在中枢神经系统损伤修复中的应用

研究表明，MSC 注射到颅内能够迁移到脑和脊髓的各个部分，因此非常适合于作为治疗神经干细胞（CNS）疾病的载体。还有实验证实，MSC 在体内能分化成具有形态学和表型特征的神经胶质细胞与神经元，并且已经证实部分分化细胞具有神经元的电生理特征[32-34]。这一发现无疑对解决 NSC 来源有限这一难题具有重要意义。MSC 中以其多向分化潜能及可作为细胞治疗和基因治疗载体这一特点，日益引起国内外学者的关注[35]。

（一）MSC 移入缺血损伤脑区

血脑屏障（blood-brain barrier）是一道天然的生理屏障，它将 CNS 与其他的系统隔离开。组成血脑屏障的内皮细胞之间形成紧密连接，成为外周细胞入脑的主要障碍[36-38]。虽然可以采用侧脑室或脑实质内定位注射的方法，但是临床操作复杂，推广难。近期，有研究发现抑制 MSC 的 Rho/ROCK 通路能够增强 MSC 的迁移能力，促进 MSC 穿过血脑屏障，而且这种促进作用可能与 MSC 细胞骨架改变和 MSC 引起人脑微血管内皮细胞（human brain microvascular endothelial cell，HBMEC）单层上紧密连接的破坏有关。在神经退行性疾病和炎症反应过程中，趋化因子是介导细胞定向迁移的决定性因素。研究表明，正常的 MSC 能够表达多种细胞因子、趋化因子及其受体，并且肿瘤坏死因子-α（TNF-α）能够上调这些因子的表达。脑缺血后 1~7 天，单核细胞趋化蛋白-1（MCP-1）、巨噬细胞炎性蛋白-1（MIP-1）和白介素-8（IL-8）水平增高，这些细胞因子与 MSC 表达的受体，如 IL-1R、IL-4R、IL-8R 和趋化因子受体等相互作用，促进 MSC 向缺血脑区迁移。移植的 MSC 治疗神经损伤时的时间窗较大多数神经保护剂更长[39,40]。

目前研究的 MSC 治疗 CNS 损伤，主要包括脑损伤、脊髓离断。由车祸等暴力引起的急性损伤，患者群偏年轻化，对这部分人群的生活会产生毁灭性的影响，因此 MSC 如能起到改善功能、促进修复的作用将会给许多濒临绝望的人们带来生活的勇气和希望[41-43]。Jarocha 等[40]报道称某脊髓离断患者经静脉和 Th2-3 髓鞘注射自体 BM-MSC，1 次/3~4 个月，共 5 个疗程后，相较于脊髓离断后 6h，美国脊髓损伤委员会（American Spinal Injury Association，ASIA）评分（该评分分为 A、B、C、D、E 五级，A 是完全损伤，E 是正常）从 A 变成了 C/D。患者能自主控制整个躯干，膀胱感觉功能得到恢复。另外，在脊髓损伤的慢性期，De Almeida 等研究发现 MSC 依然对脊髓的功能恢复有帮助，MSC 作用组的脊髓微结构更完整，营养因子分泌更多，但其修复途径不是分化成神经胶质细胞而是通过旁分泌的方式对损伤部位进行修复[44-47]。Donega 等[48]提出鼻腔给予 MSC 能改善新生小鼠脑由缺氧导致的运动和认知障碍，并认为 CXCL10 在 MSC 向病灶迁移过程中起到了重要的作用。

（二）MSC 分化替代损伤的神经细胞

MSC 能够表达多种神经细胞的标志物，虽然这种表达是微量的，Boido 等[49]认为

MSC 有向神经细胞分化的倾向。近年来关于 MSC 在体外被诱导分化为神经细胞的研究较多，而且有一些令人兴奋的结果[49-51]。有研究将人骨髓间充质干细胞（hBM-MSC）与分化的人 NSC 用膜隔开进行非接触的共培养，发现约 40%的 hBM-MSC 表达 B 微管蛋白-Ⅲ，约 60%表达神经胶质酸性蛋白（glial fibrillary acidic protein，GFAP），该研究认为 hBM-MSC 分泌的一些因子形成的微环境能够诱导 MSC 向神经细胞分化，并认为 MSC 有望被用作神经损伤的替代细胞。Villanueva 等[37]发现 MSC 移植入新生小鼠的脑内能够分化为星形胶质细胞。后来又有人将 MSC 先用含神经生长因子（nerve growth factor，NGF）培养基培养，后移植入短暂大脑中动脉闭塞（middle cerebral artery occlusion，MCAO）模型大鼠脑内，大鼠的运动功能明显恢复，且免疫组化显示，大部分 MSC 位于缺血区，只有一小部分 MSC 表达神经元标志物 NeuN、MAP2 和星形胶质细胞标志物 GFAP，但是这些细胞是球形的，并没有整合到周围的环境中，因此作者认为这些细胞并不参与神经传导。因此，MSC 替代损伤的神经细胞并不能完全解释其移植后能促进脑功能的恢复，需进一步研究[49,50]。

人胎盘间充质干细胞（hP-MSC）被诱导成 NSC 后移植到大鼠脊髓损伤的局部，可使脊髓损伤模型大鼠运动及感觉功能有所恢复，而且没有移植排斥反应发生。P-MSC 具有与 BM-MSC 类似的功能和特性，两者的多向分化潜能也类似，且 P-MSC 的增殖能力更强。相对于其他来源的 MSC，P-MSC 容易获得，对供者无损伤，不涉及伦理问题，作为细胞移植的种子细胞来源更具优势[38,51-53]。另外，hP-MSC 负载于人羊膜上，用于治疗桡神经损伤取得良好效果。有实验证实了 hP-MSC 具有向神经元样细胞分化的能力，诱导成 NSC 后，能够趋化至脊髓损伤神经断端，修复脊髓损伤，神经电生理检测及 BBB 运动功能评分（Basso, Beattie, and Bresnahan Locomotor Rating Scale, BBB Scale）证实模型大鼠运动及感觉功能有所恢复。该研究成果为 P-MSC 作为一种新的种子细胞应用于脊髓损伤的细胞治疗提供了实验数据和理论依据[54-56]。

（三）MSC 分泌神经营养因子

神经营养因子是近年来研究人员非常关注的特异蛋白分子。神经营养因子不仅调节发育过程中的神经元存活，而且能够阻止成年神经损伤后的神经元死亡，调节突触可塑性和神经递质传递等许多神经系统功能[55,57]。它包括 NGF 家族、胶质细胞源性神经营养因子（glial cell line-derived neurotrophic factor，GDNF）家族和其他 NGF。NGF 不仅具有 CNS 营养因子的作用，同时能促进受损神经元再生，改善神经元的病理状态，而且对缺血缺氧导致的神经损伤具有重要的保护和修复作用[58-60]。BM-MSC 与宿主脑相互作用导致自分泌和旁分泌神经营养因子增加，这有助于神经功能的康复。研究发现，用创伤性损伤大鼠脑提取物培养人 BM-MSC，则 BM-MSC 表达脑源性神经营养因子（brain-derived neurotrophic factor，BDNF）、碱性成纤维细胞生长因子（bFGF）、NGF、血管内皮生长因子（VEGF）显著增加，而且这种增加既取决于培养时间的长短，也取决于脑创伤组织提取时间的早晚，因此可以认为 BM-MSC 是通过产生多种细胞因子和营养因子发挥其神经保护作用的。Dimos 等研究发现，BM-MSC 移植入 MCAO 模型大鼠脑后，在缺血后的早期阶段（4~5 天）局部脑血流量和 BBB 评分恢复至接近正常水平，且这种作用与 BM-MSC 的剂量有关：在移植大鼠检测到转化生长因子-β（TGF-β），神经营养因子家族的表达，显示移植 MSC 后局部脑血流量和 BBB 评分的早期恢复可介导脑卒中动物功能的康复。NGF 和 BDNF 均可通过跨胞浆膜上的 Trk 受体与 P75 糖蛋白对神经元起作用。NGF 由 TrkA 受体介导，P75 可增强 TrkA 信号通路[61-63]。催化型 Trk 是介导 BDNF 生物学活性的受体，该受体介导将 BDNF 逆行转运至神经元胞体中。神经营养因子与 Trk 蛋白结合后启动一系列反应，激活 Raf-MAPK-ERK 通路，使酪氨酸残基磷酸化，产生信号传递；提高自由基清除剂[如过氧化物酶、超氧化物歧化酶（SOD）、谷胱甘肽过氧化酶]的活性，稳定细胞内 Na^+/K^+ 浓度比；改变兴奋性氨基酸（excitatory amino acid, EAA）NMDA 受体对 EAA 的反应性，缓解 EAA 的毒性；诱导钙结合蛋白表达，影响钙通道及钙排除系统的表达和活化，使细胞内钙浓度保持稳定，发挥神经保护作用[64,65]。

人脐带间充质干细胞（hUC-MSC）除分泌神经生长相关细胞外基质（ECM）外，还能够分泌多种 NGF，包括 BDNF、神经营养因子 3（neurotrophin 3）、神经营养因子 4/5、GDNF 等，通过体外实验证实 hUC-MSC 制备的条件培养基能够促进 NSC 增殖及背根神经节轴突生长。已有研究证实，将 hUC-MSC 注射到神经损伤段内只有极少数细胞可以自然分化为 NSC，因此一些研究者认为干细胞修复神经损伤的主要机制不是干细胞的分化，而是分泌一些 NGF 和 ECM，构建一个适合轴突再生的微环境，从而促进神经的再生[66-68]。

（四）MSC 促进内源性神经祖细胞的增殖分化

存在于脑室区/脑室下区的神经祖细胞可在发育的大脑中迁移，但在健康成年脑中未见前脑神经元的生成，不是由于缺乏合适的神经祖细胞，而是缺乏细胞有丝分裂后的营养和迁移支持[69]。脑缺血时脑室区/脑室下区神经祖细胞迁移到受损的 CNS 组织，显示一种限制和修复损伤的适应性反应。MSC 可能为缺血损伤的脑组织提供促分化因子，由此加强成年个体 CNS 内来自 NSC 和神经祖细胞的细胞增殖、分化和迁移[70]。在研究神经祖细胞迁移到受损神经组织时发现，干细胞因子（SCF）mRNA 及其蛋白在受损脑区高度表达，SCF 受体 c-Kit 在干细胞的体内和体外实验中均显示有表达，认为 SCF/c-Kit 通路有助于 NSC 迁移到缺血脑区。Li 等经静脉输注 hBM-MSC 治疗大鼠脑缺血，显示在缺血侧脑室下带 BrdU 阳性细胞增殖，且这些细胞部分表达 BDNF 和 NGF，并伴有缺血周边区凋亡细胞减少和神经功能康复[71,72]。

未分化的 MSC 移植在脊髓损伤中已显示出有积极的治疗作用。脊髓损伤两天后，移植的成体 BM-MSC 表现出神经保护作用。相比对照组，研究组损伤部位髓鞘明显增加[73]。组织学分析显示，随着轴突延伸，MSC 数量与层粘连蛋白表达增加，且排列方向与脊髓方向一致[38,39]。但使用 BBB 评分功能分析表明，MSC 治疗组和对照组之间没有显著差异。结果表明，MSC 表达层粘连蛋白并有助于突起排列。如上所述，MSC 也可分化成髓鞘细胞。成年大鼠局部脱髓鞘病变后 3 天，未分化的 MSC 有助于提高整个病变的电传导速度。用含有交感特征的髓鞘形成的 MSC 治疗脊髓轴突髓鞘，提示伤口内有 MSC 分化和髓鞘形成。在大鼠脊髓脱髓鞘病变中，直接注射和静脉内给予 MSC，组织学检查提示髓鞘发生呈剂量依赖性。这项研究还表明了 MSC 静脉内给药的有效性，且创伤小[65]。MSC 也有助于脑损伤模型的恢复。在原位环境中，联合移植 MSC 和 NSC 到海马，大多数的 NSC 分化成少突胶质细胞。与此相反，单独 NSC 移植大部分分化为星形胶质细胞。体内研究结果亦是如此。核磁共振成像（MRI）和皮质血流量（cerebral blood flow, CBF）研究表明，干细胞脑内注射显著减少心室扩张并帮助维持靠近损伤部位的 CBF。相较于注射生理盐水，MSC 治疗的大鼠的神经功能缺损评分和莫里斯水迷宫测试表现出显著的功能改进[74,75]。

二、MSC 在周围神经损伤修复中的应用

周围神经损伤（PNI）的发病率逐年上升，且 PNI 修复的过程非常复杂，神经损伤后生长效果不理想。目前主要采用组织工程的方法来修复 PNI。同 MSC 在 CNS 中的功能相似，MSC 在 PNI 中可以分化为神经系细胞，还能分泌酸性钙结合蛋白 S100、GFAP、NGF 等，为临床 PNI 修复提供了新的思路[76,77]。

PNI 的再生和修复过程既复杂又漫长，其在断端进行一系列的分子和细胞变化，称为沃勒变性。施万细胞（Schwann cell, SC）下调结构蛋白，如蛋白质 O（PO）、髓鞘基本蛋白、髓磷脂相关蛋白[78]。与此同时，这些细胞上调可以促进轴突生长的蛋白及因子，如 NGF、BDNF、胶质源性神经营养因子、bFGF、神经营养因子 3 等。随着 SC 和巨噬细胞联合将崩解的产物消除，SC 可以沿着基底膜管纵行排列，相互结合，形成一条 Büngner 带。从损伤轴突近端终末发出新的轴突芽，再生的轴突芽能沿 Büngner 带生长，与靶细胞重新建立关系并延伸达到远端靶器官[77]。

脂肪间充质干细胞（AD-MSC）是从脂肪组织中分离的一类多能干细胞，具有来源广泛、取材方便、便于自体移植等优点，因此日益受到研究者的关注[32,46]。目前，在特定的条件下人们已成功地将其诱导分化为脂肪、成骨、软骨等多种细胞。近几年报道 AD-MSC 在一定的微环境中能够向神经细胞分化。QIAN 等利用特定的诱导剂作用于 AD-MSC 后，能够检测到神经元标志物 β-微管蛋白Ⅲ（β-tubulin Ⅲ）和微管相关蛋白 2（microtubule-associated protein 2, MAP2），因此，人们将 AD-MSC 作为种子细胞，应用到组织工程修复 PNI。研究者将 AD-MSC 与纤维蛋白生物导管复合并移植入大鼠 10mm 坐骨神经缺损处，术后 2 周观察，AD-MSC 导管移植组轴突再生的距离和近端 SC 迁移的距离都明显高于空白导管组，进而证实 AD-MSC 可以促进周围神经的再生[79,80]。Aberg 等将从脂肪中分离获得的 AD-MSC，培养在含有 β-巯基乙醇、视黄酸、bFGF、血小板源生长因子（PDGF）、神经调节蛋白-β1、表皮生长因子（EGF）等成分的诱导液中，能分化为表达神经生长因子受体（nerve growth factor receptor, NGFR）p75 和整合蛋白 B4 等标志物的 SC 样细胞。同时将 AD-MSC 和分化 AD-MSC 分别植入硅胶导管修复 10mm 缺损坐骨神经处，手术 6 个月后，大鼠行走轨迹（walking track）、神经传导速度（nerve conduction velocity, NCV）、髓鞘纤维密度均比单一的硅胶导管组有显著的提高[81]。Benraiss 等将 AD-MSC 通过尾静脉注入鼠体内后研究其对损伤的坐骨神经的修复，结果发现，坐骨神经功能指数（sciatic nerve functional index, SFI）、神经纤维再生数量均较磷酸盐缓冲液（phosphate buffer saline, PBS）组（对照组）明显提高。在不同的微环境下，AD-MSC 可以对 PNI 进行较好的修复，其来源丰富、取材方便、是临床构建组织工程化神经最理想的种子细胞之一[82,83]。

BM-MSC 来源于发育早期的中胚层和外胚层，在体内或体外特定的诱导条件下，可以向脂肪、成骨、肌肉及神经等多种组织细胞分化[84,85]。BM-MSC 具有较强的分化潜能和可自体移植等优点，亦成为 PNI 修复的理想种子细胞。研究者利用含 β-巯基乙醇、视黄酸、bFGF 等的诱导液培养 BM-MSC，通过免疫组化发现，诱导过的细胞可以分化为表达 P0、NGFR p75、GFAP、L1、O4、S100 等标志物的 SC 样细胞[42]。随后将诱导过的细胞注入反渗透导管中一起移植入 10mm 缺损坐骨神经处，8 周后在近端和远端均等检测到再生的神经纤维，且 SFI 具有极明显的提高，进而证实 BM-MSC 促进损伤坐骨神经的再生与修复。Mori 等[71]研究坐骨神经损伤后在肌肉中注射 BM-MSC 的其修复作用，结果发现，缺损部位再生神经纤维和脊髓前角神经元数量均较对照组明显提高，再生髓鞘厚度也显著增加。将 BM-MSC 与胶原-壳聚糖神经支架复合后修复大鼠 15mm 坐骨神经缺损，术后 4 周观察，实验大鼠未出现明显的全身炎症反应和排斥反应，之后通过多种研究方法综合评估其修复效果，显示桥接段的远端能够检测到表达 NF160 和 S100 阳性蛋白的再生神经纤维，并且脊髓前角运动神经元和背根神经节感觉神经元数量比空白组显著增加。Ulrich 等[86]将从大鼠骨髓中分离获得的 BM-MSC 植入纤维蛋白胶导管，修复 12mm 鼠坐骨神经缺损，结果表明其疗效显著优于单纯导管组。

20 世纪 90 年代初，研究者在羊水中发现一种体外增殖能力强、可以向多种细胞分化、没有免疫排斥、致瘤性低的干细胞——羊水干细胞（amniotic stem cell，AFSC），其比成体干细胞更原始，全能性更接近胚胎干细胞（ESC）[3,7]。目前 AFSC 作为种子细胞在心血管疾病、肝脏疾病等医学领域具有重要的价值。Yang 等建立 SD（Sprague-Dawley）大鼠坐骨神经损伤模型并静脉注射 AFSC 来研究其促进损伤神经再生的作用，结果发现神经再生标志物基质细胞衍生因子-1α（SDF-1α）及其受体 CXCR4 的表达均显著增高，SFI 和再生髓鞘数量明显增加，神经肌肉动作电位的潜伏期短于对照组[87,88]。通过免疫组化检测出 AFSC 能够表达 BDNF、睫状神经营养因子（ciliary neurotrophic factor，CNTF）、S100 等蛋白。最终证实 AFSC 能够促进周围神经的再生。Wetzig 等[89]报道 AFSC 与高压氧联合移植入受损的坐骨神经处，S100 蛋白的表达上调，并且 SFI、踝关节角度均较其他组明显增高，进而表明 AFSC 与高压氧联合应用可促进轴突的再生，恢复局部神经功能。

NSC、AD-MSC、BM-MSC、髓系抑制性细胞（MDSC）、诱导多能干细胞（iPSC）、DPSC 及 AFSC 对 PNI 的再生与修复均具有明显的促进作用[83]。虽然这些干细胞各自都存在优缺点，并且它们修复神经缺损的作用机制目前仍不清楚，但是随着对干细胞研究的深入，这些困难终究会得到解决[56,61]。因此需进一步研究干细胞对神经损伤的修复效果，发现和阐明干细胞的作用机制，为干细胞移植应用于修复 PNI 提供更坚实的理论基础[90]。

第三节 间充质干细胞与神经损伤修复的临床研究

已有充分临床证据证明采用 MSC 治疗神经损伤有良好的效果，应加强临床推广[91,92]。MSC 的临床研究已经在许多国家开展，随着 MSC 基础研究的推进及相关技术的日益成熟，我国也批准了多项临床试验，走入了 MSC 核心技术研发的舞台[31,93]。

颅脑损伤（TBI）不仅具有较高的患病率和致死率，而且是继发性癫痫发病的一个重要原因，严重影响患者的生活质量。在我国，TBI是40岁以下成年人创伤性死亡的主要原因[31,41,45]。即使有幸存活下来，一部分TBI也会存在严重的神经功能缺失和行为能力障碍，给家庭和社会造成极大的负担。人MSC（hMSC）被认为是一个很好的治疗脑损伤的备选细胞。在实验和临床研究中BM-MSC是目前使用最广泛的种子细胞。然而抽取骨髓是一个高度侵入性的过程，存在疼痛和感染的风险，而且BM-MSC分化的潜能和扩增能力都会随着年龄的增加而降低[94]。因此寻找BM-MSC的替代细胞已经成为一个研究方向。来自胎盘、胎膜、羊水或胎儿组织MSC在细胞数量、扩张潜力和分化能力上都较成体组织来源的MSC更高。同时它们当中的一些细胞已经被用来研究治疗神经退行性疾病和神经创伤性疾病[95]。尽管胎盘来源的某些MSC已经被用来研究对实验动物某些神经系统疾病模型的治疗作用，但是目前仍不清楚围生期组织来源的各种MSC对神经系统疾病的治疗作用是否相同，所以相关的临床研究应用较少。已开展的试验侧重于对MSC治疗神经损伤修复的有效性及安全性进行评价[49,96]。

脑源性神经营养因子（BDNF）是一种蛋白质，在脑组织内合成，广泛分布于CNS内，对神经细胞的生长发育、分化等起着十分重要的作用，同时可以改善神经细胞的病理状态、防止神经细胞受损而死亡、促进受损神经细胞的再分化和再生长发育等，也是成熟神经中枢和周围神经系统的神经细胞生存及发挥正常生理功能所必需的营养物质[81,84]。临床研究发现，BDNF可通过自分泌、靶源性等方式与神经细胞TrkB受体结合，发挥生物学等作用，对神经细胞的分化、发育等起到关键支持功能，并在记忆和学习等功能中承担着不可缺少的重要角色。胶质细胞源性神经营养因子（GDNF）由Lin等[93]首次从大鼠神经胶质细胞系B49的培养液中分离、培养、纯化所得，并于近年用于临床。目前已在多种神经细胞和神经相关细胞的培养中发现GDNF的表达，具有靶源性神经营养因子的功能，对损伤神经细胞的修复功能有调节作用，在交感神经和感觉神经的生存方面具有一定的支持作用，促进神经细胞的分化，对轴突的伸展方向有一定的决定作用。另外有临床研究表明，GDNF对多种神经元有营养支持功能，对缺血性带来的损害有保护作用[91]。

临床研究当中，对不同MSC疗效的比较，应该着重于体外神经分化潜能，以及比较它们的形态、体外扩增能力、免疫表型和多能性，哪种细胞更容易分化成NSC，它就更适合作为治疗TBI的种子细胞。细胞移植治疗TBI，除了细胞替代以外，移植细胞提供的旁分泌功能也起到一定的作用[4,37,91]。MSC被移植到脑损伤区域后可以提高损伤局部的神经营养因子浓度，这对受损细胞的存活和分化具有重要的意义。诸多实验使用人细胞因子抗体芯片检测多种来源MSC的细胞因子表达[5,16]。同时在诱导MSC向神经干分化的过程中，去检测对神经系统发育和重建具有关键作用的5个神经营养因子。它们是BDNF、NGF、神经营养因子3、GDNF和CNTF。这一结果将在一定程度上对TBI损伤修复中种子细胞的选择提供一定的参考，尤其是在神经营养因子分泌方面。同时这也反映了NSC诱导过程对不同细胞分泌神经营养因子能力的影响。另外，移植入体内干细胞的存活能力也是干细胞移植治疗需要考虑的重要方面。MSC的移植治疗效果往往被植入细胞的大量损失所限制，细胞死亡是由损伤部位的恶劣微环境所引发的。抗凋亡能力强的MSC更适合体内移植治疗TBI[39]。

MSC 移植包括自体移植和异体移植。自体移植是指从患者体内提取，一般需要传代 3~6 次才能获得临床治疗需要的细胞，准备时间长达 4 周以上，此特点使异体甚至异种移植成为趋势。研究发现，人来源的 AD-MSC 对大鼠缺血缺氧性脑病具有较好的治疗效果，且与同种来源 AD-MSC 的治疗效果无明显差异，为临床异体移植提供了理论支持[96]。MSC 注射途径主要有颅内定位注射、鞘内注射和静脉注射 3 种。颅内定位注射 MSC 的归巢细胞数量多，但存在不能大剂量注射、易引起继发性损伤等缺点。鞘内注射相对风险较小，但存在操作烦琐、容易导致脊髓损伤等不足[27,86]。静脉注射安全、快捷，可大剂量注射，但存在干细胞归巢数量少、部分细胞被肺毛细血管网捕获的缺点。临床上需要结合患者的特点，选择合适的注射途径。初步的临床试验给予了我们很多提示，但目前仍存在一些亟待解决的问题，如最佳注射途径、治疗的时间窗、作用机制、与肿瘤的关系等，以及受到法律和伦理的限制，临床研究需要更多支持与动力才能继续深入[19,94]。

总之，MSC 因取材方便、体外扩增速度快、多向分化潜能及体内外均可稳定表达外源性目的基因、排斥反应小等优势，已成为治疗 TBI、SCI、帕金森病、脑卒中、新生儿缺氧缺血性脑病、溶酶体贮积病等累及神经系统疾病的种子细胞[97]。根据初步的临床报告，MSC 对这些疾病的治疗疗效令人鼓舞。随着对其作用机制研究的不断深入，在其细胞治疗和基因工程中将有广阔的应用前景。

（陈旭义　牛学刚）

参 考 文 献

[1] Reynolsd BA, Weiss S. Generation of neurons and astrocytes from isolated cells of the adult mammalian central nervous system. Science, 1992, 255:1707-1710.

[2] Iwashita Y, Blakemore WF. Areas of demyelination do not attract significant numbers of schwann cells transplanted into normal white matter. Glia, 2015, 31(3):232-240.

[3] Jin YQ, Liu W, Hong TH, et al. Efficient Schwann cell purification by differential cell detachment using multiplex collagenase treatment. Journal of Neuroscience Methods, 2008, 170(1):140-148.

[4] Tang YY, Guo WX, Lu ZF, et al. Ginsenoside Rg1 promotes the migration of olfactory ensheathing cells via the PI3K/Akt pathway to repair rat spinal cord injury. Biological & Pharmaceutical Bulletin, 2017, 40(10):1630-1637.

[5] Keyvanfouladi N, Raisman G, Li Y. Functional repair of the corticospinal tract by delayed transplantation of olfactory ensheathing cells in adult rats. Journal of Neuroscience the Official Journal of the Society for Neuroscience, 2003, 23(28):9428-9234.

[6] Savaskan NE, Weinmann O, Heimrich B, et al. High resolution neurochemical gold staining method for myelin in peripheral and central nervous system at the light- and electron-microscopic level. Cell & Tissue Research, 2009, 337(2):213.

[7] Cho SW, Kim IK, Kang JM, et al. Evidence for in vivo growth potential and vascular remodeling of tissue-engineered artery. Tissue Eng Part A, 2009, 15(4):901-912.

[8] Doucette R. Glial influences on axonal growth in the primary olfactory system. Glia, 2010, 3(6):433-449.

[9] Han MH, Piao YJ, Guo DW, et al. The role of Schwann cells and macrophages in the removal of myelin during Wallerian degeneration. Acta Histochemica et Cytochemica Official Journal of the Japan Society of Histochemistry & Cytochemistry, 2010, 22(2):161-172.

[10] Costa CCD, Laan LJWVD, Dijkstra CD, et al. The role of the mouse macrophage scavenger receptor in myelin phagocytosis. European Journal of Neuroscience, 2010, 9(12):2650-2657.

[11] Alexanian AR. An efficient method for generation of neural-like cells from adult human bone marrow-derived mesenchymal stem cells. Regenerative Medicine, 2010, 5(6):891-900.

[12] Sasaki M, Honmou O, Akiyama Y, et al. Transplantation of an acutely isolated bone marrow fraction repairs demyelinated adult rat spinal cord axons. Glia, 2010, 35(1):26-34.

[13] Blakemore WF, Crang AJ, Curtis R. The interaction of Schwann cells with CNS axons in regions containing normal astrocytes. Acta Neuropathologica, 1986, 71(3-4):295-300.

[14] Bachelin C, Zujovic V, Buchet D, et al. Ectopic expression of polysialylated neural cell adhesion molecule in adult macaque Schwann cells promotes their migration and remyelination potential in the central nervous system. Brain, 2010, 133(2):406-420.

[15] Lavdas AA, Franceschini I, Dubois-Dalcq M, et al. Schwann cells genetically engineered to express PSA show enhanced migratory potential without impairment of their myelinating ability *in vitro*. Glia, 2010, 53(8):868-878.

[16] Papastefanaki F, Chen J, Lavdas AA, et al. Grafts of Schwann cells engineered to express PSA-NCAM promote functional recovery after spinal cord injury. Brain, 2007, 130(8):2159-2174.

[17] Hess JL, Glatt SJ. How might ZNF804A variants influence risk for schizophrenia and bipolar disorder? A literature review, synthesis, and bioinformatic analysis. American Journal of Medical Genetics Part B: Neuropsychiatric Genetics, 2013, 165(1):28-40.

[18] Stratton JA, Kumar R, Sinha S, et al. Purification and characterization of Schwann cells from adult human skin and nerve. Eneuro, 2017, 4(3). pii: ENEURO.0307-16.2017.

[19] Coleman J, Yost A, Goren R, et al. Nonthermal atmospheric pressure plasma decontamination of protein-loaded biodegradable nanoparticles for nervous tissue repair. Plasma Medicine, 2011, 1(3-4):215-230.

[20] Nuria G, Pérez-López V, Sanz-Jaka JP, et al. Age-dependent depletion of human skin‐derived progenitor cells. Stem Cells, 2010, 27(5):1164-1172.

[21] Langford LA, Porter S, Bunge RP. Immortalized rat Schwann cells produce tumours *in vivo*. Journal of Neurocytology, 1988, 17(4):521-529.

[22] Lakatos A, Franklin RJ, Barnett SC. Olfactory ensheathing cells and Schwann cells differ in their *in vitro* interactions with astrocytes. Glia, 2015, 32(3):214-225.

[23] Imaizumi T, Lankford KL, Burton WV, et al. Xenotransplantation of transgenic pig olfactory ensheathing cells promotes axonal regeneration in rat spinal cord. Nature Biotechnology, 2000, 18(9):949-953.

[24] Meyer M, Matsuoka I, Wetmore C, et al. Enhanced synthesis of brain-derived neurotrophic factor in the lesioned peropheral nerve. Journal of Cell Biology, 1992, 119(1):45-54.

[25] Gilmore SA, Duncan D. On the presence of peripheral-like nervous and connective tissue within irradiated spinal cord Anatomical Record Advances in Integrative Anatomy & Evolutionary Biology, 2010, 160(4):675-689.

[26] Ramóncueto A, Cordero MI, Santosbenito FF, et al. Functional recovery of paraplegic rats and motor axon regeneration in their spinal cords by olfactory ensheathing glia. Neuron, 2000, 25(2):425-435.

[27] Funk D, Fricke C, Schlosshauer B. Aging Schwann cells *in vitro*. European Journal of Cell Biology, 2007, 86(4):207-219.

[28] Ruitenberg MJ, Plant GW, Hamers FP, et al. *Ex vivo* adenoviral vector-mediated neurotrophin gene transfer to olfactory ensheathing glia: effects on rubrospinal tract regeneration, lesion size, and functional recovery after implantation in the injured rat spinal cord. Journal of Neuroscience, 2003, 23(18):7045-7058.

[29] Joannides AJ, Chandran S. Human embryonic stem cells: an experimental and therapeutic resource for neurological disease. Journal of the Neurological Sciences, 2008, 265(1):84-88.

[30] Lankford KL, Sasaki M, Radtke C, et al. Olfactory ensheathing cells exhibit unique migratory, phagocytic, and myelinating properties in the X-irradiated spinal cord not shared by Schwann cells. Glia, 2010, 56(15):1664-1678.

[31] Sasaki M, Lankford KL, Zemedkun M, et al. Identified olfactory ensheathing cells transplanted into the transected dorsal funiculus bridge the lesion and form myelin. Journal of Neuroscience the Official Journal of the Society for Neuroscience, 2004, 24(39):8485-8493.

[32] Au E, Richter MW, Vincent AJ, et al. SPARC from olfactory ensheathing cells stimulates Schwann cells to promote neurite outgrowth and enhances spinal cord repair. Journal of Neuroscience, 2007, 27(27):7208-7221.

[33] Fouad K, Schnell L, Bunge MB, et al. Combining Schwann cell bridges and olfactory-ensheathing glia grafts with chondroitinase promotes locomotor recovery after complete transection of the spinal cord. Journal of Neuroscience the Official Journal of the Society for Neuroscience, 2005, 25(5):1169.

[34] O'Donovan KJ, Ma K, Guo H, et al. B-RAF kinase drives developmental axon growth and promotes axon regeneration in the injured mature CNS. Journal of Experimental Medicine, 2014, 211(5):801-814.

[35] Rogers C, Moody SA, Casey E. Neural induction and factors that stabilize a neural fate. Birth Defects Research Part C Embryo Today Reviews, 2010, 87(3):249-262.

[36] Maden M. Retinoic acid in the development, regeneration and maintenance of the nervous system. Nature Reviews Neuroscience, 2007, 8(10):755.

[37] Villanueva S, Glavic A, Ruiz P, et al. Posteriorization by FGF, Wnt, and retinoic acid is required for neural crest induction. Developmental Biology, 2002, 241(2):289-301.

[38] Jessell TM. Neuronal specification in the spinal cord: inductive signals and transcriptional codes. Nature Reviews Genetics, 2000, 1(1):20-29.

[39] O'Donovan KJ, Ma K, Guo H, et al. B-RAF kinase drives developmental axon growth and promotes axon regeneration in the injured mature CNS. Journal of Experimental Medicine, 2014, 211(5):801-814.

[40] Jarocha D, Lukasiewicz E, Majka M. Adventage of mesenchymal stem cells [MSC] expansion directly from purified bone marrow CD105plus and CD271plus cells. Folia Histochemica Et Cytobiologica, 2008, 46(3):307-314.

[41] Le DN, Calloni GW, Dupin E. The stem cells of the neural crest. Cell Cycle, 2008, 7(8):1013-1019.

[42] Nagoshi N, Shibata S, Nakamura M, et al. Neural crest-derived stem cells display a wide variety of characteristics. Journal of Cellular Biochemistry, 2010, 107(6):1046-1052.

[43] Li XJ, Du ZW, Zarnowska ED, et al. Specification of motoneurons from human embryonic stem cells. Nature Biotechnology, 2005, 23(2):215.

[44] De Almeida SM, Teive HA, Brandi I, et al. Fatal *Bacillus cereus* meningitis without inflammatory reaction in cerebral spinal fluid after bone marrow transplantation. Transplantation, 2003, 76(10):1533-1534.

[45] Jha BS, Rao M, Malik N. Motor neuron differentiation from pluripotent stem cells and other intermediate proliferative precursors that can be discriminated by lineage specific reporters. Stem Cell Reviews, 2015, 11(1):194.

[46] Niknejad H, Mirmasoumi M, Torabi B, et al. Near-IR absorbing quantum dots might be usable for growth factor-based differentiation of stem cells. Journal of Medical Hypotheses & Ideas, 2015, 9(1):24-28.

[47] Rathjen J, Rathjen PD. Formation of neural precursor cell populations by differentiation of embryonic stem cells *in vitro*. The Scientific World Journal, 2014, 2:690.

[48] Donega V, Nijboer CH, Velthoven CTJV, et al. Assessment of long-term safety and efficacy of intranasal mesenchymal stem cell treatment for neonatal brain injury in the mouse. Pediatric Research, 2015, 78(5):520-526.

[49] Boido M, Rupa R, Garbossa D, et al. Embryonic and adult stem cells promote raphespinal axon outgrowth and improve functional outcome following spinal hemisection in mice. European Journal of Neuroscience, 2010, 30(5):833-846.

[50] Perrin FE, Boniface G, Serguera C, et al. Grafted human embryonic progenitors expressing neurogenin-2 stimulate axonal sprouting and improve motor recovery after severe spinal cord injury. PLoS One, 2010, 5(12):e15914.

[51] Sun LT, Yamaguchi S, Hirano K, et al. Nanog co-regulated by Nodal/Smad2 and Oct4 is required for pluripotency in developing mouse epiblast. Developmental Biology, 2014, 392(2):182-192.

[52] Llewellyn KJ, Nalbandian A, Lan NW, et al. Myogenic differentiation of VCP disease-induced pluripotent stem cells: a novel platform for drug discovery. PLoS One, 2017, 12(6):e0176919.

[53] Malaekehnikouei B, Gholami L, Asghari F, et al. Viral vector mimicking and nucleus targeted nanoparticles based on dexamethasone polyethylenimine nanoliposomes: preparation and evaluation of transfection efficiency. Colloids & Surfaces B Biointerfaces, 2018, 165:252.

[54] Kaji K, Norrby K, Paca A, et al. Virus-free induction of pluripotency and subsequent excision of reprogramming factors. Nature, 2009, 458(7239):771-775.

[55] Lee CH, Kim JH, Lee HJ, et al. The generation of iPS cells using non-viral magnetic nanoparticle based transfection. Biomaterials, 2011, 32(28):6683-6691.

[56] Chen W, Tsai PH, Hung Y, et al. Nonviral cell labeling and differentiation agent for induced pluripotent stem cells based on Mesoporous Silica Nanoparticles. Acs Nano, 2013, 7(10):8423.

[57] Okita K, Nakagawa M, Hong H, et al. Generation of mouse induced pluripotent stem cells without viral vectors. Science, 2008, 322(5903):949-953.

[58] Khalil IA, Kogure K, Akita H, et al. Uptake pathways and subsequent intracellular trafficking in nonviral gene delivery. Pharmacological Reviews, 2006, 58(1):32-45.

[59] Ma H, Diamond SL. Nonviral gene therapy and its delivery systems. Current Pharmaceutical Biotechnology, 2001, 2(1):1-17.

[60] Malgrange B, Borgs L, Grobarczyk B, et al. Using human pluripotent stem cells to untangle neurodegenerative disease mechanisms. Cellular & Molecular Life Sciences, 2011, 68(4):635-649.

[61] Dimos JT, Rodolfa KT, Niakan KK, et al. Induced pluripotent stem cells generated from patients with ALS can be differentiated into motor neurons. Science, 2008, 321(5893):1218-1221.

[62] Nissant A, Pallotto M. Integration and maturation of newborn neurons in the adult olfactory bulb-from synapses to function. European Journal of Neuroscience, 2011, 33(6):1069-1077.

[63] Soldner F, Hockemeyer D, Beard C, et al. Parkinson's disease patient-derived induced pluripotent stem cells free of viral reprogramming factors. Cell, 2009, 136(5):964-977.

[64] Weiss S, Dunne C, Hewson J, et al. Multipotent CNS stem cells are present in the adult mammalian spinal cord and ventricular neuroaxis. Journal of Neuroscience, 1996, 16(23):7599-7609.

[65] Lois C, Alvarezbuylla A. Proliferating subventricular zone cells in the adult mammalian forebrain can differentiate into neurons and glia. Proc Natl Acad Sci Usa, 1993, 90(5):2074-2077.

[66] Gage FH, Kempermann G, Palmer TD, et al. Multipotent progenitor cells in the adult dentate gyrus. Journal of Neurobiology, 2015, 36(2):249-266.

[67] Kitamura T, Saitoh Y, Murayama A, et al. LTP induction within a narrow critical period of immature stages enhances the survival of newly generated neurons in the adult rat dentate gyrus. Molecular Brain, 2010, 3(1):13.

[68] Ihunwo AO, Tembo LH, Dzamalala C. The dynamics of adult neurogenesis in human hippocampus. Neural Regeneration Research, 2016, 11(12):1869-1883.

[69] Kornack DR, Rakic P. Continuation of neurogenesis in the hippocampus of the adult macaque monkey. Proceedings of the National Academy of Sciences of the United States of America, 1999, 96(10):5768-5773.

[70] Gould E, Reeves AJ, Fallah M, et al. Hippocampal neurogenesis in adult old world primates. Proc Natl Acad Sci USA, 1999, 96(9):5263-5267.

[71] Mori K, Kaneko YS, Nakashima A, et al. Subventricular zone under the neuroinflammatory stress and Parkinson's disease. Cellular & Molecular Neurobiology, 2012, 32(5):777-785.

[72] Kim JB, Zaehres H, Wu G, et al. Pluripotent stem cells induced from adult neural stem cells by reprogramming with two factors. Nature, 2008, 454(7204):646-650.

[73] Park IH, Zhao R, West JA, et al. Reprogramming of human somatic cells to pluripotency with defined factors. Nature, 2008, 451(7175):141-146.

[74] Song J, Crowther AJ, Olsen RH, et al. A diametric mode of neuronal circuitry-neurogenesis coupling in the adult hippocampus via parvalbumin interneurons. Neurogenesis, 2014, 1(1):150-154.

[75] Moga DE, Shapiro ML, Morrison JH. Bidirectional redistribution of AMPA but not NMDA receptors after perforant path simulation in the adult rat hippocampus in vivo. Hippocampus, 2010, 16(11):990-1003.

[76] Nomura H, Zahir T, Kim H, et al. Extramedullary chitosan channels promote survival of transplanted neural stem and progenitor cells and create a tissue bridge after complete spinal cord transection. Tissue Engineering Part A, 2008, 14(5):649-665.

[77] Rice AC, Khaldi A, Harvey HB, et al. Proliferation and neuronal differentiation of mitotically active cells following traumatic brain injury. Experimental Neurology, 2003, 183(2):406-417.

[78] Bye N, Carron S, Han X, et al. Neurogenesis and glial proliferation are stimulated following diffuse traumatic brain injury in adult rats. Journal of Neuroscience Research, 2011, 89(7):986-1000.

[79] Vedammai V, Gardner B, Okun MS, et al. Increased precursor cell proliferation after deep brain stimulation for Parkinson's disease: a human study. PLoS One, 2014, 9(3):e88770.

[80] Kulbatski I, Mothe AJ, Nomura H, et al. Endogenous and exogenous CNS derived stem/progenitor cell approaches for neurotrauma. Current Drug Targets, 2005, 6(1):11-16.

[81] Aberg MA, Aberg ND, Hedbäcker H, et al. Peripheral infusion of IGF-I selectively induces neurogenesis in the adult rat hippocampus. Journal of Neuroscience the Official Journal of the Society for Neuroscience, 2000, 20(8):2896-2903.

[82] Benraiss A, Chmielnicki E, Lerner K, et al. Adenoviral brain-derived neurotrophic factor induces both neostriatal and olfactory neuronal recruitment from endogenous progenitor cells in the adult forebrain. Journal of Neuroscience, 2001, 21(17):6718-6731.

[83] Liu X, Zhou X, Yuan W. The angiopoietin1-Akt pathway regulates barrier function of the cultured spinal cord microvascular endothelial cells through Eps8. Experimental Cell Research, 2014, 328(1):118-131.

[84] Chohan MO, Bragina O, Kazim SF, et al. Enhancement of neurogenesis and memory by a neurotrophic Peptide in mild to moderate traumatic brain injury. Neurosurgery, 2015, 76(2):201-215.

[85] Rosa AI, Gonçalves J, Cortes L, et al. The angiogenic factor angiopoietin-1 is a proneurogenic peptide on subventricular zone stem/progenitor cells. Journal of Neuroscience the Official Journal of the Society for Neuroscience, 2010, 30(13):4573-4584.

[86] Ulrich H, Nascimento ICD, Bocsi J, et al. Immunomodulation in stem cell differentiation into neurons and brain repair. Stem Cell Reviews & Reports, 2015, 11(3):474-486.

[87] Yang DY, Sheu ML, Su HL, et al. Dual regeneration of musele and nerve by intravenous administration of human amniotic fluid-derived mesenchymal stem cells regulated by stromal cell-derived factor-1a in a sciatic nerve injury mode. J Neurosurg, 2012, 116(6):1357-1367.

[88] Caminiti SP, Presotto L, Baroncini D, et al. Axonal damage and loss of connectivity in nigrostriatal and mesolimbic dopamine pathways in early Parkinson's disease. Neuroimage Clinical, 2017, 14(C):734-740.

[89] Wetzig A, Mackay-Sim A, Murrell W. Characterization of olfactory stem cells. Cell Transplantation, 2011, 20(11-12):1673-1691.

[90] Kim SJ, Son TG, Kim K, et al. Interferon-γ promotes differentiation of neural progenitor cells via the JNK pathway. Neurochemical Research, 2007, 32(8):1399-1406.

[91] Teng YD, Liao WL, Choi H, et al. Physical activity-mediated functional recovery after spinal cord injury: potential roles of neural stem cells. Regenerative Medicine, 2006, 1(6):763-776.

[92] Matthews MA, St OMF, Faciane CL, et al. Axon sprouting into segments of rat spinal cord adjacent to the site of a previous transection. Neuropathol Appl Neurobiol, 2010, 5(3):181-196.

[93] Meletis K, Barnabé-Heider F, Carlén M, et al. Spinal cord injury reveals multilineage differentiation of ependymal cells. PLos Biology, 2008, 6(7):e182.

[94] Schöniger S, Caprile T, Yulis CR, et al. Physiological response of bovine subcommissural organ to endothelin 1 and bradykinin. Cell & Tissue Research, 2009, 336(3):477-488.

[95] Attar A, Kaptanoglu E, Aydın Z, et al. Electron microscopic study of the progeny of ependymal stem cells in the normal and injured spinal cord.. Surgical Neurology, 2005, 64(64 Suppl 2):S28-S32.

[96] Kojima A, Tator CH. Intrathecal administration of epidermal growth factor and fibroblast growth factor 2 promotes ependymal proliferation and functional recovery after spinal cord injury in adult rats. Journal of Neurotrauma, 2002, 19(2):223-238.

[97] Ayusosacido A, Roy NS, Schwartz TH, et al. Long-term expansion of adult human brain subventricular zone precursors. Neurosurgery, 2008, 62(1):223.

第十三章 间充质干细胞在炎性肠病治疗中的应用

第一节 炎性肠病

炎性肠病（IBD）是发生在胃肠道的病因不明的慢性非特异性炎性疾病，主要包括溃疡性结肠炎（UC）和克罗恩病（CD）两种类型。常表现为腹痛、腹泻、黏液脓血便、肠梗阻、肠穿孔甚至癌变。因病情反复、迁延，严重影响患者日常生活质量。其发病机制尚未完全阐明，目前认为主要是肠黏膜屏障被破坏，免疫细胞及其分泌的细胞因子失衡，导致对肠道微生物的免疫应答失调，引起慢性肠道黏膜损伤。IBD 发生时，肠黏膜组织内存在大量炎性细胞浸润，其中 T 细胞介导的免疫应答紊乱起关键作用。CD 患者血循环中主要是辅助性 T 细胞（Th1/Th17）活化增加，调节性 T 细胞（Treg）减少；UC 患者血循环中主要是 Th2 活化增加，Treg 细胞减少[1-4]。IBD 治疗的最终目标是达到深度缓解，包括症状缓解和内镜下黏膜愈合，维持最大的肠道功能，减少长期并发症，维持正常的生活质量。目前的治疗主要着眼于控制活动性炎症，调节免疫紊乱。常用的治疗药物包括氨基水杨酸制剂、糖皮质激素、免疫抑制剂和生物制剂。但是仍有部分患者对现有治疗方式出现抵抗或无反应，甚至一部分难治性的 IBD 患者最终需要手术治疗，严重影响其生活质量。这促使人们去研究新的治疗方式，目前细胞疗法被认为是一种新的治疗策略。

第二节 间充质干细胞治疗炎性肠病的可能机制

间充质干细胞（MSC）是一种具有自我复制能力和多向分化潜能的成体干细胞，具有归巢、组织修复和免疫调节功能，近年来有众多的研究表明 MSC 在治疗 IBD 方面具有独特的优势。目前认为有以下几种可能的机制。

1）选择性地迁移到受损组织和炎症部位。体内示踪技术显示，在多项动物实验中[5-8]，MSC 移植进入结肠炎动物体内后，可以逐渐向炎症部位迁移、滞留，并在该部位发挥作用。MSC 的迁移和归巢作用与其表达的受体、黏附分子及其与内皮细胞相互作用相关，如 CCR2、CXCR4、血管细胞黏附分子-1（VCAM-1）、基质金属蛋白酶-2（MMP-2）等分子[9-14]。

2）促进肠道受损上皮的再生和血管生成。肠上皮屏障功能障碍在 IBD 的发生发展过程中起着重要的作用[15]。在炎症或组织损伤的条件下，MSC 选择性地迁移到受损部位，不以分化为肠道上皮细胞为目的，而是通过分泌血管内皮生长因子（VEGF）和转化生长因子-β1（TGF-β1）来促进血管新生、抑制受损组织凋亡和纤维化，促进组织存活及修复[16-20]。经葡聚糖硫酸钠（DSS）诱导的结肠炎大鼠静脉注射 MSC 后，MSC 迁移至受损的结肠固有层，上调了 α-平滑肌肌动蛋白的表达，促进上皮损伤愈合[17]。经 2,4,6-三硝基苯磺酸（TNBS）诱导的结肠炎小鼠经静脉移植 MSC 后，受损组织肠上皮细胞的增殖和肠道干细胞的分化增加[20,21]。

3）调控内质网应激，改善肠道炎症。内质网应激可以导致未折叠蛋白质活化，打破内质网平衡，若不及时恢复该平衡，未折叠蛋白质会促进炎症和细胞凋亡[12,22-25]。Banerjee 等[26]研究发现，经 DSS 诱导的结肠炎小鼠远端结肠内质网的应激相关标志物免疫球蛋白结合蛋白（immunoglobulin-binding protein，BiP）、蛋白激酶 R 样内质网激酶（PERK）、蛋白二硫化物异构酶（protein disulfide isomerase，PDI）表达显著增加，脐带间充质干细胞（UC-MSC）移植后，BiP、PDI 的表达较对照组显著降低。

4）上调紧密连接蛋白的表达，减少 MMP 的表达，以改善肠黏膜屏障功能。肠上皮细胞是肠黏膜屏障的主要组织结构基础，紧密连接是细胞间最重要的连接方式，参与维持上皮细胞的完整性。IBD 发病时肠黏膜屏障功能异常，肠腔内的抗原物质向肠黏膜固有层移位，激活固有层免疫细胞，产生大量炎性细胞因子及介质，进一步损伤肠黏膜屏障功能，这使得肠上皮细胞紧密连接发生变异、减少或缺失，导致细胞间隙通透性增加。Lin 等[27]利用伊文思蓝染色液研究发现，DSS 诱导的结肠炎小鼠肠道通透性增加，UC-MSC 移植可以降低其肠道通透性。进一步研究发现，DSS 诱导的结肠炎小鼠紧密连接蛋白 occludin、claudin-1 和 ZO-1 表达下降，但是 UC-MSC 处理后其表达会显著增加。

基质金属蛋白酶（MMP）是一组重要的裂解细胞外基质（ECM）成分的蛋白分解酶。生理状况下，MMP 表达和活性受到精确调控，保持正常的生理功能，其表达调控失衡将引起病理变化。IBD 患者过表达的 MMP 通过改变 ECM 成分、降解内皮细胞间的紧密连接、增加血管内皮的通透性，影响血管生成。MMP-2、MMP-9 是由结肠上皮和中性粒细胞释放的明胶酶，参与Ⅳ型胶原的降解，而Ⅳ型胶原是 ECM 的重要成分。Heimesaat 等[28]研究显示，MMP-2、MMP-9 参与 IBD 肠道功能紊乱。在 DSS 诱导的结肠炎小鼠中，结肠 MMP-2、MMP-9 的活性上调，UC-MSC 处理后，两者的活性均显著下降，差异有统计学意义[26]。

5）MSC 促进炎症部位浸润的 T 细胞和巨噬细胞向抗炎表型转化[8,29,30]，调节免疫反应。在 IBD 发病过程中，免疫细胞及其分泌的细胞因子起到非常重要的作用。Th17 可介导慢性炎症和自身免疫系统疾病的发生，Treg 有抑制自身免疫的功能，Th17/Treg 细胞平衡是维持肠道免疫稳态的重要因素。Th17 数量增加、Treg 数量减少或功能异常均可导致 IBD 的发生发展。在动物实验中，与对照组比较，UC-MSC 处理的结肠炎小鼠脾、肠系膜淋巴结中 Treg 增加，Th17 减少[31]。

巨噬细胞是一种异质性、可塑性较强的免疫细胞，不同来源的不同细胞因子及不同组织微环境对其增殖、分化、激活和极化可产生不同的影响，使其呈现出不同表型，发挥不同生物学功能。多项研究显示，MSC 可以旁分泌可溶性细胞因子增加结肠炎小鼠 M2 型巨噬细胞，减少 M1 型巨噬细胞[32-35]。此外，体内外实验均显示，MSC 可以减少促炎因子干扰素-γ（IFN-γ）、白介素-6（IL-6）、IL-1β、IL-17 的表达，显著增加抗炎因子 IL-10、TGF-β 等的表达[8]。

6）MSC 通过 FAS 配体/FAS 结合诱导 T 细胞凋亡，重建免疫耐受。Akiyama 等[36]在动物实验中证实 MSC 可以通过 FAS 配体/FAS 通路介导 T 细胞凋亡，凋亡 T 细胞随后触发巨噬细胞产生高水平的 TGF-β，进一步导致 CD4$^+$CD25$^+$Foxp3$^+$ Treg 数量增加，产生免疫耐受。

第三节　间充质干细胞治疗炎性肠病的基础及临床研究

近几年，大量的基础及临床试验证明了不同来源的 MSC 治疗实验性结肠炎的安全性和有效性，其中研究最多的是骨髓间充质干细胞（BM-MSC）和脂肪间充质干细胞（AD-MSC）。Liang 等[8]的研究发现，MSC 进入结肠炎动物体内后，可以选择性地迁移至腹部，并在该部位发挥免疫调节、组织修复等作用，改善结肠炎症及组织损伤。Chen 等证实 MSC 可以促进肠道上皮细胞的增殖和肠道干细胞的分化，下调 Th1、Th17 介导的自身免疫和炎症反应[IL-2、肿瘤坏死因子-α（TNF-α）、IFN-γ、T-bet、IL-6、IL-17、RORγt]，上调 Th2 活性（IL-4、IL-10、GATA-3）[21,37]，诱导 CD4$^+$CD25$^+$Foxp3$^+$ Treg 细胞的活化及 M2 型巨噬细胞的活化[38]。2015 年，Yang 等[39]提出 BM-MSC 来源的外泌体可以通过显著降低结肠炎动物损伤部位 NF-κB p65、TNF-α、诱导型一氧化氮合酶（iNOS）、环加氧酶 2（COX2）的蛋白质和 mRNA 水平，增加 IL-1β、IL-10 的表达，调节氧化应激，减少 Caspase-3、Caspase-8 和 Caspase-9 抑制细胞凋亡来缓解肠道炎症。然而 2015 年 Nam 等[40]提出，BM-MSC 对由葡聚糖硫酸钠（DSS）诱导的结肠炎动物的治疗作用有限，并不能引起有统计学意义的临床或组织学改善。

另外多项研究显示，对 MSC 进行预处理可以增强其在实验性结肠炎中的治疗作用。UC-MSC 经聚肌胞苷酸（poly(I:C)）预处理活化 Toll 样受体 3（TLR3）可以使 UC-MSC 向炎症部位迁移的能力增强，前列腺素 E2（PGE2）的表达增加，改变（Th1/Th17）/Treg 细胞平衡，对结肠炎的治疗效果明显[41,42]。同样的，胞壁酰二肽预处理活化 UC-MSC 的 NOD2 受体可以增强其免疫调节作用[43]。另外，一些生物活性试剂，如一氧化氮（NO）、IL-1β、IFN-γ 或 TNF-α 均可以激发 MSC 的功能，改变它们的迁移、分化和免疫功能[10,44-47]。Chen 等[48]将能表达 IFN-γ 的质粒转染至 UC-MSC，体外实验发现其较正常的 UC-MSC 明显抑制 T 细胞的增殖，将 IFN-γ-MSC 注射至结肠炎动物体内，发现 IFN-γ-MSC 增加了 Treg 和 Th2 细胞的数量，减少了 Th1 和 Th17 细胞的数量，上调了吲哚胺 2,3-双加氧酶（IDO）的表达，抑制了结肠中炎性细胞因子的产生，更加有效地缓解了 DSS 诱导的结肠炎。Fan 等[47]提出 IL-1β 预处理可以上调 MSC CXCR4 的表达，增强 MSC 的治疗作用。同样的，Liu 等[49]在动物体内证实过表达 CXCR4 的 MSC 向炎症部位迁移的能力增强，可抑制损伤部位的炎症反应和 STAT3 的磷酸化，对结肠炎的治疗效果更明显。另外，有研究显示[50]，来自人脐血的血小板溶解产物同样能增强 AD-MSC 的治疗效果。近期，Mao 等[51]将 UC-MSC 来源的外泌体经尾静脉注射至结肠炎小鼠体内，发现外泌体可以归巢至脾、结肠部位，缓解结肠炎症，调节细胞因子的表达，抑制巨噬细胞的浸润。研究人员[52,53]将羊膜来源的 MSC 或其条件培养基注射入结肠炎动物体内，显著减少了炎症部位单核/巨噬细胞的浸润。同时在体外实验中，羊膜来源 MSC 的条件培养基可以显著抑制 NF-κB 的活化。15-LOX-1 是炎症反应的重要调节因子，可以调节 IL-6/STAT3 信号通路，Mao 等[54]的研究显示，UC-MSC 可以显著减少结肠炎小鼠 15-LOX-1、IL-6 和磷酸化 STAT3 的表达。2015 年 Stavely 等[55]用豚鼠 BM-MSC/AD-MSC 治疗由 TNBS 诱导的豚鼠结肠炎，发现 BM-MSC/AD-MSC 均可以通过分泌 TGF-β1 在减少肠神经元损失和改变神经元亚群方面发挥神经保护作用。Ryska 等[56]将 AD-MSC 局部应用于大鼠 CD 肠皮瘘模型，结果显示 AD-MSC 治疗组较 PBS 治疗组瘘管愈合率更高（$P=0.033$）。在慢性实验性结肠炎小鼠中，在疾病初期静脉给予 BM-MSC 可以通过长时间的免疫抑制预防疾病进展，加速结肠炎的恢复[57]。

据报道，在临床试验中，同种异体的 MSC 已用于 31 例难治性克罗恩病（CD）或溃疡性结肠炎（UC）的治疗。约有 60%的患者达到临床反应，约 40%的患者达到了临床缓解。首个研究 MSC 临床应用安全性的报道来自荷兰[58]，9 例难治性 CD 患者静脉输注自体 BM-MSC（1~2）$\times 10^6$cells/kg×2 次，在治疗后第 6 周，2 例患者达到内镜下改善，3 例患者达到临床改善，另外 3 例患者由于疾病恶化需要进行手术。但是对患者的结肠组织研究发现，$CD4^+$ T 细胞浸润减少，Treg 细胞浸润增加，TNF-α、IL-1β、IL-10 和 IL-6 减少。另一项安全性研究[59]显示，12 例难治性 CD 患者接受单次 BM-MSC（剂量 2×10^6cells/kg、5×10^6cells/kg、10×10^6cells/kg）治疗后第 9 周，5 例患者达到临床反应，另外 5 例患者疾病加重，2 例患者发生严重不良反应（可能与治疗相关）。Liang 等[60]研究了异体 BM-MSC 移植是否能加强传统 IBD 治疗的效果。该研究纳入了 7 例难治性 IBD 患者，结果显示 BM-MSC 可以增强传统治疗的免疫抑制作用。2013 年，Mayer 等[61]发表了其研究团队利用胎盘来源的 MSC（PDA-001）治疗难治性 CD 患者的Ⅰ期临床研究。12 例活动性、中重度、治疗无效的 CD 患者被分成高剂量组（CD 活动指数更高）和低剂量组，分别接受高剂量和低剂量的 PDA-001 治疗，结果显示低剂量组患者全部达到临床反应，3 例患者达到临床缓解。高剂量组 2 例患者达到临床反应，均未达到临床缓解；可见短时间的轻中度头痛、恶心、发烧等不良反应。2014 年，该团队发表了其Ⅰb/Ⅱa 期研究结果。研究纳入的 50 例 CD 患者均表现为内镜下活动性炎症，粪便钙卫蛋白水平上升，对传统治疗无效。试验过程中，患者联合稳定剂量的免疫抑制剂或生物制剂治疗。研究结果显示，PDA-001 治疗组 36%患者达到临床反应，14%达到临床缓解，对照组无人达到临床反应。试验过程中 1 例患者出现了过敏反应，Ⅱa 期试验中，严重不良反应的发生率为 32%，对照组为 7%[62]。2014 年 Forbes 等[63]进行的一项Ⅱ期临床研究显示，生物制剂治疗无效的 15 例 CD 患者经过异体 AD-MSC 静脉输注治疗后，疾病活动指数下降，12 例患者达到临床反应，8 例患者达到临床缓解，7 例患者达到内镜下改善，1 例患者出现严重不良反应（可能与治疗无关）。为了研究 MSC 的致瘤性，进行了一项长达 5 年的随访研究，103 例经 MSC 治疗的 IBD 患者与对照组患者相比，肿瘤的发生率无统计学差异[64]。

第四节 间充质干细胞对炎性肠病相关瘘管的局部治疗

克罗恩病（CD）患者在疾病的进程中会出现多种临床并发症，包括与疾病相关的肛周病变、肛管病变，包括肛裂、溃疡和狭窄、肛周瘘管、脓肿及癌症。CD 肛瘘可伴有局部疼痛、肛周溢液及脓肿形成等情况，严重影响患者的生活质量，是临床上一治疗难题。研究表明，诊断 CD 后 20 年内有 23%~26%的患者可形成肛瘘[65,66]。肛瘘病程长，易反复。有相当比例 CD 肛瘘患者对现有的治疗药物（包括抗生素、免疫抑制剂及生物制剂等）无反应[67]。为防止脓肿形成，部分患者需要进行手术放置挂线引流，更有甚者需要进行大便转流来缓解肛周疾病。CD 肛瘘治疗的最终目标是在保留括约肌功能的同时治愈瘘管。尽管有多种治疗手段，但文献报道 CD 肛瘘的长期缓解率仅为 37.0%[68]。目前应用 MSC 治疗 CD 肛瘘的临床研究越来越多，已有多项针对 MSC 治疗 CD 并发肛瘘的Ⅰ～Ⅲ期临床试验完成。研究显示，MSC 有良好的免疫调节功能，能够促进 CD 肛瘘的愈合，减少复发，无明显不良反应，能保留肛门功能，不影响患者生活质量。

  首个将 AD-MSC 应用于 CD 肛瘘治疗的病例报道来自西班牙，García-Olmo 等[69]将 AD-MSC 局部注射在 CD 患者直肠黏膜靠近直肠阴道瘘内口的部位，结果显示，治疗后直肠阴道瘘外口逐渐完全闭合。尽管该患者很快又出现了肛周瘘管，但随访 3 个月，经 AD-MSC 治疗的直肠阴道瘘未出现复发或不良反应。随后该研究团队进行了一项前瞻性、单中心Ⅰ期临床试验[70]，研究纳入了 4 例 CD 肛瘘患者（共 9 道瘘管），局部注射 3×10^6 cells 自体 AD-MSC，观察治疗的安全性及有效性。75%瘘管在治疗 8 周后完全愈合，25%瘘管不完全愈合。在整个试验及为期 2 年的随访过程中，参加试验人员均未出现明显的不良反应。2013 年，de la Portilla 等[71]发表了一项异体 AD-MSC 治疗 CD 肛瘘的多中心Ⅰ/Ⅱa 临床试验。该研究纳入了 24 例患者，瘘管内注射 2×10^7 cells AD-MSC，12 周后若未愈合，再次给予 4×10^7 cells AD-MSC 治疗，随访观察至首次给药后 24 周。结果显示，56.3%患者瘘管达到完全愈合，且 6 个月的随访过程中未见治疗相关不良反应。2009 年，该研究团队发表了该项目的Ⅱb 期临床试验结果[72]。研究共纳入 49 例复杂肛瘘患者，随机分为两组（纤维蛋白胶组、纤维蛋白胶加 2×10^7 cells AD-MSC 组），治疗后 8 周若瘘管未完全愈合，进行 2 次治疗（纤维蛋白胶加 4×10^7 cells AD-MSC）。结果显示，治疗后第 8 周，纤维蛋白胶加 2×10^7 cells AD-MSC 组 71%患者瘘管愈合，纤维蛋白胶组 16%患者瘘管愈合（$P<0.001$）。同样的，纤维蛋白胶加 AD-MSC 治疗组患者的生活质量评分较纤维蛋白胶组更高。2012 年发表了该研究的长期随访结果[73]，纤维蛋白胶加 AD-MSC 组共 21 例患者，纤维蛋白胶组共 13 例患者，随访期（38 个月、42.6 个月）内未见治疗相关不良反应。纤维蛋白胶加 AD-MSC 组 12 例完全愈合的患者，有 7 例未再复发。该研究再次证实了 MSC 治疗的安全性，但是长期随访中仍有部分患者复发。同年，该研究团队完成了一项大型多中心、随机、单盲Ⅲ期临床试验[74]。200 例成年患者随机分配到接受 2×10^7 cells AD-MSC 组（A 组，64 名患者）、纤维蛋白胶加 2×10^7 cells AD-MSC 组（B 组，60 名患者）、纤维蛋白胶组（C 组，59 名患者）。如果治疗后瘘管在 12 周内没有愈合，则使用第 2 剂（A 组、B 组均增加为 4×10^7 cells AD-MSC）。该试验证明 AD-MSC 单独治疗或者联合纤维蛋白胶治疗均是安全有效的，治疗后第 6 个月瘘管的愈合率达到 40%，1 年达到 50%，与单独使用纤维蛋白胶无统计学差异。2015 年，García-Olmo 等[75]研究了 MSC 对瘘管患者进行挽救治疗的效果。该研究纳入的研究对象均为经历过多次手术治疗仍无法解决瘘管且有症状复发的患者，共 10 例。自体 AD-MSC 局部注射至瘘管壁后 8 周进行评估，6 例患者出现临床反应，瘘管完全停止化脓，3 名患者出现部分反应，瘘管化脓明显减少。治疗后 1 年，6 名患者瘘管完全愈合，瘘管外口完全再上皮化。2 例肛门

失禁患者症状有所改善,且试验过程中未出现治疗相关不良反应。该项研究再次证明MSC治疗肛瘘安全有效,并且对难治性患者仍有治疗作用,可减少大便失禁的风险,提高患者的生活质量。2015年,该研究团队发表了AD-MSC治疗妇女CD肛瘘后对妊娠是否有影响的临床研究[76]。研究共纳入了5例CD年轻女性(1例有肛瘘合并直肠阴道瘘,2例有肛瘘,2例有直肠阴道瘘),所有患者均接受了2次治疗,治疗结束后分别受孕。研究结果显示,AD-MSC对生育力、妊娠期、新生儿体重或新生儿的身体状况没有任何影响。2016年在《柳叶刀》(Lacent)上发表的一项随机、双盲、多中心、安慰剂对照临床研究显示,一种新型的AD-MSC(CX601)能够显著将肛瘘患者的24周缓解率从34%提高到50%,并体现有较高的安全性[77]。2018年,该研究团队发表了该研究的Ⅲ期临床试验的长期随访结果,显示试验组患者的52周完全缓解率为56.3%,对照组为38.6%($P=0.010$),且两组不良事件发生率无明显差异[78]。2017年,Dietz等[79]对12例肛瘘患者进行了一项为期6个月的临床试验,应用可吸收的生物基质包被自体MSC,在治疗后6个月,10/12(83%)患者达到临床愈合,影像学表现有反应,无治疗相关严重不良反应。

2012年,Cho等[80]进行了AD-MSC治疗CD瘘管的剂量逐步上升的Ⅰ期临床试验,证明AD-MSC治疗CD瘘管安全有效。2013年,该项目的Ⅱ期临床研究显示,AD-MSC治疗后8周的愈合率约82%,26例愈合患者在长达1年的随访中,未出现治疗相关不良反应,且88%患者维持完全愈合,仅3例患者复发[81]。2015年,Cho等[82]利用AD-MSC治疗CD患者瘘管的Ⅱ期临床研究的长期随访结果显示,26例完全愈合的CD瘘管患者中21例维持完全愈合。

2011年,意大利科学家Ciccocioppo[83]发表了一项自体BM-MSC治疗CD瘘管的临床试验。该项试验共纳入了10例难治性CD患者,接受自体BM-MSC瘘管注射治疗,每4周注射一次,直到临床症状改善或BM-MSC不能继续使用。试验过程中通过核磁共振成像(MRI)或手术、内镜检测至首次注射后1年。结果显示,70%患者瘘管完全愈合,30%患者瘘管不完全愈合,所有患者直肠黏膜愈合。在治疗过程中黏膜和循环中Treg细胞显著增加且维持稳定至随访结束。2015年,该团队发表了BM-MSC治疗CD瘘管的长期随访结果。在试验结束后2年,患者平均CD活动指数显著上升,然后逐渐减少,在5年随访期结束时,患者再次获得缓解。瘘管的无复发生存率为分别为1年88%,2年50%,之后的4年37%,无手术累计生存率是1年100%,3年75%,5年63%,或无药物的累计生存率是1年88%,2年25%直到随访结束,未见不良事件记录[84]。

2015年,Molendijk等[85]对异体BM-MSC治疗难治性CD肛瘘的作用进行研究。21例难治性CD肛瘘患者随机分成4组:$1×10^7$cells组5例、$3×10^7$cells组5例、$9×10^7$cells组5例及安慰剂组6例。治疗后6周,第1组60.0%肛瘘患者开始愈合,第2组80.0%,第3组20.0%,安慰剂组16.7%(第2组与安慰剂组$P=0.08$)。治疗后第12周瘘管正在愈合的比例各组分别为40.0、80.0%、20.0%、33.3%。这样的结果一直持续到第24周,此时第1组80.0%患者的瘘管可见愈合。在治疗后第6周,第1组44.4%的瘘管已经愈合,第2组85.7%,第3组28.6%,安慰剂组22.2%(第2组与安慰剂组$P=0.04$)。治疗后第12周,第一组33.3%患者的瘘管已经愈合,第2组85.7%,第3组28.6%,安慰剂组33.3%。直到治疗后第24周,第一组瘘管的愈合率增加到66.7%。

2016 年，Lightner 等[86]在《胃肠病学》(*Gastroenterology*)上发表了其研究团队利用 AD-MSC 治疗 CD 难治性肛瘘的早期结果。他们将 AD-MSC 移植到人工合成的瘘管塞，然后将瘘管塞放置在肛瘘瘘管内发挥治疗作用。该研究共纳入 7 例难治性 CD 肛瘘患者（药物、手术治疗失败），将 AD-MSC 黏附包被在瘘管塞内。挂线手术后 6 周，将包被 AD-MSC 的瘘管塞手术放置在瘘管部位，未出现治疗相关围手术期不良反应。在术后 6 个月，经 MRI 证实，6/7 患者达到完全愈合。该研究进一步证实 MSC 在治疗难治性肛瘘中的潜在作用。我们在广州中山大学附属第六医院的试点临床试验中，连续 6 例 CD 肛瘘患者纳入研究，随机分为两组，分别接受 $1×10^6$cells 或 $5×10^6$cells 胎盘来源的 MSC 凝胶瘘管内注射的治疗。结果显示，治疗后 6 周、12 周、24 周，观察到的瘘管愈合率分别为 44.4%、50%、75%，且试验过程中未出现治疗相关不良反应。

综上所述，MSC 具有归巢、组织修复和免疫调节功能，大量基础及临床试验显示，MSC 对大部分 IBD 患者进行静脉滴注或局部注射治疗是安全有效的，甚至对难治性 CD 患者可以起到挽救治疗的作用，极大地改善了患者的生活质量。但是 MSC 临床应用的安全性仍需要进一步关注，尤其是 MSC 的来源、运输、使用剂量及治疗途径等。总之，MSC 治疗 IBD 是有潜力的，需要更多的临床研究数据证实。

（侯慧星　曹晓沧）

参 考 文 献

[1] Bandzar S, Gupta S, Platt MO. Crohn's disease: a review of treatment options and current research. Cellular Immunology, 2013, 286:45-52.

[2] Chao K, Zhang S, Yao J, et al. Imbalances of CD4(+) T-cell subgroups in Crohn's disease and their relationship with disease activity and prognosis. Journal of Gastroenterology and Hepatology, 2014, 29:1808-1814.

[3] Sisakhtnezhad S, Alimoradi E, Akrami H. External factors influencing mesenchymal stem cell fate *in vitro*. European Journal of Cell Biology, 2017, 96:13-33.

[4] Strober W, Fuss IJ. Proinflammatory cytokines in the pathogenesis of inflammatory bowel diseases. Gastroenterology, 2011, 140:1756-1767.

[5] Devine SM, Cobbs C, Jennings M, et al. Mesenchymal stem cells distribute to a wide range of tissues following systemic infusion into nonhuman primates. Blood, 2003, 101:2999-3001.

[6] Tanaka F, Tominaga K, Ochi M, et al. Exogenous administration of mesenchymal stem cells ameliorates dextran sulfate sodium-induced colitis via anti-inflammatory action in damaged tissue in rats. Life Sciences, 2008, 83:771-779.

[7] Bruck F, Belle L, Lechanteur C, et al. Impact of bone marrow-derived mesenchymal stromal cells on experimental xenogeneic graft-versus-host disease. Cytotherapy, 2013, 15:267-279.

[8] Liang L, Dong C, Chen X, et al. Human umbilical cord mesenchymal stem cells ameliorate mice trinitrobenzene sulfonic acid (TNBS)-induced colitis. Cell Transplantation, 2011, 20:1395-1408.

[9] Belema-Bedada F, Uchida S, Martire A, et al.. Efficient homing of multipotent adult mesenchymal stem cells depends on FROUNT-mediated clustering of CCR2. Cell Stem Cell, 2008, 2:566-575.

[10] Cheng Z, Ou L, Zhou X, et al. Targeted migration of mesenchymal stem cells modified with CXCR4 gene to infarcted myocardium improves cardiac performance. Molecular Therapy: the Journal of the American Society of Gene Therapy, 2008, 16:571-579.

[11] Ko IK, Kim BG, Awadallah A, et al. Targeting improves MSC treatment of inflammatory bowel disease. Molecular therapy: the Journal of the American Society of Gene Therapy, 2010, 18:1365-1372.

[12] Christian R, Virginia E, Marisa K, et al. MMP-2, MT1-MMP, and TIMP-2 are essential for the invasive capacity of human mesenchymal stem cells: differential regulation by inflammatory cytokines. Blood, 2007, 109(9):4055-4063.

[13] Ruster B, Gottig S, Ludwig RJ, et al. Mesenchymal stem cells display coordinated rolling and adhesion behavior on endothelial cells. Blood, 2006, 108:3938-3944.

[14] Wynn RF, Hart CA, Corradi-Perini C, et al. A small proportion of mesenchymal stem cells strongly expresses functionally active CXCR4 receptor capable of promoting migration to bone marrow. Blood, 2004, 104:2643-2645.

[15] Bouma G, Strober W. The immunological and genetic basis of inflammatory bowel disease. Nature Reviews Immunology, 2003, 3:521-533.

[16] Prockop DJ. "Stemness" does not explain the repair of many tissues by mesenchymal stem/multipotent stromal cells (MSCs). Clin Pharmacol Ther, 2007, 82:241-243.

[17] Hayashi Y, Tsuji S, Tsujii M, et al. Topical implantation of mesenchymal stem cells has beneficial effects on healing of experimental colitis in rats. The Journal of Pharmacology and Experimental Therapeutics, 2008, 326:523-531.

[18] Caplan AI. Why are MSCs therapeutic? New data: new insight. The Journal of Pathology, 2009, 217:318-324.

[19] Kachgal S, Putnam AJ. Mesenchymal stem cells from adipose and bone marrow promote angiogenesis via distinct cytokine and protease expression mechanisms. Angiogenesis, 2011, 14:47-59.

[20] Ferrand J, Noel D, Lehours P, et al. Human bone marrow-derived stem cells acquire epithelial characteristics through fusion with gastrointestinal epithelial cells. PLoS One, 2011, 6:e19569.

[21] Chen QQ, Yan L, Wang CZ, et al. Mesenchymal stem cells alleviate TNBS-induced colitis by modulating inflammatory and autoimmune responses. World Journal of Gastroenterology, 2013, 19:4702-4717.

[22] Eri RD, Adams RJ, Tran TV, et al. An intestinal epithelial defect conferring ER stress results in inflammation involving both innate and adaptive immunity. Mucosal Immunology, 2011, 4:354-364.

[23] Garg AD, Kaczmarek A, Krysko O, et al. ER stress-induced inflammation: does it aid or impede disease progression? Trends in Molecular Medicine, 2012, 18:589-598.

[24] Osorio F, Tavernier SJ, Hoffmann E, et al. The unfolded-protein-response sensor IRE-1alpha regulates the function of CD8alpha+ dendritic cells. Nature Immunology, 2014, 15:248-257.

[25] Shenderov K, Riteau N, Yip R, et al. Cutting edge: endoplasmic reticulum stress licenses macrophages to produce mature IL-1beta in response to TLR4 stimulation through a Caspase-8- and TRIF-dependent pathway. Journal of Immunology, 2014, 192:2029-2033.

[26] Banerjee A, Bizzaro D, Burra P, et al. Umbilical cord mesenchymal stem cells modulate dextran sulfate sodium induced acute colitis in immunodeficient mice. Stem Cell Research & Therapy, 2015, 6:79.

[27] Lin Y, Lin L, Wang Q, et al. Transplantation of human umbilical mesenchymal stem cells attenuates dextran sulfate sodium-induced colitis in mice. Clinical and Experimental Pharmacology & Physiology, 2015, 42:76-86.

[28] Heimesaat MM, Dunay IR, Fuchs D, et al. The distinct roles of MMP-2 and MMP-9 in acute DSS colitis. European Journal of Microbiology & Immunology, 2011, 1:302-310.

[29] Liang J, Zhang HY, Wang DD, et al. Allogeneic mesenchymal stem cell transplantation in seven patients with refractory inflammatory bowel disease. Gut, 2012, 61(3):468-469.

[30] Markovic BS, Nikolic A, Gazdic M, et al. Pharmacological inhibition of Gal-3 in mesenchymal stem cells enhances their capacity to promote alternative activation of macrophages in dextran sulphate sodium-induced colitis. Stem Cells International, 2016, 2016(3):2640746.

[31] Li L, Liu S, Xu Y, et al. Human umbilical cord-derived mesenchymal stem cells downregulate inflammatory responses by shifting the Treg/Th17 profile in experimental colitis. Pharmacology, 2013, 92:257-264.

[32] Song WJ, Li Q, Ryu MO, A et al. TSG-6 released from intraperitoneally injected canine adipose tissue-derived mesenchymal stem cells ameliorate inflammatory bowel disease by inducing M2 macrophage switch in mice. Stem Cell Research & Therapy, 2018, 9:91.

[33] Song WJ, Li Q, Ryu MO, et al. TSG-6 secreted by human adipose tissue-derived mesenchymal stem cells ameliorates DSS-induced colitis by Inducing M2 macrophage polarization in mice. Scientific Reports, 2017, 7:5187.

[34] Anderson P, Souza-Moreira L, Morell M, et al. Adipose-derived mesenchymal stromal cells induce immunomodulatory macrophages which protect from experimental colitis and sepsis. Gut, 2013, 62:1131-1141.

[35] Park HJ, Kim J, Saima FT, et al. Adipose-derived stem cells ameliorate colitis by suppression of inflammasome formation and regulation of M1-macrophage population through prostaglandin E2. Biochemical and Biophysical Research Communications, 2018, 498:988-995.

[36] Akiyama K, Chen C, Wang D, et al. Mesenchymal-stem-cell-induced immunoregulation involves FAS-ligand-/FAS-mediated T cell apoptosis. Cell Stem Cell, 2012, 10:544-555.

[37] Kong QF, Sun B, Bai SS, et al. Administration of bone marrow stromal cells ameliorates experimental autoimmune myasthenia gravis by altering the balance of Th1/Th2/Th17/Treg cell subsets through the secretion of TGF-beta. Journal of Neuroimmunology, 2009, 207:83-91.

[38] Gonzalez MA, Gonzalez-Rey E, Rico L, et al. Adipose-derived mesenchymal stem cells alleviate experimental colitis by inhibiting inflammatory and autoimmune responses. Gastroenterology, 2009, 136:978-989.

[39] Yang J, Liu XX, Fan H, et al. Extracellular vesicles derived from bone marrow mesenchymal stem cells protect against experimental colitis via attenuating colon inflammation, oxidative stress and apoptosis. PLoS One, 2015, 10:e0140551.

[40] Nam YS, Kim N, Im KI, et al. Negative impact of bone-marrow-derived mesenchymal stem cells on dextran sulfate sodium-induced colitis. World Journal of Gastroenterology, 2015, 21:2030-2039.

[41] Fuenzalida P, Kurte M, Fernandez-O'ryan C, et al. Toll-like receptor 3 pre-conditioning increases the therapeutic efficacy of umbilical cord mesenchymal stromal cells in a dextran sulfate sodium-induced colitis model. Cytotherapy, 2016, 18:630-641.

[42] Qiu Y, Guo J, Mao R, et al. TLR3 preconditioning enhances the therapeutic efficacy of umbilical cord mesenchymal stem cells in TNBS-induced colitis via the TLR3-Jagged-1-Notch-1 pathway. Mucosal Immunology, 2017, 10:727-742.

[43] Kim HS, Shin TH, Lee BC, et al. Human umbilical cord blood mesenchymal stem cells reduce colitis in mice by activating NOD2 signaling to COX2. Gastroenterology, 2013, 145:1392-1403.

[44] Fuseler JW, Valarmathi MT. Modulation of the migration and differentiation potential of adult bone marrow stromal stem cells by nitric oxide. Biomaterials, 2012, 33:1032-1043.

[45] Du WJ, Reppel L, Leger L, et al. Mesenchymal stem cells derived from human bone marrow and adipose tissue maintain their immunosuppressive properties after chondrogenic differentiation: role of HLA-G. Stem Cells and Development, 2016, 25:1454-1469.

[46] Castro-Manrreza ME, Montesinos JJ. Immunoregulation by mesenchymal stem cells: biological aspects and clinical applications. Journal of Immunology Research, 2015, 2015(2):394917.

[47] Fan H, Zhao G, Liu L, et al. Pre-treatment with IL-1beta enhances the efficacy of MSC transplantation in DSS-induced colitis. Cellular & Molecular Immunology, 2012, 9:473-481.

[48] Chen Y, Song Y, Miao H, et al. Gene delivery with IFN-gamma-expression plasmids enhances the therapeutic effects of MSCs on DSS-induced mouse colitis. Inflammation Research: Official Journal of the European Histamine Research Society, 2015, 64:671-681.

[49] Liu X, Zuo D, Fan H, et al. Over-expression of CXCR4 on mesenchymal stem cells protect against experimental colitis via immunomodulatory functions in impaired tissue. Journal of Molecular Histology, 2014, 45:181-193.

[50] Forte D, Ciciarello M, Valerii MC, et al. Human cord blood-derived platelet lysate enhances the therapeutic activity of adipose-derived mesenchymal stromal cells isolated from Crohn's disease patients in a mouse model of colitis. Stem Cell Research & Therapy, 2015, 6:170.

[51] Mao F, Wu Y, Tang X, et al. Exosomes derived from human umbilical cord mesenchymal stem cells relieve inflammatory bowel disease in mice. BioMed Research International, 2017, 2017(4):5356760.

[52] Miyamoto S, Ohnishi S, Onishi R, et al. Therapeutic effects of human amnion-derived mesenchymal stem cell transplantation and conditioned medium enema in rats with trinitrobenzene sulfonic acid-induced colitis. Am J Transl Res, 2017, 9(3):940-952.

[53] Onishi R, Ohnishi S, Higashi R, et al. Human amnion-derived mesenchymal stem cell transplantation ameliorates dextran sulfate sodium-induced severe colitis in rats. Cell Transplantation, 2015, 24:2601-2614.

[54] Mao F, Xu M, Zuo X, et al. 15-lipoxygenase-1 suppression of colitis-associated colon cancer through inhibition of the IL-6/STAT3 signaling pathway. FASEB Journal: Official Publication of the Federation of American Societies for Experimental Biology, 2015, 29:2359-2370.

[55] Stavely R, Robinson AM, Miller S, et al. Allogeneic guinea pig mesenchymal stem cells ameliorate neurological changes in experimental colitis. Stem Cell Research & Therapy, 2015, 6:263.

[56] Ryska O, Serclova Z, Mestak O, et al. Local application of adipose-derived mesenchymal stem cells supports the healing of fistula: prospective randomised study on rat model of fistulising Crohn's disease. Scandinavian Journal of Gastroenterology, 2017, 52:543-550.

[57] Lee HJ, Oh SH, Jang HW, et al. Long-term effects of bone marrow-derived mesenchymal stem cells in dextran sulfate sodium-induced murine chronic colitis. Gut and Liver, 2016, 10:412-419.

[58] Duijvestein M, Vos AC, Roelofs H, et al. Autologous bone marrow-derived mesenchymal stromal cell treatment for refractory luminal Crohn's disease: results of a phase I study. Gut, 2010, 59:1662-1669.

[59] Dhere T, Copland I, Garcia M, et al. The safety of autologous and metabolically fit bone marrow mesenchymal stromal cells in medically refractory Crohn's disease – a phase 1 trial with three doses. Alimentary Pharmacology & Therapeutics, 2016, 44:471-481.

[60] Liang J, Zhang H, Wang D, et al. Allogeneic mesenchymal stem cell transplantation in seven patients with refractory inflammatory bowel disease. Gut, 2012, 61:468-469.

[61] Mayer L, Pandak William M, Melmed Gil Y, et al. Safety and tolerability of human placenta-derived cells (PDA001) in treatment-resistant crohn's disease: a phase 1 study. Inflamm Bowel Dis, 2013, 19(4):754-760.

[62] Melmed GY, Pandak WM, Casey K, et al. Human placenta-derived cells (PDA-001) for the treatment of moderate-to-severe Crohn's disease: a phase 1b/2a study. Inflammatory Bowel Diseases, 2015, 21:1809-1816.

[63] Forbes GM, Sturm MJ, Leong RW, et al. A phase 2 study of allogeneic mesenchymal stromal cells for luminal Crohn's disease refractory to biologic therapy. Clinical Gastroenterology and Hepatology: the Official Clinical Practice Journal of the American Gastroenterological Association, 2014, 12:64-71.

[64] Knyazev OV, Parfenov AI, Konoplyannikov AG, et al. Safety of mesenchymal stromal cell therapy for inflammatory bowel diseases: results of a 5-year follow-up. Ter Arkh, 2015, 87(2):39-44.

[65] Schwartz DA, Loftus EV, Tremaine WJ, et al. The natural history of fistulizing Crohn's disease in Olmsted County, Minnesota. Gastroenterology, 2002, 122:875-880.

[66] Hellers G, Bergstrand O, Ewerth S, et al. Occurrence and outcome after primary treatment of anal fistulae in Crohn's disease. Gut, 1980, 21:525-527.

[67] West RL, van der Woude CJ, Hansen BE, et al. Clinical and endosonographic effect of ciprofloxacin on the treatment of perianal fistulae in Crohn's disease with infliximab: a double-blind placebo-controlled study. Alimentary Pharmacology & Therapeutics, 2004, 20:1329-1336.

[68] Molendijk I, Nuij VJ, van der Meulen-de Jong AE, et al. Disappointing durable remission rates in complex Crohn's disease fistula. Inflammatory Bowel Diseases, 2014, 20:2022-2028.

[69] Garcia-Olmo D, Garcia-Arranz M, Garcia LG, et al. Autologous stem cell transplantation for treatment of rectovaginal fistula in perianal Crohn's disease: a new cell-based therapy. International Journal of Colorectal Disease, 2003, 18:451-454.

[70] Garcia-Olmo D, Garcia-Arranz M, Herreros D, et al. A phase I clinical trial of the treatment of Crohn's fistula by adipose mesenchymal stem cell transplantation. Diseases of the Colon and Rectum, 2005, 48:1416-1423.

[71] de la Portilla F, Alba F, Garcia-Olmo D, et al. Expanded allogeneic adipose-derived stem cells (eASCs) for the treatment of complex perianal fistula in Crohn's disease: results from a multicenter phase I/IIa clinical trial. International Journal of Colorectal Disease, 2013, 28:313-323.

[72] Garcia-Olmo D, Herreros D, Pascual I, et al. Expanded adipose-derived stem cells for the treatment of complex perianal fistula: a phase II clinical trial. Diseases of the Colon and Rectum, 2009, 52:79-86.

[73] Guadalajara H, Herreros D, De-La-Quintana P, et al. Long-term follow-up of patients undergoing adipose-derived adult stem cell administration to treat complex perianal fistulas. International Journal of Colorectal Disease, 2012, 27:595-600.

[74] Herreros MD, Garcia-Arranz M, Guadalajara H, et al. Autologous expanded adipose-derived stem cells for the treatment of complex cryptoglandular perianal fistulas:a phase III randomized clinical trial (FATT 1: fistula Advanced Therapy Trial 1) and long-term evaluation. Diseases of the Colon and Rectum, 2012, 55:762-772.

[75] Garcia-Olmo D, Guadalajara H, Rubio-Perez I, et al. Recurrent anal fistulae: limited surgery supported by stem cells. World Journal of Gastroenterology, 2015, 21:3330-3336.

[76] Sanz-Baro R, Garcia-Arranz M, Guadalajara H, et al. First-in-human case study: pregnancy in women with Crohn's perianal fistula treated with adipose-derived stem cells: a safety study. Stem Cells Translational Medicine, 2015, 4:598-602.

[77] Panés J, García-Olmo D, Van Assche G, et al. Expanded allogeneic adipose-derived mesenchymal stem cells (Cx601) for complex perianal fistulas in Crohn's disease: a phase 3 randomised, double-blind controlled trial. The Lancet, 2016, 388:1281-1290.

[78] Panes J, Garcia-Olmo D, Van Assche G, et al. Long-term efficacy and safety of stem cell therapy (Cx601) for complex perianal fistulas in patients with Crohn's disease. Gastroenterology, 2018, 154:1334-1342.

[79] Dietz AB, Dozois EJ, Fletcher JG, et al. Autologous mesenchymal stem cells, applied in a bioabsorbable matrix, for treatment of perianal fistulas in patients with Crohn's disease. Gastroenterology, 2017, 153:59-62.

[80] Cho YB, Lee WY, Park KJ, et al. Autologous adipose tissue-derived stem cells for the treatment of Crohn's fistula: a phase I clinical study. Cell Transplantation, 2013, 22:279-285.

[81] Lee WY, Park KJ, Cho YB, et al. Autologous adipose tissue-derived stem cells treatment demonstrated favorable and sustainable therapeutic effect for Crohn's fistula. Stem Cells, 2013, 31:2575-2581.

[82] Cho YB, Park KJ, Yoon SN, et al. Long-term results of adipose-derived stem cell therapy for the treatment of Crohn's fistula. Stem Cells Translational Medicine, 2015, 4:532-537.

[83] Ciccocioppo R, Bernardo ME, Sgarella A, et al. Autologous bone marrow-derived mesenchymal stromal cells in the treatment of fistulising Crohn's disease. Gut, 2011, 60:788-798.

[84] Ciccocioppo R, Gallia A, Sgarella A, et al. Long-term follow-up of Crohn disease fistulas after local injections of bone marrow-derived mesenchymal stem cells. Mayo Clinic Proceedings, 2015, 90:747-755.

[85] Molendijk I, Bonsing BA, Roelofs H, et al. Allogeneic bone marrow-derived mesenchymal stromal cells promote healing of refractory perianal fistulas in patients with Crohn's disease. Gastroenterology, 2015, 149:918-927.

[86] Lightner AL, Dozois E, Fletcher JG, et al. Su1163 early results using an adipose derived mesenchymal stem cells coated fistula plug for the treatment of refractory perianal fistulizing Crohn's disease. Gastroenterology, 2016, 150:S483-S484.

第十四章 间充质干细胞在肝硬化治疗中的应用

肝硬化的转归通常是慢性肝衰竭，除肝移植外目前还没有有效的治疗手段，但肝移植手术风险高、肝源有限，多数患者不能得到有效的救治。间充质干细胞（MSC）是来源于发育早期中胚层的一类多能干细胞，具有增殖能力高、多向分化潜能的特性，属于成体干细胞的一种。由于 MSC 能够被诱导分化为肝细胞，同时 MSC 具有调节免疫反应、促进血管新生等功能，在肝硬化的治疗中显示出有广阔的应用前景。在本章节中，我们将详细描述 MSC 治疗肝硬化的临床前和临床研究，同时列举目前在线注册的 MSC 治疗肝硬化的临床试验。

第一节 肝硬化治疗现状

肝硬化是由一种或多种病因长期或反复作用导致的弥漫性肝损害。在我国大多数为肝炎后肝硬化，少部分为酒精性肝硬化和血吸虫性肝硬化。病理组织学上有广泛的肝细胞坏死、残存肝细胞结节性再生、结缔组织增生与纤维隔形成，导致肝小叶结构破坏和假小叶形成，肝脏逐渐变形、变硬而发展为肝硬化。

肝硬化的临床表现分为代偿期和失代偿期[1]。代偿期一般属 Child-Pugh A 级，可有肝炎临床表现，亦可隐匿起病。可有轻度乏力、腹胀、肝脾轻度肿大、轻度黄疸、肝掌、蜘蛛痣等表现。失代偿期一般属 Child-Pugh B、C 级，有肝功能受损及门脉高压症候群，具体表现有：①全身症状，如乏力、消瘦、面色晦暗、尿少、下肢水肿；②消化道症状，如食欲减退、腹胀、胃肠功能紊乱甚至吸收不良综合征、肝源性糖尿病，可出现多尿、多食等症状；③出血倾向及贫血，如齿龈出血、鼻衄、紫癜、贫血；④内分泌障碍，如蜘蛛痣、肝掌、皮肤色素沉着、女性月经失调、男性乳房发育、腮腺肿大；⑤低蛋白血症，如双下肢水肿、尿少、腹腔积液、肝源性胸腔积液；⑥门脉高压，如腹腔积液、胸腔积液、脾大、脾功能亢进、门脉侧支循环建立、食管-胃底静脉曲张，腹壁静脉曲张。肝硬化是由组织结构紊乱而致的肝功能障碍。

肝硬化目前尚无根治办法，主要预防手段为及早发现和阻止病程进展。治疗手段主要包括：①支持治疗。静脉输入高渗葡萄糖液以补充热量，输液中可加入维生素 C、胰岛素、氯化钾等。注意维持水、电解质、酸碱平衡。病情较重者可输入白蛋白、新鲜血浆。②药物治疗。可给予保肝、降酶、退黄等治疗，如葡醛内酯、维生素 C。必要时静脉输液治疗，如给予促肝细胞生长素、还原型谷胱甘肽、甘草酸类制剂等。门脉压力升高时给予药物如普萘洛尔、消心痛、钙通道阻滞剂等。③手术治疗。食管-胃底静脉曲张可采用内镜套扎手术；对于静脉曲张破裂高危或出血的患者，采取外科手术治疗，包括门-腔静脉分流术，门-奇静脉分流术和脾切除术等。

由肝硬化造成的终末期肝病，除对症治疗外，肝移植仍是最有效的手段。原位肝移植虽较为有效，但是因为费用高、肝源不足、免疫排斥等因素，不能大规模应用于临床。人工肝是肝功能衰竭治疗领域的研究热点，但其临床效果还没有得到确认[2]。在美国，姑息治疗是终末期肝病患者的主要治疗手段，采用姑息治疗能够减少患者的整体住院费用[3]。

MSC 主要存在于骨髓、脂肪、脐血、脐带或胎盘中。由于其具有体外扩增迅速、低免疫原性、遗传背景稳定等特点，同时可避免胚胎干细胞（ESC）的免疫排斥、伦理学争论等方面的问题，被认为在细胞治疗组织工程等领域具有广泛的临床应用前景。目前已有大量研究证实 MSC 能够阻止或减缓肝硬化进展[4,5]。

第二节　间充质干细胞治疗肝硬化的临床前研究

骨髓间充质干细胞（BM-MSC）能够在体内外分化为肝细胞，相比骨髓造血干细胞（BM-HSC），BM-MSC 分化为肝细胞的能力更强[6]。体外实验发现，当用肝细胞生长因子（HGF）、烟酰胺及地塞米松同时刺激人骨髓间充质干细胞（hBM-MSC）时，分化细胞中白蛋白（albumin）和甲胎蛋白（α-fetoprotein，AFP）的基因表达水平接近人的成熟肝细胞[7]。另有研究将大鼠 BM-MSC 在体外用 HGF 刺激 2 周后，通过尾静脉注射到肝损伤大鼠体内，发现 MSC 能够使大鼠血白蛋白水平恢复，抑制转氨酶活性及肝纤维化，这说明 BM-MSC 在体内可以分化为肝细胞，同时缓解肝纤维化程度[8]。Lam 等[9]将大鼠 BM-MSC 通过门静脉移植到由 D-氨基半乳糖诱导的急性肝损伤大鼠体内，发现移植的 MSC 能够较活跃地增殖、分化并提高肝细胞的增殖活性。同时他们发现，MSC 向肝细胞的定向分化是通过白介素-6（IL-6）/gp130 介导的 STAT3 信号通路的激活实现的。在用 BM-MSC 治疗肝硬化过程中，激素可能发挥一定的作用，因为雌二醇预处理能够促进 BM-MSC 增殖并降低 H_2O_2 诱导的细胞凋亡，进而提高 BM-MSC 治疗肝硬化的疗效[10]。另外，BM-MSC 治疗肝硬化可能与抑制肝星状细胞增殖有关，并且 BM-MSC 可能通过增加基质金属蛋白酶-1（MMP-1）的产生，抑制肝星状细胞中 MMP-1 和 2 的表达，从而起到抗纤维化的作用[11]。随后进一步深入研究发现 BM-MSC 也可能通过干扰转化生长因子-β（TGF-β）/Smad 信号转导通路，诱导肝星状细胞的凋亡，从而阻止肝纤维化的进展[12]。

相比于 BM-MSC，围产期 MSC 如脐血间充质干细胞（UCB-MSC）、脐带间充质干细胞（UC-MSC）和胎盘间充质干细胞（P-MSC）来源广泛、采集对供者无痛苦、免疫原性低的特点越来越受到国内外学者的重视。

研究发现，从脐血中分离的有核细胞在体外培养时加入生长因子和分化因子，21 天后有 50% 的细胞表达白蛋白，并且这些表达白蛋白的细胞能够在体外增殖。另外，在肝损伤免疫缺陷小鼠体内，脐血细胞能够在肝定植并分化为功能性肝细胞，因此脐血细胞可以作为肝祖细胞的来源[13]。将未分类的人脐血单个核细胞输入到免疫缺陷小鼠体内，发现肝内移植的脐血单个核细胞并不与小鼠肝细胞融合，可以独自分化为成熟的肝细胞，再次证明移植的脐血干细胞能够在体内通过转分化机制分化为肝细胞[14]。

将从人脐血中分离的 MSC 在含有 HGF 的培养基中诱导分化，发现培养 1 周后大部分白蛋白染色呈阳性，21 天后出现肝细胞样形态，在 28 天后为成熟肝细胞，6 周后能够摄取低密度脂蛋白（low density lipoprotein，LDL），而未分化的细胞却不能表达白蛋白和摄取 LDL [15]。Hong 等[16]将人脐血间充质干细胞（hUCB-MSC）培养在和 BM-MSC 相似的培养基中，发现 UCB-MSC 可以分化为表达白蛋白、AFP、细胞角蛋白 18 和 19 的肝细胞样细胞，这提示 UCB-MSC 在体外一定条件下可以分化为肝细胞[16]。另外，Jung 等[17]将 UCB-MSC 移植到经四氯化碳诱导的肝硬化大鼠体内，发现移植后大鼠胰的岛素抵抗明显得到改善，移植的 UCB-MSC 能够增加胰岛素信号通路的活性，改善肝硬化大鼠的糖代谢。

脐带外包被羊膜，内含黏液性的结缔组织，该结缔组织内有脐动脉和脐静脉。血管周围被黏蛋白样组织所包裹，后者称为华通氏胶（Wharton's jelly），它富含透明质酸，形成了成纤维细胞周围的水凝胶结构。目前的研究显示，脐带组织中的 MSC 主要分布在 2 个位置：脐血管周[18]和华通氏胶[19]。

UC-MSC 同样能够分化为肝细胞，Zhao 等[20]利用 HGF 和碱性成纤维细胞生长因子（bFGF）刺激人脐带间充质干细胞（hUC-MSC），MSC 呈现出高度分化的特性，分化的 MSC 在形态上具有肝细胞样特征，同时在基因和蛋白质水平表达肝细胞特有的标记，发挥分泌白蛋白、摄取 LDL 和产生尿素的功能。分化的 MSC 不表达主要组织相容性复合体 II（MHC II）类抗原，不会诱导淋巴细胞增殖。将分化的 MSC 通过尾静脉注入肝损伤免疫缺陷小鼠后，肝组织切片发现有人类特有的白蛋白表达。此外，将 hUC-MSC 分离鉴定后输注到经四氯化碳诱导的肝损伤小鼠体内，检测到人 AFP 和细胞角蛋白 18 的表达。此外，移植的 hUC-MSC 能够抑制肝细胞的凋亡并促进其增殖，降低血中转氨酶水平，减轻肝细胞变性[21]。Ren 等[22]将 hUC-MSC 输注到肝损伤的非肥胖糖尿病/重症联合免疫缺陷小鼠体内，发现 MSC 能够分化为肝细胞样细胞，但没有促进肝窦发生毛细血管化和小静脉化。

P-MSC 包括胎盘羊膜 MSC、绒毛膜 MSC 和底蜕膜 MSC。对 P-MSC 进行分离培养的历史并不长，最早报道于 2004 年[23,24]。与 BM-MSC 类似，P-MSC 也可用于修复组织，且具备免疫调节功能[25]。目前还未见到 P-MSC 治疗肝硬化的临床研究报道，但已发表的动物实验结果表明，P-MSC 对肝硬化具有治疗作用。南京医科大学的 Zhang 等[26]用从人胎盘羊膜中提取的 MSC 治疗经四氯化碳诱导的肝硬化小鼠模型，发现 P-MSC 能够迁移至肝组织并表达肝细胞的特异标记分子、人白蛋白和 AFP。另外，肝硬化小鼠肝纤维化的程度明显减轻，肝功能（血清丙氨酸氨基转移酶和天冬氨酸氨基转移酶水平）得到明显改善，肝星形细胞的激活被显著抑制，肝细胞凋亡减少，肝的再生能力增强。韩国国立江原大学的 Kim 研究团队[27]用胎盘绒毛膜来源的 MSC 治疗经四氯化碳诱导的肝硬化大鼠模型，发现 MSC 能显著抑制肝纤维化发展。动物实验的结果显示，P-MSC 能够迁移到由四氯化碳损伤的肝组织，并且显著降低 α-平滑肌肌动蛋（α-SMA）和 I 型胶原蛋白的表达水平，提升白蛋白和 MMP-9 的表达水平，同时使受损肝细胞摄取/分泌靛青绿（一种用作测试肝排泄能力的染料）的能力显著增强。Kim 研究团队进一步深入发现，自噬是 P-MSC 发挥治疗肝硬化作用的关键机制。他们发现，MSC 能显著降低肝硬化组织中凋亡相关激酶 Caspase 3/7 的活性，并能上调缺氧诱导因子-1α（hypoxia-inducible factor-1α，HIF-1α）及自噬、存活和再生因子的表达。另外，将胎盘绒毛膜 MSC 与经四氯化碳处理的原代肝细胞共培养，能显著减少坏死细胞数量和增加自噬信号。同时，HIF-1α 能通过自噬的机制促进受损肝细胞的再生[28]。

在 MSC 对肝硬化发挥治疗作用的机制方面，Xu 等[29]发现 MSC 能显著增加患者血清中抑制炎症反应的调节性 T 细胞（Treg）的水平，同时降低促进炎症反应的 Th17 细胞的水平，进而上调 Treg/Th17 值，抑制肝硬化中炎症反应。更进一步，在 mRNA 水平上，Treg 相关转录因子（Foxp3）表达上调，Th17 相关转录因子（RORγt）表达下调。另外，在 MSC 输注的早期阶段，血清中具有抑制炎症反应功能的 TGF-β 的水平显著升高，而具有促进炎症反应功能的 IL-17、肿瘤坏死因子-α（TNF-α）和 IL-6 的水平显著下降。

第三节　间充质干细胞治疗肝硬化的临床研究

随着对骨髓间充质干细胞（BM-MSC）分化为肝细胞的深入研究，利用自体 BM-MSC 治疗肝硬化的临床研究正在世界各地开展。Mohamadnejad 等[30]对 4 例失代偿期肝硬化患者进行自体 BM-MSC 治疗，体外培养 BM-MSC，并用流式细胞仪按照 MSC 表面标记蛋白对细胞进行分类后，经外周静脉将 MSC 输入患者体内，结果提示治疗安全可行，患者的终末期肝病评分显著提高，生活质量得到明显提升。另外，Kharaziha 等[31]研究了 8 例肝硬化患者，取髂骨 BM-MSC 通过外周静脉或门静脉输入患者体内，随访 24 周发现所有患者均较好地耐受治疗，肝功能得到明显改善，白蛋白水平上升，血肌酐、胆红素水平和国际标准化比值下降。为了研究由乙型肝炎病毒（HBV）引起的肝功能衰竭患者自体 BM-MSC 移植的有效性和长期预后，Peng 等[32]对 53 例肝硬化及无肝硬化的乙型肝炎肝功能衰竭患者进行了经肝动脉的自体 BM-MSC 移植，同时以 105 例患者作为对照。结果显示，治疗 2~3 周后，相比对照组，移植组患者血清白蛋白、总胆红素、凝血酶原时间及终末期肝病评分（model for end-stage liver disease score，MELD）显著改善。随访 192 周后两组患者肝癌的发生率及病死率差异无统计学意义。这说明自体 BM-MSC 移植治疗乙型肝炎肝功能衰竭具有良好的短期效果和安全性，但是远期结果不显著。

广州中山大学第三附属医院于 2017 年发表论文公布了一项研究异体 BM-MSC 治疗乙型肝炎相关慢加急性肝功能衰竭的随机对照临床试验。该随机对照临床试验共纳入 110 例患者，其中对照组 54 例，按照标准方案进行治疗；实验组 56 例，在标准治疗方案的基础上静脉输注异体 BM-MSC，每周输注一次，共输注 4 次。随访 24 周，研究结束时对照组的生存率为 55.6%，而输注 BM-MSC 的试验组患者生存率高达 73.2%，二者之间的差异有统计学意义。试验期间未发现干细胞输注相关不良反应，说明干细胞静脉输注安全性良好。随访结束时，试验组和对照组患者均无恶性肿瘤发生，排除了 MSC 致瘤的风险。试验组患者血清总胆红素的水平和肝病终末期评分的改善程度明显优于对照组。试验组患者严重感染的发生率（16.1%）显著低于对照组（33.3%），因多器官衰竭和严重感染造成的死亡率（17.9%）显著低于对照组（37.0%）。由此可见，外周静脉输注异体 BM-MSC 对乙型肝炎相关终末期肝病患者是安全的，并且能够通过改善患者肝功能和降低严重感染发生率来减少患者的死亡率[33]。

目前脐带间充质干细胞（UC-MSC）治疗肝硬化的临床研究也在逐步展开。将 hUC-MSC 每周 3 次通过外周静脉输注到 30 例乙型肝炎失代偿肝硬化患者体内，另外 15 例作为阴性对照组。随访 40 周后，治疗组患者白蛋白水平上升和腹水程度有所改善，并发症的发生率明显降低,说明 UC-MSC 治疗乙型肝炎肝硬化具有明显疗效[34]。于双杰等[35]给予 60 例乙型肝炎失代偿肝硬化患者综合治疗联合静脉输注 UC-MSC,每月 1 次,共输注 3 次,120 例乙型肝炎失代偿肝硬化患者作为对照组,仅给予综合治疗。结果表明,与对照组相比,试验组在回输 UC-MSC 12 周及 24 周后胆碱酯酶和白细胞水平显著升高($P<0.05$);12 周、24 周和 96 周时球蛋白水平明显升高($P<0.05$),同时碱性磷酸酶(ALP)水平也有明显升高($P<0.05$);Child-Pugh 评分在 12 周开始降低($P<0.05$)。但两组间丙氨酸转氨酶、天冬氨酸转氨酶、总胆红素、白蛋白、总胆固醇、甘油三酯、肾功能指标和凝血酶原活动度等指标的变化差异无统计学意义($P>0.05$)。研究期间没有发生明显不良反应。

UC-MSC 对于熊脱氧胆酸部分反应的原发性胆汁性肝硬化患者亦有一定疗效。每次间隔 4 周,共给予原发性胆汁性肝硬化患者 3 次 UC-MSC 治疗,大部分患者的症状如疲劳、瘙痒明显缓解;血清 ALP 和 γ-谷氨酰转移酶水平显著降低。但血清丙氨酸氨基转移酶、天冬氨酸转氨酶、总胆红素、白蛋白、凝血酶原时间、国际标准化比值、免疫球蛋白 M 水平无明显变化[36]。

UC-MSC 对自身免疫性肝炎相关肝硬化也能发挥治疗作用,这一研究是由南京鼓楼医院完成的[37]。总计 26 例患者通过外周静脉输注异体 MSC,其中 23 例接受 UC-MSC,2 例接受 UCB-MSC,1 例接受 BM-MSC。MSC 输注后的随访期间,有 3 例患者死于肝硬化相关的并发症,2 例患者接受肝移植。在 MSC 输注后的 6 个月、1 年和 2 年,患者血清丙氨酸转氨酶的水平均有所下降,但下降程度无统计学意义。在 6 个月和 1 年随访时,患者血清总胆红素水平明显下降。在 6 个月、1 年和 2 年随访时,患者血清白蛋白水平均有所改善,2 年时血清白蛋白的平均水平显著高于基线值。MSC 移植后 6 个月患者凝血酶原时间降低。MELD 在随访 6 个月、1 年和 2 年时均有所改善。26 例肝硬化患者在 MSC 输注后 24h 内均无严重不良反应发生。这一研究说明异体 MSC 静脉输注治疗由自身免疫性肝炎造成的肝硬化是安全和有效的,但未来还需要更大规模的随机对照临床试验进一步验证异体 MSC 的有效性。

解放军第 302 医院王福生院士团队[38]应用 UC-MSC 对乙型肝炎相关慢加急肝衰竭进行治疗性研究,24 例慢加急肝衰竭患者接受 UC-MSC 输注 3 次,每次间隔 4 周,研究期间未发生不良反应。与对照组相比,UC-MSC 输注显著提高患者生存率,改善终末期肝病评分,患者血清白蛋白水平、胆碱酯酶水平、凝血酶原活动度和血小板计数增加,血清总胆红素、丙氨酸氨基转移酶水平显著降低。

在一篇尚未发表的病例报道中,王伟强等应用胎盘绒毛膜来源的 MSC 治疗了一例 41 岁男性酒精性肝硬化腹水患者。P-MSC 通过静脉输注,每次输注剂量为 1×10^7cells/kg 体重,共输注 3 次,每次间隔 1 个月。治疗结束后,B 超显示患者腹水深度从 10.37cm 下降到 1.9cm(大量腹水转变为少量腹水),糖类抗原 125 的血清浓度从 156U/mL 下降到 30U/mL(正常范围:0~35U/mL),白蛋白的血清浓度从 25.3g/L 恢复至 44.9g/L(正常范围:33~55g/L),患者精神状态明显好转。治疗后 1 年随访,各项临床指标仍维持在治疗结束时的状态。这提示 P-MSC 对肝硬化具有良好的治疗作用,但更确切的结论还需要大规模随机对照临床研究结果进行证实。

第四节 间充质干细胞治疗肝硬化的临床试验

在中国临床试验注册中心注册开展的 MSC 治疗肝硬化的临床试验有两项，如表 14-1 所示。

表 14-1 在中国临床试验注册中心注册开展的 MSC 治疗肝硬化的临床试验

注册题目	注册编号	干细胞种类	承担单位	实施时间	临床分期
干细胞移植治疗肝硬化的疗效分析	ChiCTR-TNRC-11001488	脐带间充质干细胞	武警总医院，中国	2011-09-15 至 2013-05-01	Ⅱ 期
自体骨髓干细胞移植治疗肝硬化的临床研究	ChiCTR-ONRC-12001892	骨髓间充质干细胞	解放军第 452 医院，中国	2010-09-07 至 2015-09-07	Ⅰ 期+Ⅱ 期

在美国国立卫生研究院（NIH）临床试验网站（ClinicalTrials.gov）注册的 MSC 治疗肝硬化的临床试验有 10 项，如表 14-2 所示。

表 14-2 在 ClinicalTrials.gov 网站注册的间充质干细胞治疗肝硬化的临床试验

注册题目	注册编号	干细胞种类	承担单位	完成情况	临床分期
Human Umbilical Cord Mesenchymal Stem Cells Transplantation for Patients With Decompensated Liver Cirrhosis	NCT01342250	脐带间充质干细胞	Shenzhen Beike Bio-Technology Co., Ltd，中国	完成	Ⅰ 期+Ⅱ 期
Human Umbilical Cord-Mesenchymal Stem Cells for Hepatic Cirrhosis	NCT02652351	脐带间充质干细胞	Shenzhen Hornetcorn Bio-technology Co., Ltd，中国	进行中	Ⅰ 期
Mesenchymal Stem Cells Transplantation for Liver Cirrhosis Due to HCV Hepatitis	NCT02705742	自体脂肪间充质干细胞	Saglik Bilimleri Universitesi Gulhane Tip Fakultesi，土耳其	进行中	Ⅰ 期+Ⅱ 期
Clinical Trial Study About Human Adipose-Derived Stem Cells in the Liver Cirrhosis	NCT02297867	脂肪间充质干细胞	Gwo Xi Stem Cell Applied Technology Co., Ltd，中国台湾	进行中	Ⅰ 期
Dose Finding Study to Assess Safety and Efficacy of Stem Cells in Liver Cirrhosis	NCT01591200	骨髓间充质干细胞	Centre for Liver Research & Diagnostics，印度	完成	Ⅱ 期
REVIVE (Randomized Exploratory Clinical Trial to Evaluate the Safety and Effectiveness of Stem Cell Product in Alcoholic Liver Cirrhosis Patient)	NCT01875081	自体骨髓间充质干细胞	Pharmicell Co., Ltd，韩国	完成	Ⅱ 期
Improvement of Liver Function in Liver Cirrhosis Patients After Autologous Mesenchymal Stem Cell Injection:a Phase Ⅰ-Ⅱ Clinical Trial	NCT00420134	自体间充质干细胞诱导分化的肝祖细胞	Research Center for Gastroenterology and Liver Diseases，伊朗	完成	Ⅰ 期+Ⅱ 期
Efficacy of In vitro Expanded Bone Marrow Derived Allogeneic Mesenchymal Stem Cell Transplantation Via Portal Vein or Hepatic Artery or Peripheral Vein in Patients With Wilson Cirrhosis	NCT01378182	骨髓间充质干细胞	Department of Gastroenterology; Gulhane Military Medical Academy，土耳其	完成	Ⅰ 期
Transplantation of Autologous Mesenchymal Stem Cell in Decompensate Cirrhotic Patients With Pioglitazone	NCT01454336	自体骨髓间充质干细胞	Royan Institute，伊朗	完成	Ⅰ 期
Injectable Collagen Scaffold™ Combined With hUC-MSC Transplantation for Patients With Decompensated Cirrhosis	NCT02786017	脐带间充质干细胞	南京鼓楼医院，中国	进行中	Ⅰ 期+Ⅱ 期

（王伟强　林美光）

参 考 文 献

[1] Heidelbaugh JJ, Sherbondy M. Cirrhosis and chronic liver failure: part Ⅱ. Complications and treatment. Am Fam Physician, 2006, 74(5):767-776.

[2] 贾继东. 终末期肝病的诊疗现状. 中华肝脏病杂志, 2007, 15(6):401-402.

[3] Patel AA, Walling AM, May FP, et al. Palliative care and health care utilization for patients with end-stage liver disease at the end of life. Clin Gastroenterol Hepatol, 2017, 15(10):1612-1619.

[4] 张立婷. 间充质干细胞在肝硬化基础研究和临床应用中的进展. 临床肝胆病杂志, 2016, 32(6):1196-1198.

[5] Kim G, Eom YW, Baik SK, et al. Therapeutic effects of mesenchymal stem cells for patients with chronic liver diseases: systematic review and meta-analysis. J Korean Med Sci, 2015, 30(10):1405-1415.

[6] Sato Y, Araki H, Kato J, et al. Human mesenchymal stem cells xenografted directly to rat liver are differentiated into human hepatocytes without fusion. Blood, 2005, 106(2):756-763.

[7] Chivu M, Dima SO, Stancu CI, et al. *In vitro* hepatic differentiation of human bone marrow mesenchymal stem cells under differential exposure to liver-specific factors. Transl Res, 2009, 154(3):122-132.

[8] Oyagi S, Hirose M, Kojima M, et al. Therapeutic effect of transplanting HGF-treated bone marrow mesenchymal cells into CCl4-injured rats. J Hepatol, 2006, 44(4):742-748.

[9] Lam SP, Luk JM, Man K, et al. Activation of interleukin-6-induced glycoprotein 130/signal transducer and activator of transcription 3 pathway in mesenchymal stem cells enhances hepatic differentiation, proliferation, and liver regeneration. Liver Transpl, 2010, 16(10):1195-1206.

[10] 颜小明, 张立婷, 骆菁怡, 等. 雌二醇预处理的人骨髓间充质干细胞对实验性小鼠肝硬化模型的疗效对比研究, 临床肝胆病杂志, 2015, 31(3):424-430.

[11] 张立婷, 王珊, 李俊峰, 等. 人骨髓间充质干细胞上清液对体外肝星状细胞增殖周期及MMP-1表达的影响. 临床肝胆病杂志, 2012, 28(11):836-838.

[12] Zhang LT, Fang XQ, Chen QF, et al. Bone marrow-derived mesenchymal stem cells inhibit the proliferation of hepatic stellate cells by inhibiting the transforming growth factor β pathway. Mol Med Rep, 2015, 12(5):7227-7232.

[13] Kakinuma S, Tanaka Y, Chinzei R, et al. Human umbilical cord blood as a source of transplantable hepatic progenitor cells. Stem Cells, 2003, 21(2):217-227.

[14] Newsome PN, Johannessen I, Boyle S, et al. Human cord blood-derived cells can differentiate into hepatocytes in the mouse liver with no evidence of cellular fusion. Gastroenterology, 2003, 124(7):1891-900.

[15] Lee OK, Kuo TK, Chen WM, et al. Isolation of multipotent mesenchymal stem cells from umbilical cord blood. Blood, 2004, 103(5):1669-1675.

[16] Hong SH, Gang EJ, Jeong JA, et al. *In vitro* differentiation of human umbilical cord blood-derived mesenchymal stem cells into hepatocyte-like cells. Biochem Biophys Res Commun, 2005, 330(4): 1153-1161.

[17] Jung KH, Uhm YK, Lim YJ, et al. Human umbilical cord blood-derived mesenchymal stem cells improve glucose homeostasis in rats with liver cirrhosis. Int J Oncol, 2011, 39(1):137-443.

[18] Covas DT, Siufi JL, Silva AR, et al. Isolation and culture of umbilical vein mesenchymal stem cells. Braz J Med Biol Res, 2003, 36(9):1179-1183.

[19] Wang HS, Hung SC, Peng ST, et al. Mesenchymal stem cells in the Wharton's jelly of the human umbilical cord. Stem Cells, 2004, 22(7):1330-1337.

[20] Zhao Q, Ren H, Li X, et al. Differentiation of human umbilical cord mesenchymal stromal cells into low immunogenic hepatocyte-like cells. Cytotherapy, 2009, 11(4):414-426.

[21] Yan Y, Xu W, Qian H, et al. Mesenchymal stem cells from human umbilical cords ameliorate mouse hepatic injury *in vivo*. Liver Int, 2009, 29(3):356-365.

[22] Ren H, Zhao Q, Cheng T, et al. No contribution of umbilical cord mesenchymal stromal cells to capillarization and vanularization of hepatic sinusoids accompanied by hepatic differentiation in carbon tetrachloride-induced mouse liver fibrosis. Cytotherapy, 2010, 12(3):371-383.

[23] In't APS, Scherjon SA, der Keur CK, et al. Isolation of mesenchymal stem cells of fetal or maternal origin from human placenta. Stem Cells, 2004, 22(7):1338-1345.

[24] Fukuchi Y, Nakajima H, Sugiyama D, et al. Human placenta-derived cells have mesenchymal stem/progenitor cell potential. Stem Cells, 2004, 22(5):649-658.

[25] Zhang Y, Li CD, Jiang XX, et al. Comparison of mesenchymal stem cells from human placenta and bone marrow. Chin Med J (Engl), 2004, 117(6):882-887.

[26] Zhang D, Jiang M, Miao D. Transplanted human amniotic membrane-derived mesenchymal stem cells ameliorate carbon tetrachloride-induced liver cirrhosis in mouse. PLoS One, 2011, 6(2):e16789.

[27] Lee MJ, Jung J, Na KH, et al. Anti-fibrotic effect of chorionic plate-derived mesenchymal stem cells isolated from human placenta in a rat model of CCl(4)-injured liver: potential application to the treatment of hepatic diseases. J Cell Biochem, 2010, 111(6):1453-1463.

[28] Jung J, Choi JH, Lee Y, et al. Human placenta-derived mesenchymal stem cells promote hepatic regeneration in CCl4-injured rat liver model via increased autophagic mechanism. Stem Cells, 2013, 31(8):1584-1596.

[29] Xu L, Gong Y, Wang B, et al. Randomized trial of autologous bone marrow mesenchymal stem cells transplantation for hepatitis B virus cirrhosis: regulation of Treg/Th17 cells. J Gastroenterol Hepatol, 2014, 29(8):1620-1628.

[30] Mohamadnejad M, Alimoghaddam K, Mohyeddin-Bonab M, et al. Phase 1 trial of autologous bone marrow mesenchymal stem cell transplantation in patients with decompensated liver cirrhosis. Arch Iran Med, 2007, 10(4):459-466.

[31] Kharaziha P, Hellström PM, Noorinayer B, et al. Improvement of liver function in liver cirrhosis patients after autologous mesenchymal stem cell injection: a phase Ⅰ-Ⅱ clinical trial. Eur J Gastroenterol Hepatol, 2009, 21(10):1199-1205.

[32] Peng L, Xie DY, Lin BL, et al. Autologous bone marrow mesenchymal stem cell transplantation in liver failure patients caused by hepatitis B: short-term and long-term outcomes. Hepatology, 2011, 54(3):820-828.

[33] Lin BL, Chen JF, Qiu WH, et al. Allogeneic bone marrow-derived mesenchymal stromal cells for hepatitis B virus-related acute-on-chronic liver failure: a randomized controlled trial. Hepatology, 2017, 66(1):209-219.

[34] Zhang Z, Lin H, Shi M, et al. Human umbilical cord mesenchymal stem cells improve liver function and ascites in decompensated liver cirrhosis patients. J Gastroenterol Hepatol, 2012, 27 Suppl 2:112-20.

[35] 于双杰, 陈黎明, 吕飒, 等. 人脐带间充质干细胞治疗失代偿性乙型肝炎肝硬化的安全性与疗效. 中华肝脏病杂志, 2016, 24(1):51-55.

[36] Wang L, Li J, Liu H, et al. Pilot study of umbilical cord-derived mesenchymal stem cell transfusion in patients with primary biliary cirrhosis. J Gastroenterol Hepatol, 2013, 28 Suppl 1:85-92.

[37] Liang J, Zhang H, Zhao C, et al. Effects of allogeneic mesenchymal stem cell transplantation in the treatment of liver cirrhosis caused by autoimmune diseases. Int J Rheum Dis, 2017, 20(9):1219-1226.

[38] Shi M, Zhang Z, Xu R, et al. Human mesenchymal stem cell transfusion is safe and improves liver function in acute-on-chronic liver failure patients. Stem Cells Transl Med, 2012, 1(10):725-731.

第十五章 间充质干细胞在糖尿病治疗中的应用

第一节 糖尿病的概述

糖尿病是由于胰岛素分泌缺乏和/或胰岛素抵抗引发的以高血糖为表现形式的糖、脂肪、蛋白质代谢紊乱综合征。近年来，随着人民生活水平的提高及生活方式的改变，以及人口老龄化，糖尿病发病率逐渐上升，2013年我国糖尿病患患者数量高达1.2亿，糖尿病前期人数达4亿[1]，严重威胁着人们的身体健康和生活质量。更加令人担忧的是，我国一项流行病学调研发现血糖控制达标标率只有32%[2]。长期的高血糖状态会影响体内各种激素代谢、能量消耗、营养摄取，引发糖尿病肾病、糖尿病大血管病变（心肌梗死、脑梗死）等一系列并发症，严重影响人们的身体健康和生活质量[3]，甚至导致患者残疾或死亡。有效控制和治疗糖尿病已成为一项刻不容缓的战略任务。

胰岛β细胞功能衰竭和胰岛素抵抗是导致糖尿病发生发展的主要机制，目前的抗糖尿病药物并不是针对糖尿病发病的关键环节，只能解除或缓解症状，延缓疾病进展，不能从根本上治愈该病。而占总发病率90%以上的是2型糖尿病[4]，其主要病理生理特征是由胰岛素抵抗、胰岛细胞功能进行性下降所致的血糖升高与胰岛失能[5]。因此，针对2型糖尿病的主要治疗手段是减轻胰岛素抵抗、降低血糖及恢复胰岛功能。胰岛素治疗虽能控制症状、延缓和减少并发症的发生，但不能使糖尿病彻底治愈，长期使用不仅会对外源性胰岛素抵抗、失敏，还会产生诸如失明、肾衰竭、肥胖等严重的并发症，并有增加肿瘤发生的可能性。此外，每天注射胰岛素本身也会给患者带来极大的痛苦，明显降低患者生活质量，缩短其期望寿命。同时人为注射胰岛素并不能发挥精确调节血糖的功能，也阻止不了各种器官的功能损伤。20世纪初，就有人提出胰腺移植可以治疗糖尿病的设想，但因供体缺乏和移植后免疫排斥问题制约了其广泛应用。而2型糖尿病因其突出的胰岛素抵抗的病理机制，仅靠补充外源性胰岛素包括胰岛移植亦不能从根本上治疗此疾病。由于无法完全模拟生理状态下胰岛素的调节，加之胰岛素抵抗，糖尿病患者会随着疾病进程，用药量越来越大，随之而来的是不可避免的并发症的发生[6]。减少胰岛素抵抗，阻止细胞功能衰退，增加细胞数量，最大限度地挽救和恢复胰岛功能，成为治疗糖尿病的新思路，有望从根本上改善胰岛功能，有效控制血糖并防止并发症的发生。

间充质干细胞（MSC）所具有的独特生物学特性、极强的自我更新能力、多向分化潜能及可分泌多种细胞因子，特别是近年来MSC的基础与临床前应用研究取得了长足的进展，为MSC治疗糖尿病带来了希望。MSC治疗能够使部分糖尿病患者停用胰岛素或减少胰岛素用量，干细胞治疗糖尿病的有效性及安全性也得到了有效验证[7]。

第二节　间充质干细胞治疗糖尿病的机制

尽管糖尿病分为以胰岛 β 细胞受到细胞介导的自身免疫性破坏为特征的胰岛素绝对缺乏的 1 型，以及起始于胰岛素抵抗与胰岛功能障碍的 2 型糖尿病，但是到了后期均表现为由胰岛 β 细胞损害导致胰岛素分泌绝对或相对不足，因此重建患者体内的功能性胰岛 β 细胞总量是治疗糖尿病的理想目标。基于 MSC 具有的自我更新和不断增殖、多向分化潜能的特点，早期的研究试图从两个方面解决重建功能性胰岛 β 细胞的问题：①体外采用适宜的诱导条件，将 MSC 诱导分化为胰岛素分泌细胞，用于治疗糖尿病。在现有的条件下，无论如何体外诱导，其效率还无法办法满足临床治疗所需[7,8]。②通过系统或胰岛局部直接移植的 MSC，归巢到受损胰岛的局部，分化为胰岛 β 细胞。然而，随着基础与临床研究的进一步深入，发现 MSC 治疗后，虽然胰岛功能明显提高，但动物模型中受损的胰岛中并没有发现移植 MSC 的存在[9]。越来越多研究显示，MSC 发挥降低血糖、恢复胰岛功能的效应主要通过以下几个方面：①MSC 通过其分泌特性分泌了多种细胞因子，改善了胰岛素靶组织的胰岛素抵抗，促进了组织对糖的吸收，从而改善了体内高血糖状态，减轻了代谢紊乱对胰岛的进行性损害，缓解或再生了胰岛 β 细胞；②MSC 通过分泌多种细胞因子改善了微环境，诱导了受损胰岛内 α 细胞向 β 细胞的转化，从而实现了胰岛 β 细胞的原位再生，改善了体内的高血糖状态；③MSC 通过免疫抑制作用，抑制了 T 细胞介导的对新生 β 细胞的免疫反应，改善了微环境的营养，促进了胰岛的修复与再生。

一、MSC 改善外周组织胰岛素抵抗

2 型糖尿病的病理基础是肥胖引起机体低浓度的慢性炎症，从而诱发胰岛素抵抗，以及由此引起的糖与脂代谢紊乱诱发的胰岛分泌胰岛素能力的进行性下降，因此治疗的策略除了修复胰岛功能以外，更重要的是改善外周靶组织对胰岛素的吸收。Si 等[10]于 2012 年报道用高脂高糖饮食加链脲佐菌素（streptozotocin，STZ）注射建立 2 型糖尿病大鼠模型，并注射同种异体骨髓间充质干细胞（BM-MSC），7 天后大鼠血糖较输注前显著下降，治疗效果持续 2 周，而后血糖再次升高。另外，第 21 天再次输注后大鼠血糖持续下降，且下降幅度大于第一次输注。在进一步对机制进行解析时发现，BM-MSC 通过活化骨骼肌、脂肪和肝胰岛素受体底物-1（insulin receptor substrate-1，IRS-1）-Akt-GLUT 信号通路来改善外周组织胰岛素抵抗而达到降低血糖的目的。这个实验结果为 MSC 治疗糖尿病提供了新的理论基础，即 MSC 可能通过分泌效应来改善胰岛素抵抗，从而发挥降糖效应。

在新的发现提出后，陆续的研究结果证实了这个理论。Hao 等[11]用经高脂高糖饮食加 STZ 注射建立的 2 型糖尿病小鼠模型，证实了连续注射 BM-MSC，每周 1 次，至第 6 次后小鼠的血糖逐步降到接近正常水平，与炎症相关的胰岛素抵抗的指标也伴随血糖接近正常而恢复正常。结果不仅进一步验证了 MSC 通过改善胰岛素抵抗来降低血糖的理论，而且通过多次注射方式可以发挥持续的治疗效应，从而彻底修复外周靶组织受损的胰岛素吸收能力。

胰岛素抵抗被认为与系统低浓度慢性炎症密切相关,而在胰岛素抵抗的发生发展中,脂肪组织中 M1 型炎性状态的巨噬细胞分泌的诸如肿瘤坏死因子-α（TNF-α）、白介素-1β（IL-1β）等细胞因子是重要触发因素,而巨噬细胞 M2 型,一种抗炎状态的巨噬细胞,已经被证实可以发挥改善胰岛素抵抗的作用[12]。脐带间充质干细胞（UC-MSC）被用于治疗 2 型糖尿病大鼠,发挥的改善胰岛素抵抗从而降糖的效应明显促进了脂肪组织中巨噬细胞的极化（M1 型向 M2 型的转变）[13]。近期研究显示,BM-MSC 通过抑制炎症小体 NLRP3 的表达降低全身慢性低度炎症来改善大鼠外周靶组织胰岛素抵抗。巨噬细胞 M2 型改善胰岛素抵抗的证据新近也在由高糖高脂饮食诱导的肥胖小鼠动物模型中得到证实,这个研究采用体外诱导的同种异体巨噬细胞 M2 型静脉回输的方法,成功降低了小鼠的体重,改善了脂肪组织中的胰岛素抵抗。

二、MSC 改善胰岛 β 细胞功能

MSC 通过改善胰岛 β 细胞功能来降低血糖是目前普遍认可的理论假设,这个理论在 MSC 治疗糖尿病的基础与临床试验中均得到了证实。但 MSC 是否通过定向分化为胰岛素分泌细胞来达到改善血糖的目的目前尚有争议。支持的观点是：①MSC 具有多向分化潜能,在特定适宜的条件下可以分化为胰岛素分泌细胞,移植到动物模型体内评测[14],可以增加胰岛素和 C 肽分泌；②MSC 具有趋化特性,在另外的糖尿病动物模型中,找到了输注的 MSC 归巢到受损胰腺组织的证据[15]；③MSC 具有促进细胞增殖的能力,有研究证实了移植 MSC 可促进动物模型胰岛的再生,结果证明 MSC 可促进胰岛细胞的增殖,增加胰岛素分泌细胞的数量[16]。然而到目前为止,还没有很确切的证据能证明胰岛功能恢复与 MSC 归巢到受损胰岛后直接分化成胰岛素分泌细胞相关[17]。

（一）MSC 促进胰岛 β 细胞再生

越来越多的证据表明,MSC 通过自分泌与旁分泌效应,分泌细胞因子与生长因子参与修复过程[18]。MSC 输注后,无论是大鼠还是小鼠糖尿病模型,无论是 1 型还是 2 型动物模型,1 次或多次的治疗,均发现受损胰岛 β 细胞从结构到胰岛数目都可以再生修复到接近正常的状态[19]。同时用培养 MSC 的条件培养液输注治疗,也可以达到类似的修复效果,尽管治疗效应要弱些,但是从另外一个方面证实了 MSC 通过分泌效应发挥治疗作用。MSC 分泌的细胞因子包括血管内皮生长因子（VEGF）、胰岛素样生长因子-1（IGF-1）、血小板源生长因子-BB（PDGF-BB）、血管生成素-1（Ang-1）,在修复与再生胰岛中均扮演了重要角色[20]。

（二）MSC 促进 α 细胞 β 化再生胰岛

研究发现,胰岛中的 α 细胞与 β 细胞,在 β 细胞完全破坏的极端情况下,大约 1% 的 α 细胞会转分化为 β 细胞。这一现象为人们治疗糖尿病,寻求实现 α 细胞 β 化原位再生胰岛,提供了新的治疗思路[21]。

Cheng 等[22]研究发现,在 STZ 注射后的小鼠糖尿病模型中,胰岛 α 细胞在 12h 出现上皮间质化的过程,在此过程中发现许多的同时具有 α 细胞与 β 细胞特点的细胞。采用

α细胞谱系追踪技术，研究者发现，α细胞部分转变为β细胞。进一步的研究解析了巨噬细胞M2型，通过转化生长因子-β（TGF-β）通路，促进了α细胞的β化。这个研究小组的另外一个实验结果（未发表）显示，BM-MSC单次输注，明显促进了STZ诱导的糖尿病小鼠胰岛功能的恢复。单次MSC输注，小鼠血糖并不能恢复到正常水平，并且2周后血糖又开始升高。研究者采用α细胞与β细胞谱系追踪技术，验证了部分α细胞转化为β细胞，并发现MSC的分泌效应通过激活巨噬细胞M2极化来实现。

（三）MSC修复受损的胰岛β细胞

MSC的免疫调节作用在受损胰岛的修复中同样发挥了重要作用。MSC具有非人白细胞抗原（HLA）限制性的免疫抑制特性，因此能够针对1型糖尿病患者及动物模型中机体对自身胰岛β细胞的免疫攻击发挥治疗作用，这已经被多项研究所证实。MSC可抑制淋巴细胞的增殖，可通过使T细胞从Th1向Th2转换，使效应T细胞分泌的细胞因子从炎性细胞因子向抗炎因子转换。MSC还可以增加调节性T细胞（Treg）比例，进而发挥抗炎作用。

MSC的输注可明显逆转糖尿病的高血糖状态已经被广泛证实，其修复胰岛功能、重建胰岛α细胞与β细胞结构与MSC的免疫调节功能密切相关[23-25]。新近的一项多次经STZ注射诱导小鼠糖尿病模型的研究显示，经过STZ注射后胰腺中的β细胞没有完全被破坏，但激活了自体T细胞攻击，残存的β细胞逐步遭到破坏。在注射STZ后25天，接受MSC回输。结果发现，小鼠的高血糖状态被逆转，胰岛功能与结构被修复，但是没有发现MSC源的胰岛素分泌细胞，而在回输后7天和65天在第二淋巴器官发现回输的MSC。进一步解析的结果显示，MSC治疗糖尿病的效应与修复Th1和Th2的平衡进而改造胰岛微环境，从而保护了一部分胰岛β细胞免受持续的损伤有关。

细胞的自噬能力是维持细胞稳态、修复损伤的重要生理功能，自噬作用清除细胞内受损的细胞器与毒性蛋白，起到"去腐生新"的作用。在许多疾病发生发展过程中，细胞自身的自噬能力受到影响，无法及时清除受损的细胞器与毒性蛋白。在糖尿病发病的早期，一些诸如免疫、慢性炎症、糖毒性、脂毒性等因素逐渐影响胰岛β细胞分泌胰岛素的能力，最后彻底破坏胰岛功能，糖尿病患者只能靠补充外源胰岛素维持正常生理功能。Zhao等[26]研究发现，BM-MSC的分泌效应在体外可以提高胰岛细胞（INS-1）的自噬能力，抵抗高糖削弱细胞分泌胰岛素的作用。在高糖高脂加STZ注射的2型糖尿病大鼠模型中，输注BM-MSC，高血糖被明显调节，胰岛功能显著提升，胰岛素及C肽分泌能力得到修复。进一步对胰岛 细胞解析发现，β细胞的与自噬相关的蛋白表达水平明显增强，细胞内自噬小体明显增多。

第三节　间充质干细胞治疗糖尿病的临床试验进展

第一次使用MSC的临床试验早在1995年就完成了，15例血液肿瘤患者成为自体BM-MSC的受者[27]。在动物模型中已经证实了MSC治疗可以降低血糖，主要通过修复和再生胰岛功能与改善胰岛素抵抗等起作用。MSC的临床试验研究逐步开展起来，目前干细胞治疗糖尿病的临床试验研究已在美国国立卫生研究院（NIH）临床试验网站（ClinicalTrials.gov）上注册的高达 186 项，其中 MSC 治疗糖尿病的临床试验有 26 个。已发表的研究论文有 13 篇，研究结果显示，MSC 可以通过改善胰岛素 β 细胞功能或/和改善外周靶组织胰岛素抵抗来有效改善糖尿病患者的高血糖状态。

一、MSC 治疗 1 型糖尿病

2007 年 Voltarelli 等[28]第一次在 JAMA 期刊报道了干细胞移植治疗 1 型糖尿病的临床试验研究结果。方案首先采用药物抑制免疫系统，防止免疫系统对移植细胞产生攻击，然后施以自体非清髓造血干细胞移植（HSCT）试图阻止 β 细胞的进一步损伤与修复 β 细胞的功能。15 例不伴有糖尿病酮症酸中毒的新诊断 1 型糖尿病患者（病程小于 6 周）静脉输注干细胞后，14 例患者停用外源性胰岛素，其中 1 例患者停用胰岛素时间为 25 个月，4 例患者为 21 个月，7 例患者为 6 个月，另外 2 例患者分别为 1 个月和 5 个月；静脉输注干细胞后 6 个月 C 肽曲线下面积较输注前明显增大，且维持 24 周不变，提示可改善胰岛 β 细胞功能。另外，13 例患者的糖化血红蛋白（HbA1c）稳定在 7%以下。结果显示了干细胞治疗 1 型糖尿病的有效性、安全性、可行性，开创性地提出治疗 1 型糖尿病的有效策略。治疗相关不良反应，包括 1 例晚期出现肺炎，2 例发展为晚期内分泌失调，多数出现中性粒细胞减少、恶心、呕吐等，主要由免疫抑制所致。随后，针对第一次临床试验研究所存在的一些限制，Couri 等[29]设计了另外的长期临床试验研究。入选的 15 例新发 1 型糖尿病患者和另外 8 例患者，在干细胞治疗后长达 58 个月的时间内，20 例患者停用胰岛素，治疗效应平均 32 个月，最长持续 52 个月。研究结果进一步支持了干细胞治疗 1 型糖尿病的可靠性与有效性，只是仍然没有随机对照、小样本研究。

HSCT 治疗 1 型糖尿病的发现为 MSC 治疗 1 型糖尿病提供了实践基础。随后一个随机对照临床试验研究，入组了 29 例平均年龄为 17.6 岁的 1 型糖尿病患者，观察了 UC-MSC 的长期治疗效应[30]。试验组在接受常规胰岛素治疗的同时，施以平均 $2.6×10^7$cells± $1.2×10^7$cells UC-MSC 细胞治疗。结果发现，治疗后治疗组的 C 肽水平从治疗前的 0.8ng/mL±0.074ng/mL 提升到 1.4ng/mL±0.09ng/mL，平均 HbA1c 水平从治疗前的 6.8%±0.57%下降到 6.1%±0.67%。4 例患者脱离胰岛素治疗，7 例患者日胰岛素用量下降一半的疗效持续时间为 20~22 个月。结果证实，UC-MSC 治疗 1 型糖尿病可以达到一个稳定、长期的效果。

另外的一个使用 BM-MSC 的单盲随机临床试验，入组 20 例患者，患者年龄 24 岁±2 岁，病程小于 3 个月。治疗组接受了平均 $2.75×10^6$cells/kg 自体 BM-MSC，C 肽和 HbA1c 12 月后均明显得到改善[31]。

二、MSC 治疗 2 型糖尿病的临床试验

2009 年，Bhansali 等[32]首次发表了使用 BM-MSC 治疗 2 型糖尿病的试验结果，自此共有 4 个研究采用 BM-MSC 治疗 2 型糖尿病，涉及 189 例患者（包括 82 例对照）[33-35]，患者的平均年龄为 53.7 岁±3.5 岁。干细胞治疗后 12 月，C 肽水平明显增高，从治疗前的 1.6ng/mL±0.5ng/mL 增加到 2.5ng/mL±1.13ng/mL，HbA1c 水平从治疗前的 7.86%±0.68%下

降到 7.125%±0.30%。其中 18 例患者脱离胰岛素治疗，47 例患者的胰岛素日用量减量超过 50%，8 例患者报道存在包括恶心、腹痛等轻微的不良反应。结果显示了 BM-MSC 治疗 2 型糖尿病的良好安全性与有效性。

另外使用 UC-MSC 的一项研究显示[36]，22 例患者入组，平均年龄为 52.9 岁±10.5 岁，2 型糖尿病史 8.7 年±4.3 年，分别经胰背动脉、静脉输注共平均 $1×10^6$cells/kg 体重。基础 C 肽水平为 1.29ng/mL±0.83ng/mL，治疗后 6 个月升高到 1.95ng/mL±1.3ng/mL，12 月略微下降到 1.86ng/mL±1.0ng/mL。平均 HbA1c 水平是 8.20%±1.69%，治疗后 6 个月下降到 6.91%±0.96%，至 12 个月时略微增高到 7.0%±0.6%。平均的胰岛素需求量从基础的 0.49U/kg±0.22U/kg 减少到 12 个月后的 0.23U/kg±0.19U/kg。结果证实了 UC-MSC 治疗 2 型糖尿病的有效性。

2012 年，由中国医学科学院血液学研究所牵头的一个 2 型糖尿病的小样本临床试验研究[37]，所有接受移植的 10 名患者，胰岛素的日用量由 63.7IU±18.7IU 降低至 34.7IU±13.4IU（$P<0.01$），C 肽水平明显增多，由 4.1ng/mL±3.7ng/mL 上升至 5.6ng/mL±3.8ng/mL（$P<0.05$）。未见诸如发热、寒战、肝损伤和其他不良反应，同时改进了肾、心脏功能。

第四节 问题与展望

MSC 治疗糖尿病的基础和临床研究的开展为糖尿病患者点燃了新的希望，为未来彻底治愈糖尿病提供了一种新模式、新思路，也为研究人员及医生提供了治疗糖尿病的新的突破口。同时，MSC 治疗糖尿病就像治疗其他疾病一样，仍存在一些重要的科学问题尚未解决。

一、如何应用 MSC 治疗

MSC 种子细胞的来源，移植的剂量、次数，何种移植途径能够获得最好的治疗效果，到目前都没有最佳答案。

（一）种子细胞的来源

尽管干细胞种类繁多，但本研究只涉及 MSC。临床研究用的 MSC 细胞来源，包括自体骨髓、自体脂肪、异体脐带和羊膜[38-40]。采自异体脐带、羊膜的细胞时，需说明供体的年龄、性别，供体必须经医院检验科等检验证明乙型肝炎病毒（HBV）抗原、抗丙型肝炎病毒（HCV）抗体、抗人类免疫缺陷病毒（HIV）抗体、梅毒抗体、霉菌均为阴性，必要时需说明供体的既往病史、家族史等临床资料。来源于自体骨髓的 MSC，具有取材简单方便，安全性高。低免疫原性，移植时不存在组织配型和排斥反应问题，在体外可快速大量扩增到所需数量，便于操作等优点，是干细胞移植的首选来源。但是骨髓会随着年龄衰老，长期高血糖刺激等因素会使得自体的干细胞多数不能满足临床需求[41,42]。取自分娩后废弃的脐带华通氏胶（Wharton's jelly）及羊膜中的 MSC，在形态学、分化功能和表面标志物等生物学表型方面具有典型的 MSC 特性，是临床试验研究新的种子细胞来源[43,44]。进一步研究发现，其基因表达特征更像胚胎干细胞（ESC），如表达 ESC 基因标记的 NANOG、DNMT3B 和 GABRB3。另外发现其具有分泌细胞因子的能力[45,46]。更为值得注意的是，虽然为同种异体种子细胞，但具有不同于其他异种干细胞的免疫原特性。

（二）MSC 治疗的次数与剂量

糖尿病涉及免疫系统的损害、慢性炎症触发的靶组织的胰岛素抵抗及持续的胰岛功能的受损，加上长期慢性炎症与糖代谢紊乱合并的其他并发症，利用 MSC 进行单次移植治疗，疗效并不理想，治疗后血糖改善只能维持 4 周多的时间[11,47]。采用多次输注 MSC 治疗 2 型糖尿病小鼠的动物实验结果显示，在每周 1 次、连续治疗 6 周以后，高血糖状况得到明显改善，接近正常水平，并且胰岛素抵抗及胰岛功能同步得到改善。另外，一项还在进行中的研究发现，采用每周 1 次、连续 6 个月的 MSC 输注治疗，晚期 2 型糖尿病大鼠的血糖稳定在正常水平，包括脂肪肝、心功能减弱、白内障、肾功能受损等并发症修复到正常，胰岛正常分泌。在一项还在进行的实验中研究者发现，对于 2 型糖尿病大鼠晚期合并皮肤、肾、眼、心脏等多种并发症，经过 MSC 每周 1 次、连续 6 个月的输注治疗，大鼠的糖代谢功能、胰岛分泌胰岛素的能力稳定恢复到正常水平，同时相关并发症均得以修复。结果提示，经过 MSC 多次的输注，可明显达到修复胰岛、降低血糖的目的。当然，最佳治疗次数需要通过大样本临床研究来确定。

关于 MSC 治疗糖尿病剂量的选择上，像治疗其他疾病一样，没有一个确切的标准。基于 MSC 来源不同，治疗路径不同，活性与纯度不同，以及 MSC 质量标准匮乏，患者的病情有差异，因此很难能得出结论。我们在糖尿病动物模型中，探索了 MSC 治疗剂量与疗效的关系，1×10^6 cells/kg、1×10^7 cells/kg、1×10^8 cells/kg 体重三个浓度，未发现明显差异。无论如何，大多数临床前及临床试验的剂量选择在 $(1\sim2)\times10^6$ cells/kg 体重。

（三）MSC 的质量

虽然国际细胞治疗协会（ISCT）于 2006 年制定了 MSC 鉴定标准（定义），但是这些标准并没有评价与质量相关的指标，而且 MSC 具有多功能属性，很难用单一指标来界定 MSC 的质量。基于目前对 MSC 治疗糖尿病机制的了解，MSC 发挥的治疗效应主要表现在两个方面：分泌细胞因子和免疫抑制能力[48]。其分泌的细胞因子容易量化及检测，而免疫抑制能力不容易量化及检测。因为免疫抑制能力的检测涉及免疫细胞的应答能力，而不同个体免疫细胞的应答能力不同，甚至同一个体在不同状态下，免疫细胞的应答能力也不同；更关键的是 MSC 的免疫抑制有多重机制，更难以统一量化。

供者的年龄对 MSC 的功能和特性有着重要的影响。BM-MSC 中端粒的长度会随着年龄与病损加重而缩短，出现细胞衰老等，其分化和增殖能力下降明显，而且伴随有克隆形成单位的减少和衰老细胞的增多。甚至大龄个体来源的 MSC 失去了促进组织器官干祖细胞更新增殖的能力。

MSC 在体内"龛"生理微环境下，保持相对的稳定状态，参与修复时以非对称方式复制子代发挥作用，从而保持自身的稳定。在临床应用中，大多数研究需要对细胞进行体外扩增培养，这就不可避免会出现复制衰老，也就是说培养代数越高，DNA 复制导致的细胞衰老会越多。需要解决体外培养过程如何更有效地模拟体内微环境，尽量避免复制衰老过早出现，在有限的培养代数内维持干细胞的治疗效能。

（四）MSC 的治疗路径

目前，从已有的文献看，MSC 治疗糖尿病的主要路径包括：单独系统输注、单独胰岛局部给药及两者结合给药三种模式。基于 MSC 来源不同，活性与纯度不同及 MSC 质量标准匮乏，患者的病情存在差异，无法得到正确的结论。理论上，随着 MSC 治疗糖尿病的优质作用机制逐步明晰，似乎系统输注是更好的选择，当然需要大样本临床试验研究来证实。

二、MSC 治疗糖尿病的优势作用机制

虽然干细胞可通过向胰岛细胞分化及其旁分泌作用修复受损的胰岛细胞，改善胰岛细胞功能，但这些机制主要是通过对 1 型糖尿病研究获得的，上述机制是否也在 2 型糖尿病的治疗中发挥作用还需大量研究证实。另外，新近的研究发现，胰岛中的 α 细胞与 β 细胞之间，在损伤条件下可以相互转化，一些研究也证实了胰岛损伤早期 α 细胞经历了一个上皮细胞间质化而后转分化为 β 细胞的过程[49]。另外有研究发现，临床病例中存在 α 细胞与 β 细胞呈双阳性的过程，糖尿病过程中 β 细胞转分化为 α 细胞[50]。推测 MSC 也许通过其分泌效应，促进 α 细胞分化为 β 细胞而再生修复胰岛，需要进一步研究证实。

三、进行大样本多中心随机双盲临床试验

目前各研究中心 MSC 来源不同，获得的 MSC 的活性、纯度、质量可能存在差异，从而使各中心临床疗效的评价难以进行比较。干细胞输注目前主要有 3 种：外周静脉输注、胰腺动脉内输注或 2 种方法联合应用。虽然有人认为胰腺动脉内输注可促进干细胞归巢至胰腺，但目前尚无 2 种输注途径头对头比较的研究数据。多数临床试验存在一定的干扰因素或混杂因素。例如，受试者的入选标准是否存在酮症酸中毒病史，骨髓干细胞移植是否联合其他细胞治疗，是否联合应用大剂量免疫抑制剂或高压氧等其他医疗措施等。这些均可影响对骨髓干细胞本身确切治疗作用的客观评价；对于 2 型糖尿病患者，胰岛素剂量减少大于等于 50% 作为首要终点是否合适，如何能够更准确地评估干细胞移植治疗 2 型糖尿病的疗效。目前这类临床试验的样本量均非常小，随机对照研究仅有 1 项，且远期疗效和安全性尚未可知，作用机制和影响临床疗效的相关因素也尚待确定。因此，MSC 移植治疗糖尿病目前还只能是试验性研究，而不能作为常规的临床实践。

（郝好杰　陈惠华　易　军）

参 考 文 献

[1] Yang W, Lu J, Weng J, et al. Prevalence of diabetes among men and women in China. N Engl J Med, 2010, 362(12):1090-1101.

[2] Rong C, Linong J, Liming C, et al. Glycemic control rate of T2DM outpatients in China: a multi-center survey. Med Sci Monit, 2015, 21:1440-1446.

[3] Pearson-Stuttard J, Blundell S, Harris T, et al. Diabetes and infection: assessing the association with glycaemic control in population-based studies. Lancet Diabetes Endocrinol, 2015, 15:379-384.

[4] Kahn SE, Haffner SM, Heise MA, et al. Glycemic durability of rosiglitazone, metformin, or glyburide monotherapy. N Engl J Med, 2006, 355(23):2427-2443.

[5] Cersosimo E, Triplitt C, Mandarino LJ, et al. Pathogenesis of type 2 diabetes mellitus. In: De Groot LJ, Chrousos G, Dungan K, et al. Endotext [Internet]. South Dartmouth (MA): MDText.com, Inc, 2015.

[6] Bailey CJ. The challenge of managing coexistent type 2 diabetes and obesity. BMJ, 2011, 342:d1996.

[7] El-Badawy A, El-Badri N. Clinical efficacy of stem cell therapy for diabetes mellitus: a meta-analysis. PLoS One, 2016, 11(4):e0151938.

[8] Zang L, Hao H, Liu J, et al. Mesenchymal stem cell therapy in type 2 diabetes mellitus. Diabetol Metab Syndr, 2017, 9:36.

[9] Kadam SSSM, Nair PD, Bhonde RR. Reversal of experimental diabetes in mice by transplantation of neo-islets generated from human amnion-derived mesenchymal stromal cells using immuno-isolatory macrocapsules. Cytotherapy, 2010, 12(8):982-991.

[10] Si Y, Zhao Y, Hao H, et al. Infusion of mesenchymal stem cells ameliorates hyperglycemia in type 2 diabetic rats: identification of a novel role in improving insulin sensitivity. Diabetes, 2012, 61:1616-1625.

[11] Hao H, Liu J, Shen J, et al. Multiple intravenous infusions of bone marrow mesenchymal stem cells reverse hyperglycemia in experimental type 2 diabetes rats. Biochem Biophys Res Commun, 2013 436(3):418-423.

[12] Zhang Q, Hao H, Xie Z, et al. M2 macrophages infusion ameliorates obesity and insulin resistance by remodeling inflammatory/macrophages' homeostasis in obese mice. Mol Cell Endocrinol, 2017, 443:63-71.

[13] Xie Z, Hao H, Tong C, et al. Human umbilical cord-derived mesenchymal stem cells elicit macrophages into an anti-inflammatory phenotype to alleviate insulin resistance in type 2 diabetic rats. Stem Cells, 2016, 34(3):627-639.

[14] Kadam S, Muthyala S, Nair P, et al. Human placenta-derived mesenchymal stem cells and islet-like cell clusters generated from these cells as a novel source for stem cell therapy in diabetes. Rev Diabet Stud, 2010, 7(2):168-182.

[15] Choi JB, Uchino H, Azuma K, et al. Little evidence of transdifferentiation of bone marrow-derived cells into pancreatic beta cells. Diabetologia, 2003, 46(10):1366-1374.

[16] Donath MY, Shoelson SE. Type 2 diabetes as an inflammatory disease. Nat Rev Immunol, 2011, 11:98-107.

[17] Fujisaka S, Usui I, Bukhari A, et al. Regulatory mechanisms for adipose tissue M1 and M2 macrophages in diet-induced obese mice. Diabetes, 2009, 58(11):2574-2582.

[18] Kim HY, Hwang JI, Moon MJ, et al. A novel long-acting glucagon-like peptide-1 agonist with improved efficacy in insulin secretion and beta-cell growth. Endocrinol Metab, 2014, 29(3):320-327.

[19] Sasaki S, Miyatsuka T, Matsuoka TA, et al. Activation of GLP-1 and gastrin signalling induces in vivo repro-gramming of pancreatic exocrine cells into beta cells in mice. Diabetologia, 2015, 58(11):2582-2591.

[20] Zang L, Hao H, Liu J, et al. Mesenchymal stem cell therapy in type 2 diabetes mellitus. Diabetol Metab Syndr, 2017, 9:36.

[21] Piran R, Lee S H, Li C R, et al. Pharmacological induction of pancreatic islet cell transdifferentiation:relevance to type Ⅰ diabetes. Cell Death & Disease, 2014, 5(7):e1357.

[22] Cheng Y, Kang H, Shen J, et al. Beta-cell regeneration from vimentin+/MafB+ cells after STZ-induced extreme beta-cell ablation. Sci Rep, 2015, 5:11703.

[23] Ezquer F, Ezquer M, Contador D, et al. The antidiabetic effect of mesenchymal stem cells is unrelated to their transdifferentiation potential but to their capability to restore Th1/Th2 balance and to modify the pancreatic microenvironment. Stem Cells, 2012, 30(8):1664-1674

[24] Ezquer F, Ezquer M, Simon V, et al. The antidiabetic effect of MSCs is not impaired by insulin prophylaxis and is not improved by a second dose of cells. PLoS One, 2011, 6:e16566.

[25] Ezquer FE, Ezquer ME, Parrau DB, et al. Systemic administration of multipotent mesenchymal stromal cells reverts hyperglycemia and prevents nephropathy in type 1 diabetic mice. Biol Blood Marrow Transplant, 2008, 14(6):631-640.

[26] Zhao K, Hao H, Liu J, et al. Bone marrow-derived mesenchymal stem cells ameliorate chronic high glucose-induced β-cell injury through modulation of autophagy. Cell Death Dis, 2015, 6:e1885.

[27] Lazarus HM, Haynesworth SE, Gerson SL, et al. *Ex vivo* expansion and subsequent infusion of human bone marrow-derived stromal progenitor cells: implications for therapeutic use. Bone Marrow Transplant, 1995, 16:557-564.

[28] Voltarelli JC, Couri CE, Stracieri AB, et al. Autologous nonmyeloablative hematopoietic stem cell transplantation in newly diagnosed type 1 diabetes mellitus. JAMA, 2007, 297:1568-1576.

[29] Couri CE, Oliveira MC, Stracieri AB, et al. C-peptide levels and insulin independence following autologous nonmyeloablative hematopoietic stem cell transplantation in newly diagnosed type 1 diabetes mellitus. JAMA, 2009, 301:1573-1579.

[30] Hu J, Yu X, Wang Z, et al. Long term effects of the implantation of Wharton's jelly-derived mesenchymal stem cells from the umbilical cord for newly-onset type 1 diabetes mellitus. Endocr J, 2013, 60(3):347-357.

[31] Carlsson PO, Schwarcz E, Korsgren O, et al. Preserved beta-cell function in type 1 diabetes by mesenchymal stromal cells. Diabetes, 2014, 64(2):587-592.

[32] Bhansali A, Asokumar P, Walia R, et al. Efficacy and safety of autologous bone marrow-derived stem cell transplantation in patients with type 2 diabetes mellitus: a randomized placebo-controlled study. Cell Transplant, 2013, 23(9):1075-1085.

[33] Bhansali A, Upreti V, Khandelwal N, et al. Efficacy of autologous bone marrow-derived stem cell transplantation in patients with type 2 diabetes mellitus. Stem Cells Dev, 2009, 18(10):1407-1416.

[34] Hu J, Li C, Wang L, et al. Long term effects of the implantation of autologous bone marrow mononuclear cells for type 2 diabetes mellitus. Endocr J, 2012, 59(11):1031-1039.

[35] Wu Z, Cai J, Chen J, et al. Autologous bone marrow mononuclear cell infusion and hyperbaric oxygen therapy in type 2 diabetes mellitus: an open-label, randomized controlled clinical trial. Cytotherapy, 2014, 16(2):258-265.

[36] Liu X, Zheng P, Wang X, et al. A preliminary evaluation of efficacy and safety of Wharton's jelly mesenchymal stem cell transplantation in patients with type 2 diabetes mellitus. Stem Cell Res Ther, 2014, 5(2):57.

[37] Jiang R, Han Z, Zhuo G, et al. Transplantation of placenta-derived mesenchymal stem cells in type 2 diabetes: a pilot study. Front Med, 2011, 5(1):94-100.

[38] Picinich SC, Meshra PJ, Prasun JM, et al. The therapeutic potential of mesenchymal stem cells. Expert Opin, Biol Ther, 2007, 7:965-973.

[39] Secco M, Zucconi E, Natassia M, et al. Multipotent stem cells from umbilical cord: cord is richer than blood! Stem Cells, 2008, 26:146-150.

[40] Wang HS, Hung SC, Peng ST, et al. Mesenchymal stem cells in the Wharton's jelly of the human umbilical cord. Stem Cells, 2004, 22:1330-1337.

[41] Si YL, Zhao YL, Hao HJ, et al. MSCs: biological characteristics, clinical applications and their outstanding concerns. Ageing Res Rev, 2011, 10:93-103.

[42] Yan J, Tie G, Wang S, et al. Type 2 diabetes estricts multioncy of mesenchymal stem cells and impairs their capacity to augment postischemic neovascularization in db/db mice. J Am Heart Assoc, 2012, 1(6):e002238.

[43] Deryl L, Mark LT, Weiss, et al. Concise review: Wharton's jelly-derived cells are a primitive stromal cell population. Stem Cells, 2008, 26:591-599.

[44] Qiao C, Xu W, Zhu W, et al. Human mesenchymal stem cells isolated from the umbilical cord. Cell Biology International, 2008, 32:8-15.

[45] Weiss ML, Anderson C, Medicetty S, et al. Immune properties of human umbilical cord Wharton's jelly-derived cells. Stem Cells, 2008, 26:2865-2874.

[46] Cho PS, Messina DJ, Erica L. et al. Immunogenicity of umbilical cord tissue derived cells. Blood, 2008, 111:430-438.

[47] Banerjee M, Kumar A, Bhonde RR. Reversal of experimental diabetes by multiple bone marrow transplantation. Biochem Biophys Res Commun, 2005, 328(1):318-325.

[48] Domínguez-Bendala J, Lanzoni G, Inverardi L, et al. Concise review: mesenchymal stem cells for diabetes. Stem Cells Transl Med, 2012, 1(1):59-63.

[49] Cheng Y, Kang H, Shen J, et al. Beta-cell regeneration from vimentin[+]/MafB[+] cells after STZ-induced extreme beta-cell ablation. Sci Rep, 2015, 5:11703.

[50] Chakravarthy H, Gu X, Enge M, et al. Converting adult pancreatic islet α cells into β cells by targeting both Dnmt1 and Arx. Cell Metab, 2017, 25(3):622-634.

第十六章 间充质干细胞在骨关节炎疾病治疗中的应用

第一节 间充质干细胞与骨关节炎

一、骨关节炎概述

骨关节炎（OA）是一种常见的慢性退行性关节疾病。其主要侵害关节软骨、骨和滑膜组织，导致关节疼痛、肿胀、变形。OA 在中老年群体中发病率极高，主要表现为关节疼痛、肿胀，骨质增生，并影响患者的活动能力。轻者表现为关节疼痛、膝关节活动障碍，重者可导致患者关节畸形和残疾，给患者的工作、生活都造成极大的不便。

OA 可以分为两类：原发性 OA 和继发性 OA。原发性 OA 多发生于中老年，无明确的全身或局部诱因，与遗传和体质因素有一定的关系。继发性 OA 可发生于青壮年，可继发于创伤、炎症、关节不稳定、慢性反复的积累性劳损或先天性疾病等。

由于关节表面透明的软骨组织无神经、无血管，因此软骨自我修复和再生能力很差，一旦受损很难自行恢复。同时，目前对 OA 尚缺乏理想的手段促进软骨再生，大部分患者只能采取关节腔冲洗、骨髓刺激和消炎镇痛等治疗，主要作用是止痛和改善功能。这些疗法都不能完全阻断 OA 的病理进程，最终还需人工关节置换才能解除病痛，治疗费用昂贵，使用寿命有限，届时可能需要再次置换。

目前临床上采用多种方法修复关节软骨损伤，如微骨折术、自体或异体软骨移植术、自体软骨细胞或基质诱导的自体软骨细胞移植术，以及基因疗法和干细胞疗法等[1]。然而，关节软骨受损后再生修复能力有限，药物治疗效果较差，各种外科治疗如软骨移植术及微骨折术虽能暂时缓解疼痛，但远期疗效并不理想；组织工程性自体软骨细胞或基质诱导的自体软骨细胞移植研究的发展，为退变关节组织的修复或再生提供了可能，但主要局限在于无法治疗较大面积的软骨缺损[2]。

近年来，随着对软骨的结构、生化组成及代谢变化进一步认识，再加以对软骨细胞培养、OA 动物模型的深入研究，认为 OA 是多种因素联合作用的结果，主要有：①软骨基质合成和分解代谢失调；②软骨下骨板损害使软骨失去缓冲作用；③关节内局灶性炎症。目前缺乏有效的治疗方法。例如，软骨细胞移植法不仅受到细胞来源的限制，而且提取细胞的供体部位出现的软骨缺损会形成新的 OA。

近年来，采用间充质干细胞（MSC）治疗 OA 成为研究热点，应用前景广阔。MSC 具有强大的自我更新、增殖能力及多向分化潜能，可替代受损细胞；可通过旁分泌和自分泌合成多种生物活性分子如胰岛素样生长因子-1（IGF-1）、转化生长因子-β（TGF-β）、血管内皮生长因子（VEGF）等，激活细胞和血管再生途径，对软骨起到营养作用；具有归巢作用，在体内微环境作用下主动迁移至软骨缺血或受损部位进行修复重建[3]；具有免疫抑制和抗炎作用，即表达主要组织相容性复合体Ⅰ（MHCⅠ）来逃避宿主的免疫清除，并通过抑制T淋巴细胞活性和自然杀伤细胞分化实现抗炎作用[4]，这些特点使MSC逐渐成为治疗OA的理想种子细胞。

二、MSC的生物特性

MSC是一种存在于多种组织（如骨髓、脐带、胎盘、脂肪等许多组织中）的具有自我复制能力和多向分化潜力的非造血干细胞（HSC）的成体干细胞。

Friedenstein等[5]于1968年首次发现在骨髓里存在一群不均一的细胞群体，可支持造血和分化为骨细胞，在1974年体外培养获得这群细胞，这类贴壁培养的干细胞呈漩涡状生长[2]；1988年将这类干细胞命名为"骨髓基质干细胞"[6,7]。Caplan教授在1991年进一步把这类干细胞命名为"间充质干细胞"。"间充质干细胞"的命名逐渐被广泛接受和使用。也有学者认为间充质干细胞的英文名称为mesenchymal stromal cell，但是更多学者习惯采用mesenchymal stem cell。

2006年国际细胞治疗协会（ISCT）提出人MSC（hMSC）最基本定义：①MSC在标准培养条件下呈贴壁生长；②MSC必须表达CD105、CD73和CD90，不表达CD45、CD32、CD14或CD11b、CD79α或CD19及HLA-DR表面标记；③MSC在体外可以分化为成骨细胞、脂肪细胞及软骨细胞[8]。

MSC可以分化为成骨、脂肪、软骨细胞[9,10]，同时也具有向内、外胚层组织细胞分化的能力，科学家成功地在体外诱导它向神经和肝细胞方向分化[11,12]，同时MSC也可利用其多向分化潜能向内胚层分化，从而得到肝实质样细胞及胰岛样细胞等，用于临床严重器官损伤移植，其应用价值很大。

MSC作为一种多能干细胞，在临床上已经作为一种新的疾病治疗方法而进行了广泛研究，如心脏疾病、肝脏疾病、糖尿病及相关并发症、克罗恩病（CD）、多发性硬化（MS）、移植物抗宿主病（GVHD）、骨/软骨疾病、脊髓损伤、脑部疾病、肺部疾病和肿瘤等[13]。20世纪70年代，Friedenstein等[5]从全骨髓培养物中分离得到MSC，随后MSC得到广泛研究。1995年，Lazarus等[14]在《骨髓移植》（*Bone Marrow Transplantation*）上报道了首项骨髓间充质干细胞（BM-MSC）用于骨髓移植患者治疗的临床研究。

虽然多数临床研究采用的MSC均来自骨髓，但由于抽取骨髓对供者创伤较重，骨髓受病毒感染程度较高，骨髓中MSC含量低，以及细胞数量和增殖/分化能力随年龄增加显著下降等，许多研究者正在研究其他材料来替代骨髓，而且已有研究表明，脂肪组织、脐血和脐带来源的MSC具有典型MSC的表面标志物和与BM-MSC类似的抑制T细胞增殖的能力，能替代BM-MSC[15,16]。虽然不同来源的MSC具有类似的生物学活性，并在许多免疫和非免疫疾病中显示出良好的治疗前景，但仍有一些重要问题尚需解决，如MSC的最佳输注剂量、输注途径、输注后细胞迁移和存活[17]。

因此，阐明MSC治疗作用的机制至关重要。虽然有关MSC治疗作用的确切机制还未完全阐明，但目前研究表明主要涉及四个方面：①归巢效应。MSC倾向归巢至损伤的组织部位，炎症介导的MSC归巢作用经研究表明涉及几种重要的细胞迁移相关分子，如趋化因子、黏附分子和基质金属蛋白酶（MMP）。②分化潜能。MSC具有分化为多种细胞类型的潜能，如脂肪细胞、成骨细胞、软骨细胞、肌细胞和神经元样细胞。③分泌营养因子。越来越多研究表明，MSC 的治疗作用很大程度上依赖于它分泌营养因子和免疫调节的能力。MSC 归巢至待修复的损伤部位之后，将与局部刺激因子相互作用，这些因子包括炎性细胞因子、Toll 样受体（TLR）的配体和缺氧条件，缺氧将刺激 MSC 产生大量用于组织再生的具有多种功能的生长因子。这些因子在血管新生和阻止细胞凋亡中发挥关键作用，如 VEGF、IGF-1、碱性成纤维细胞生长因子（bFGF）、肝细胞生长因子（HGF）、白介素-6（IL-6）和 CCL2。④免疫调节。体内外研究均表明，MSC 抑制 T 细胞、B 细胞、树突细胞（DC）、巨噬细胞和 NK 细胞的过度免疫反应[18]。

第二节　间充质干细胞治疗骨关节炎的研究进展

骨关节炎（OA）是一种常见的关节软骨退行性疾病，其病理变化包括软骨磨损、滑膜炎症、骨赘生成及软骨下骨病变等，常导致疼痛和行动不便[19]。由于关节软骨并无血管和神经，受损软骨组织难以自我修复[20]。MSC 具有良好的自我复制能力和多向分化潜能，有治愈多种退行性疾病的潜能。多项研究证实[21]，骨髓间充质干细胞（BM-MSC）、脐带间充质干细胞（UC-MSC）、胎盘间充质干细胞（P-MSC）、脂肪间充质干细胞（AD-MSC）等可有效治疗 OA。

MSC 移植治疗 OA 通常需选用一定的载体支架。而载体支架的选择直接影响种子细胞的种植、迁移和增殖。理想的 MSC 载体支架应具备以下特点[22]：①能均匀搭载并保留细胞；②支持血管快速内生；③便于观察新骨形成；④新骨形成后能被吸收和替代利于骨改建；⑤可提高骨的骨传导性桥接；⑥可以保留分化细胞的功能；⑦组织相容性好；⑧易于加工制作。目前常用的软骨组织工程支架材料按来源可分为人工合成材料和天然材料两大类。人工合成材料主要有聚乳酸（poly lactic acid，PLA）、聚羟基乙酸共聚物（poly glycolic acid，PGA）、聚羟基乙酸聚乳酸（poly lactic glycolic acid，PLGA）、羟基磷灰石、聚磷酸酯等。天然材料主要有Ⅰ型或Ⅱ型胶原、纤维蛋白凝胶、透明质酸、脱钙骨基质、明胶等。

一、MSC 治疗 OA 的动物实验研究

关于动物的体内实验，文献中采用多种不同动物模型，其中家兔前足的前交叉韧带切断术（anterior cruciate ligament transaction，ACLT）是经典的制造家兔 OA 模型的方法之一。这种经 3~8 周造模成功的家兔模型，具有和人相似的生化及病理改变。此外，也有大鼠、马等动物模型。通过动物实验研究发现，关节腔直接注射 MSC、关节腔注射体外预处理过的 MSC、注射载体支架材料混悬的 MSC 均能有效缓解 OA 病情。

（一）单纯使用 MSC 治疗 OA

2003 年就有研究 MSC 治疗软骨缺损的报道[23]，但经关节腔注射 MSC 治疗 OA 的研究开始较晚。Frisbie 等[24]在 2009 年采用关节腔注射体外扩增的自体来源 AD-MSC 和 BM-MSC 治疗关节镜下诱导的马 OA，未发生不良反应，但也未改善 OA 情况。Toghraie 等[25]采用 ACLT 诱导的家兔 OA 模型，证实关节腔注射生理盐水混悬的脂肪垫来源 MSC

能缓解 OA 病情，修复 OA 中损伤的软骨，表明了 MSC 治疗 OA 有一定的效果。

（二）使用体外预处理的 MSC 治疗 OA

将 MSC 向软骨细胞分化相关基因转染至 MSC，可提高其向软骨细胞分化的能力。Lee 等[26]研究发现，在 ACLT 诱导的大鼠 OA 模型中，实验组关节腔注射经纤维蛋白凝胶混悬且转染 *SOX5*、*SOX6*、*SOX9* 基因的 MSC 进行治疗，对照组只注射纤维蛋白凝胶治疗，8 周后对比治疗效果，发现实验组症状获得明显缓解。采用生物活性物质体外预处理 MSC，可提高其增殖分化能力。Van Pham 等[27]研究发现，使用富血小板血浆（platelet rich plasma，PRP）预处理 MSC 能大大提高其增殖和向软骨细胞分化的能力，从而提高其治疗 OA 的效果，具体操作方法为在体外先将 MSC 向软骨细胞诱导分化，再经关节腔进行注射治疗。Ude 等[28]对经 ACLT 诱导的山羊 OA 模型进行实验，治疗组分别向关节腔注射体外经诱导向软骨细胞分化的 AD-MSC 或 BM-MSC，对照组注射含 10%胎牛血清（FBS）的 DMEM/F12 培养基治疗，6 周后发现与对照组相比，AD-MSC 与 BM-MSC 治疗组均能明显降低国际软骨修复协会（International Cartilage Repair Society，ICRS）组织学评分，修复受损软骨。

（三）使用 MSC 与载体支架混悬液治疗 OA

MSC 与可注射载体支架混悬液治疗 OA 的研究较多。常用的载体支架分别为透明质酸（HA）、纤维蛋白凝胶等高分子支架和血清、PRP 等富含营养物质的液体。Kuroda 等[29]对家兔经 ACLT 诱导形成 OA，治疗组关节腔注射 HA 溶液混悬 MSC，对照组注射单纯 HA 溶液，结果发现与对照组相比，治疗组软骨退变明显被抑制。Lee 和 Im[30]采用以纤维蛋白凝胶为支架的 MSC 治疗经 ACLT 诱导的大鼠 OA 模型，取得良好疗效。ter Huurne 等[31]将由绿色荧光蛋白（GFP）标记的 MSC 用 4%小鼠血清混悬，关节腔注射治疗经胶原诱导的小鼠 OA 模型，结果显示可明显缓解软骨退变、滑膜增厚和骨赘形成等 OA 相关病理变化。Desando 等[32]在经 ACLT 诱导的家兔 OA 模型中用 4%家兔血清混悬的 MSC 治疗，取得明显疗效。

二、MSC 治疗 OA 的临床试验研究

关于人体移植临床试验，相关文献报道较少，多采用 BM-MSC 和 AD-MSC。临床上采用关节腔注射 MSC/PRP 混悬液、MSC/PRP/HA 混悬液和 MSC/纤维蛋白凝胶混悬液治疗 OA 较为常见。

（一）BM-MSC 移植治疗 OA

Centeno 等[33]进行了关于自体 BM-MSC 移植治疗 OA 的研究，抽取 1 例双侧膝关节炎患者的椎体和髂骨骨髓，分离培养获取自体 BM-MSC，分 3 次，每次间隔 1 周，注射 2.24×10^7 cells 至关节腔内，6 个月后功能评价指数（functional rating index，FRI）由 21 降至 9，视觉模拟量表（visual analog scale，VAS）疼痛评分由 4 分降为 0.38 分，核磁共振成像（MRI）定量分析结果显示关节软骨和半月板体积明显增加。Davatchi 等[34]将体外培

养的自体 BM-MSC 通过关节腔注射治疗 4 例严重膝关节炎患者单侧膝关节，发现治疗侧膝关节的疼痛明显减轻，VAS 评分降低，关节摩擦音明显改善，关节活动度提高。Orozco 等[35]将 4×10^7cells 经体外培养的自体 BM-MSC 注射到 12 例膝关节炎患者的关节腔内，1 年后所有患者的膝关节相关功能指数均得到改善，MRI 检查 T2WI 的横向弛豫时间图定量显示关节软骨面积损伤降低至 27%，并且其间 11 例患者膝关节软骨质量提高。Vega 等[36]在细胞移植治疗膝关节炎中使用了异体 BM-MSC，其将 30 例患者随机分为异体 BM-MSC 注射治疗和透明质酸注射对照组两组，1 年后 BM-MSC 注射组的膝关节功能指数得到显著改善，伴随膝关节软骨质量的提高，治疗组的损伤软骨区域面积显著降低。

（二）关节腔注射 MSC 与载体支架混悬液治疗 OA

关节腔注射 MSC/PRP 混悬液治疗 OA 的研究较多。Bui 等[37]对 21 例 OA 患者实施关节腔注射腹部脂肪来源 MSC/PRP 混悬液，随访 6 个月，所有患者病情均明显好转，MRI 检查显示软骨变厚。Koh 等[38]对 18 例 OA 患者实施关节腔注射脂肪垫来源 MSC/PRP 混悬液治疗，随访 24.3 个月，所有患者临床症状均得到明显改善；将 25 例 OA 患者分为治疗组与对照组，治疗组关节腔注射脂肪垫来源 MSC/PRP 混悬液，对照组关节腔注射单纯 PRP，随访 16.4 个月，均未发现任何不良反应，与治疗前相比，治疗后两组患者关节 Lysholm 评分、Tegner 运动评分均提高，VAS 评分降低，治疗组疗效优于对照组，但无统计学差异。

Koh 和 Choi[39]对 30 例 OA 患者进行关节腔注射 MSC/PRP 混悬液，与治疗前相比，87.5%患者软骨缺损得到修复或未进一步缺损，且治疗 24 个月后的疗效明显优于治疗后 12 个月。一项临床研究[40]显示，对 35 例 OA 患者进行关节腔注射 MSC/PRP 混悬液，随访 24 个月，发现患者国际膝关节文献委员会膝关节评估表（简称 IKDC 评分）和 Tegner 运动评分均得到明显提高，且 94%的患者对疗效满意，但经关节镜评价软骨缺损发现，76%患者仍有软骨缺损，体重指数（body mass index，BMI）和大软骨缺损是影响 MSC 治疗效果的主要因素。有对照研究[41]显示，对试验组 21 例患者实施高位胫骨截骨术（high tibial osteotomy，HTO）与关节腔注射 MSC/PRP 混悬液联合治疗，对对照组 23 例患者实施 HTO 与关节腔注射单纯 PRP 联合治疗，随访 24 个月，试验组疼痛缓解 OA 病情改善程度明显优于对照组，经关节镜评价软骨缺损发现，试验组 50%患者软骨缺损处存在纤维软骨修复，而对照组仅 10%患者存在类似现象。

Pak[42]对 2 例 OA 患者实施关节腔注射 MSC/PRP/HA 混悬液联合微量地塞米松治疗，3 个月后 MRI 检查显示，2 例患者关节软骨均较治疗前明显增多，疼痛明显减轻，关节活动度明显增强。Pak 等[43]对 91 例 OA 患者进行关节腔注射 MSC/PRP/HA 混悬液，1 个月和 3 个月后平均 VAS 评分从治疗前 26.62±0.32 分别减小至 6.55±0.32 和 4.43±0.41，随访 30 个月，未发现任何不良反应。Kim 等[44]对 54 例 OA 患者进行研究，其中 A 组（17 例）实施关节腔注射纤维蛋白凝胶混悬 MSC 治疗，B 组（37 例）实施关节腔注射无支架材料 MSC 治疗，随访 28.6 个月，发现两组 IKDC 评分和 Tegner 运动评分均无显著性差异，但与治疗前相比均显著提高，经关节镜评价软骨缺损发现第 1 组 58%患者、第 2 组 23%患者的关节软骨达到正常水平。

第三节 间充干细胞治疗骨关节炎的机制

MSC 对骨关节炎（OA）的治疗作用一般认为可通过以下 3 种途径实现：①免疫调节作用。通过分泌抗炎因子，抑制关节腔内巨噬细胞活性，降低炎性细胞因子释放，从而抑制 OA 炎症。②分化潜能。MSC 可分化为软骨细胞，修复受损软骨组织；③促进软骨生长。通过分泌营养因子促进软骨祖细胞增殖，并通过抑制软骨细胞发生炎症、凋亡、肥大化、纤维化和去分化等来抑制软骨组织被破坏，减轻软骨受损。

一、MSC 的免疫调节作用

MSC 的免疫调节作用是其治疗 OA 的重要机制之一。MSC 能被炎性细胞因子激活而分泌前列腺素 E2（PGE2）、吲哚胺 2,3-双加氧酶（IDO）、一氧化氮（NO）等直接或间接抑制免疫细胞。与 OA 发生发展较为密切的免疫细胞为滑膜巨噬细胞。在 OA 过程中滑膜巨噬细胞较为活跃，分泌 IL-1β 和 TGF-α 等炎性细胞因子导致软骨细胞凋亡[45]。Schelbergen 等[46]研究发现，MSC 能通过与巨噬细胞相互作用来抑制巨噬细胞的活化，抑制其分泌 IL-1β、TGF-α 等炎性细胞因子，从而抑制 OA 中软骨退化，经干扰素-γ（IFN-γ）和 IL-1β 刺激的 MSC 的培养上清能提高巨噬细胞精氨酸酶、IDO 和诱导型一氧化氮合酶（iNOS）的表达而使巨噬细胞由 M1 型转化为 M2 型。

MSC 还分泌丰富的趋化因子，如 RANTES、基质细胞衍生因子-1α（SDF-1α）、fractalkine、巨噬细胞炎性蛋白-1α（MIP-1α）、单核细胞趋化蛋白（MCP-1、MCP-2）等[47]。这些趋化因子都能吸引淋巴细胞、树突细胞（DC）、单核细胞、巨噬细胞等，募集这些细胞趋化到损伤和炎症部位，参与机体的损伤修复。MSC 分泌的趋化因子 SDF-1α 还能吸引更多 MSC 到损伤炎症部位，进一步促进机体组织的修复。例如，辅助性 Th2 细胞、调节性 T 细胞（Treg）、巨噬细胞和单核细胞分泌的 IL-10 是抗炎因子，能抑制巨噬细胞和 Th1 细胞等分泌炎性细胞因子[IFN-γ、IL-2、IL-3、肿瘤坏死因子-α（TNF-α）]，还能促进 B 细胞增殖及分泌抗体。

二、MSC 向软骨细胞分化

早期研究发现 MSC 具有分化为成骨细胞、软骨细胞、脂肪细胞、肌细胞和神经细胞等多种组织细胞的潜能。软骨组织作为无血管组织，自我修复能力十分有限，因此需要由 MSC 向软骨细胞分化，以起到修复缺损的目的。鉴于这一特性，学者对 MSC 向软骨细胞分化进行了研究。

（一）体外诱导分化

Wasim 等[48]将来源于人髌下脂肪垫的 MSC 聚集体接种于软骨诱导培养物上，三维空间低氧条件培养下，14 天后成功诱导培养出能够分泌来自软骨细胞的蛋白聚糖、多功能蛋白聚糖和Ⅱ、Ⅸ、Ⅹ、Ⅺ型胶原的软骨细胞。Johnstone 等[49]取 5 月龄兔 BM-MSC，经离心使细胞形成聚集体，在培养基中加入地塞米松和 TGF-β1 培养，7 天后检测到细胞分泌Ⅱ和X型胶原。Mackay 等[50]采用相同的培养体系诱导人骨髓间充质干细胞（hBM-MSC）分化为软骨细胞获得成功。Sekiya 等[51]在上述体系中添加了骨形成蛋白（bone morphogenetic protein，BMP-6 或 BMP-2）后，发现可以增强 TGF-β3 对软骨的诱导作用。Worsterf 等[52]研究发现 TGF 与 IGF-1 有协同诱导 MSC 向软骨分化的作用。Hegewaid 和 Ringe[53]取马骨髓的 MSC 经体外离心形成微聚体，以 TGF-β1、透明质酸（HA）、自体滑膜液为诱导因子，诱导其分化为软骨细胞，结果显示单纯 TGF-β1 诱导有高蛋白聚糖表达，TGF-β1 与 HA 共同诱导不能增加蛋白聚糖的表达，自体滑膜液与 HA 诱导有低蛋白聚糖表达，所有诱导组均有软骨特有的Ⅱ型胶原高表达。以上研究说明诱导 MSC 分化为软骨细胞是一个多因子参与的复杂过程。

（二）体内诱导分化

体外 MSC 经诱导向软骨分化获得成功，使研究进一步转向动物体内实验。Katayama 和 Wakitani [54]把 TGF 家族成员软骨衍生形态发生蛋白-1（cartilage derived morphogenetic protein-1，CDMP-1）通过脂质转染到自体 BM-MSC，将 MSC 移植到兔膝关节软骨缺损处，结果缺损处被透明软骨填充，深部区域显示重建的软骨下骨。霍建忠[55]利用壳聚糖负载有 TGF 基因的质粒转染修饰 BM-MSC 用于修复兔关节软骨缺损，术后 6 周还可检测到 TGF-β，12 周后软骨缺损外完全被透明软骨组织覆盖修复，结构与正常软骨非常相近。Mason 等[56]以逆转录病毒为载体将 BMP-7 转染入骨膜来源的兔 MSC，体外培养扩增后植入聚乙醇酸支架，移植到兔膝关节软骨缺损处，12 周后，从大体形态、组织学和免疫组化等方面均证实有完全再生的骨与软骨。Madry 和 Orth[57]通过质粒把 IGF-1 和成纤维细胞生长因子-2（FGF-2）基因转染到 NIH 3T3 细胞中，并把细胞移植到兔股骨滑车凹陷骨软骨缺损处，3 周后对关节软骨修复程度进行定性定量评估，结果显示在此基因的作用下软骨加速修复。

MSC 在适当诱导条件下能向软骨细胞分化。治疗软骨缺损时，将 MSC 与支架材料填补于软骨缺损处，MSC 能向软骨细胞分化而修复缺损软骨[58,59]。但关节腔注射 MSC 治疗 OA 时，荧光标记的 MSC 分布在滑膜和内侧半月板中，而软骨组织中并未检测到被标记的细胞，因此治疗 OA 时 MSC 在体内是否能向软骨细胞分化需进一步研究[60-62]。

三、MSC 促进软骨细胞再生

现有实验数据的结果均提示 MSC 对软骨细胞的再生有促进作用。MSC 可抑制 OA 中软骨细胞发生炎症、凋亡、肥大化、纤维化和去分化等而保护软骨细胞并促进软骨细胞增殖。Platas 等[63]报道 MSC 的体外培养液能通过抑制转录因子 NF-κB 实现促进软骨细胞的增殖并抑制 IL-1β 诱导的软骨细胞表达 MMP、NO、PGE2 和趋化因子（CXCL）等炎性介质。vanBuul 等[64]研究发现，IFN-γ 和 TNF-α 刺激的 MSC 的培养上清能抑制 OA 患者软骨和滑膜组织的炎症。Manferdini 等[65]将 MSC 与 OA 患者软骨细胞和滑膜细胞共培养，发现能抑制软骨细胞和滑膜细胞表达 IL-1β、TNF-α、IL-6、CXCL1/生长相关癌基因 α、CXCL8/IL-8、CXCL2/MCP-1、CXCL3/MIP-1α 与 CXCL5/RANTES 等炎性细胞因子，该抗炎作用不通过拮抗 IL-10、IL-11 受体和 FGF-2、IDO 和半乳凝素-1（galectin-1）

抗体等常见免疫调节因子介导,而是通过环加氧酶2(COX2)/PGE2信号转导通路实现,在此过程中软骨细胞和滑膜炎症程度越严重,抗炎效果越明显。Maumus等[66]将MSC与OA患者软骨细胞共培养,发现MSC能抑制软骨细胞纤维化、肥大化而使其保持透明软骨表型,并能抑制喜树碱诱导的软骨细胞凋亡,共培养后软骨细胞分泌的TGF-β数量明显减少,而MSC分泌的HGF数量明显增多,中和HGF时,抗软骨细胞纤维化的作用消失,推测MSC抗软骨细胞纤维化的作用可能通过HGF实现。

第四节 问题与展望

尽管在MSC在治疗OA的研究中显示出有良好治疗前景,但仍有一些重要问题尚需解决,如治疗的最佳时机、治疗的MSC最佳剂量、治疗的最佳途径(是否需要载体,选择何种载体)、输注后细胞迁移和存活时间等。

目前来看,MSC治疗OA的研究仍处于基础实验阶段,对于人体在体的试验依然比较少。MSC具有多向分化潜能,MSC定向分化的机制目前尚未完全明确。现阶段研究表明,MSC分化方向的确定需要多转录因子参与调控,而不是由单一因素刺激完成。MSC与软骨细胞共培养能够很好地诱导其向软骨细胞定向分化,为后续其治疗OA提供了必要的技术支持。在技术手法运用方面,细胞分离、扩增、分化,MSC与载体支架混合及无支架的MSC植入关节腔,自体MSC移植等治疗技术已成为成熟的技术方法。但由于OA的发生机制多样,对软骨细胞修复对OA有多大改善,是能够将已发生的OA逆转性地恢复到发病以前,还是能够阻止其进一步发展,亦或只是起到预防OA损伤的作用,这些问题都需要经行大范围的人体在体试验才能得出结论。对于人体在体试验中要面对的生物力学因素,生理微环境的不同,营养状态的改变等问题,都有待其他学者做进一步的研究和临床分析。

(马步鹏 吴志宏)

参 考 文 献

[1] Lubis AM, Lubis VK. Adult bone marrow stem cells in cartilage therapy. Acta Medica Indonesiana, 2012, 44(1):62.

[2] Steinert AF, Ghivizzani SC, Rethwilm A, et al. Major biological obstacles for persistent cell-based regeneration of articular cartilage. Arthritis Research & Therapy, 2007, 9(3):213.

[3] Fong EL, Chan CK, Goodman SB. Stem cell homing in musculoskeletal injury. Biomaterials, 2011, 32(2):395

[4] Singer NG, Caplan AI. Mesenchymal stem cells: mechanisms of inflammation. Annu Rev Pathol, 2010, 6(1):457-478.

[5] Friedenstein AJ, Petrakova KV, Kurolesova AI, et al. Heterotopic of bone marrow. Analysis of precursor cells for osteogenic and hematopoietic tissues. Transplantation, 1968, 6(2):230-247.

[6] Friedenstein AJ, Deriglasova UF, Kulagina NN, et al. Precursors for fibroblasts in different populations of hematopoieticcells as detected by the *in vitro* colony assay method. Exp Hematol, 1974, 2(2):83-92.

[7] Owen M, Friedenstein AJ. Stromal stem cells:marrow-derived osteogenic precursors. Ciba Found Symp, 1988, 136(136):42-60.

[8] Dominici M, Blanc KL, Mueller I, et al. Minimal criteria for defining multipotent mesenchymal stromal cells. The International Society for Cellular Therapy position statement. Cytotherapy, 2006, 8(4):315.

[9] Meirelles LD, Nardi NB. Methodology, biology and clinical applications of mesenchymal stem cells. Front Biosci, 2009, 14:4281-4298.

[10] Pittenger MF, Mackay AM, Beck SC, et al. Multilineage potential of adult human mesenchymal stem cells. Science, 1999, 284:143-147.

[11] Jang S, Cho HH, Cho YB, et al. Functional neural differentiation of human adipose tissue-derived stem cells using bFGF and forskolin. BMC Cell Biol, 2010, 11:25.

[12] Campard D, Lysy PA, Najimi M, et al. Native umbilical cord matrix stem cells express hepatic markers and differentiate into hepatocyte-like cells. Gastroenterology, 2008, 134:833-848.

[13] Wang SH, Qu XB, Zhao CH. Clinical applications of mesenchymal stem cells. Journalof Hematology & Oncology, 2012, 5:19.

[14] Lazarus HM, Haynesworth SE, Gerson SL, et al. *Ex vivo* expansion and subsequent infusion of human bone marrow-derived stromal progenitor cells (mesenchymal progenitor cells): implications for therapeutic use. Bone Marrow Transplant, 1995, 16(4):557-564.

[15] Yoo KH, Jang IK, Lee MW, et al. Comparison of immunomodulatory properties of mesenchymal stem cells derived from adult human tissues. Cellular Immunology, 2009, 259:150-156.

[16] Koc ON, Peters C, Aubourg P, et al. Bone marrow-derived mesenchymal stem cells remain host-derived despite successful hematopoietic engraftment after allogeneic transplantation in patients with lysosomal and peroxisomal storage diseases. Exp Hematol, 1999, 27:1675-1681.

[17] Karp JM, Leng teo GS. Mesenchymal stem cell homing: the devil is in the details. Cell Stem Cell, 2009, 4:206-216.

[18] Uccelli A, Moretta L, Pistoia V. Mesenchymal stem cells in health and disease. Nat Rev Immunol, 2008, 8(9):726-736.

[19] Jo CH, Lee YG, Shin WH, et al. Intra-articular injection of mesenchymal stem cells for the treatment of osteoarthritis of the knee: a proof-of-concept clinical trial. Stem Cells, 2014, 32(5):1254.

[20] 王鑫, 荀文隆, 徐小龙, 等. 骨髓间充质干细胞归巢促小鼠骨折愈合的实验研究. 国际骨科学杂志, 2015, (3):218-223.

[21] Liu Y, Buckley CT, Almeida HV, et al. Infrapatellar fat pad-derived stem cells maintain their chondrogenic capacity in disease and can be used to engineer cartilaginous grafts of clinically relevant dimensions. Tissue Engineering Part A, 2014, 20(21-22):3050.

[22] 张文志, 孔荣, 方诗元, 等. 兔自体骨髓间充质干细胞体内复合移植的成骨研究. 临床骨科杂志. 2002, 5(z):81.

[23] Nathan S, Das DS, Thambyah A, et al. Cell-based therapy in the repair of osteochondral defects: a novel use for adipose tissue. Tissue Engineering, 2003, 9(4):733-744.

[24] Frisbie DD, Kisiday JD, Kawcak CE, et al. Evaluation of adipose-derived stromal vascular fraction or bone marrow-derived mesenchymal stem cells for treatment of osteoarthritis. Journal of Orthopaedic Research Official Publication of the Orthopaedic Research Society, 2009, 27(12):1675-1680.

[25] Toghraie FS, Chenari N, Gholipour MA, et al. Treatment of osteoarthritis with infrapatellar fat pad derived mesenchymal stem cells in Rabbit. Knee, 2011, 18(2):71-75.

[26] Lee JM, Im GI. SOX trio-co-transduced adipose stem cells in fibrin gel to enhance cartilage repair and delay the progression of osteoarthritis in the rat. Biomaterials, 2012, 33(7):2016-2024.

[27] Van Pham PV, Bui HT, Ngo DQ, et al. Activated platelet-rich plasma improves adipose-derived stem cell transplantation efficiency in injured articular cartilage. Stem Cell Research & Therapy, 2013, 4(4):91.

[28] Ude CC, Sulaiman SB, Minhwei N, et al. Cartilage regeneration by chondrogenic induced adult stem cells in osteoarthritic sheep model. PLoS One, 2014, 9(6):e98770.

[29] Kuroda K, Kabata T, Hayashi K, et al. The paracrine effect of adipose-derived stem cells inhibits osteoarthritis progression. BMC Musculoskeletal Disorders, 2015, 16(1):236.

[30] Lee JM, Im GI. SOX trio-co-transduced adipose stem cells in fibrin gel to enhance cartilage repair and delay the progression of osteoarthritis in the rat. Biomaterials, 2012, 33(7):2016-2024.

[31] Ter HM, Schelbergen R, Blattes R, et al. Antiinflammatory and chondroprotective effects of intraarticular injection of adipose-derived stem cells in experimental osteoarthritis. Arthritis & Rheumatology, 2012, 64(11):3604.

[32] Desando G, Cavallo C, Sartoni F, et al. Intra-articular delivery of adipose derived stromal cells attenuates osteoarthritis progression in an experimental rabbit model. Arthritis Research & Therapy, 2013, 15(1):R22.

[33] Centeno CJ, Busse D, Kisiday J, et al. Increased knee cartilage volume in degenerative joint disease using percutaneously implanted, autologous mesenchymal stem cells. Pain Physician, 2008, 11(3):343-353.

[34] Davatchi F, Abdollahi BS, Mohyeddin M, et al. Mesenchymal stem cell therapy for knee osteoarthritis. Preliminary report of four patients. Int J Rheum Dis, 2011, 14(2):211-215.

[35] Orozco L, Munar A, Soler R, et al. Treatment of knee osteoarthritis with autologous mesenchymal stem cells. Transplantation Journal, 2013, 95(12):1535-1541.

[36] Vega A, Martin-Ferrero MA, Del Canto F, et al. Treatment of knee osteoarthritis with allogeneic bone marow mesenchymal Stem cells. Transplantation, 2015, 99(8):1681-1690.

[37] Bui HT, Duong TD, Nguyen NT, et al. Symptomatic knee osteoarthritis treatment using autologous adipose derived stem cells and platelet-rich plasma: a clinical study. Biomedical Research & Therapy, 2014, 1(1):2-8.

[38] Koh YG, Jo SB, Kwon OR, et al. Mesenchymal stem cell injections improve symptoms of knee osteoarthritis. Arthroscopy: the journal of Arthroscopic & Related Surgery: Official Publication of the Arthroscopy Association of North America and the International Arthroscopy Association, 2013, 29(4):748.

[39] Koh YG, Choi YJ. Infrapatellar fat pad-derived mesenchymal stem cell therapy for knee osteoarthritis. Knee, 2012, 19(6):902.

[40] Koh YG, Choi YJ, Kwon OR, et al. Second-look arthroscopic evaluation of cartilage lesions after mesenchymal stem cell implantation in osteoarthritic knees. American Journal of Sports Medicine, 2015, 43(1):176-185.

[41] Koh YG, Kwon OR, Kim YS, et al. Comparative outcomes of open-wedge high tibial osteotomy with platelet-rich plasma alone or in combination with mesenchymal stem cell treatment: a prospective study. Arthroscopy-the Journal of Arthroscopic & Related Surgery, 2014, 30(11):1453-1460.

[42] Pak J. Regeneration of human bones in hip osteonecrosis and human cartilage in knee osteoarthritis with autologous adipose-tissue-derived stem cells: a case series. Journal of Medical Case Reports, 2011, 5(1):296.

[43] Pak J, Chang JJ, Lee JH, et al. Safety reporting on implantation of autologous adipose tissue-derived stem cells with platelet-rich plasma into human articular joints. BMC Musculoskeletal Disorders, 2013, 14(1):1-8.

[44] Kim YS, Choi YJ, Suh DS, et al. Mesenchymal stem cell implantation in osteoarthritic knees: is fibrin glue effective as a scaffold? Am J Sports Med, 2015, 42(7):176-185..

[45] Clouet J, Vinatier C, Merceron C, et al. From osteoarthritis treatments to future regenerative therapies for cartilage. Drug Discovery Today, 2009, 14(19):913-925..

[46] Schelbergen RF, Van DS, Ter HM, et al. Treatment efficacy of adipose-derived stem cells in experimental osteoarthritis is driven by high synovial activation and reflected by S100A8/A9 serum levels. Osteoarthritis & Cartilage, 2014, 22(8):1158-1166.

[47] Samsonraj RM, Rai B, Sathiyanathan P, et al. Establishing criteria for human mesenchymal stem cell potency. Stem Cells, 2015, 33(6):1878-1891.

[48] Wasim SK, Adetola BA, Timothy EH. Hypoxic conditions increase hypoxia inducible transcription factor 2a and enhance chondrogenesis in stem cells from the infrapatellar fat Pad of osteoarthritis patients. Arthritis Research & Therapy, 2007, 9:55.

[49] Johnstone B, Hering TM, Caplan AJ, et al. *In vitro* chondmgenesis of bone marrow-derived mesenchymal progenitor cell. Exp Cel Res, 1998, 238:265-272.

[50] Mackay AM, Beck SC, Murphy JM, et al. The chondrogenic differentiation of cultured mesenchymal stem cells from marrow. Tissue Engineering, 1998, 4:415-428.

[51] Sekiya I, Colter DC, Prockop DJ. BMP-6 enhances chondrogenesis in a subpopulation of human bone stromal cells. Biochem Biophys Res Commun, 2001, 284:411-418.

[52] Worsterf AA, Brower-Toland BD, Fortier LA, et al. Chondro cytic differentiation of mdsenchym al stem cells sequentially exposed to transforming growth factot-p1 in monolayer and insulin-like growth factor 1 in a three-dimensional matr. CIrthop Res, 2001, 19:738-749.

[53] Hegewald AA, Ringe G. Hyalurnic acid and autologous synovial fluid induci chondrogenic differentiation if equine mesenchymal stem cells: a preliminary study. Tissue Cel, 2004, 36(6):431-438.

[54] Katayama R, Wakitani S. Repair of articular cartilage defects in rabbits using CDM P1 gene ansfected autologous mesenchymal cells derived from hone marrow. Rheumatology, 2004, 43:980-985.

[55] 霍建忠. 转化生长因子-B 基因修饰骨髓间充质干细胞复合壳聚糖修复兔关节软骨缺损. 中华风湿病学杂志, 2005, 9(7):393-396.

[56] Mason JM, Breitbart AS, Barcia M, et al. Cartilage and bone regeneration using gene enhanced tissue engineering. Clin OrhopRelat Res, 2000, 379:171-178.

[57] Madry H, Orth P. Acceleration of articular cartilage repair by combined gene transfer of human insulin-like growth factor 1 and fibmblast factor 2 in vivo. Arch Orthop Trauma Surg, 2010, 130(10):1311-1322.

[58] Xie X, Wang Y, Zhao C, et al. Comparative evalution of MSCs from bone marrow and adipose tissue seeded in PRP derived scaffold for cartilage regeneration . Biomaterials, 2012, 33(29):7008-7018.

[59] Soto M, Uchida K, Nakajima H, et al. Direct transplantation of mesenchymal stem cells into the knee joints of Hartley strain guinea pigs with spontaneous osteoarthritis. Arthtitis Res Ther, 2012, 14(1):R13.

[60] Kuroda K, Kabata T, Hayashi K, et al. The paracrine effect of adipose derived stem cells inhibits osteoarthritis progression,BMC Musculoskelet Disord, 2015, 16:236.

[61] ter Huurne M, Schelbergen R, Blattes R, et al. Antiinflammatory and chodroprotective effect of intraarticular injection of adipose derived stem cells in experimental osteoarthritis. Arthritis Rheum, 2012, 64(11):3064-3013.

[62] Desando G, Gavallo C, Sartoni F, et al. Intraarticular delivery of adipose derived stromal cells attenuates osteoarthritis progression in an experimental rabbit madel. Arthtitis Res Ther, 2013, 15(1)R22.

[63] Platas J, Guillén MI, Pérez del Caz MD, et al. Conditioned media from adipose-tissue-derived mesenchymal stem cells downregulate degradative mediators induced by interleukin-1β in osteoarthritic chondrocytes. Mediators of Inflammation, 2013, (1-2):357014.

[64] Buul GMV, Villafuertes E, Bos PK, et al. Mesenchymal stem cells secrete factors that inhibit inflammatory processes in short-term osteoarthritic synovium and cartilage explant culture. Osteoarthritis & Cartilage, 2012, 20(10):1186-1196.

[65] Manferdini C, Maumus M, Gabusi E, et al. Adipose-derived mesenchymal stem cells exert antiinflammatory effects on chondrocytes and synoviocytes from osteoarthritis patients through prostaglandin E2. Arthritis & Rheumatism, 2013, 65(5):1271-1281.

[66] Maumus M, Manferdini C, Toupet K, et al. Adipose mesenchymal stem cells protect chondrocytes from degeneration associated with osteoarthritis. Stem Cell Research, 2013, 11(2):834.

第十七章 间充质干细胞在皮肤修复与再生中的应用

第一节 皮肤创面修复的一般过程

皮肤是人体最大的器官，分为表皮层及真皮层。表皮层位于皮肤最外层，发挥抵御病原体和保湿的功能。其主要由角质细胞构成，同时包含少量黑素细胞、朗格汉斯细胞和默克尔细胞[1]。真皮层位于表皮层之下，包含大量编织状的细胞外基质（ECM），发挥减震和抗拉的作用。同时，皮肤包含多种附件结构，如汗腺、皮脂腺和毛囊。它们分别发挥排泄、调节体温、滋润皮肤和保护皮肤的功能。因此皮肤组织结构的完整性对维持皮肤正常的生理功能十分重要[2]。皮肤是机体的重要屏障，易于遭受各种物理、化学及生物伤害。创面修复是机体面对损伤的主要修补机制，能够防止感染，恢复组织完整性及功能。不幸的是，在成人皮肤，创面修复方式趋向于快速愈合，虽然可防止感染，但影响创面愈合质量，最终形成瘢痕[3]。瘢痕组织与原有皮肤结构完全不同，且难以再生附件。理想的创面治疗手段应该满足以下三方面需求：①加速愈合；②改善创面愈合质量；③促进附件再生。皮肤修复是一个复杂过程，按照其发生特征，主要分为出血期、炎症期、增殖期及改建期。同时，这4个时期又相互交叠，甚至创面的不同部位呈现不同的修复期。

一、止血期

止血期发生在皮肤损伤后的数秒至数分钟内。起初创面出血，流出的血小板与血管内皮下层接触，形成内皮下血栓。血小板参与整个创面的愈合过程。它们不仅能够止血，而且释放多种细胞因子、激素和趋化因子以启动下个愈合阶段。血管活性物质如儿茶酚胺和5-羟色胺通过内皮上的特异性受体发挥作用，引起血管收缩，继而导致血管舒张和通透性增强，白细胞、红细胞和血浆蛋白得以进入创面。血小板与受损内皮下胶原的GpⅡb-Ⅲa受体相互作用，形成初始血栓。活化的血小板脱颗粒引发内外凝血级联反应。纤维蛋白聚合促使血栓成熟，并作为创面愈合相关修复细胞（白细胞、角质形成细胞和成纤维细胞）的重要网络支架。血小板释放众多的趋化因子和细胞因子，如血小板源生长因子（PDGF）、转化生长因子-β（TGF-β）、成纤维细胞生长因子（FGF）、血小板因子4（platelet factor 4, PF4）、血清素、前列腺素（PG）、前列环素、促凝血素和组胺，吸引炎症细胞进入创面[4,5]。在修复开始数分钟内，大量炎症细胞（主要是中性粒细胞和巨噬细胞）涌入创面，进入炎症期。

二、炎症期

炎症期伴随着大量中性粒细胞、巨噬细胞和淋巴细胞涌入创面。中性粒细胞在PDGF和TGF-β驱动下最先迁移到创口部位[6,7]，以创伤后1~2天为最多，其通过嗜菌杀灭机制杀灭细菌,去除入侵病原体和失活组织碎片,保护创面免受感染。另外,中性粒细胞能够活化角质形成细胞、成纤维细胞和免疫细胞[8,9]。在修复的数小时内,创缘角质细胞过表达基质金属蛋白酶(MMP)和纤维蛋白溶酶[10,11],促进角质细胞早期的增殖和迁移。

随着中性粒细胞进入创面,大量的淋巴细胞开始出现,尽管淋巴细胞在炎症期的具体作用尚不明确,但它通过分泌细胞因子,如白介素-2(IL-2)、一氧化氮合酶(NOS)、FGF[12],促进创面愈合。同时,淋巴细胞能够帮助清除创面中老化中性粒细胞。在创伤后的2~3天,伴随着中性粒细胞凋亡,巨噬细胞被中性粒细胞释放的细胞因子和生长因子趋化到创口。在炎症期的初期,巨噬细胞和中性粒细胞均能通过 Toll 样受体(TLR)识别入侵病原体和受损组织,并且在激活这些受体时,促炎细胞因子的表达进一步增加[13],从而招募更多的免疫细胞聚集在创口。当进入炎症期的后期,巨噬细胞分化为抗炎表型,表达抗炎细胞因子如血管内皮生长因子(VEGF)和 IL-10,同时分泌大量 TGF-β,招募创缘成纤维细胞进入创面[8,9,11,14]。目前认为,巨噬细胞为创面炎症细胞群落的优势细胞,并参与创面修复的整个过程。

炎症期是创面修复的关键时期,如果中断或延长(即超过3周),可导致慢性创面,愈合受损,并最终形成更多瘢痕。一些重要因素,如高细菌负荷(大于 10^5cells/g 组织)、反复创伤和伤口持久存在异物均可导致炎症期的延长。这不仅影响愈合速度,还可决定纤维化程度。因此,过度的炎症反应将导致创面延迟愈合或不愈合和严重的纤维化[15,16]。此时,再上皮化进程加速,创缘角质细胞过表达 MMP 和纤维蛋白溶酶,以降解纤维蛋白凝块[10,17],形成角质细胞增殖、迁移的空间。一旦重建上皮,基底膜将重新形成。

三、增殖期

增殖期以出现再上皮化的形式呈现。另外,增殖期还包括血管新生及胞外基质形成来填补创面缺损。上皮化由角质形成细胞的增殖及向创缘的迁移完成。同时毛囊及顶泌汗腺膨出部干细胞分化为角质形成细胞并重新构建基底层,同时逐渐向创缘迁移,其主要通过分泌蛋白酶降解周围 ECM 完成迁移[18]。创伤后2~3天,成纤维细胞被 PDGF 和 TGF-β 招募到创面[19],开始在创面分泌并沉积Ⅲ型胶原、纤连蛋白、透明质酸和糖胺聚糖(GAG)(透明质酸酶、4-硫酸软骨素、硫酸软骨素 B 和硫酸乙酰肝素)。在成纤维细胞增殖过程中,胶原不断产生,胶原水平在大约3周内不断上升,最后达到胶原降解与合成相平衡的状态。创面中的胶原量与创面的张力密切相关。另外,成纤维细胞在 TGF-β 作用下分化为肌成纤维细胞,该细胞表达 α-SMA,能够收缩创口,并能够对创面的机械环境发生物理和生物化学反应[20,21]。另外,受损内皮细胞、角化细胞、成纤维细胞和巨噬细胞释放 VEGF 和 FGF-2,促进血管内皮细胞大量增殖,形成高血管化的肉芽组织[22]。血供能够为损伤皮肤提供营养及氧气,并促使参与修复细胞的增殖、迁移及分化。

四、改建期

创伤后3周,胶原合成与降解达到稳态平衡,创面重建开始。该过程可持续数月,甚至数年。在该阶段,肌成纤维细胞发生凋亡,以防止胶原过度沉积。基底膜中的硫酸乙酰肝素被纤连蛋白和透明质酸所替代[18]。Ⅲ型胶原逐渐被Ⅰ型胶原取代。该过程受到MMP和其组织抑制剂[金属蛋白酶组织抑制物（TIMP）]的表达严格控制与调节[23]。MMP负责降解胶原蛋白网络，而TIMP通过与相应的MMP结合来抑制其发挥作用[24]。GAG稳定降解，直到达到正常真皮中的含量。理想情况是改建后的真皮组织与原有皮肤相同，这一现象只在胚胎发育的早、中期才能观察到[25,26]。在成体皮肤修复中，修复后皮肤与原有皮肤组织明显不同[27]，即瘢痕性愈合。其由大量胶原纤维构成，且胶原纤维之间相互交联。最为显著的特征是，修复后皮肤的附件结构，如毛囊及汗腺不能重建，这将导致修复后皮肤生理功能缺失。

五、皮肤创面的病理性修复

创面愈合是多种细胞、结构蛋白、生长因子及蛋白水解酶之间相互作用的过复杂程。修复过程顺序发生，若这一自然级联愈合过程中断或延迟，将破坏后续愈合进程，导致创面愈合延迟，或者过度愈合。前者形成慢性创面，后者形成增生性瘢痕（hyperplastic scar）或瘢痕疙瘩（keloid）。

持续6周以上不愈合的创面称为慢性创面[28]。姜玉峰等在2010年开展多中心、横断面临床数据调研，统计了17家医院在2007~2008年收治的2513例慢性创面患者，慢性创面发生率为1.7‰[29]，其中糖尿病和创伤是导致慢性创面的主要原因。慢性创面按照发生分为三种类型：血管型、糖尿病型和压力型[30,31]。尽管病因不同，但慢性创面存在病理生理上的共同点，即持续慢性炎症、低氧及微生物感染[30,32]。由于慢性创面患者存在基础疾病或感染，大量的免疫细胞被招募到创面，并产生活性氧（ROS）及炎性细胞因子，诱导创面多种细胞产生蛋白水解酶，如MMP和其他蛋白酶，降解TIMP、VEGF和ECM[32,33]。另外，异常表达的基质糖蛋白进一步减弱ECM的稳定性，抑制角质细胞的迁移，影响再上皮化[30,34]。在慢性创面，成纤维细胞分化发生障碍，肌成纤维细胞数量减少，影响肉芽组织形成，因此创面愈合能力受损。

纤维增生性修复是创面发生的异常炎症，导致修复创面过度纤维化，包括增生性瘢痕和瘢痕疙瘩。其可能与致伤因素形成的创面局部机械应力相关，其主要发生在烧伤或深度创伤时[35]。瘢痕疙瘩形成除受到局部炎症影响外，还受到遗传及环境因素的影响，易发生在肤色较深和具有遗传倾向的人群[36,37]。临床上，HTS和瘢痕疙瘩表现为隆出正常皮肤，质地较硬，呈红色，瘙痒，且瘢痕疙瘩超出创缘生长[38]。通常HTS能够自发萎缩，二者均缺乏有效的治疗手段。

第二节　间充质干细胞参与皮肤修复的机制

近年来，干细胞逐渐成为创面治疗领域的热点。不同来源的干细胞，包括胚胎干细胞（ESC）、间充质干细胞（MSC）、诱导多能干细胞（iPSC）已相继开展临床前及临床研究[39]。其中，MSC是具有多向分化潜能的成体干细胞，广泛分布在多种人体组织中，如骨、软骨、肌肉、脂肪、肝、胰腺、胎盘、脐带及羊膜组织[40]。MSC具有以下基本生物学特征：①能够体外贴壁培养；②CD105、CD73、CD90阳性，CD34、CD19、CD 45、

CD11a、HLA-DR 阴性；③在体外能够分化为成骨、脂肪及软骨细胞[41]。除此之外，MSC 具低免疫原性，使其成为创面修复领域的明星细胞。

越来越多的研究证据显示 MSC 在创面治疗中有积极作用，它能够参与创面修复的整个过程，并发挥多种生物学效力，主要分为以下几个方面（图 17-1）。

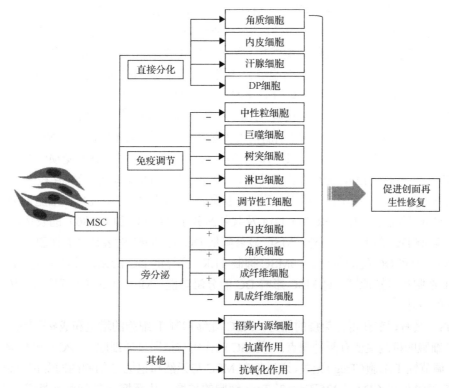

图 17-1　MSC 促进皮肤创面愈合的机制

MSC 主要通过直接分化、免疫调节、旁分泌及其他相关机制如招募内源性干/祖细胞、抗菌及抗氧化作用来改善创面局部的微环境，缩短炎症期，加速进入增殖期，改善创面修复细胞的生物学功能，并改变创面基质构成，以及抑制创面胶原沉积，实现创面的加速和少瘢痕性修复

一、MSC 直接分化为修复细胞类型

目前，体内外均已发现 MSC 可分化为其他细胞的证据，并分为单向分化和多向分化。通过体外定向诱导，MSC 可以分化为角质细胞[42]、内皮细胞[43]、汗腺细胞[44]或真皮乳头（dermal papilla，DP）细胞[45]。通过体内示踪技术，发现 MSC 在创伤环境下多向分化的证据。2006 年，付小兵院士研究团队[46]将 5-溴脱氧尿苷标记的 MSC 移植入创面后发现，在创面修复的 7 天、14 天及 28 天分别有 3.34%、3.46%、2.94%的移植 MSC 分化为内皮细胞，并有 0~1.49%的移植 MSC 参与表皮重构[46]。另外，2008 年 Sasaki 等[47]通过系统性输注 GFP-MSC 证实，MSC 能够在创面分化为角质细胞、内皮细胞和血管周细胞。这些证据都说明，MSC 能够通过分化为其他类型细胞参与创面修复。

二、MSC 的免疫调节作用

创面中的炎性细胞因子[INF-γ、肿瘤坏死因子-α（TNF-α）和 IL-1]和免疫细胞均能够活化 MSC，激发其免疫调节功能。MSC 能够调节不同免疫细胞亚群的增殖和分泌。

MSC 具有调节固有免疫的能力。MSC 分泌高水平 IL-6，激活 STAT3 转录因子，使中性粒细胞寿命延长，同时减弱中性粒细胞呼吸爆发程度，所以 MSC 作用后，并未明显改变中性粒细胞对创面环境的影响[48,49]。另外，MSC 抑制中性粒细胞脱颗粒，从而阻止炎性细胞因子，如防御素-α 发挥细胞溶解作用。在对巨噬细胞的调节方面，MSC 能够促进巨噬细胞由 M1 型向 M2 型的转化，多种介质，如吲哚胺 2,3-双加氧酶（IDO）、前列腺素 E2（PGE2）和 MSC 源性 IL-4 与 IL-10 均参与该过程[50-53]。MSC 分泌的 TGF-β1 与 PGE2 发挥协同作用，抑制巨噬细胞的促炎细胞因子（如 IL-1β、IL-6、TNF-α 和 IFN-γ）产生[50,52]。同时，抗炎细胞因子 IL-10 上调，可阻止中性粒细胞进入受损组织[53,54]。但是，MSC 并不影响巨噬细胞的吞噬功能，意味着其碎片清除功能仍然存在[52,55]。MSC 对树突细胞（DC）的作用体现在以下四个方面：①抑制 CD14 阳性单核细胞向 DC 的分化。②MSC 降低了 HLA-DR 和 CD1a 及共刺激分子（CD80 和 CD86）的表达，并下调 IL-12，影响 DC 成熟[56]。另外，MSC 显著降低 DC 上 CD83 的表达（T 细胞活化的共刺激受体），导致 DC 的幼稚化。③MSC 抑制 TLR4 诱导的 DC 活化，影响其向 $CD4^+$ T 和 $CD8^+$ T 细胞提呈抗原[36]。④MSC 影响 DC 的分泌功能。MSC 上调 DC 表达 IL-10，下调 TNF-α 的表达[57]。

MSC 具有调节获得性免疫的能力。MSC 能够抑制 T 细胞的增殖和成熟[58-62]。关于其抑制 T 细胞的机制主要有两种观点：①MSC 的作用与抗原呈递细胞（APC）和 $CD4^+CD2^+$ $Foxp3^+$ 调节性 T 细胞(Treg)无关，通过抑制 MSC 与天然和记忆 T 细胞的同源抗原识别[61]；②MSC 通过促进 $CD4^+CD25^+Foxp3^+$ Treg 细胞的扩增，从而抑制 T 细胞活性[58]。另外，MSC 分泌的 PGE2 能够促进 Th1 细胞向 Th2 细胞转变[63]。MSC 对 B 细胞的作用取决于培养条件和所属物种。MSC 能够抑制 B 细胞增殖及其抗体（IgA、IgG 和 IgM）的分泌，并减弱其化学趋化能力[64]。

三、MSC 改善修复细胞功能

目前，MSC 的旁分泌效应被认为是其发挥作用的主要途径。大量的科学证据表明，MSC 分泌的调节分子主要分为以下几类：①生长因子，包括血管内皮生长因子（VEGF）、血小板源生长因子（PDGF）、碱性成纤维细胞生长因子（bFGF）、表皮生长因子（EGF）、肝细胞生长因子（HGF）和基质细胞衍生因子-1（SDF-1）等；②细胞因子，包括 IL-4、IL-6 和 IL-8；③趋化因子和受体，包括 CCL2、CCL7、CXCR4、CXCR5 和 CXCR6；④其他因子，包括 MMP 和 TIMP 等[47,65]。

MSC 通过旁分泌机制调节多种修复细胞，包括上皮细胞、内皮细胞、角质形成细胞、成纤维细胞和内源性干/祖细胞的存活、增殖、迁移和基因表达[66]。血供的重建是伤口愈合的基础，MSC 通过旁分泌信号在血管新生中发挥重要作用。血管新生是一个复杂过程。MSC 分泌的生物活性分子参与血管形成的整个过程。VEGF 是血管新生中重要的启动及调节细胞因子[67,68]。IL-6 和 IL-8 可增强内皮细胞迁移[69,70]。SDF-1 参与内皮细胞存活、血管树形成及周细胞和平滑肌细胞募集的过程[71,72]。Ang-1 抑制周细胞凋亡并招募周细胞促进新血管的成熟[73]。MSC 还表达旁分泌因子以提高血管稳定性和发挥血管保护作用[74,75]，如肾上腺髓质素[76]。

此外，MSC 通过旁分泌效应能够逆转功能受损细胞的生物学活性。MSC 的条件培养基能够通过上调 ERK 信号通路，从而逆转糖尿病微环境下角质细胞的功能障碍[77]，并能够通过抗氧化应激策略抵抗糖尿病微环境诱导的成纤维细胞衰老[78]。

四、MSC 促进创面少瘢痕性修复

瘢痕的形成是由成纤维细胞生成过量 ECM 沉积在创面引起的，不仅影响皮肤修复后的外观，而且缺乏毛囊和神经末梢等附属结构。MSC 可分泌具有抗纤维化特性的各种细胞因子和生长因子，包括 HGF、IL-10 和肾上腺髓质素[65,79,80]。MSC 主要通过调节成纤维细胞的 ECM 分泌和成纤维细胞的分化两个层面来促进创面少瘢痕性愈合。HGF 作用下，成纤维细胞表达 TGF-β1[81]、Ⅰ型胶原[82]和Ⅲ型胶原[81]下调，并促进 TGF-β 信号通路下游转录分子 Smad3 的核移出，进一步降低促纤维化相关靶基因的表达[83]。HGF 还刺激成纤维细胞中 MMP-1、MMP-3 和 MMP-13 表达的上调[84]，从而促进 ECM 的更新。

在伤口愈合的增殖阶段，伤口边缘的成纤维细胞增殖并迁移到伤口中部，产生肉芽组织，开始改造伤口基质以产生新的真皮组织。MSC 能够显著抑制成纤维细胞向肌成纤维细胞分化。肌成纤维细胞是组织纤维化中的"效应"细胞，其能合成更多的 ECM 成分，比成纤维细胞呈现更强的收缩能力。创面中成纤维细胞的来源主要分为三类：①创面原有成纤维细胞增殖、迁移；②损伤血管内皮细胞可发生上皮间质转化（epithelial-mesenchymal transition，EMT）成为肌成纤维细胞参与愈合[85]；③持续的炎症或是肉芽组织形成障碍，血液中的纤维细胞在 TGF-β1 和 T 细胞的作用下被招募到创面并分化为肌成纤维细胞[86]。MSC 通过分泌 HGF[81]和 PGE2，一方面调节创面 TGF-β1 和 TGF-β3 的平衡，另一方面抑制 EMT[87]过程进而抑制肌成纤维细胞形成。

五、MSC 的其他促修复机制

干细胞在创面修复中发挥着关键性作用。MSC 通过 VEGF、SDF-1-CXCR4 轴、炎性细胞因子（IL-6、IL-8）[70]和生长因子受体轴（如 HGF-c-met 轴、PDGF[88]）从骨髓、外周血和周围组织中募集内源性干/祖细胞，包括 MSC 和内皮祖细胞（EPC）来促进修复。

除此之外，MSC 具有抗氧化及抗菌的效应。MSC 能够分泌大量的抗氧化调节因子，如胰岛素样生长因子（IGF）、血小板源生长因子（PDGF）、超氧化物歧化酶（SOD）、肝细胞生长因子（HGF）、粒细胞集落刺激因子（G-CSF）、粒细胞-巨噬细胞集落刺激因子（GM-CSF）、IL-12 和 IL-6，它们能够改善创面局部微环境，维持创面细胞的活力[89]。MSC 具有显著的抗菌作用。一方面 MSC 通过分泌免疫调节因子促进免疫细胞的杀伤和吞噬作用[90]，另一方面通过分泌 LL37，一种具有广谱微生物防御性质的肽，直接破坏细菌细胞膜，杀死细菌[91]。

此外，MSC 还以外泌体的方式发挥促修复作用。外泌体是细胞分泌的双脂膜小囊泡，

包含蛋白质、DNA/RNA 和微 RNA（miRNA），参与细胞间通讯。MSC 来源外泌体在疾病治疗中的优势被逐渐证实[92]。就皮肤修复而言，MSC 外泌体能够通过以下三种方式干预创面修复：①抑制创面炎症。MSC 外泌体来源 miR-181 通过抑制 TLR4 产生抑制烧伤创面炎症[93,94]，let7b 通过 TLR4/NF-κB/STAT3/Akt 信号通路促进巨噬细胞 M1 表型向 M2 表型的转换。②促进成纤维细胞增殖、迁移，新生血管形成[95]；③抑制成纤维细胞分化。MiR-21、MiR-23a、MiR-125b 和 MiR-145 能够通过抑制 TGF-β2/Smad2 信号通路发挥作用[96]。

第三节　影响间充质干细胞促修复能力的因素

通过上一节的介绍，我们可知 MSC 可以通过多种方式参与皮肤组织的修复过程，并在创面愈合中发挥重要作用，影响创面愈合的结局。机体本身是一个巨大的干细胞库，多种组织中均已发现 MSC。因此，促进内源性 MSC 迁移，从而发挥损伤修复作用对于组织修复再生是十分必要的。因此，理想的修复模式是皮肤损伤后，来自于临近及远端组织的内源性 MSC 被招募到创面，发挥修复作用[97]。新的证据显示，在损伤状态下，如急性烧伤[98]、骨折[99]和缺氧[100]，骨髓中少量的 MSC 进入外周血，并循环到损伤部位。然而，尚缺乏有效的内源性 MSC 的动员策略。MSC 表达 CCR7（SLC/CCL21 的受体），在 MSC 向创面迁移过程中发挥重要作用。皮内注射 SLC/CCL21 显著增加 MSC 在伤口中的募集，加速伤口修复[101]。受损组织释放 SDF-1α，通过 SDF-1α 受体 CXCR4 诱导经静脉注射的 MSC 向创面迁移[102]。此外，静脉注射 HMGB1 通过增强 SDF-1α 受体 CXCR4 的表达，也能够诱导内源性骨髓中 PDGFRα$^+$ MSC 进入创面[103]。中国传统医学中的电针[104]及创面的负压伤口治疗（negative pressure wound therapy，NPWT）[105]分别通过增加组织中 SDF-1 的水平和营造低氧环境来发挥动员内源性 MSC 的作用。尽管如此，动员后的 MSC 数量仍然不能满足损伤创面进行修复的需求，更为有效地调动内源性 MSC 的策略需要进一步深入挖掘。因此，施以外源性 MSC 仍然是目前促进皮肤修复的常用手段。

外源性 MSC 常规通过系统性输注和局部注射两种方式进行促修复的治疗。有证据显示系统性输注的 MSC 在胰腺损伤模型[106]、心脏[107]、肾[108]和肝[109]中均表现出仅有小于 3%的长期移植率，同样通过局部注射给皮肤伤口部位输送 MSC 同样表现出低植入率[47]。因此，改善 MSC 在损伤部位植入率可进一步提升干细胞的治疗效应。相关的改善策略主要分为以下几个方面。

首先，选择适当的 MSC 施予方式。对于皮肤修复，MSC 进入体内的路径主要分为系统性输注和局部注射两种。相较之下，后者比前者具有更高的创面保留率。即使如此，经创面直接注射的 MSC 在创面低氧、炎症、高氧化应激的条件下很快凋亡和死亡。干细胞与由生物材料构建的三维支架相结合形成组织工程化皮肤在再生医学中发挥的作用越来越突出[110]。生物材料可以支持细胞保持活力并增强治疗效果。因此，MSC 与生物支架的联合移植能够优化 MSC 植入率，以及提升 MSC 疗效。美国食品药品监督管理局（FDA）批准的商品化纤维蛋白密封剂在基础和临床试验中得到广泛的应用，其可以作为 MSC 进行体内植入的载体[111]。此外，基于 ECM 大分子的生物支架也得到广泛研究。

混合 MSC 的胶原基质能够显著缩短创面愈合的早期阶段，并在大鼠[68]、小鼠[67]、裸鼠[112]及兔[113]糖尿病模型的创面得到证实。另外，MSC 混合改性胶原支架，包括胶原-壳聚糖支架[114]、端胶原支架[115]及脱细胞生物支架[116]同样具有良好的促修复作用。

其次，对移植 MSC 进行预处理。MSC 预处理的方式分为以下三种：①低氧预处理。尽管严重缺氧可致细胞死亡，但短期暴露于缺氧环境（缺氧预处理）对细胞具有保护作用[117]。缺氧培养能够显著降低 MSC 体外培养的死亡率[118]，与缺氧诱导因子-1α（HIF-1α），血管生成素-1（Ang-1）、VEGF、红细胞生成素（EPO）、Bcl-2 和 Bcl-xL 的促生长因子与促血管生成因子的表达增加有关[118]。低氧预处理还能够诱导 MSC 的自噬过程，防止凋亡，提高创面局部的细胞植入率[119]。②高氧预处理。高氧（100%氧）预处理能够减少 MSC 中半胱天冬酶 1、3、6、7 和 9 的表达并增加存活基因如 Akt 的表达，从而提高MSC 活力并促进其增殖[84]。③细胞因子预处理。低水平的炎性细胞因子预处理同样能够增强 MSC 的生物学功能。TGF-α 预处理 MSC 后，其迁移能力增强，分泌 VEGF 明显增多[120]。高迁移率族蛋白 1（HMGB1）是一种炎性细胞因子，由损伤坏死的细胞被动释放，HMGB1 能够促进 MSC 的迁移[121]。另外，PDGF-BB 预处理 MSC，通过活化 Akt 和 ERK 信号通路上调其 VEGF 的表达[122]。此外，Netrin-1 是新型血管生成因子，预处理 MSC 能够显著增加血液灌注评分和血管密度，以及改善缺血肢体的功能[123]。

再次，基因改造 MSC，可以靶向性增强 MSC 的生物学功能。MSC 进行 Akt 和血管生成素-1（血管生成中的重要调节剂）双表达系统改造后，细胞存活率明显提高[124]。另外，热激蛋白 20（Hsp-20）[125]、分泌性卷曲相关蛋白 2（sFRP2）、Wnt 信号调节因子[126]、survivin[127]、血红素加氧酶（HO-1）[128]、GSK-3β[129]、ERBB4[130]、CCR1[131]和 SDF-1[132,133]过表达，均证实能够提高 MSC 的体内植入率。此外，miR-1 和 miR-210 分别通过调节半胱天冬酶 9、Bcl-2 与 Bax[134]及抗氧化 c-Met 途径[135]促进 MSC 在体内的存活。

最后，MSC 的联合移植策略。与单独 MSC 移植相比，辛伐他汀联合 MSC 移植，能够 4 倍提高 MSC 存活率，主要通过抑制氧化应激和炎症反应来实现[136]。另外，罗苏伐他汀可以增加 Akt、ERK 磷酸化，进一步影响下游 FoxO3a 磷酸化和减少促凋亡蛋白的表达，从而提高移植脂肪间充质干细胞（AD-MSC）的存活率，并进一步改善心脏功能和减少纤维化[137]。

另外，值得一提的是 MSC 自身的分化功能。如前所述，MSC 能够通过分化为内皮细胞、角质细胞等修复细胞参与创面愈合。因此，促使 MSC 的定向分化也是增强其修复功能的策略之一。MSC 具有多向分化潜能，鉴定中已经将多向分化视为其常规检测标准。细胞所处微环境对干细胞的分化具有导向性作用。其中细胞因子及小分子微环境起主要作用。PCGF[138]和 VEGF 具有促进 MSC 向内皮细胞分化的能力[139]。Rho 激酶信号通路抑制剂（Y-27632）能够促进 MSC 向角质细胞分化[140]。EGF、角质细胞生长因子（KGF）[141]和汗腺细胞培养[142]上清均具有促进 MSC 向汗腺细胞分化的能力。另外，胞外基质微环境对细胞分化方向也具有积极引导作用。ECM 的固有弹性、纳米形态、蛋白质构成和机械应变力都是诱导干细胞定向分化的独立参数[143-146]。更有趣的是，干细胞还具有机械记忆性，即细胞对之前所处微环境中机械参数的敏感性[147]，这样就保证了干细胞所处体内机械环境在体外的可复制性。此外，表观遗传调控已经成为 MSC 分化的重要机制[148]。

低甲基化试剂（hypomethylating agent，HMA）如 5-氮胞苷、组蛋白脱乙酰酶（histone deacetylase，HDAC）抑制剂如曲古抑菌素 A（trichostatin A，TSA）或丙戊酸（valproate，VPA）已经被用来调节 MSC 的分化[149-152]，获得特异性的目的细胞。

第四节　间充质干细胞治疗皮肤损伤的临床研究

MSC 在临床前研究中，展示出其具有优越的创面治疗效果，并且其相关促愈合机制在逐渐得到阐明，显示出其在创面治疗中的应用前景。目前，美国国立卫生研究院（NIH）临床试验网站（ClinicalTrials.gov）中可查到的利用 MSC 进行创面治疗的临床研究越来越多。

早在 2003 年，Badiavas 和 Falanga[97]首先就将自体骨髓间充质干细胞（BM-MSC）应用于 3 例慢性创面患者（创伤 1 年以上），并且这些患者前期对生物工程皮肤及自体皮肤移植均反应不良。经过 BM-MSC 治疗后，创面呈少瘢痕性愈合，并在创面组织中检测到移植的 BM-MSC，这是应用 MSC 进行创面治疗的第一个临床案例。在此之后，Ichioka 等[153]在 2005 年报道，对传统创面治疗手段反应不良的慢性腿部溃疡患者，创面持续 1 年仍然未愈，利用自体骨髓浸渍胶原基质（包含 BM-MSC）治疗，创面完全愈合。同样 Kirana 等[154]报道使用自体 BM-MSC 移植可促进伴有下肢溃疡的糖尿病患者创面愈合。另外，异体来源 MSC 也具有促愈合作用。Conget 等[155]在 2010 年使用同种异体 BM-MSC 治疗 2 例隐性营养不良的大疱性表皮松解症患者，可促进创面Ⅶ型胶原蛋白沉积并加速创面的再上皮化过程。

除了上述这些临床案例之外，还有几项大型临床试验。2008 年，Yoshikawa 等[156]利用在胶原海绵上培养的自体 BM-MSC 治疗 20 例由各种病因引发的慢性创面，证明在所有患者中均有效。此后，Prochazka 等先后进行了两项临床研究，发现自体 BM-MSC 移植可以保留需要截断的患肢，并改善与创面愈合相关的物理指征，如脚趾压力、脚趾指数和 $TcpO_2$[157,158]。Lu 等[159]进行一项双盲、随机、对照试验，用于比较 BM-MSC 与骨髓单个核细胞（BMMC）治疗糖尿病的严重肢体缺血和足部溃疡患者的疗效。在 41 例 2 型糖尿病患者接受移植后，BM-MSC 组的溃疡愈合率显著高于 BMMC 组（$P=0.022$），且愈合速度明显加快。2012 年，Kirana 等[160]的研究显示，22 例慢性创面患者中 18 例在 MSC 移植后，创面条件改善，愈合加快。

MSC 对烧伤创面也具有明显的治疗作用。Rasulov 等[161]在 2005 年通过将同种异体 BM-MSC 移植到烧伤创面，成功地治疗了患有广泛皮肤灼伤（Ⅰ-Ⅱ-ⅢAB 度皮肤烧伤，总面积 40%，ⅢB 度面积 30%）的患者。Bey 等[162]在 2010 年通过使用 BM-MSC 来修复由严重放射性灼伤造成的伤口，具有显著的治疗效果。

MSC 与组织工程支架联合移植在慢性创面患者中也收到理想的治疗效果。2006 年，Vojtassak 等[163]使用由自体皮肤成纤维细胞组成的自体生物移植物在可生物降解的胶原膜（Coladerm）上与自体 BM-MSC 组合来进行糖尿病相关慢性创面的治疗。前后分别进行 3 次联合治疗，患者的创面面积显著减小，并逐渐愈合，创面真皮血管密度增加，伤口床的真皮厚度增厚。另外，2011 年，Ravari 等[164]使用自体 BM-MSC 联合血小板、纤维蛋白胶和胶原基质治疗 8 例糖尿病相关慢性创面。联合治疗 4 周后，3 例患者创面完全愈合，另 5 例创面面积显著缩小。

创面治疗相关的临床试验，主要以采用自体 BM-MSC 为主。AD-MSC 可通过抽取患者自体脂肪组织来提取，因此 AD-MSC 易于获得，并且能够大量扩增，能够满足患者大面积创面对细胞数量的需求。但是，关于 AD-MSC 相关临床试验仅见一篇报道。Lee 等[165]在 2012 年初步研究了 AD-MSC 移植对患有严重肢体缺血的糖尿病患者的安全性和有效性。在随访期间，未见相关不良反应及并发症，且 66.7%的患者临床症状得到明显改善。此外，脐带间充质干细胞（UC-MSC）来源方便、有较强的增值能力及极低的免疫原性，使其成为再生医学中的研究热点，但是目前尚缺乏利用 UC-MSC 进行创面治疗的临床数据。由付小兵带领的解放军 301 医院创面治疗中心利用 UC-MSC 来源的干细胞凝胶，成功治疗了对传统治疗手段反应不良的慢性创面患者，临床治愈率达到 83.33%，并未见不良反应（未发表数据）。干细胞凝胶治疗慢性创面的多中心临床试验正在进行中。尚未见其他组织来源 MSC，如羊膜 MSC 和脐血间充质干细胞（UCB-MSC），进行创面治疗的临床试验报道。

关于利用 MSC 进行创面治疗的临床试验需要在以下两个方面进一步改进：①现有的"大规模试验"远远不足以将基于 MSC 的治疗形成标准化方案推广到临床，需要更多双盲、随机、对照设计的大规模临床试验，并且需要更长的随访时间；②基于不同组织来源 MSC 的异质性，需要根据临床研究比较不同来源 MSC（如 BM-MSC 与 AD-MSC、自体与同种异体）的治疗效率、输注策略、患者的个体差异对细胞治疗的反应性等问题。

皮肤再生，包括附件的再生是创面愈合的最高境界。MSC 促进人皮肤附件再生的临床试验也已开展。2009 年，盛志勇院士率先在国际上将由 BM-MSC 分化的汗腺样细胞移植到深部烧伤的切痂创面。手术后 2~12 个月，所有接受移植的病例，发汗功能均恢复[166]。这些结果不仅表明 MSC 的移植可以促进伤口愈合，而且 MSC 在皮肤附件重建方面具有极大优势。

旁分泌效应是 MSC 发挥治疗功效的主要方式。MSC 释放的各种调节性因子，刺激伤口中主要细胞类型的增殖、迁移和分化。目前，关于利用 MSC 分泌产物直接进行创面治疗的临床研究仅见一例报道。肛周瘘管男性病例，先后经过 7 次外科清创手术，溃疡进一步恶化、扩大，并伴有疼痛。皮内注射 MSC 分泌产物并于创面上覆盖混合有 MSC 分泌产物的抗菌凝胶，最后用绷带隔离固定。瘘管在治疗 4 天后封闭愈合[167]。目前，还需要更多的利用 MSC 分泌产物进行创面治疗的临床数据。

（李美蓉）

参 考 文 献

[1] Kirby GT, Mills SJ, Cowin AJ, et al. Stem cells for cutaneous wound healing. Biomed Res Int, 2015, 2015(2):285869.

[2] Breitkreutz D, Mirancea N, Nischt R. Basement membranes in skin: unique matrix structures with diverse functions? Histochem Cell Biol, 2009, 132:1-10.

[3] Martin P. Wound healing-aiming for perfect skin regeneration. Science, 1997, 276:75-81.

[4] Reinke JM, Sorg H. Wound repair and regeneration. Eur Surg Res, 2012, 49:35-43.

[5] Piccin A, Di Pierro AM, Canzian L, et al. Platelet gel: a new therapeutic tool with great potential. Blood Transfus, 2016, 15(4):333-340.

[6] Ng LG, Qin JS, Roediger B, et al. Visualizing the neutrophil response to sterile tissue injury in mouse dermis reveals a three-phase cascade of events. J Invest Dermatol, 2011, 131:2058-2068.

[7] Tzeng DY, Deuel TF, Huang JS, et al. Platelet-derived growth factor promotes human peripheral monocyte activation. Blood, 1985, 66:179-183.

[8] Baum CL, Arpey CJ. Normal cutaneous wound healing: clinical correlation with cellular and molecular events. Dermatol Surg, 2005, 31:674-686.

[9] Wilgus TA, Roy S, McDaniel JC. Neutrophils and wound repair: positive actions and negative reactions. Adv Wound Care (New Rochelle), 2013, 2:379-388.

[10] Saarialho-Kere UK, Pentland AP, Birkedal-Hansen H, et al. Distinct populations of basal keratinocytes express stromelysin-1 and stromelysin-2 in chronic wounds. J Clin Invest, 1994, 94:79-88.

[11] Singer AJ, Clark RA. Cutaneous wound healing. N Engl J Med, 1999, 341:738-746.

[12] Schaffer M, Barbul A. Lymphocyte function in wound healing and following injury. Br J Surg, 1998, 85:444-460.

[13] Ruzehaji N, Mills SJ, Melville E, et al. The influence of Flightless I on Toll-like-receptor-mediated inflammation in a murine model of diabetic wound healing. Biomed Res Int, 2013, (2013):1-9.

[14] Sindrilaru A, Scharffetter-Kochanek K. Disclosure of the culprits: macrophages-versatile regulators of wound healing. Adv Wound Care (New Rochelle), 2013, 2:357-368.

[15] Louis H, Van Laethem JL, Wu W, et al. Interleukin-10 controls neutrophilic infiltration, hepatocyte proliferation, and liver fibrosis induced by carbon tetrachloride in mice. Hepatology, 1998, 28:1607-1615.

[16] Titos E, Claria J, Planaguma A, et al. Inhibition of 5-lipoxygenase induces cell growth arrest and apoptosis in rat Kupffer cells: implications for liver fibrosis. FASEB J, 2003, 17:1745-1747.

[17] Romer J, Bugge TH, Pyke C, et al. Impaired wound healing in mice with a disrupted plasminogen gene. Nat Med, 1996, 2:287-292.

[18] Clark RA. Biology of dermal wound repair. Dermatol Clin, 1993, 11:647-666.

[19] Ballas CB, Davidson JM. Delayed wound healing in aged rats is associated with increased collagen gel remodeling and contraction by skin fibroblasts, not with differences in apoptotic or myofibroblast cell populations. Wound Repair Regen, 2001, 9:223-237.

[20] Bond JS, Duncan JA, Mason T, et al. Scar redness in humans: how long does it persist after incisional and excisional wounding? Plast Reconstr Surg, 2008, 121:487-496.

[21] Gottrup F. A specialized wound-healing center concept: importance of a multidisciplinary department structure and surgical treatment facilities in the treatment of chronic wounds. Am J Surg, 2004, 187:38S-43S.

[22] Stoltz RA, Conners MS, Gerritsen ME, et al. Direct stimulation of limbal microvessel endothelial cell proliferation and capillary formation *in vitro* by a corneal-derived eicosanoid. Am J Pathol, 1996, 148:129-139.

[23] Birkedal-Hansen H, Moore WG, Bodden MK, et al. Matrix metalloproteinases: a review. Crit Rev Oral Biol Med, 1993, 4:197-250.

[24] Emmert-Buck MR, Emonard HP, Corcoran ML, et al. Cell surface binding of TIMP-2 and pro-MMP-2/TIMP-2 complex. FEBS Lett, 1995, 364:28-32.

[25] Armstrong JR, Ferguson MW. Ontogeny of the skin and the transition from scar-free to scarring phenotype during wound healing in the pouch young of a marsupial, Monodelphis domestica. Dev Biol, 1995, 169:242-260.

[26] Cowin AJ, Brosnan MP, Holmes TM, et al. Endogenous inflammatory response to dermal wound healing in the fetal and adult mouse. Dev Dyn, 1998, 212:385-393.

[27] Whitby DJ, Ferguson MW. Immunohistochemical localization of growth factors in fetal wound healing. Dev Biol, 1991, 147:207-215.

[28] Markova A, Mostow EN. US skin disease assessment: ulcer and wound care. Dermatol Clin, 2012, 30:107-111.

[29] Jiang Y, Huang S, Fu X, et al. Epidemiology of chronic cutaneous wounds in China. Wound Repair Regen, 2011, 19:181-188.

[30] Demidova-Rice TN, Hamblin MR, Herman IM. Acute and impaired wound healing: pathophysiology and current methods for drug delivery, part 1: normal and chronic wounds: biology, causes, and approaches to care. Adv Skin Wound Care, 2012, 25:304-314.

[31] Medina A, Scott PG, Ghahary A, et al. Pathophysiology of chronic nonhealing wounds. J Burn Care Rehabil, 2005, 26:306-319.

[32] Eming SA, Krieg T, Davidson JM. Inflammation in wound repair: molecular and cellular mechanisms. J Invest Dermatol, 2007, 127:514-525.

[33] Stadelmann WK, Digenis AG, Tobin GR. Physiology and healing dynamics of chronic cutaneous wounds. Am J Surg, 1998, 176:26S-38S.

[34] Raja Sivamani K, Garcia MS, Isseroff RR. Wound re-epithelialization: modulating keratinocyte migration in wound healing. Front Biosci, 2007, 12:2849-2868.

[35] Zhu Z, Ding J, Shankowsky HA, et al. The molecular mechanism of hypertrophic scar. J Cell Commun Signal, 2013, 7:239-252.

[36] Al-Attar A, Mess S, Thomassen JM, et al. Keloid pathogenesis and treatment. Plast Reconstr Surg, 2006, 117:286-300.

[37] Halim AS, Emami A, Salahshourifar I, et al. Keloid scarring: understanding the genetic basis, advances, and prospects. Arch Plast Surg, 2012, 39:184-189.

[38] Ehrlich HP, Desmouliere A, Diegelmann RF, et al. Morphological and immunochemical differences between keloid and hypertrophic scar. Am J Pathol, 1994, 145:105-113.

[39] Rennert RC, Rodrigues M, Wong VW, et al. Biological therapies for the treatment of cutaneous wounds: phase III and launched therapies. Expert Opin Biol Ther, 2013, 13:1523-1541.

[40] Caplan AI, Correa D. The MSC: an injury drugstore. Cell Stem Cell, 2011, 9:11-15.

[41] Dominici M, Le Blanc K, Mueller I, et al. Minimal criteria for defining multipotent mesenchymal stromal cells. The International Society for Cellular Therapy position statement. Cytotherapy, 2006, 8:315-317.

[42] Sun TJ, Tao R, Han YQ, et al. Wnt3a promotes human umbilical cord mesenchymal stem cells to differentiate into epidermal-like cells. Eur Rev Med Pharmacol Sci, 2015, 19:86-91.

[43] Oswald J, Boxberger S, Jorgensen B, et al. Mesenchymal stem cells can be differentiated into endothelial cells in vitro. Stem Cells, 2004, 22:377-384.

[44] Zhang C, Chen Y, Fu X. Sweat gland regeneration after burn injury: is stem cell therapy a new hope? Cytotherapy, 2015, 17:526-535.

[45] Wu M, Sun Q, Guo X, et al. hMSCs possess the potential to differentiate into DP cells *in vivo* and in vitro. Cell Biol Int Rep, 2012, 19:e00019.

[46] Fu X, Fang L, Li X, et al. Enhanced wound-healing quality with bone marrow mesenchymal stem cells autografting after skin injury. Wound Repair Regen, 2006, 14:325-335.

[47] Sasaki M, Abe R, Fujita Y, et al. Mesenchymal stem cells are recruited into wounded skin and contribute to wound repair by transdifferentiation into multiple skin cell type. J Immunol. 2008, 180(4):2581-2587.

[48] Badiavas EV, Ford D, Liu P, et al. Long-term bone marrow culture and its clinical potential in chronic wound healing. Wound Repair Regen, 2007, 15:856-865.

[49] Raffaghello L, Bianchi G, Bertolotto M, et al. Human mesenchymal stem cells inhibit neutrophil apoptosis: a model for neutrophil preservation in the bone marrow niche. Stem Cells, 2008, 26:151-162.

[50] Dayan V, Yannarelli G, Billia F, et al. Mesenchymal stromal cells mediate a switch to alternatively activated monocytes/macrophages after acute myocardial infarction. Basic Res Cardiol, 2011, 106:1299-1310.

[51] Francois M, Romieu-Mourez R, Li M, et al. Human MSC suppression correlates with cytokine induction of indoleamine 2,3-dioxygenase and bystander M2 macrophage differentiation. Mol Ther, 2012, 20:187-195.

[52] Maggini J, Mirkin G, Bognanni I, et al. Mouse bone marrow-derived mesenchymal stromal cells turn activated macrophages into a regulatory-like profile. PLoS One, 2010, 5:e9252.

[53] Nemeth K, Leelahavanichkul A, Yuen PS, et al. Bone marrow stromal cells attenuate sepsis via prostaglandin E(2)-dependent reprogramming of host macrophages to increase their interleukin-10 production. Nat Med, 2009, 15:42-49.

[54] Machado Cde V, Telles PD, Nascimento IL. Immunological characteristics of mesenchymal stem cells. Rev Bras Hematol Hemoter, 2013, 35:62-67.

[55] Frangogiannis NG. Regulation of the inflammatory response in cardiac repair. Circ Res, 2012, 110:159-173.

[56] Jiang XX, Zhang Y, Liu B, et al. Human mesenchymal stem cells inhibit differentiation and function of monocyte-derived dendritic cells. Blood, 2005, 105:4120-4126.

[57] Aggarwal S, Pittenger MF. Human mesenchymal stem cells modulate allogeneic immune cell responses. Blood, 2005, 105:1815-1822.

[58] Bartholomew A, Sturgeon C, Siatskas M, et al. Mesenchymal stem cells suppress lymphocyte proliferation *in vitro* and prolong skin graft survival *in vivo*. Exp Hematol, 2002, 30:42-48.

[59] Bernardo ME, Zaffaroni N, Novara F, et al. Human bone marrow derived mesenchymal stem cells do not undergo transformation after long-term in vitro culture and do not exhibit telomere maintenance mechanisms. Cancer Res, 2007, 67:9142-9149.

[60] Glennie S, Soeiro I, Dyson PJ, et al. Bone marrow mesenchymal stem cells induce division arrest anergy of activated T cells. Blood, 2005, 105:2821-2827.

[61] Krampera M, Glennie S, Dyson J, et al. Bone marrow mesenchymal stem cells inhibit the response of naive and memory antigen-specific T cells to their cognate peptide. Blood, 2003, 101:3722-3729.

[62] Prevosto C, Zancolli M, Canevali P, et al. Generation of $CD4^+$ or $CD8^+$ regulatory T cells upon mesenchymal stem cell-lymphocyte interaction. Haematologica, 2007, 92:881-888.

[63] Bouffi C, Bony C, Courties G, et al. IL-6-dependent PGE2 secretion by mesenchymal stem cells inhibits local inflammation in experimental arthritis. PLoS One, 2010, 5:e14247.

[64] Corcione A, Benvenuto F, Ferretti E, et al. Human mesenchymal stem cells modulate B-cell functions. Blood, 2006, 107:367-372.

[65] Chen L, Tredget EE, Wu PY, et al. Paracrine factors of mesenchymal stem cells recruit macrophages and endothelial lineage cells and enhance wound healing. PLoS One, 2008, 3:e1886.

[66] Chen L, Tredget EE, Liu C, et al. Analysis of allogenicity of mesenchymal stem cells in engraftment and wound healing in mice. PLoS One, 2009, 4:e7119.

[67] Egana JT, Fierro FA, Kruger S, et al. Use of human mesenchymal cells to improve vascularization in a mouse model for scaffold-based dermal regeneration. Tissue Eng Part A, 2009, 15:1191-1200.

[68] Kim CH, Lee JH, Won JH, et al. Mesenchymal stem cells improve wound healing *in vivo* via early activation of matrix metalloproteinase-9 and vascular endothelial growth factor. J Korean Med Sci, 2011, 26:726-733.

[69] Barcelos LS, Duplaa C, Krankel N, et al. Human $CD133^+$ progenitor cells promote the healing of diabetic ischemic ulcers by paracrine stimulation of angiogenesis and activation of Wnt signaling. Circ Res, 2009, 104:1095-1102.

[70] Yew TL, Hung YT, Li HY, et al. Enhancement of wound healing by human multipotent stromal cell conditioned medium: the paracrine factors and p38 MAPK activation. Cell Transplant, 2011, 20:693-706.

[71] Di Rocco G, Gentile A, Antonini A, et al. Enhanced healing of diabetic wounds by topical administration of adipose tissue-derived stromal cells overexpressing stromal-derived factor-1: biodistribution and engraftment analysis by bioluminescent imaging. Stem Cells Int, 2010, (2011):304562.

[72] Galiano RD, Tepper OM, Pelo CR, et al. Topical vascular endothelial growth factor accelerates diabetic wound healing through increased angiogenesis and by mobilizing and recruiting bone marrow-derived cells. Am J Pathol, 2004, 164:1935-1947.

[73] Qin D, Trenkwalder T, Lee S, et al. Early vessel destabilization mediated by Angiopoietin-2 and subsequent vessel maturation via Angiopoietin-1 induce functional neovasculature after ischemia. PLoS One, 2013, 8:e61831.

[74] Kato J, Tsuruda T, Kita T, et al. Adrenomedullin: a protective factor for blood vessels. Arterioscler Thromb Vasc Biol, 2005, 25:2480-2487.

[75] Lozito TP, Taboas JM, Kuo CK, et al. Mesenchymal stem cell modification of endothelial matrix regulates their vascular differentiation. J Cell Biochem, 2009, 107:706-713.

[76] Renault MA, Roncalli J, Tongers J, et al. The Hedgehog transcription factor Gli3 modulates angiogenesis. Circ Res, 2009, 105:818-826.

[77] Li M, Zhao Y, Hao H, et al. Mesenchymal stem cell-conditioned medium improves the proliferation and migration of keratinocytes in a diabetes-like microenvironment. Int J Low Extrem Wounds, 2015, 14:73-86.

[78] Li M, Zhao Y, Hao H, et al. Umbilical cord-derived mesenchymal stromal cell-conditioned medium exerts *in vitro* antiaging effects in human fibroblasts. Cytotherapy, 2017, 19:371-383.

[79] Li L, Zhang S, Zhang Y, et al. Paracrine action mediate the antifibrotic effect of transplanted mesenchymal stem cells in a rat model of global heart failure. Mol Biol Rep, 2009, 36:725-731.

[80] Li L, Zhang Y, Li Y, et al. Mesenchymal stem cell transplantation attenuates cardiac fibrosis associated with isoproterenol-induced global heart failure. Transpl Int, 2008, 21:1181-1189.

[81] Mou S, Wang Q, Shi B, et al. Hepatocyte growth factor suppresses transforming growth factor-beta-1 and type III collagen in human primary renal fibroblasts. Kaohsiung J Med Sci, 2009, 25:577-587.

[82] Schievenbusch S, Strack I, Scheffler M, et al. Profiling of anti-fibrotic signaling by hepatocyte growth factor in renal fibroblasts. Biochem Biophys Res Commun, 2009, 385:55-61.

[83] Inagaki Y, Higashi K, Kushida M, et al. Hepatocyte growth factor suppresses profibrogenic signal transduction via nuclear export of Smad3 with galectin-7. Gastroenterology, 2008, 134:1180-1190.

[84] Kanemura H, Iimuro Y, Takeuchi M, et al. Hepatocyte growth factor gene transfer with naked plasmid DNA ameliorates dimethylnitrosamine-induced liver fibrosis in rats. Hepatol Res, 2008, 38:930-939.

[85] McAnulty RJ. Fibroblasts and myofibroblasts: their source, function and role in disease. Int J Biochem Cell Biol, 2007, 39:666-671.

[86] Abe R, Donnelly SC, Peng T, et al. Peripheral blood fibrocytes: differentiation pathway and migration to wound sites. J Immunol, 2001, 166:7556-7562.

[87] Yang J, Dai C, Liu Y. A novel mechanism by which hepatocyte growth factor blocks tubular epithelial to mesenchymal transition. J Am Soc Nephrol, 2005, 16:68-78.

[88] Zhao L, Liu X, Zhang Y, et al. Enhanced cell survival and paracrine effects of mesenchymal stem cells overexpressing hepatocyte growth factor promote cardioprotection in myocardial infarction. Exp Cell Res, 2016, 344:30-39.

[89] Kim WS, Park BS, Kim HK, et al. Evidence supporting antioxidant action of adipose-derived stem cells: protection of human dermal fibroblasts from oxidative stress. J Dermatol Sci, 2008, 49:133-142.

[90] Mei SH, Haitsma JJ, Dos Santos CC, et al. Mesenchymal stem cells reduce inflammation while enhancing bacterial clearance and improving survival in sepsis. Am J Respir Crit Care Med, 2010, 182:1047-1057.

[91] Krasnodembskaya A, Song Y, Fang X, et al. Antibacterial effect of human mesenchymal stem cells is mediated in part from secretion of the antimicrobial peptide LL-37. Stem Cells, 2010, 28:2229-2238.

[92] Lai RC, Chen TS, Lim SK. Mesenchymal stem cell exosome: a novel stem cell-based therapy for cardiovascular disease. Regen Med, 2011, 6:481-492.

[93] Zhang B, Shen L, Shi H, et al. Exosomes from human umbilical cord mesenchymal stem cells: identification, purification, and biological characteristics. Stem Cells Int, 2016, 2016(19):1929536.

[94] Ti D, Hao H, Tong C, et al. LPS-preconditioned mesenchymal stromal cells modify macrophage polarization for resolution of chronic inflammation via exosome-shuttled let-7b. J Transl Med, 2015, 13:308.

[95] Shabbir A, Cox A, Rodriguez-Menocal L, et al. Mesenchymal stem cell exosomes induce proliferation and migration of normal and chronic wound fibroblasts, and enhance angiogenesis in vitro. Stem Cells Dev, 2015, 24:1635-1647.

[96] Li X, Liu L, Yang J, et al. Exosome derived from human umbilical cord mesenchymal stem cell mediates MiR-181c attenuating burn-induced excessive inflammation. EBio Medicine, 2016, 8:72-82.

[97] Badiavas EV, Falanga V. Treatment of chronic wounds with bone marrow-derived cells. Arch Dermatol, 2003, 139:510-516.

[98] Mansilla E, Marin GH, Drago H, et al. Bloodstream cells phenotypically identical to human mesenchymal bone marrow stem cells circulate in large amounts under the influence of acute large skin damage: new evidence for their use in regenerative medicine. Transplant Proc, 2006, 38:967-969.

[99] Alm JJ, Koivu HM, Heino TJ, et al. Circulating plastic adherent mesenchymal stem cells in aged hip fracture patients. J Orthop Res, 2010, 28:1634-1642.

[100] Rochefort GY, Delorme B, Lopez A, et al. Multipotential mesenchymal stem cells are mobilized into peripheral blood by hypoxia. Stem Cells, 2006, 24:2202-2208.

[101] Sasaki M, Abe R, Fujita Y, et al. Mesenchymal stem cells are recruited into wounded skin and contribute to wound repair by transdifferentiation into multiple skin cell type. J Immunol, 2008, 180:2581-2587.

[102] Lu MH, Hu CJ, Chen L, et al. miR-27b represses migration of mouse MSCs to burned margins and prolongs wound repair through silencing SDF-1a. PLoS One, 2013, 8:e68972.

[103] Aikawa E, Fujita R, Kikuchi Y, et al. Systemic high-mobility group box 1 administration suppresses skin inflammation by inducing an accumulation of PDGFRalpha(+) mesenchymal cells from bone marrow. Sci Rep, 2015, 5:11008.

[104] Liu L, Yu Q, Hu K, et al. Electro-acupuncture promotes endogenous multipotential mesenchymal stem cell mobilization into the oeripheral blood. Cell Physiol Biochem, 2016, 38:1605-1617.

[105] Shou K, Niu Y, Zheng X, et al. Enhancement of bone-marrow-derived mesenchymal stem cell angiogenic capacity by NPWT for a combinatorial therapy to promote wound healing with large defect. Biomed Res Int, 2017, (3):7920265.

[106] Lee RH, Seo MJ, Reger RL, et al. Multipotent stromal cells from human marrow home to and promote repair of pancreatic islets and renal glomeruli in diabetic NOD/scid mice. Proc Natl Acad Sci USA, 2006, 103:17438-17443.

[107] Iso Y, Spees JL, Serrano C, et al. Multipotent human stromal cells improve cardiac function after myocardial infarction in mice without long-term engraftment. Biochem Biophys Res Commun, 2007, 354:700-706.

[108] Burst VR, Gillis M, Putsch F, et al. Poor cell survival limits the beneficial impact of mesenchymal stem cell transplantation on acute kidney injury. Nephron Exp Nephrol, 2010, 114:e107-e116.

[109] di Bonzo LV, Ferrero I, Cravanzola C, et al. Human mesenchymal stem cells as a two-edged sword in hepatic regenerative medicine: engraftment and hepatocyte differentiation versus profibrogenic potential. Gut, 2008, 57:223-231.

[110] Chai C, Leong KW. Biomaterials approach to expand and direct differentiation of stem cells. Mol Ther, 2007, 15:467-480.

[111] Falanga V, Iwamoto S, Chartier M, et al. Autologous bone marrow-derived cultured mesenchymal stem cells delivered in a fibrin spray accelerate healing in murine and human cutaneous wounds. Tissue Eng, 2007, 13:1299-1312.

[112] Lee SH, Lee JH, Cho KH. Effects of human adipose-derived stem cells on cutaneous wound healing in nude mice. Ann Dermatol, 2011, 23:150-155.

[113] O'Loughlin A, Kulkarni M, Creane M, et al. Topical administration of allogeneic mesenchymal stromal cells seeded in a collagen scaffold augments wound healing and increases angiogenesis in the diabetic rabbit ulcer. Diabetes, 2013, 62:2588-2594.

[114] Liu P, Deng Z, Han S, et al. Tissue-engineered skin containing mesenchymal stem cells improves burn wounds. Artif Organs, 2008, 32:925-931.

[115] Nambu M, Kishimoto S, Nakamura S, et al. Accelerated wound healing in healing-impaired db/db mice by autologous adipose tissue-derived stromal cells combined with atelocollagen matrix. Ann Plast Surg, 2009, 62:317-321.

[116] Liu S, Zhang H, Zhang X, et al. Synergistic angiogenesis promoting effects of extracellular matrix scaffolds and adipose-derived stem cells during wound repair. Tissue Eng Part A, 2011, 17:725-739.

[117] Li JH, Zhang N, Wang JA. Improved anti-apoptotic and anti-remodeling potency of bone marrow mesenchymal stem cells by anoxic pre-conditioning in diabetic cardiomyopathy. J Endocrinol Invest, 2008, 31:103-110.

[118] Hu X, Yu SP, Fraser JL, et al. Transplantation of hypoxia-preconditioned mesenchymal stem cells improves infarcted heart function via enhanced survival of implanted cells and angiogenesis. J Thorac Cardiovasc Surg, 2008, 135:799-808.

[119] Wang L, Hu X, Zhu W, et al. Increased leptin by hypoxic-preconditioning promotes autophagy of mesenchymal stem cells and protects them from apoptosis. Sci China Life Sci, 2014, 57:171-180.

[120] Yu Y, Yin Y, Wu RX, et al. Hypoxia and low-dose inflammatory stimulus synergistically enhance bone marrow mesenchymal stem cell migration. Cell Prolif, 2017, 50(1).

[121] Xue D, Zhang W, Chen E, et al. Local delivery of HMGB1 in gelatin sponge scaffolds combined with mesenchymal stem cell sheets to accelerate fracture healing. Oncotarget, 2017, 8(26):42098-42115.

[122] Xu B, Luo Y, Liu Y, et al. Platelet-derived growth factor-BB enhances MSC-mediated cardioprotection via suppression of miR-320 expression. Am J Physiol Heart Circ Physiol, 2015, 308:H980-H989.

[123] Ke X, Liu C, Wang Y, et al. Netrin-1 promotes mesenchymal stem cell revascularization of limb ischaemia. Diab Vasc Dis Res, 2016, 13:145-156.

[124] Jiang S, Haider H, Idris NM, et al. Supportive interaction between cell survival signaling and angiocompetent factors enhances donor cell survival and promotes angiomyogenesis for cardiac repair. Circ Res, 2006, 99:776-784.

[125] Wang X, Zhao T, Huang W, et al. Hsp20-engineered mesenchymal stem cells are resistant to oxidative stress via enhanced activation of Akt and increased secretion of growth factors. Stem Cells, 2009, 27:3021-3031.

[126] Mirotsou M, Zhang Z, Deb A, et al. Secreted frizzled related protein 2 (Sfrp2) is the key Akt-mesenchymal stem cell-released paracrine factor mediating myocardial survival and repair. Proc Natl Acad Sci USA, 2007, 104:1643-1648.

[127] Fan L, Lin C, Zhuo S, et al. Transplantation with survivin-engineered mesenchymal stem cells results in better prognosis in a rat model of myocardial infarction. Eur J Heart Fail, 2009, 11:1023-1030.

[128] Tang YL, Tang Y, Zhang YC, et al. Improved graft mesenchymal stem cell survival in ischemic heart with a hypoxia-regulated heme oxygenase-1 vector. J Am Coll Cardiol, 2005, 46:1339-1350.

[129] Cho J, Zhai P, Maejima Y, et al. Myocardial injection with GSK-3beta-overexpressing bone marrow-derived mesenchymal stem cells attenuates cardiac dysfunction after myocardial infarction. Circ Res, 2011, 108:478-489.

[130] Ranganath SH, Levy O, Inamdar MS, et al. Harnessing the mesenchymal stem cell secretome for the treatment of cardiovascular disease. Cell Stem Cell, 2012, 10:244-258.

[131] Huang J, Zhang Z, Guo J, et al. Genetic modification of mesenchymal stem cells overexpressing CCR1 increases cell viability, migration, engraftment, and capillary density in the injured myocardium. Circ Res, 2010, 106:1753-1762.

[132] Tang J, Wang J, Guo L, et al. Mesenchymal stem cells modified with stromal cell-derived factor 1 alpha improve cardiac remodeling via paracrine activation of hepatocyte growth factor in a rat model of myocardial infarction. Mol Cells, 2010, 29:9-19.

[133] Li L, Chen X, Wang WE, et al. How to improve the survival of transplanted mesenchymal stem cell in ischemic heart? Stem Cells Int, 2016, (2016):9682757.

[134] Huang F, Li ML, Fang ZF, et al. Overexpression of MicroRNA-1 improves the efficacy of mesenchymal stem cell transplantation after myocardial infarction. Cardiology, 2013, 125:18-30.

[135] Xu J, Huang Z, Lin L, et al. miR-210 over-expression enhances mesenchymal stem cell survival in an oxidative stress environment through antioxidation and c-Met pathway activation. Sci China Life Sci, 2014, 57:989-997.

[136] Yang YJ, Qian HY, Huang J, et al. Combined therapy with simvastatin and bone marrow-derived mesenchymal stem cells increases benefits in infarcted swine hearts. Arterioscler Thromb Vasc Biol, 2009, 29:2076-2082.

[137] Zhang Z, Li S, Cui M, et al. Rosuvastatin enhances the therapeutic efficacy of adipose-derived mesenchymal stem cells for myocardial infarction via PI3K/Akt and MEK/ERK pathways. Basic Res Cardiol, 2013, 108:333.

[138] Tancharoen W, Aungsuchawan S, Pothacharoen P, et al. Differentiation of mesenchymal stem cells from human amniotic fluid to vascular endothelial cells. Acta Histochem, 2017, 119:113-121.

[139] Alaminos M, Perez-Kohler B, Garzon I, et al. Transdifferentiation potentiality of human Wharton's jelly stem cells towards vascular endothelial cells. J Cell Physiol, 2010, 223:640-647.

[140] Li Z, Han S, Wang X, et al. Rho kinase inhibitor Y-27632 promotes the differentiation of human bone marrow mesenchymal stem cells into keratinocyte-like cells in xeno-free conditioned medium. Stem Cell Res Ther, 2015, 6:17.

[141] Xu Y, Hong Y, Xu M, et al. Role of keratinocyte growth factor in the differentiation of sweat gland-like cells from human umbilical cord-derived mesenchymal stem cells. Stem Cells Transl Med, 2016, 5:106-116.

[142] Yang S, Ma K, Feng C, et al. Capacity of human umbilical cord-derived mesenchymal stem cells to differentiate into sweat gland-like cells: a preclinical study. Front Med, 2013, 7:345-353.

[143] Hadden WJ, Choi YS. The extracellular microscape governs mesenchymal stem cell fate. J Biol Eng, 2016, 10:16.

[144] Fu J, Wang YK, Yang MT, et al. Mechanical regulation of cell function with geometrically modulated elastomeric substrates. Nat Methods, 2010, 7:733-736.

[145] Flanagan LA, Rebaza LM, Derzic S, et al. Regulation of human neural precursor cells by laminin and integrins. J Neurosci Res, 2006, 83:845-856.

[146] Saha S, Ji L, de Pablo JJ, et al. Inhibition of human embryonic stem cell differentiation by mechanical strain. J Cell Physiol, 2006, 206:126-137.

[147] Yang C, Tibbitt MW, Basta L, et al. Mechanical memory and dosing influence stem cell fate. Nat Mater, 2014, 13:645-652.

[148] Kim HJ, Kwon YR, Bae YJ, et al. Enhancement of human mesenchymal stem cell differentiation by combination treatment with 5-azacytidine and trichostatin A. Biotechnol Lett, 2016, 38:167-174.

[149] Lee S, Park JR, Seo MS, et al. Histone deacetylase inhibitors decrease proliferation potential and multilineage differentiation capability of human mesenchymal stem cells. Cell Prolif, 2009, 42:711-720.

[150] Furumatsu T, Ozaki T. Epigenetic regulation in chondrogenesis. Acta Med Okayama, 2010, 64:155-161.

[151] Nakatsuka R, Nozaki T, Uemura Y, et al. 5-Aza-2'-deoxycytidine treatment induces skeletal myogenic differentiation of mouse dental pulp stem cells. Arch Oral Biol, 2010, 55:350-357.

[152] Rosca AM, Burlacu A. Effect of 5-azacytidine: evidence for alteration of the multipotent ability of mesenchymal stem cells. Stem Cells Dev, 2011, 20:1213-1221.

[153] Ichioka S, Kouraba S, Sekiya N, et al. Bone marrow-impregnated collagen matrix for wound healing: experimental evaluation in a microcirculatory model of angiogenesis, and clinical experience. Br J Plast Surg, 2005, 58:1124-1130.

[154] Kirana S, Stratmann B, Lammers D, et al. Wound therapy with autologous bone marrow stem cells in diabetic patients with ischaemia-induced tissue ulcers affecting the lower limbs. Int J Clin Pract, 2007, 61:690-692.

[155] Conget P, Rodriguez F, Kramer S, et al. Replenishment of type VII collagen and re-epithelialization of chronically ulcerated skin after intradermal administration of allogeneic mesenchymal stromal cells in two patients with recessive dystrophic epidermolysis bullosa. Cytotherapy, 2010, 12:429-431.

[156] Yoshikawa T, Mitsuno H, Nonaka I, et al. Wound therapy by marrow mesenchymal cell transplantation. Plast Reconstr Surg, 2008, 121:860-877.

[157] Prochazka V, Gumulec J, Chmelova J, et al. Autologous bone marrow stem cell transplantation in patients with end-stage chronical critical limb ischemia and diabetic foot. Vnitr Lek, 2009, 55:173-178.

[158] Prochazka V, Gumulec J, Jaluvka F, et al. Cell therapy, a new standard in management of chronic critical limb ischemia and foot ulcer. Cell Transplant, 2010, 19:1413-1424.

[159] Lu D, Chen B, Liang Z, et al. Comparison of bone marrow mesenchymal stem cells with bone marrow-derived mononuclear cells for treatment of diabetic critical limb ischemia and foot ulcer: a double-blind, randomized, controlled trial. Diabetes Res Clin Pract, 2011, 92(1):26-36.

[160] Kirana S, Stratmann B, Prante C, et al. Autologous stem cell therapy in the treatment of limb ischaemia induced chronic tissue ulcers of diabetic foot patients. Int J Clin Pract, 2012, 66(4):384-393.

[161] Rasulov MF, Vasilchenkov AV, Onishchenko NA, et al. First experience of the use bone marrow mesenchymal stem cells for the treatment of a patient with deep skin burns. Bull Exp Biol Med, 2005, 139:141-144.

[162] Bey E, Prat M, Duhamel P, et al. Emerging therapy for improving wound repair of severe radiation burns using local bone marrow-derived stem cell administrations. Wound Repair Regen, 2010, 18(1):50-58.

[163] Vojtassak J, Danisovic L, Kubes M, et al. Autologous biograft and mesenchymal stem cells in treatment of the diabetic foot. Neuro Endocrinol Lett, 2006, 27 (Suppl 2):134-137.

[164] Ravari H, Hamidi-Almadari D, Salimifar M, et al. Treatment of non-healing wounds with autologous bone marrow cells, platelets, fibrin glue and collagen matrix. Cytotherapy, 2011, 13(6):705-711.

[165] Lee HC, An SG, Lee HW, et al. Safety and effect of adipose tissue-derived stem cell implantation in patients with critical limb ischemia: a pilot study. Circ J, 2012, 76:1750-1760.

[166] Sheng Z, Fu X, Cai S, et al. Regeneration of functional sweat gland-like structures by transplanted differentiated bone marrow mesenchymal stem cells. Wound Repair Regen, 2009, 17:427-435.

[167] Julianto I, Rindastuti Y. Topical delivery of mesenchymal stem cells "Secretomes" in wound repair. Acta Med Indones, 2016, 48:217-220.

第十八章 间充质干细胞组织工程

第一节 间充质干细胞组织工程的概述

一、组织工程与再生医学

随着生命科学的飞速发展，目前组织工程、干细胞研究已经成为 21 世纪生命科学研究的热点和前沿领域。组织工程是利用生命科学、医学、工程学原理与技术，在体外或体内利用细胞、生物材料、细胞因子等构建具有生物功能的人工取代物，从而修复或替代受损的组织和器官，最终实现组织修复或再生的一门技术。再生医学是指利用生物学及工程学的理论方法创造丢失或功能受损的组织和器官，使其具备正常组织和器官的结构与功能。具体讲是应用生命科学、材料科学、临床医学和工程学等学科的原理和方法，研究和开发用于替代、修复、重建或再生人体各种组织、器官的理论和技术的新型学科。我国组织工程与再生医学研究起步于 20 世纪 90 年代，随着组织工程与再生医学领域科学技术成果的不断涌现，组织工程与再生医学产品在医疗及产业领域的应用越来越广泛。其中，组织工程的研究进展经历了从结构组织的组织工程化构建到具有复杂功能器官的组织工程的构建与应用，再到与再生医学的融合，使其研究范围和应用领域更加广泛[1]。

其中，组织工程与再生医学的理论研究主要分为三个方面：种子细胞、生物材料及组织工程支架。干细胞的发现及其进一步的研究，为组织工程与再生医学提供了丰富的种子细胞，特别是 MSC 相关方面的研究；生物材料主要在研发新型生物材料、与生长因子相结合及模拟干细胞微环境等方面不断发展；此外，组织工程支架的出现，也为组织工程与再生医学的发展提供了新思路。通过组织脱细胞、干细胞分化及 3D 打印等关键技术，越来越多的组织工程产品出现在市场上，从而满足了临床日益增长的需求（图 18-1）。

二、MSC 作为种子细胞的优势特性及局限性

目前，组织工程与再生医学研究的基础和关键之一是选择种子细胞。而干细胞作为种子细胞越来越受到人们的重视，它将有望解决组织工程研究中种子细胞的来源问题，是组织工程研究中理想的种子细胞。干细胞主要包括胚胎干细胞（ESC）和成体干细胞，其中 ESC 虽然具有高度的全能性，具有发育成我们想要的人类所有细胞类型的潜能，但其来源、获取和相关研究受到了较多的伦理及法律限制，因此影响了其在工程领域的发展及应用；而在许多种类的成体干细胞中，MSC 凭借其天然的优势特性，成为组织工程、再生医学和细胞移植等研究领域较为理想的种子细胞类型，其中广泛应用于组织工程的几种 MSC 类型包括：骨髓间充质干细胞（BM-MSC）、脂肪间充质干细胞（AD-MSC）、脐带间充质干细胞（UC-MSC）、胎盘间充质干细胞（P-MSC）等。这些 MSC 拥有作为

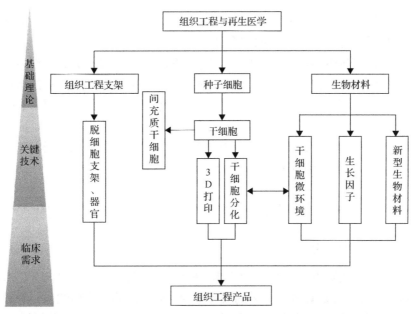

图 18-1 组织工程与再生医学的理论研究（彩图请扫封底二维码）

理想种子细胞相关的各种天然优势特性，如来源广泛，取材方便，不受伦理学限制，并且增殖能力强，能够在体外大量增殖和分化等。近年来的研究表明，MSC 除具有多向分化潜能外[2]，还具有免疫调节作用，不表达或仅表达可以忽略水平的主要组织相容性复合体（MHC）类分子，因此不容易被宿主免疫细胞识别，可成功逃脱免疫系统的免疫排斥反应，在受体中可以长期存在（图 18-2）。

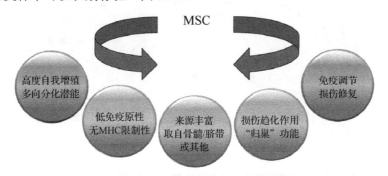

图 18-2 间充质干细胞拥有作为理想种子细胞相关的各种天然优势特性（彩图请扫封底二维码）

如今，虽然 MSC 作为种子细胞拥有许多优势，并且相关研究已取得较大进展。但实际上到目前为止，相关研究才刚刚起步，人们对组织工程种子细胞，尤其是对 MSC 的了解仍存在许多未知的东西，许多相关的技术和理论问题还需要进一步解决，其中阻碍 MSC 在组织工程中发展和临床应用的主要限制因素主要包括以下几点：①用来作为种子细胞修复骨、软骨组织时，细胞内分子学特性还未彻底弄清楚，并且体外环境容易诱导其分化成其他细胞类型。②不同种属之间，不同组织部位来源的 MSC 分化潜能存在较大差异，需要探究选用哪一种或哪几种效果较好的 MSC 来源。③探明 MSC 分化和调控的确切机制,研究多种调控因素间的相互作用,找到使其向不同组织转化的最佳条件。④MSC 作为种子细胞应用于临床的安全性、可靠性。

总之,作为目前研究最广的一种主要成体干细胞,MSC 既具有适合作为种子细胞的优势特性,也具有相关的局限性。但无论如何,MSC 仍是当前研究最多的组织工程种子细胞之一。因此,目前还需要进一步提高相关技术水平,解决相关问题,着眼于其在临床医学的应用,为 MSC 在组织工程、再生医学、细胞治疗等领域得到广泛应用与发展奠定良好的基础。相信随着组织工程种子细胞尤其是针对 MSC 的研究不断向更深和更广扩展,人类对组织工程种子细胞的了解也将逐渐深入而且更加全面。

第二节 间充质干细胞在组织工程中的应用

一、基于 MSC 功能的组织工程

MSC 是属于中胚层的一类多能干细胞,是具有自我复制能力和多向分化潜能的成体干细胞,可以诱导分化成体外和体内的成骨细胞、软骨细胞、脂肪细胞、肌细胞和神经系统细胞等,具有广泛的临床应用前景。除此之外,MSC 还存在于多种组织中,如骨髓、脐血和脐带、胎盘和脂肪等组织,可以作为种子细胞广泛应用于组织工程,因此基于 MSC 对于受损组织定向分化功能为基础的组织工程具有非常广阔的应用前景(图 18-3)。

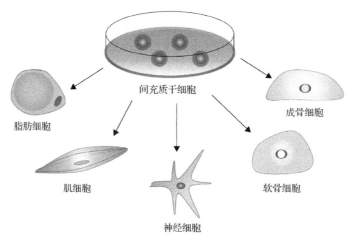

图 18-3 间充质干细胞属于多能干细胞(彩图请扫封底二维码)

(一)组织工程皮肤

皮肤是人体最大的器官,起着保持人体动态平衡的功能。深部烧伤和各种皮肤损伤等相关疾病是常见的临床皮肤病之一[3],于是使用皮肤移植物进行皮肤移植成为最有效的治疗手段和方法。然而,皮肤移植物的需求量远远大于目前可获得的量,于是现代皮肤组织工程中的皮肤接种应运而生[4]。近年来组织工程皮肤已经取得了重大进展,目前已经开发了多种不同类型的组织工程皮肤替代品,这些组织工程皮肤替代品在慢性伤口修复、伤口愈合、皮肤复原和瘢痕重塑等方面都发挥着重要作用。与此同时,种子细胞

作为影响皮肤组织工程临床应用的重要因素，研究者们正在不断积极推行优质种子细胞类型的选择研究[5]。目前，由于MSC具有自我更新能力，表现出多能性，可高度增殖，且免疫原性低等，并在伤口修复和组织再生中发挥关键作用，因此现在许多研究工作重点关注MSC的应用，旨在为其临床开发提供一条新思路。

BM-MSC可以分化成表皮细胞并修复皮肤缺陷[6]。此外，骨髓和脐带来源的MSC可以在体外被诱导分化为皮肤成纤维细胞[7,8]。同时，在来自MSC的毛囊、汗腺和皮脂腺附着细胞的培养与分化方面也取得了进展，通过直接培养BM-MSC与正常乳腺细胞，可以成功诱导BM-MSC向汗腺细胞的表型转化[9]。AD-MSC可以从脂肪抽吸物中分离，并具有与BM-MSC相似的多向分化能力。此外，AD-MSC衍生的脂肪组织应用广泛，与从100mL骨髓中提取的BM-MSC数量相比，从100g的脂肪组织中获得的AD-MSC是其300倍[10,11]。因此，AD-MSC可能是皮肤组织工程中理想的种子细胞。目前已知AD-MSC直接或间接地参与皮肤新生血管形成和皮肤愈合过程，并且其皮肤年轻化功能也处于积极的研究过程中。此外，AD-MSC也是皮肤再生的强大来源，它们不仅能够提供细胞元件，还提供许多细胞因子[12]。最近一项研究表明，AD-MSC可以代替皮肤成纤维细胞进行皮肤重建，并通过分泌相关细胞因子及利用AD-MSC的脂肪形成潜力，可以成功地设计出更加成熟的三层皮肤成分[13]（包括表皮、真皮和含脂肪细胞的皮下组织）。与此同时，人脐带间充质干细胞（hUC-MSC）也是皮肤组织工程较理想的种子细胞，它是具组织再生及修复功能的干细胞的重要来源。最近对其可塑性研究也已经进展到重建了部分皮肤附属物，如汗腺等。此研究表明，UC-MSC通过模拟汗腺的发生和发展过程从而分化为汗腺细胞，这可能是汗腺重建的途径之一，并且是预期的可以产生含有汗腺结构的组织工程全层皮肤的种子细胞。因此，汗腺上皮细胞的离体培养具有潜在的及其重要的生物学和临床意义，这不仅为进一步研究病毒和疾病的发病机制提供了良好的实验体系，也为研究组织工程皮肤附属物的发生和发展提供了指导[14]。总之，UC-MSC在未来很可能成为细胞移植和皮肤组织工程的重要干细胞来源[15]。

目前，虽然皮肤的再生与移植技术在多年前就已成功实现，但随着人们对手术效果要求的提高，以及人口老龄化导致糖尿病发病率增加，现有的技术已不能完全满足临床需求。于是组织工程皮肤已成为有力的替代品。目前，烧伤，糖尿病造成的下肢、足部皮肤损伤已经可以使用组织工程学皮肤进行治疗。同时，由羊膜和动物小肠开发的材料越来越多地被外科医生用来恢复损伤的组织并支持身体自身的愈合过程，如二度烧伤、慢性压力性溃疡、糖尿病性皮肤溃疡和深层皮肤撕裂等伤口的愈合。

（二）组织工程骨

通过体外观察发现，MSC可以分化为成骨细胞和软骨细胞。与此同时，研究人员也多次尝试使用扩增的MSC进行体内组织修复[16,17]。在各种动物模型中，已经使用MSC来修复节段性骨缺损[18,19]，如将MSC接种在支架上，然后植入小鼠体内，随后观察骨形成。目前已经开发了许多支架作为递送MSC的载体来协助MSC用于组织工程治疗，生物材料领域的最新进展也已经确定了使用无孔到多孔的生物惰性材料（如陶瓷或钛）、可再吸收及骨传导生物材料（如羟基磷灰石和磷酸三钙）等众多天然或合成的生物材料[20,21]。

此外，诱导和增强 MSC 分化成成骨细胞用于骨再生也是非常重要的一个方面。因为成骨细胞在骨的初始形成、维持骨骨化和骨折修复过程中扮演着极其重要的角色，所以 MSC 在骨的初始形成和维持过程中也起着非常关键的作用。其中，软骨内成骨是一种类型的骨愈合机制，涉及 MSC 分化成软骨细胞，然后软骨持续生长与退化、钙化，最后重塑成骨[22]；而骨膜内成骨是另一种类型的骨愈合机制，涉及 MSC 直接分化为成骨细胞或未分化的骨形成细胞[23]。除此之外，MSC 分化为成骨细胞还涉及许多旁分泌和自分泌信号之间的复杂相互作用，需要启动允许完全成骨分化的相关分子机制[24]。其中许多 MSC 分泌因子在促进或调节 MSC 分化为成骨细胞的新骨形成中发挥重要作用，如骨形成蛋白（BMP）、转化生长因子-β（TGF-β）、基质细胞衍生因子-1（SDF-1）、胰岛素样生长因子-1（IGF-1）、组蛋白去甲基化酶 JMJD3、CDK1、岩藻多糖、Runx2、PDZ 结合域的转录共刺激因子（transcriptional co-activator with PDZ-binding motif，TAZ）等。针对这些因素可知通过增加成骨细胞信号加速骨愈合可能是治疗复杂骨折并最大限度提高愈合程度的有效方法。此外，MSC 分泌的广泛生物活性分子还有助于创造最佳的再生微环境[25]。该领域正在迅速发展，并具有潜在的临床应用前景。

（三）组织工程软骨

关节软骨是一种高度特化的组织减震器，使滑膜关节能够以较低摩擦力进行衔接，由于其无血管、无神经及有较低的有丝分裂活动，关节软骨具有有限的自我修复潜力[26]，一旦损伤就很难恢复其全部功能及正常的生理状态。软骨损失会导致疼痛、身体残疾、运动限制和其他疾病。目前已经提出了各种再生软骨的治疗策略，但最佳治疗策略尚未确定。软骨是应用组织工程技术最早构建的组织，可以通过向细胞提供机械、生物和化学支持来帮助软骨再生。使用支架作为底物来支持干细胞或自体软骨细胞已经取得了良好成果[27]。

MSC 作为一种多能细胞，可以提供一种不受可用性和供体位点发病率限制的软骨修复技术。同时将其与快速发展的组织工程相结合，通过在体外形成工程化组织构建体并将其植入体内，可以产生功能性组织替代物（包括软骨）。一方面，生物力学在软骨形成中的重要性是众所周知的，MSC 在体外分化获得透明质酸结构需要在三维环境中进行接种和培养，三维环境可以根据周围细胞的几何形状和物理化学特性采用多种形式。目前相关研究通过提高三维结构材料的机械和生物化学性能，使其作用更加高效，可以增强 MSC 的增殖能力和活力。因此，三维环境是 MSC 向软骨分化的重要影响因素[28]。另一方面，由于体外扩增的关节软骨细胞（articular chondrocyte，ACh）的去分化受限，如果初始种群的 ACh 与 MSC 共培养，该共培养体系不仅可以促进软骨的形成，而且还可以提高软骨细胞的活力。同时相关体外实验也表明，MSC 在共培养环境中相比于单独培养的确可以支持软骨细胞再分化及增殖[29]。

虽然最近在软骨组织工程方面取得了较大进展，但细胞渗漏、细胞活力差、细胞分化不良、宿主组织融合不足、细胞分布不匀、正常软骨去分化等常见问题在组织工程中仍然存在。经过长期的发展，人们清楚地意识到，种子细胞将是限制软骨组织工程走向临床应用最主要的因素之一，这也是软骨组织工程经过长期的发展仍未进入大规模临床应用的主要原因[30]。因此，MSC 凭借其天然优势应用于软骨组织工程，会为软骨组织工程带来新的希望。

（四）人工血管

目前心血管疾病（CVD）发病率为全球第一，全球慢性疾病中导致死亡的主要疾病是缺血性 CVD，如脑卒中和心肌梗死[31]。其中血管支架和人工血管置换的市场需求规模巨大。在接受冠脉搭桥手术的患者中，有 30%的病例无法进行自体血管切取，需要采用人工合成的血管替代物，并且自体血管的 5 年血管通畅率最多只有 40%~50%。而组织工程血管用于治疗心肌梗死可以避免这些问题，同时还可以防止心力衰竭、缩短恢复时间[32]。最新进展表明，3D 生物血管打印技术未来将主要应用于 CVD 领域，市场前景广阔。在适当条件下培养的 MSC 可以分化为血管平滑肌细胞、肌节肌细胞（骨骼和心脏）和内皮细胞，3D 生物打印的血管与受体自身血管完全融为一体，在结构和功能方面，脂肪间充质干细胞（AD-MSC）可以有序分化为内皮细胞、平滑肌细胞等血管组织。

目前，利用干细胞和生物材料生成治疗性血管似乎是一个强有力的选择。于是，涉及心血管再生的成体干细胞治疗是 CVD 最有希望的治疗策略。同时，由于其较好的免疫调节特性和血管修复能力，MSC 是用于心血管再生的最佳候选治疗性干细胞。然而，MSC 在应用中也存在一些治疗缺陷，包括其在治疗缺血性病变时存活率非常低等[33]。目前，一种有希望的治疗策略是在移植前使用干细胞调节剂预处理 MSC，即在移植前使用细胞因子和天然化合物，通过旁分泌机制诱导细胞内信号转导或预刺激促进移植细胞的存活[34,35]。另一种是基于组织工程的治疗策略，比如细胞外基质（ECM）或水凝胶相关的生物材料制成的细胞-蛋白-支架结构，及利用基于细胞贴片和 3D 打印的组织工程，通过细胞-细胞之间或细胞-支架之间的交流及相互作用增加细胞存活率[36]，从而在细胞体外或体内进一步提高 MSC 应用于心血管再生的组织工程治疗策略。

二、基于增强 MSC 功能的生物工程策略

结合 MSC 的再生潜力和免疫调节作用，很容易想象这些细胞可能会成为治疗退变性和炎性疾病的有力工具[37-40]。目前已经将这种方法应用于动物模型中，用来修复和再生各种组织[41,42]。然而，任何组织的重建不仅需要修复细胞（即 MSC），而且需要适合的支架，从而支持植入的 MSC 增殖及其与特定的生长因子和细胞因子相互作用。因此，再生医学已经成为可通过结合细胞生物学、组织工程及临床医学，并通过细胞、生物材料和信号分子来更新组织的一门学科[43]。细胞和支架之间的相互作用，基质表面的细胞黏附，细胞增殖、成熟和分化，以及 ECM 的产生都是决定细胞疗法成功的重要因素。

目前，干细胞治疗是再生医学或组织修复最有希望的治疗策略。然而，移植细胞极低的存活率，以及内源性干细胞在受损伤位点显著募集不足和激活不够是干细胞治疗面临的主要挑战[44,45]。与此同时，细胞命运的选择取决于对复杂的细胞外信号的反应[46,47]，其中 ECM 在细胞的命运选择中提供了许多可溶性和固定性因子，对细胞生理及功能的发挥起着至关重要的作用[48]。因此，从改善细胞微环境方面入手，通过有效因子的释放来增强 MSC 功能的相关生物工程策略可进一步提高干细胞的治疗效果[49]。

（一）仿生组织

组织再生期间需要可以持续释放并能改善局部再生能力的因子，如生长因子和ECM[50]。然而，这些分子很容易快速降解，因此它们会迅速失去其功能和临床疗效[51]。此外，细胞生长过程也受到细胞和 ECM 中非可溶性成分之间相互作用的影响[52]。由于这些原因，将信号分子或功能组分固定到 ECM 可适当稳定这些高反应性活性分子，增加生物化学刺激物的局部浓度，并增加工程化 ECM 的生物活性。越来越多的研究使用较短的合成肽来模拟全长生长因子的生物学特性并用其来替代亲本蛋白[53,54]。例如，IGF-1 被认为是组织再生中必需的生物化学刺激物。IGF-1 的 C 结构域（IGF-1C）（12 个氨基酸序列）被证明是 IGF-1 的活性区[55]。因此，IGF-1C 已被用作 IGF-1 的替代物并应用于水凝胶生物材料，作为组织工程和再生医学的生物仿生材料。将 AD-MSC 接种到 IGF-1C 固定的壳聚糖（chitosan，CS）水凝胶上，细胞的增殖、凋亡抵抗和旁分泌作用显著增强。同时将 AD-MSC 和 CS-IGF-1C 水凝胶共移植到缺血器官中，这种仿生基质可以为 AD-MSC 的生存创造良好的人工微环境，进一步促进受损器官的功能性及结构性的恢复。目前就有研究表明，在体内将 CS-IGI-1C 水凝胶和 AD-MSC 共移植到缺血性肾中可以改善肾功能，可能是急性肾损伤有希望的治疗方法[56]。

（二）生物缓释材料

一氧化氮（NO）作为一种气体信号分子，已被证明是必不可少的调节细胞/分子功能的分子，并在许多生理系统中发挥重要作用，如心血管、呼吸道和免疫系统，以及与组织再生和癌症相关的发展过程[57,58]。在组织再生方面，NO 可参与调节细胞增殖、分化和迁移，是起着重要作用的气体信号分子[59]。同时，它通过增强自我更新能力从而有利于组织再生并抑制间质祖细胞的病态分化[60,61]。因此外源性 NO 能够提高 NO 水平，并被认为是组织再生的良好策略。除此之外，NO 也是组织修复不同阶段的关键介质，包括胶原沉积、成纤维细胞迁移[62]和血管生成[63]。

外源性 NO 的释放需要特定的生物材料，为细胞再生应用而设计的生物材料为载体提供了控制 NO 释放的支撑结构。同时，具有控制 NO 释放功能的工程化基质已被用作递送干细胞的载体，利用合成生物材料作为体内载体用于干细胞的移植已被证明是一种有希望提高干细胞移植率和存活率的策略，能够进一步提高移植细胞的治疗效果。例如，一种多肽水凝胶（NapFF-NO）能通过经 β-半乳糖苷酶催化的可控方式释放 NO，这种外源性 NO 可显著提高 MSC 的移植率和旁分泌效应，激活 MSC 中 VEGF/VEGFR2 途径进行血管重建并改善心脏功能，在小鼠模型中可与 MSC 共移植来治疗心肌梗死[64]。因此，这些生物缓释材料不仅为细胞锚定提供支架、改善细胞功能，还为细胞移植提供了支持性的利基[65,66]，在再生医学和组织工程中有巨大的潜力。

三、从 MSC 为载体的肿瘤靶向治疗

肿瘤是人类发展中最具破坏性的疾病之一，但有效的治疗方法仍然存在。目前可用的放射疗法和化学疗法仅在有限类型的癌症中具有一定的效果，但在许多情况下，都伴

有严重的不良反应[67]，于是现在迫切需要创新肿瘤治疗方法。

MSC 作为干细胞的一种，在机体大部分组织中广泛存在，并且较易获得。它不仅在再生医学和治疗自身免疫疾病方面具有巨大的潜力，可以治疗许多不同种类的疾病，包括肝纤维化、糖尿病、移植物抗宿主病（GVHD）和克罗恩病（CD）等[68]，而且 MSC 具有免疫调节作用及主动向损伤部位迁移的特性，从而参与伤口修复[69]。由于肿瘤被认为是"永不愈合的创伤"，因此 MSC 可以响应肿瘤发出的信号，连续地被招募并成为肿瘤微环境的组成部分，在体内特异性迁移到肿瘤中[70]。由于 MSC 具有特异性迁移到肿瘤部位的能力，其可以作为递送抗肿瘤药物的理想肿瘤载体[71,72]。总之，MSC 作为运载抗癌药物或基因的载体在抗肿瘤领域还是很有研究价值的。

目前，肿瘤治疗中的一个难题是靶向药物的递送[73]。一型干扰素（IFN-α）是临床上广泛应用的抗肿瘤药物，由于其半衰期较短，在治疗肿瘤时应用剂量往往较高，给患者带来严重不良反应。于是，MSC 被基因工程化用来表达各种抗肿瘤因子，构建了能高效、持续分泌功能性 IFN-α 的 MSC（MSC-IFN-α）[74]，在临床上被用于治疗白血病和黑色素瘤。另外值得注意的是，MSC 产生的 IFN-α 尽管远低于临床应用剂量，但其治疗效果明显优于高剂量 IFN-α 的治疗效果。由此可见，MSC 是一种理想的肿瘤靶向治疗及药物缓释载体，为临床治疗肿瘤提供了重要参考。

此外，其他相关研究还表明：肿瘤细胞处于复杂的微环境中，包括如内皮细胞、免疫细胞、肿瘤相关巨噬细胞（tumor-associated macrophage，TAM）和 MSC 等细胞的环境中[75]。这些非癌基质细胞具有不同的促进肿瘤生长的能力[76]。其中，经 M1 型巨噬细胞活化的 MSC 便可通过免疫抑制效应协助肿瘤逃脱免疫监视，从而促进癌症发展[77]。总之，MSC 无论是作为肿瘤靶向治疗的载体，还是作为促进癌症发展的非癌基质细胞，都与癌症发生、发展及治疗密切相关，是癌症相关研究中不可忽略的重要因素之一。

第三节 发展趋势预测与展望

目前，我国科技部发布的"十三五"规划将"干细胞与再生医学"列为重点发展领域，肯定了"干细胞与再生医学为疾病治疗开辟了全新道路"，"十三五"期间将"重点加强干细胞的应用及基础研究和转化研究，强化干细胞、生物医用材料与组织工程的交叉融合，引导我国生物医用材料产业的技术升级和细胞治疗等新治疗手段的规范化临床应用"。MSC 作为一种重要的成体干细胞，在干细胞与组织工程及再生医学的研究中扮演着重要的角色。现在，随着基因工程、胚胎工程、细胞工程等各种生物技术的快速发展，按照一定的目的，在体外人工分离、培养 MSC 已成为可能，利用 MSC 构建各种细胞、组织、器官作为移植器官的来源，这将成为 MSC 应用的主要方向。

组织工程的提出、建立和发展标志着医学将走出组织器官移植的范畴，步入制造组织和器官的新时代。从某种意义上讲，它的发展水平已成为一个国家医学发展水平的标志之一。目前组织工程研究已经涉及生命科学几乎所有相关研究领域，同时研发的产品也蕴含着巨大的应用前景。2003 年以来，组织工程的发展更是突飞猛进，特别是材料学、先进制造技术（包括 3D 打印等）的发展，基因工程的介入，移植免疫的进展和 MSC 研究的突破为组织工程的持续发展和不断完善注入了新的活力,并出现了一些新的研究动向。例如,一些发达国家在组织工程领域的研究重点,从种子细胞、生物材料等基础领域,逐渐向组织构建领域转移,即应用干细胞特别是 MSC 进行结构性组织的构建,其中 MSC 可分泌几十上百种因子,包括生长因子、细胞因子、趋化因子、酶等,在组织和器官的生长与修复中起到了至关重要的作用,于是在此基础上将其应用于临床上组织创伤的修复;同时在再生医学领域,我们可以看到发育生物学不仅关注组织形成的问题,而且开始注意内生 MSC 在生长及再生过程中的特殊作用。如果把这些基础科学研究的成果和吸收特定内生细胞种群的新型材料结合起来,将会给用于组织修复的智能材料研究带来新的思路。此外,新的细胞源如诱导多能干细胞(iPSC)已经走上前台,使得基于细胞的自体移植治疗对于修复任何组织都有了可能。因此,随着干细胞研究的不断深入,组织工程的研究成果向临床转化的速度变得越来越快。除此之外,通过引入一组多能性相关基因进入成体细胞,或通过化学重编程或蛋白质递送产生的 iPSC 受到广泛关注。作为一种多能干细胞,iPSC 为干细胞治疗增加了一种选择。另外,通过规律成簇间隔短回文重复(clustered regularly interspaced short palindromic repeat,CRISPR)技术可以以极高的精度,高效靶向地编辑基因组。这些不断发展的基因修饰技术能让 MSC 携带药物进行疾病的治疗,正在进行的众多相关动物和人类试点研究正在为大规模临床试验治疗难治性疾病铺平道路。总之,以上这些新技术、新手段的介入使得 MSC 组织工程不断发展和完善(图 18-4)。

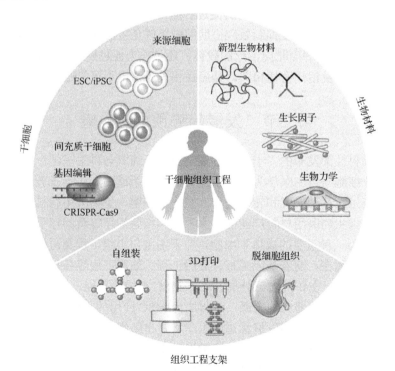

图 18-4　干细胞组织工程发展(彩图请扫封底二维码)

综上所述,我们看到了组织工程与再生医学的快速发展,与此同时,跨学科合作也为组织工程带来了更加广阔的前景。进入 21 世纪以来，我们已经见证了多个学科，包括生物学、材料科学、化学和工程学的融合，对组织工程与再生医学的发展起促进作用。MSC 作为组织工程与再生医学研究领域的应用热点，以及组织再生、疾病治疗过程的重要组成部分，受到了广泛关注。因此，对于未来，我们期望，通过再生医学的不断发展，可以创造出一个没有捐赠者的世界；同时智能材料的成功使用，也让我们可以完全按照个人要求进行组织器官打印，构建出可应用于临床的功能性组织和器官。

总之，我们相信，通过跨学科合作，并充分利用 MSC 天然的优势特性，将会推动组织工程与再生医学更好、更快地发展。

（赵向男 李宗金）

参 考 文 献

[1] Langer R. Tissue engineering: perspectives, challenges, and future directions. Tissue Engineering, 2007, 13(1):1-2.

[2] Jiang Y, Jahagirdar BN, Reinhardt RL, et al. Pluripotency of mesenchymal stem cells derived from adult marrow. Nature, 2002, 418(6893):41-49.

[3] Chrapusta A, Nessler MB, Drukala J, et al. A comparative analysis of advanced techniques for skin reconstruction with autologous keratinocyte culture in severely burned children: own experience. Postepy Dermatol Alergol, 2014, 31(3):164-169.

[4] Horch RE, Kopp J, Kneser U, et al. Tissue engineering of cultured skin substitutes. Journal of Cellular and Molecular Medicine, 2005, 9(3):592-608.

[5] Han YF, Tao R, Sun TJ, et al. Advances and opportunities for stem cell research in skin tissue engineering. European Review for Medical and Pharmacological Sciences, 2012, 16(13):1873-1877.

[6] Perng CK, Kao CL, Yang YP, et al. Culturing adult human bone marrow stem cells on gelatin scaffold with pNIPAAm as transplanted grafts for skin regeneration. Journal of Biomedical Materials Research Part A, 2008, 84(3):622-630.

[7] He L, Nan X, Wang Y, et al. Full-thickness tissue engineered skin constructed with autogenic bone marrow mesenchymal stem cells. Science in China Series C, Life Sciences, 2007, 50(4):429-437.

[8] Sheng Z, Fu X, Cai S, et al. Regeneration of functional sweat gland-like structures by transplanted differentiated bone marrow mesenchymal stem cells. Wound Repair Regen, 2009, 17(3):427-435.

[9] Han Y, Chai J, Sun T, et al. Differentiation of human umbilical cord mesenchymal stem cells into dermal fibroblasts *in vitro*. Biochemical and Biophysical Research Communications, 2011, 413(4):561-565.

[10] Danisovic L, Varga I, Polak S, et al. Comparison of *in vitro* chondrogenic potential of human mesenchymal stem cells derived from bone marrow and adipose tissue. General Physiology and Biophysics, 2009, 28(1):56-62.

[11] Oedayrajsingh-Varma MJ, van Ham SM, Knippenberg M, et al. Adipose tissue-derived mesenchymal stem cell yield and growth characteristics are affected by the tissue-harvesting procedure. Cytotherapy, 2006, 8(2):166-177.

[12] Jeong JH. Adipose stem cells and skin repair. Current Stem Cell Research & Therapy, 2010, 5(2):137-140.

[13] Trottier V, Marceau-Fortier G, Germain L, et al. IFATS collection: using human adipose-derived stem/stromal cells for the production of new skin substitutes. Stem Cells, 2008, 26(10):2713-2723.

[14] Tao R, Han Y, Chai J, et al. Isolation, culture, and verification of human sweat gland epithelial cells. Cytotechnology, 2010, 62(6):489-495.

[15] Wu KH, Zhou B, Mo XM, et al. Therapeutic potential of human umbilical cord-derived stem cells in ischemic diseases. Transplantation Proceedings, 2007, 39(5):1620-1622.

[16] Barry F, Boynton RE, Liu B, et al. Chondrogenic differentiation of mesenchymal stem cells from bone marrow: differentiation-dependent gene expression of matrix components. Experimental Cell Research, 2001, 268(2):189-200.

[17] Fibbe WE. Mesenchymal stem cells. A potential source for skeletal repair. Annals of the Rheumatic Diseases, 2002, 61(Suppl 2):ii29-ii31.

[18] Kon E, Muraglia A, Corsi A, et al. Autologous bone marrow stromal cells loaded onto porous hydroxyapatite ceramic accelerate bone repair in critical-size defects of sheep long bones. Journal of Biomedical Materials Research, 2000, 49(3):328-337.

[19] Petite H, Viateau V, Bensaid W, et al. Tissue-engineered bone regeneration. Nature Biotechnology, 2000, 18(9):959-963.

[20] Rose FR, Oreffo RO. Bone tissue engineering: hope vs hype. Biochemical and Biophysical Research Communications, 2002, 292(1):1-7.

[21] Vats A, Tolley NS, Polak JM, et al. Scaffolds and biomaterials for tissue engineering: a review of clinical applications. Clinical Otolaryngology and Allied Sciences, 2003, 28(3):165-172.

[22] Dimitriou R, Tsiridis E, Giannoudis PV. Current concepts of molecular aspects of bone healing. Injury, 2005, 36(12):1392-1404.

[23] Thompson Z, Miclau T, Hu D, et al. A model for intramembranous ossification during fracture healing. Journal of Orthopaedic Research: Official Publication of the Orthopaedic Research Society, 2002, 20(5):1091-1098.

[24] Jaiswal RK, Jaiswal N, Bruder SP, et al. Adult human mesenchymal stem cell differentiation to the osteogenic or adipogenic lineage is regulated by mitogen-activated protein kinase. The Journal of Biological Chemistry, 2000, 275(13):9645-9652.

[25] Caplan AI. Adult mesenchymal stem cells for tissue engineering versus regenerative medicine. Journal of Cellular Physiology, 2007, 213(2):341-347

[26] Oldershaw RA. Cell sources for the regeneration of articular cartilage: the past, the horizon and the future. International Journal of experimental Pathology, 2012, 93(6):389-400.

[27] Rai V, Dilisio MF, Dietz NE, et al. Recent strategies in cartilage repair: a systemic review of the scaffold development and tissue engineering. Journal of Biomedical Materials Research Part A, 2017, 105(8):2343-2354.

[28] Panadero JA, Lanceros-Mendez S, Ribelles JL. Differentiation of mesenchymal stem cells for cartilage tissue engineering: Individual and synergetic effects of three-dimensional environment and mechanical loading. Acta Biomaterialia, 2016, 33:1-12.

[29] Nazempour A, Van Wie BJ. Chondrocytes, mesenchymal stem cells, and their combination in articular cartilage regenerative medicine. Annals of Biomedical Engineering, 2016, 44(5):1325-1354.

[30] Tuli R, Li WJ, Tuan RS. Current state of cartilage tissue engineering. Arthritis Research & Therapy, 2003, 5(5):235-238.

[31] GBD 2015 Mortality and Causes of Death Collaborators. Global, regional, and national life expectancy, all-cause mortality, and cause-specific mortality for 249 causes of death, 1980-2015: a systematic analysis for the Global Burden of Disease Study 2015. Lancet, 2016, 388(10053):1459-1544.

[32] Kaushik G, Leijten J, Khademhosseini A. Concise review: organ engineering: design, technology, and integration. Stem Cells, 2017, 35(1):51-60.

[33] Li L, Chen X, Wang WE, et al. How to improve the survival of transplanted mesenchymal stem cell in ischemic heart? Stem Cells International, 2016, (2016):9682757.

[34] Cai M, Shen R, Song L, et al. Erratum: bone marrow mesenchymal stem cells (BM-MSCs) improve heart function in swine myocardial infarction model through paracrine effects. Scientific Reports, 2016, 6:31528.

[35] Dittmer J, Leyh B. Paracrine effects of stem cells in wound healing and cancer progression (Review). International Journal of Oncology, 2014, 44(6):1789-1798.

[36] Zippel N, Schulze M, Tobiasch E. Biomaterials and mesenchymal stem cells for regenerative medicine. Recent Patents on Biotechnology, 2010, 4(1):1-22.

[37] Bonassar LJ, Vacanti CA. Tissue engineering: the first decade and beyond. Journal of Cellular Biochemistry Supplement, 1998, 30-31:297-303.

[38] Caplan AI, Bruder SP. Mesenchymal stem cells: building blocks for molecular medicine in the 21st century. Trends in Molecular Medicine, 2001, 7(6):259-264.

[39] Deans RJ, Moseley AB. Mesenchymal stem cells:biology and potential clinical uses. Experimental Hematology, 2000, 28(8):875-884.

[40] Griffith LG, Naughton G. Tissue engineering-current challenges and expanding opportunities. Science, 2002, 295(5557):1009-1014.

[41] Barry FP, Murphy JM. Mesenchymal stem cells: clinical applications and biological characterization. The International Journal of Biochemistry & Cell Biology, 2004, 36(4):568-584.

[42] Devine SM. Mesenchymal stem cells: will they have a role in the clinic? Journal of Cellular Biochemistry Supplement, 2002, 38:73-79.

[43] Tuan RS, Boland G, Tuli R. Adult mesenchymal stem cells and cell-based tissue engineering. Arthritis Research & Therapy, 2003, 5(1):32-45.

[44] He N, Xu Y, Du W, et al. Extracellular matrix can recover the downregulation of adhesion molecules after cell detachment and enhance endothelial cell engraftment. Scientific Reports, 2015, 5:10902.

[45] Li Z, Han Z, Wu JC. Transplantation of human embryonic stem cell-derived endothelial cells for vascular diseases. Journal of Cellular Biochemistry, 2009, 106(2):194-199.

[46] Kim SH, Turnbull J, Guimond S. Extracellular matrix and cell signalling: the dynamic cooperation of integrin, proteoglycan and growth factor receptor. The Journal of Endocrinology, 2011, 209(2):139-151.

[47] Zouani OF, Kalisky J, Ibarboure E, et al. Effect of BMP-2 from matrices of different stiffnesses for the modulation of stem cell fate. Biomaterials, 2013, 34(9):2157-2166.

[48] Seif-Naraghi SB, Horn D, Schup-Magoffin PJ, et al. Injectable extracellular matrix derived hydrogel provides a platform for enhanced retention and delivery of a heparin-binding growth factor. Acta Biomaterialia, 2012, 8(10):3695-3703.

[49] Nie Y, Zhang SQ, Liu N, et al. Extracellular Matrix Enhances Therapeutic Effects of Stem Cells in Regenerative Medicine, Composition and Function of the Extracellular Matrix in the Human Body, Francesco Travascio, IntechOpen.

[50] Kolambkar YM, Dupont KM, Boerckel JD, et al. An alginate-based hybrid system for growth factor delivery in the functional repair of large bone defects. Biomaterials, 2011, 32(1):65-74.

[51] Grassian AR, Schafer ZT, Brugge JS. ErbB2 stabilizes epidermal growth factor receptor (EGFR) expression via Erk and Sprouty2 in extracellular matrix-detached cells. The Journal of Biological Chemistry, 2011, 286(1):79-90.

[52] Walters NJ, Gentleman E. Evolving insights in cell-matrix interactions: elucidating how non-soluble properties of the extracellular niche direct stem cell fate. Acta Biomaterialia, 2015, 11:3-16.

[53] Frith JE, Mills RJ, Cooper-White JJ. Lateral spacing of adhesion peptides influences human mesenchymal stem cell behaviour. Journal of Cell Science, 2012, 125(Pt 2):317-327.

[54] Meng Q, Man Z, Dai L, et al. A composite scaffold of MSC affinity peptide-modified demineralized bone matrix particles and chitosan hydrogel for cartilage regeneration. Scientific Reports, 2015, 5:17802.

[55] Jansen M, van Schaik FM, Ricker AT, et al. Sequence of cDNA encoding human insulin-like growth factor I precursor. Nature, 1983, 306(5943):609-611.

[56] Feng G, Zhang J, Li Y, et al. IGF-1 C domain-modified hydrogel enhances eell therapy for AKI. Journal of the American Society of Nephrology: JASN, 2016, 27(8):2357-2369.

[57] Carpenter AW, Schoenfisch MH. Nitric oxide release: part II. Therapeutic applications. Chemical Society Reviews, 2012, 41(10):3742-3752.

[58] Gao J, Zheng W, Zhang J, et al. Enzyme-controllable delivery of nitric oxide from a molecular hydrogel. Chemical Communications, 2013, 49(80):9173-9175.

[59] Fuseler JW, Valarmathi MT. Modulation of the migration and differentiation potential of adult bone marrow stromal stem cells by nitric oxide. Biomaterials, 2012, 33(4):1032-1043.

[60] Buono R, Vantaggiato C, Pisa V, et al. Nitric oxide sustains long-term skeletal muscle regeneration by regulating fate of satellite cells via signaling pathways requiring Vangl2 and cyclic GMP. Stem Cells, 2012, 30(2):197-209.

[61] Cordani N, Pisa V, Pozzi L, et al. Nitric oxide controls fat deposition in dystrophic skeletal muscle by regulating fibro-adipogenic precursor differentiation. Stem Cells, 2014, 32(4):874-885.

[62] Han G, Nguyen LN, Macherla C, et al. Nitric oxide-releasing nanoparticles accelerate wound healing by promoting fibroblast migration and collagen deposition. The American Journal of Pathology, 2012, 180(4):1465-1473.

[63] Kazakov A, Muller P, Jagoda P, et al. Endothelial nitric oxide synthase of the bone marrow regulates myocardial hypertrophy, fibrosis, and angiogenesis. Cardiovascular Research, 2012, 93(3):397-405.

[64] Yao X, Liu Y, Gao J, et al. Nitric oxide releasing hydrogel enhances the therapeutic efficacy of mesenchymal stem cells for myocardial infarction. Biomaterials, 2015, 60:130-140.

[65] Li Z, Lee A, Huang M, et al. Imaging survival and function of transplanted cardiac resident stem cells. Journal of the American College of Cardiology, 2009, 53(14):1229-1240.

[66] Li Z, Wilson KD, Smith B, et al. Functional and transcriptional characterization of human embryonic stem cell-derived endothelial cells for treatment of myocardial infarction. PLoS One, 2009, 4(12):e8443.

[67] Hainaut P, Plymoth A. Targeting the hallmarks of cancer: towards a rational approach to next-generation cancer therapy. Current Opinion in Oncology, 2013, 25(1):50-51.

[68] Ren G, Chen X, Dong F, et al. Concise review: mesenchymal stem cells and translational medicine: emerging issues. Stem Cells Translational Medicine, 2012, 1(1):51-58.

[69] Karp JM, Leng Teo GS. Mesenchymal stem cell homing: the devil is in the details. Cell Stem Cell, 2009, 4(3):206-216.

[70] Kidd S, Spaeth E, Dembinski JL, et al. Direct evidence of mesenchymal stem cell tropism for tumor and wounding microenvironments using *in vivo* bioluminescent imaging. Stem Cells, 2009, 27(10):2614-2623.

[71] Kim SM, Lim JY, Park SI, et al. Gene therapy using TRAIL-secreting human umbilical cord blood-derived mesenchymal stem cells against intracranial glioma. Cancer Research, 2008, 68(23):9614-9623.

[72] Studeny M, Marini FC, Champlin RE, et al. Bone marrow-derived mesenchymal stem cells as vehicles for interferon-beta delivery into tumors. Cancer Research, 2002, 62(13):3603-3608.

[73] Wang B, Rosano JM, Cheheltani R, et al. Towards a targeted multi-drug delivery approach to improve therapeutic efficacy in breast cancer. Expert Opinion on Drug Delivery, 2010, 7(10):1159-1173.

[74] Xu C, Lin L, Cao G, et al. Interferon-alpha-secreting mesenchymal stem cells exert potent antitumor effect *in vivo*. Oncogene, 2014, 33(42):5047-5052.

[75] Grivennikov SI, Greten FR, Karin M. Immunity, inflammation, and cancer. Cell, 2010, 140(6):883-899.

[76] Elinav E, Nowarski R, Thaiss CA, et al. Inflammation-induced cancer: crosstalk between tumours, immune cells and microorganisms. Nature Reviews Cancer, 2013, 13(11):759-771.

[77] Jia XH, Feng GW, Wang ZL, et al. Activation of mesenchymal stem cells by macrophages promotes tumor progression through immune suppressive effects. Oncotarget, 2016, 7(15):20934-20944.

第十九章　间充质干细胞基因治疗

近年来细胞疗法已经表现出巨大的潜力，其在生物医学中使用[1-3]并不是新事物，事实上已经有半个多世纪的使用历史了。首次开展细胞疗法的形式是骨髓移植，一种原始的干细胞治疗技术[4]。近年来随着基因工程与基因/药物传递方法的突破，现在可以通过基因工程技术对细胞进行更为安全和精准的操作来得到更为有效的治疗用干细胞，同时提高了其在临床中应用的可行性和适用性。

使用干细胞进行基因治疗这种新型的治疗方式获得突飞猛进的发展得益于两大基石，一是我们对干细胞的了解和驾驭能力的提高，发现了干细胞的迁移归巢能力，但同时我们也认识到要获得更好的治疗效果需要对干细胞进行改造。二是随着基因组学的发展，我们发现了越来越多的可用于治疗的功能性基因，并确证了这些基因产物的具体生物学功能，将这些基因有目的地转入干细胞可以使得本就前景广阔的干细胞治疗如虎添翼。

目前可用于基因治疗的细胞有很多种，包括分化的、未分化的祖细胞和干细胞，每一种都有其独特之处和优缺点。一般来说，分化细胞在临床应用困难较多，包括数量不足、自我更新能力不足等，而干细胞恰恰具有这方面的优势[5]。首先干细胞具有很强的自我更新能力，并可以分化成为多谱系的终末功能细胞。其次，干细胞分离扩增相对容易，十分重要的是其具有迁移至损伤和疾病部位的能力，使得其成为基因治疗最理想的细胞载体[6]。最后，干细胞在体内本就可以直接通过旁分泌，以具有抗炎、促血管新生、抗凋亡等功效的细胞因子作为桥梁，直接发挥治疗作用[7,8]。干细胞具有十分强大的基础治疗功能，基因改造时只需要在此基础上进一步强化某些特定功能就可以达到很好的治疗效果。

第一节　干细胞的来源

成体干细胞、胚胎干细胞（ESC）、诱导多能干细胞（iPSC）都已经被广泛研究，其中每种干细胞都有自己的优点和缺点。首先，ESC 可以从早期的胚胎细胞团中提取，可以产生几乎所有的细胞谱系，也是再生医学最有前景的细胞来源。然而其存在明显的伦理问题。iPSC 与 ESC 具有很多的相似之处，且没有相关的伦理问题，也显示了有很广阔的前景。但不幸的是，ESC 和 iPSC 都具有形成畸胎瘤的可能性，从而极大地制约了其临床应用[1,9-11]，成体干细胞包括造血干细胞（HSC）、神经干细胞（NSC）和间充质干细胞（MSC）。HSC 和 NSC 在治疗血液疾病和神经系统疾病方面虽具有很大的优势，但是由于其分化能力有限，直接限制了其临床应用范围，仅成为特定领域的特定工具和方法。相比之下 MSC 具有安全、分化谱系广、适应证范围大、较强的迁移归巢能力、易于扩增和获取、无伦理问题、免疫调节活性、来源广泛等特点而成为研究人员的宠儿。

第二节　间充质干细胞的归巢能力

MSC 作为一种成体干细胞，其具有的各方面优势在此不再赘述，而需要重点介绍的是 MSC 的归巢能力，因为这一点是基于 MSC 进行基因治疗的重要基础。

以 MSC 为载体的基因治疗的优势在于充分利用了干细胞的靶向迁移特性，可以将治疗物质靶向输送至病灶，由此解决基因治疗的递送靶向性难题。众所周知，肿瘤组织由恶性肿瘤细胞和部分良性细胞组成，良性组织成分包括血管、浸润其间的炎症细胞和基质成纤维细胞。基质成纤维细胞为癌症组织扩散转移的行为提供结构框架。肿瘤基质的形成过程与创伤愈合和瘢痕形成过程类似，癌性细胞为了获得足够的基质来支持生长而诱导大量的新生结缔组织生成。肿瘤细胞以类似的机制诱导某些干细胞对其的趋化，相关研究也表明，肿瘤细胞在体内处于快速增殖的状态，而引起肿瘤生长部位发生与创伤类似的病理变化，称为永不愈合的伤口，这种病理变化导致一些与细胞组织修复有关的因子释放，包括炎性细胞因子和趋化因子，在这些因子的作用下，干细胞会向肿瘤部位迁移，进而到达肿瘤的周围甚至到达肿瘤的内部，因此干细胞对肿瘤组织的趋化特性使其成为基因治疗的有效运载工具，除此之外诸多出现损伤的器官也会表现出对干细胞有强大吸引力[12-15]。

在治疗过程中如果能够增加 MSC 在肿瘤部位和损伤部位的积累效率，我们就可以进一步提高干细胞基因治疗的效果，所以对 MSC 归巢迁移机制的研究始终是近些年的热点。研究发现，MSC 会被动员到受损组织，如损伤或炎症时，细胞会释放多种细胞因子。肿瘤微环境中也含有大量炎性细胞[16]，这些炎性细胞所构建的这种微环境通过各种肿瘤和细胞分泌的可溶性因子，包括表皮生长因子（EGF）、血小板源生长因子（PDGF）、血管内皮生长因子-A（VEGF-A）、白介素-8（IL-8）、IL-6、成纤维细胞生长因子（FGF）、粒细胞集落刺激因子（G-CSF）、粒细胞-巨噬细胞集落刺激因子（GM-CSF）、单核细胞趋化蛋白-1（MCP-1）、肝细胞生长因子（HGF）、转化生长因子-β（TGF-β1）、尿激酶型纤溶酶原激活剂[17-23]等促进 MSC 聚集。全身注射 MSC 可以有效地观察到 MSC 在肿瘤部位积累，而皮下注射 MSC 通常则不能。肿瘤组织中通常含有丰富的血管，MSC 会在肿瘤内部和肿瘤边界处积累。而使用肿瘤坏死因子-α（TNF-α）刺激 MSC 将增强 MSC 对内皮细胞的黏附作用。如果通过抗体封闭血管细胞黏附分子-1（VCAM-1）和极晚期抗原（VLA-4）可以部分抑制这种黏附作用。

在 Sullivan 等[24]的一项研究中，使用小鼠关节炎模型发现，骨髓间充质干细胞（BM-MSC）在表达 CTLA4Ig 之后，可更加有效地在静脉注射之后迁移至胶原诱导的关节炎局部，并表现出了对干扰素-γ（IFN-γ）、TNF-α 等炎性细胞因子局部浓度的抑制作用。此研究为我们了解 MSC 在肿瘤之外的疾病中的归巢过程提供了很好的基础。以此为基础，可以开发出更为有效的治疗用基因工程 MSC。类似的关于桂皮鞣质 B-1（cinnamtannin B-1）和趋化因子 CXC 受体 1（CXC chemokine receptor 1）的研究，证明此类分子的过表达也可以增强 MSC 的迁移归巢能力，特别是向伤口及胶质瘤迁移[25,26]。

干细胞向肿瘤组织趋化转移的特性使其在基因治疗中可以发挥独特的作用。首先通过干细胞的运输可以在肿瘤病灶的局部表达那些无法进行全身给药的药物,如全身性注射 IFN-β,在血药浓度尚无法发挥抗肿瘤作用之前就会引起剧烈的免疫反应,而利用干细胞在肿瘤局部大量表达 IFN-β 可以在发挥抗肿瘤活性的同时有效避免上述不良反应的发生,降低全身用药产生的毒不良反应[27]。另外,干细胞可以迁移至手术无法切除甚至影像学无法发现的肿瘤转移灶,通过表达抗肿瘤分子清除转移灶,进而降低肿瘤复发的概率,特别是在胶质瘤、前列腺癌、乳腺癌等转移性较强的肿瘤治疗中具有特别的意义[28,29]。

第三节 对干细胞进行基因工程改造的方法

20 世纪 70 年代,重组 DNA 技术的研究进展标志着一个激动人心的生物学新纪元的开始。分子生物学家获得了操纵 DNA 分子的能力,这使得研究基因和利用基因成为可能,其中包括工程干细胞。然而,要获得满意的治疗效果,MSC 必须使用安全和有效的载体进行改造,不仅可以特异性地使其向靶细胞传递基因,还可以维持其表达。在方法的选择上应该考虑:①转染效率高,②在不损害宿主基因稳定性的情况下长期稳定表达,③在时间和空间上表达适当水平的治疗基因,④不刺激宿主的免疫系统或诱导细胞转化[30]。

目前经常使用的方法分为病毒载体和非病毒载体两类。这些方法都有各自的优势,同时也有难以克服的缺点,在研究和临床试验中选用何种方法则要根据具体的需要确定。

一、病毒载体

在获得较高的转染效率方面,病毒载体是具有优势的。目前常用的病毒载体主要有腺病毒、腺相关病毒、逆转录病毒、慢病毒、杆状病毒。这些病毒载体具有各自不同的特点和优势。

(一)腺病毒

在基因治疗中最经常使用的是腺病毒载体,腺病毒是无包膜的病毒,由核衣壳和双链线性 DNA 组成[31]。腺病毒载体有许多优点,具体来说,36kb 的基因组转载能力使得腺病毒载体提供了充足的空间插入大序列[32]。此外,腺病毒载体在分裂和非分裂的细胞都具有较高的转染效率,且没有明显的细胞毒效应。同时在腺病毒制备的过程中,可以通过多次传代较为容易地获得大量高滴度的病毒,这为实验研究带来了极大地方便。特别重要的是其不融入宿主基因组,对宿主细胞基因组的稳定性不产生影响,从而表现出了临床使用的安全性。其结果是使用腺病毒载体的临床试验数量不断增长[33]。

但腺病毒存在免疫原性强,外源基因无法长期表达等缺陷,同时腺病毒对干细胞的感染效率通常很低,如由于 MSC 表面缺少柯萨奇病毒-腺病毒受体,腺病毒感染 MSC 的效率低下,为了获得高的感染效率,常常需要非常高的感染复数(multiplicity of infection, MOI)。在最近的一些研究中开始使用经过尾丝改造的 5 血清型腺病毒载体,在人 MSC (hMSC)中获得非常高的感染效率。在一定程度上克服了腺病毒感染效率低的这个缺陷[34]。此外,随着 3D 培养技术的开发,研究者发现处于 3D 培养环境下的细胞,包括 MSC 表现出了与 2D 培养环境下细胞不同的生物学特性,以此为基础,Neumann 等[35]

使用纤维蛋白、藻酸盐或琼脂糖构建了 MSC 的 3D 培养环境，在培养了 28 天后，使用腺病毒载体对上述细胞进行转染，并与用 2D 环境培养的 MSC 进行了对比，发现从转染效率到外源基因的表达水平，3D 环境下的 MSC 均明显优于传统平面培养的细胞，而且这种 3D 培养体系的构建十分简单，使用的材料也都是临床级的生物材料，在未来向临床转化时具有一定的优势。

（二）逆转录病毒

目前常用的反转录病毒载体是莫氏鼠白血病病毒（Moloney murine leukemia virus，Mo2MuLV），此种病毒主要感染分裂期的细胞，感染靶细胞后，在逆转录酶作用下可将单链 RNA 转变成双链 DNA，整合酶可介导双链 DNA 整合到靶细胞基因组中，以前病毒（provirus）的形式存在，并可在细胞分裂后传递给子代细胞。逆转录病毒载体具有高效转移基因和介导目的基因整合等优点，其不足之处是：①因随机整合有引起插入突变（insertional mutagenesis）的可能；②基因转移后，随时间的延长，出现转录抑制（transcription repression），目的基因的表达水平逐渐降低甚至表达终止；③不能感染非分裂期细胞。因此近年来逆转录病毒的使用在逐渐减少。

（三）慢病毒

以人类免疫缺陷病毒（HIV）为基础的慢病毒载体可通过增加 MOI 和感染次数获得接近 90%的感染效率，同时外源基因可以通过整合进入染色体的方式获得长期稳定的表达。获得如此高的转染效率并不需要进行长时间的转染操作，这对于减少操作过程中干细胞干性的损失是有利的。另外慢病毒载体可以感染非分裂期的细胞，这也是其相比于逆转录病毒载体的一大优势。但是这同时增加了病毒使用过程中对宿主细胞的致突变作用，并有可能激活某些癌基因的表达。加之使用过程中有可能产生具有复制能力的病毒，其潜在致病性使得慢病毒载体在临床试验中的安全性受到质疑。除此之外，慢病毒载体制备成本高，无法像腺病毒那样较为容易地制备得到大量高滴度的病毒，成为实验研究中的一个障碍。

（四）腺相关病毒

腺相关病毒（adeno-associated virus，AAV）属细小病毒科，为一缺陷病毒。野生型 AAV 感染人源宿主细胞后，其中 99%的载体以游离状态存在，仅 1%发生重组[36]，并可定点整合到 19 号染色体，这使得由 AAV 所导致的插入突变变得更加可控和可预测，也方便通过进一步的研究来明确 AAV 所带来的突变风险及控制方法。AAV 载体的缺点是包装容量小于 5kb，不适于较大基因片段的插入。而其优点如下：①截至目前未发现野生型 AAV 与任何人类疾病有关；②宿主范围广，可以感染脑、肝、肺、肌肉等多种组织来源的细胞；③能够感染分裂细胞和非分裂细胞；④插入突变的危险性低，综合考虑 AAV 是一种相对安全且使用方便的病毒载体；⑤相比于其他病毒载体，AAV 在体内和体外都可以获得更高的基因表达效率，其表达效率是其他载体的 10～100 倍。因此根据目前的统计数据可以看到，自 2012 年之后使用 AAV 的研究数量显著上升并在进一步增加。此外，尽管 AAV 载体的免疫原性较低,但一项研究报道称腺癌的形成与 AAV 有关。病毒载体在已知的微 RNA（miRNA）位点附近整合参与肿瘤的发生[37]。另外,一个由 Nathwani 和他的同事进行的临床试验[38]表明,AAV 载体介导血友病 B 的基因治疗并没有导致任何急性或长期毒性,此项研究还在继续,将通过对患者更长时间的追踪对 AAV 载体进行全面评估。

（五）杆状病毒

值得注意的是,最近的研究中将杆状病毒用于感染干细胞取得了良好的效果。杆状病毒的天然宿主是昆虫细胞,研究表明杆状病毒可以有效感染多种哺乳动物细胞而并不在细胞内复制,从而以很高的 MOI 感染哺乳动物细胞后几乎不引起明显的细胞毒效应。除此之外,杆状病毒的另一优势是最大可以携带 30kb 的外源基因,这为多基因治疗载体的构建提供了可能性。截至目前,杆状病毒在 hMSC 和 ESC 中的感染效率已经被报道。在 hMSC 中,使用 200MOI 可以获得接近 90%的感染效率,外源基因可以有效表达超过一个月,同时杆状病毒的感染没有影响 MSC 的分化潜能[39,40]。在 ESC 中,Zeng 等[41]将 AAV 基因组插入杆状病毒载体中,使用这种复合载体感染 ESC,利用杆状病毒将 AAV 基因组高效转入宿主细胞,再利用 AAV 的整合特性获得了在宿主细胞中长期表达外源基因,使用 100MOI 可以获得超过 80%的感染效率。

杆状病毒的使用使我们眼前一亮,相信随着研究的深入,会不断有更加完善的载体出现,使得外源基因可以在干细胞中高效、安全、稳定地表达。

二、非病毒载体技术

病毒载体应用到临床有几个限制,如安全问题,包括致癌性、免疫原性、宿主范围广等,以及其相对较小的运载治疗 DNA 的能力。为克服上述问题,近年来非病毒载体获得了长足的发展。理想的非病毒载体应该能够克服许多原有的缺点,包括:①有针对性的呈递基因;②有效地被细胞摄取;③保证其所运载的基因不被降解的情况下有效释放。目前发现纳米颗粒可以提供一个有希望的平台,纳米颗粒具有一些病毒载体没有的优势,包括:①免疫原性较低;②更大的有效载荷;③更容易制备合成[42]。此外,纳米颗粒可用于递送多种类型的物质,包括核酸（DNA、RNA）、生物分子（如多肽、蛋白质）、小分子药物,还可以提供额外的附加功能（如加热、成像）[43]。

由于其具有巨大的潜力,大量的纳米颗粒系统已被开发,以克服非病毒方法所面临的生理障碍。具体来说,这些纳米颗粒可以由各种材料组成,包括金属、贵金属、半导体、聚合物、脂类和其他无机材料,可有各种尺寸、形状[44]。然而,这些载体中很少能通过美国食品药品监督管理局（FDA）批准进入临床试验[45]。此外,它们的输送效率相对于病毒载体仍存在一定的差距[46],所以这种运载方式虽潜力巨大,仍需要进行改进才可以广泛应用于临床。在本节中,我们将简要介绍一些最常见的纳米颗粒系统。

（一）脂质体

目前应用最为广泛的非病毒递送系统即脂质体。脂质体一般由三部分组成：阳离子头基,疏水性尾部和连接器。目前许多从脂质体为基础的具有更高效转染性能的载体已被开发。合成的阳离子脂质体，如DOTMA、DOSPA、DOTAP、DMRIE均可以对DNA进行有效封装并用于包括干细胞在内的各种哺乳动物细胞[47]。然而，尽管使用最广泛，但效率低、稳定性差、被快速清除[48]，以及炎症或抗炎反应的产生限制了其进一步应用[49]。

（二）金纳米颗粒

金纳米颗粒（gold nanoparticle，GNP）是一种最广泛用于干细胞的纳米颗粒。特别是GNP便于合成和功能化。此外，它们非常惰性且无毒。大量研究已表明，由于GNP可以很好地被包装，因此干细胞对其有较好的耐受性，而且可以通过GNP递送核酸或其他生物分子来诱导干细胞分化[50]。目前GNP的合成已经有了一系列方法。例如，最常用的方法是柠檬酸合成法，涉及使用柠檬酸三钠还原氯金酸。这个方法所得到的纳米颗粒尺寸主要由盐的浓度、温度和加入速率决定，一般的直径范围为10~25nm。若要获得范围为1~100nm或更大的颗粒也可以通过改变盐的浓度和温度来实现[51]。利用GNP传递药物或基因已经进行了不少研究，包括用GNP共价连接基因或药物、非共价连接聚乙烯亚胺（polyethylenimine，PEI）和核酸等，均已被证明可成功用于干细胞的基因工程改造中[52]。

（三）磁性纳米颗粒

近年来全球研究者对磁性纳米颗粒（magnetic nanoparticle，MNP）在干细胞工程的干细胞中应用表现出了很大的兴趣。主要由于MNP有许多独特的性能，如高生物相容性、表面易改性、磁性等，特别是磁性可以应用于核磁共振成像（MRI）增强显影并用于下一步的诱导热疗[53,54]。还有结果显示，MNP与干细胞具有很好的生物相容性，可以通过强磁场提高转染效率[55]。MNP，如最常见的磁性纳米四氧化三铁颗粒，通常是通过在基本水介质或热分解中Fe^{2+}和Fe^{3+}共沉淀合成离子，从而可以获得更均匀和更高度结晶的结构[56]。此外，将MNP与其他金属离子如Zn^{2+}、Mn^{2+}混合可以大大提高所产生的磁性纳米颗粒的磁性，这对于之后的应用十分重要，如MRI对比度增强4~14倍，可用于监测干细胞迁移，治疗癌症的热疗效果增强4倍[57]。一般来说，和GNP一样，这些MNP包被有生物相容性聚合物，如葡聚糖、葡聚糖衍生物或聚乙二醇[poly(ethylene glycol)，PEG]，从而赋予生物系统稳定性。此外，核酸、生物分子和小分子药物可以通过共价键或非共价键与颗粒结合（如PEI通过静电相互作用）。由于它们具有巨大的潜力，FDA已经批准了一些应用MNP的临床研究，其中作为MRI的增强技术是主要领域。最近关于磁性颗粒的研究集中在磁性核壳纳米颗粒（magnetic core-shell nanoparticle，MCNP）的发展，给MNP包裹一个外壳，如黄金或介孔二氧化硅，可提供额外的功能[58]。因MNP的多功能性和易改造特点，其在干细胞工程领域未来具有巨大的潜力，并将成为基因工程的有力工具。

第四节 研究进展

以MSC为基础的基因治疗已经研究了接近20年，随着周边学科，包括纳米技术、影像示踪技术、基因组工程等领域的发展，MSC基因治疗的方式和方向都在不断完善与扩大，本书的第一版曾在数年前对这方面的研究进行过梳理和总结，今天我们将进行简单的回顾并介绍近年来新的研究成果。

一、细胞来源

在之前的研究中，小鼠、大鼠、人的 MSC 使用最多，这是因为以人类临床为目标的研究始终是研究者的关注热点。但是在另外一个领域，也就是兽医领域，同样存在着巨大的治疗需求，特别是竞赛动物和城市伴侣动物在近年来也成为 MSC 基因治疗的应用领域。例如，Seo 等[59]通过分离扩增犬脂肪间充质干细胞（AD-MSC），而后使其携带 IFN-β 用于治疗黑色素瘤获得很好的效果。治疗组在肿瘤体积、生存期、免疫组化结果等方面均优于对照组。此研究证实了犬 MSC 同样可使用现有的慢病毒载体技术进行基因工程改造而用于动物临床治疗。Ahn 等[60]也进行了类似的研究，同样确证了携带 IFN-β 的犬 MSC 对犬黑色素瘤有显著治疗作用。这将为未来犬类动物肿瘤的治疗提供新的选择，特别是改变目前伴侣犬肿瘤治疗除传统化疗药物和手术外缺乏有效手段的情况。

Petersen 等[61]通过研究发现，慢病毒载体可以有效地转染马 AD-MSC，同时病毒载体的使用对干细胞的干性、多向分化潜能、增殖能力、细胞活率等没有显著的影响。另外，外源基因可以获得与人细胞中同样的长时间稳定表达效果，这就为利用马 MSC 进行基因治疗确立了良好的基础。众所周知，马作为竞赛动物经常出现关节和肌腱损伤，在之前的 30 年中，外周血和脂肪来源的干细胞已经被用于马的赛后治疗和恢复，其效果已经得到了业内公认，相信未来通过基因工程技术的改造，我们可以获得治疗活性更高的干细胞产品，并在治疗中获得更好的治疗效果。

二、工具基因的选择

肿瘤作为人类健康的第一大杀手，利用干细胞进行基因治疗也是热点研究方向。首先要介绍的是可用于治疗肿瘤的工具基因。

（一）自杀基因

将能编码某些药物敏感酶的基因转导入干细胞，产生的某些酶类将低毒或无毒的药物前体转化为细胞毒效应产物，从而杀伤肿瘤细胞。例如，把单纯疱疹病毒的胸腺嘧啶激酶（HSV-tk）基因或大肠杆菌的胞嘧啶脱氨酶（cytosine deaminase, CD）基因转染至干细胞，使干细胞表达 HSV-tk 或 CD，能使无毒的前体药物丙氧鸟苷（GCV）或 5-FU 转化为对分裂细胞有杀伤作用的代谢产物，通过直接杀伤或旁杀伤效应消除肿瘤[62]。

（二）免疫调节因子

通过激发或增强人体的抗肿瘤免疫功能来达到治疗肿瘤的目的。例如，通过干细胞将 CX3CL1 基因携带至肿瘤周围表达后可以增强细胞毒性 T 细胞（CTL）和 NK 细胞向肿瘤病灶趋化的能力，从而提高机体对肿瘤细胞的杀伤作用[63]。其他研究中，使用干细胞在肿瘤病灶局部表达 IL-2、IL-4、IL-12、IL-23 等具有免疫调节功能的细胞因子，通过调节机体肿瘤部位的免疫反应，从而达到利用机体自身免疫系统杀灭肿瘤的效果。

（三）诱导凋亡分子

通过病毒载体的感染使干细胞表达肿瘤坏死因子相关凋亡诱导配体（tumor necrosis factor related apoptosis-induced ligand，TRAIL）、IFN-α 或 IFN-β，而后发挥这些分子对肿瘤细胞的诱导凋亡作用，使干细胞周围的肿瘤细胞大量凋亡，从而达到缩小肿瘤病灶、减缓肿瘤发展的目的[64,65]。

（四）抗血管新生分子

通过 MSC 释放一些分子达到抑制肿瘤内血管新生的目的，一方面可以减少肿瘤的血供，由此使肿瘤病灶处于缺血状态而生长缓慢，另一方面可以降低肿瘤细胞通过血管转移的可能性[66]。

（五）miRNA

miRNA 作为 21 世纪的重要科学发现，使我们了解到一个全新的基因调控方式，其在发育、肿瘤发生过程中发挥着十分重要的作用，这也使得 miRNA 成为肿瘤治疗的效应分子。Katakowski 等[67]通过质粒将 miRNA-146b 基因转染入 MSC，而后在大鼠模型中用于治疗胶质瘤并获得了显著的效果。治疗过程中显著地抑制了肿瘤细胞的生长。更为有趣的是，外泌体作为 MSC 重要的旁分泌手段被单独关注，通过检测发现：过表达的 miRNA-146b 被包装入外泌体中，通过收集 MSC 所释放的外泌体可以获得一种不含细胞成分的治疗方式，这种方式虽然无法利用 MSC 向肿瘤组织迁移归巢的特性，但是由于其成分明确简单，易于通过安全性评价，短时间内更具有临床转化的可能。此外，对于很多实体肿瘤，可以通过穿刺注射的方式直接给药，也间接克服了其不具备归巢能力的缺点，在临床使用中仍然具有十分广阔的应用前景[68]。

（六）双功效分子

以蛋白质融合技术为基础，诸多具有更强生物学效力的效应分子被制备出来，在肿瘤的免疫细胞治疗方面已经取得了显著的成果，在 MSC 基因工程方面，研究者也开始考虑构建多功效的效应分子以获得更好的治疗效果。例如，Yan 等[69]的研究中设计了一个有前途的双靶向治疗系统，用于治疗非霍奇金淋巴瘤（non-Hodgkin lymphoma，NHL）。该系统基于一种新的分泌型 scfvCD20-sTRAIL 融合蛋白，其中 CD20 特异性单链抗体片段（scFv）和可溶性 TNF 相关凋亡诱导配体（TRAIL、AA 残基 114~281）与异亮氨酸拉链（Z）添加到末端（ISZ sTRAIL），将这个蛋白分子在人脐带间充质干细胞（hUC-MSC）中表达。与 sTRAIL 相比，新的蛋白 scfvCD20-sTRAIL 表现出更明显的抑制增殖作用，特别是对于 CD20 阳性 B 细胞，而对 CD20 阴性细胞的抑制作用较弱，且对正常人外周血单个核细胞（PBMC）无影响。此研究表明，scfvCD20-sTRAIL 分泌干细胞是一种有效的治疗 NHL 的方法。这种方法通过 CD20 抗体进一步强化了干细胞归巢后的靶向性和特异性，而后通过 TRAIL 有效地诱导肿瘤细胞凋亡，这种方式一方面可以提高治疗的靶向性和特异性，另一方面可以有效降低干细胞的用量，这也就为降低治疗成本和相关治疗风险提供了可能。此类研究代表了未来干细胞治疗的重要方向,随着更多的融合蛋白分子被构建,MSC基因治疗将迈上一个新的台阶。

三、MSC在肿瘤基因治疗中的进展

在本书第一版出版后的几年中,使用MSC对肿瘤进行基因治疗始终是研究的重点方向,如Zhang等[70]使用携带*IL-21*基因的慢病毒载体转染人脐血间充质干细胞(hUCB-MSC),之后裸鼠上治疗卵巢癌,在观察到对肿瘤细胞有抑制作用的同时,还观察到IL-21对脾细胞增殖有促进作用。

在Dwyer等[71]的研究中,并未选用具有治疗功能的基因,而是将钠碘转运体(sodium-iodide symporter, *NIS*)基因导入MSC而后用于成像和治疗乳腺癌。荷瘤动物接受静脉注射或瘤内注射(MSC-NIS),而后给治疗组动物腹腔注射I^{131},在之后的8周监测肿瘤体积。注射后3天,影像可见一个较弱的肿瘤图像。14天后可见清晰的肿瘤影像,其中有放射性核素积聚,但肿瘤的体积在放射线作用下相比对照组显著减小。此研究一方面通过放射性核素进一步证实了MSC的迁移归巢能力,另一方面对研究MSC注射后在体内的代谢迁移过程,以药物代谢动力学的观点研究干细胞提供了新的方法。同时也证实了通过合理控制I^{131}的剂量和给药时间可以建立有效的肿瘤体内放疗方法,尤其对转移灶的治疗有着重要的意义。

四、非肿瘤疾病的治疗进展

非肿瘤疾病的治疗在过去的几年中获得了巨大关注,特别是再生医学。

(一)自体免疫系统疾病

MSC以其突出的免疫调节活性成为自体免疫疾病的重要治疗手段,特别是一些严重、难治性自体免疫系统疾病,而如何进一步提高MSC的免疫调节功能已经受到了关注。Amari等[72]在其研究中,使用慢病毒载体将*IL-35*基因转入UC-MSC中,此后检测IL-35的表达水平和生物活性,由于IL-35可以诱导调节性T细胞(Treg)增殖和降低Th17和Th1细胞的活性,因此IL-35的局部表达可以带来更好的治疗效果。

(二)软骨损伤

Cucchiarini等[73]在研究中通过AAV病毒载体使MSC携带胰岛素样生长因子-1(*IGF-1*)基因,以提高MSC对早期软骨缺损的修复作用。在结果中可以看到,相比于对照组,MSC-IGF显示出了显著的促进软骨组织再生的作用,同时对软骨下骨的损伤修复也有一定的促进作用。这意味着将来我们可以通过基因工程的方式获取更为有效的关节炎治疗方法。

(三)神经系统疾病

Yin等[74]尝试使用慢病毒载体使MSC携带*PSPN*基因(一种神经营养因子配体),证明该基因具有促进特定神经元细胞存活的作用。对大鼠的帕金森病模型进行治疗后,

以旋转行为作为指标，可以观察到治疗组病情得到明显改善，同时高效液相层析（high performance liquid chromatography，HPLC）检测发现，治疗组中多巴胺（dopamine，DA）含量明显升高。除此之外，MSC 还被用于治疗阿尔茨海默病等退行性神经疾病。具体来说，胶质细胞源性神经营养因子（GDNF）、神经生长因子（NGF）和脑源性神经营养因子（BDNF）已在帕金森病、阿尔茨海默病和亨廷顿病的治疗中使用，并取得了初步的效果[75]。

（四）勃起障碍

针对勃起障碍（erectile dysfunction，ED）的基因治疗意在解决传统治疗方法的局限性，目前的策略包括向阴茎海绵体内注射 BM-MSC 和治疗基因，如内皮型一氧化氮合酶（eNOS）基因或血管内皮生长因子（VEGF）基因，此外还包括 FGF 基因等，通常是通过使用腺病毒载体转染 BM-MSC 从而修复和再生组织。此外，它们不诱导局部免疫反应，是稳定和安全的。因此，基因治疗是一种很有前途的 ED 治疗策略[76]。

（五）辐射损伤

Hu 等[77]在其研究中对接受了剂量为 4.5Gy 的 Co^{60} 照射小鼠注射 Trx-1 基因过表达的 MSC，之后观察到治疗组获得了更强的抗氧化效果，保护了骨髓造血干细胞（BM-HSC），促进了红细胞的生成和血红蛋白水平的提高，同时减少了炎症对重要器官（骨髓、肺、肝和小肠）的损害，最为重要的是延长了生存期。

（六）胰腺炎

Hua 的研究中使用慢病毒转染 MSC，使之表达血管生成素-1（Ang-1），利用该蛋白调节内皮细胞存活、血管稳定和血管新生的重要作用在大鼠胰腺炎模型上进行治疗尝试。治疗组可以观察到 MSC 注射后显著降低了胰腺损伤和炎症水平，血清淀粉酶和脂肪酶水平明显下降，并抑制了促炎细胞因子包括 TNF-α、IFN-γ、IL-1β 和 IL-6 的血清水平。此外，MSC 治疗还促进了胰腺血管生成。这些积极结果表明，基因工程 MSC 在胰腺炎的治疗中可以发挥重要的作用[78]。

（七）血友病

血友病是一种 X 连锁出血性疾病，血友病患者缺乏生物活性凝血因子。Watanabe 等[79]的研究中，通过慢病毒载体高效率转染小鼠 AD-MSC，使之合成分泌凝血因子Ⅸ（hFIX），使用 1~60 的 MOI，通过优化得到了最佳的 MOI 为 10，感染后 hFIX 蛋白分泌持续超过 28 天。在此期间可以有效地控制血友病的病情，获得满意的凝血活性，此方法为进一步开发长效的血友病治疗方法提供了经验。

第五节　总结与展望

在整理此部分内容的时候，笔者最直接的感受就是，随着周边学科的发展，在短短几年中，利用基因工程 MSC 治疗疾病的研究上取得了大量的成果，最为重要的是在除了肿瘤之外的更多疾病在治疗上进行尝试和探索，这得益于基因组学和蛋白质组学的发展使得我们了解更多基因表达产物的功能和其作用机制，了解更多的信号通路，这都为我们设计出更为合理、高效、安全的基因治疗策略奠定了基础。另外，干细胞独特的归巢特性使之成为目前十分理想的运载工具，上述二者结合共同打造出了一个有着广阔应用前景的技术平台。在这个平台上我们有望为诸多疾病的治疗带来新的思路。

（王　斌　李丽丽）

参 考 文 献

[1] Wei X, Yang X, Han ZP, et al. Mesenchymal stem cells: a new trend for cell therapy. Acta Pharmacol Sin, 2013, 34(6):747-754.

[2] Kim SU, de Vellis J. Stem cell-based cell therapy in neurological diseases: a review. J Neurosci Res, 2009, 87(10):2183-2200.

[3] Segers VF, Lee RT. Stem-cell therapy for cardiac disease. Nature, 2008, 451(7181):937-942.

[4] Thomas ED, Lochte HL Jr, Lu WC, et al. Intravenous infusion of bone marrow in patients receiving radiation and chemotherapy. N Engl J Med, 1957, 257(11):491-496.

[5] Wang YX, Dumont NA, Rudnicki MA, et al. Muscle stem cells at a glance. J Cell Sci, 2014, 127(Pt 21):4543-4548.

[6] Cheng Z1, Ou L, Zhou X, et al. Targeted migration of mesenchymal stem cells modified with CXCR4 gene to infarcted myocardium improves cardiac performance. Mol Ther, 2008, 16(3):571-579.

[7] Uccelli A, Moretta L, Pistoia V. Mesenchymal stem cells in health and disease. Nat Rev Immunol, 2008, 8(9):726-736.

[8] Rehman J, Li J, Orschell CM, et al. Peripheral blood "endothelial progenitor cells" are derived from monocyte/macrophages and secrete angiogenic growth factors. Circulation, 2003, 107(8):1164-1169.

[9] Robinton DA, Daley GQ. The promise of induced pluripotent stem cells in research and therapy. Nature, 2012, 481(7381):295-305.

[10] Barker N, Bartfeld S, Clevers H. Tissue-resident adult stem cell populations of rapidly self-renewing organs. Cell Stem Cell, 2010, 7(6):656-670.

[11] Martino G, Pluchino S. The therapeutic potential of neural stem cells. Nat Rev Neurosci, 2006, 7(5):395-406.

[12] Aboody KS, Brown A, Rainov NG, et al. Neural stem cells display extensive tropism for pathology in adult brain: evidence from intracranial gliomas. Proc Natl Acad Sci USA, 2000, 97(23):12846-12851.

[13] Pereboeva L, Komarova S, Mikheeva G, et al. Approaches to utilize mesenchymal progenitor cells as cellular vehicles. Stem Cells, 2003, 21(4):389-404.

[14] Ehtesham M, Kabos P, Gutierrez MA, et al. Induction of glioblastoma apoptosis using neural stem cell-mediated delivery of tumor necrosis factor-related apoptosis-inducing ligand. Cancer Res, 2002, 62(24):7170-7174.

[15] Cao F, Lin S, Xie X, et al. *In vivo* visualization of embryonic stem cell survival, proliferation, and migration after cardiac delivery. Circulation, 2006, 113(7):1005-1014.

[16] Tille JC, Pepper MS. Mesenchymal cells potentiate vascular endothelial growth factor-induced angiogenesis *in vitro*. Exp Cell Res, 2002, 280(2):179-191.

[17] Andrades JA, Han B, Becerra J, et al. A recombinant human TGF-beta1 fusion protein with collagen binding domain promotes migration, growth, and differentiation of bone marrow mesenchymal cells. Exp Cell Res, 1999, 250(2):485-498.

[18] Yu J, Ustach C, Kim HR. Platelet-derived growth factor signaling and human cancer. Biochem Mol Biol, 2003, 36(1):49-59.

[19] Wang L, Li Y, Chen X, et al. MCP-1, MIP-1, IL-8 and ischemic cerebral tissue enhance human bone marrow stromal cell migration in interface culture. Hematology, 2002, 7(2):113-117.

[20] Wang L, Li Y, Chen J, et al. Ischemic cerebral tissue and MCP-1 enhance rat bone marrow stromal cell migration in interface culture. Exp Hematol, 2002, 30(7):831-836.

[21] Rempel SA, Dudas S, Ge S, et al. Identification and localization of the cytokine SDF1 and its receptor, CXC chemokine receptor 4, to regions of necrosis and angiogenesis in human glioblastoma. Clin Cancer Res, 2000, 6(1):102-111.

[22] Honczarenko M, Le Y, Swierkowski M, et al. Human bone marrow stromal cells express a distinct set of biologically functional chemokine receptors. Stem Cells, 2006, 24(4):1030-1041.

[23] Ponte AL, Marais E, Gallay N, et al. Human bone marrow stromal cells of human bone marrow mesenchymal stem cells: comparison of chemokine and growth factor chemotactic activities. Stem Cells, 2007, 25(7):1737-1745.

[24] Sullivan C, Barry F, Ritter T, et al. Allogeneic murine mesenchymal stem cells: migration to inflamed joints in vivo and amelioration of collagen induced arthritis when transduced to express CTLA4Ig. Stem Cells Dev, 2013, 22(24):3203-3213.

[25] Fujita K, Kuge K, Ozawa N, et al. Cinnamtannin B-1 promotes migration of mesenchymal stem cells and accelerates wound healing in mice. PLoS One, 2015, 10(12):e0144166.

[26] Kim SM, Kim DS, Jeong CH, et al. CXC chemokine receptor 1 enhances the ability of human umbilical cord blood-derived mesenchymal stem cells to migrate toward gliomas. Biochem Biophys Res Commun, 2011, 407(4):741-746.

[27] Studeny M, Marini FC, Champlin RE, et al. Bone marrow-derived mesenchymal stem cells as vehicles for interferon-beta delivery into tumors. Cancer Res, 2002, 62(13):3603-3608.

[28] Martiniello-Wilks R, Stephen R. Larsen mesenchymal stem cells as suicide gene therapy vehicles for organ-confined and metastatic prostate cancer (PCa). Blood (ASH Annual Meeting Abstracts), 2007, 110:5148.

[29] Rachakatla RS, Pyle MM, Ayuzawa R, et al. Combination treatment of human umbilical cord matrix stem cell-based interferon-beta gene therapy and 5-fluorouracil significantly reduces growth of metastatic human breast cancer in SCID mouse lungs. Cancer Invest, 2008, 26(7):662-670.

[30] Kazuki Y, Oshimura M. Human artificial chromosomes for gene delivery and the development of animal models. Mol Ther, 2011, 19(9):1591-1601.

[31] Partridge KA, Oreffo RO. Gene delivery in bone tissue engineering: progress and prospects using viral and nonviral strategies. Tissue Eng, 2004, 10(1-2):295-307.

[32] Kay MA. State-of-the-art gene-based therapies: the road ahead. Nat Rev Genet, 2011, 12(5):316-328.

[33] Ginn SL, Alexander IE, Edelstein ML, et al. Gene therapy clinical trials worldwide to 2012-an update. J Gene Med, 2013, 15(2):65-77.

[34] Knaän-Shanzer S, van de Watering MJ, van der Velde I, et al. Endowing human adenovirus serotype 5 vectors with fiber domains of species B greatly enhances gene transfer into human mesenchymal stem cells. Stem Cells, 2005, 23(10):1598-1607.

[35] Neumann AJ, Schroeder J, Alini M, et al. Enhanced adenovirus transduction of hMSCs using 3D hydrogel cell carriers. Mol Biotechnol, 2013, 53(2):207-216.

[36] Inagaki K, Piao C, Kotchey NM, et al. Frequency and spectrum of genomic integration of recombinant adeno-associated virus serotype 8 vector in neonatal mouse liver. J Virol, 2008, 82(19):9513-9524.

[37] Donsante A, Miller DG, Li Y, et al. AAV vector integration sites in mouse hepatocellular carcinoma. Science, 2007, 317(5837):477.

[38] Nathwani AC, Tuddenham EG, Rangarajan S, et al. Adenovirus-associated virus vector-mediated gene transfer in hemophilia B. N Engl J Med, 2011, 365(25):2357-2365.

[39] Ho YC, Lee HP, Hwang SM, et al. Baculovirus transduction of human mesenchymal stem cell-derived progenitor cells: variation of transgene expression with cellular differentiation states. Gene Ther, 2006, 13(20):1471-1479.

[40] Ho YC, Lee HP, Hwang SM, et al. Transgene expression and differentiation of baculovirus-transduced human mesenchymal stem cells. J Gene Med, 2005, 7(7):860-868.

[41] Zeng J1, Du J, Zhao Y, et al. Baculoviral vector-mediated transient and stable transgene expression in human embryonic stem cells. Stem Cells, 2007, 25(4):1055-1061.

[42] Mintzer MA, Simanek EE. Nonviral vectors for gene delivery. Chem Rev, 2009, 109(2):259-302.

[43] Lee DE, Koo H, Sun IC, et al. Multifunctional nanoparticles for multimodal imaging and theragnosis. Chem Soc Rev, 2012, 41(7):2656-1672.

[44] Alexis F, Pridgen EM, Langer R, et al. Nanoparticle technologies for cancer therapy. Handb Exp Pharmacol, 2010, (197):55-86.

[45] Yin H, Kanasty RL, Eltoukhy AA, et al. Non-viral vectors for gene-based therapy. Nat Rev Genet, 2014, 15(8):541-555.

[46] Putnam D. Polymers for gene delivery across length scales. Nat Mater, 2006, 5(6):439-451.

[47] Mintzer MA, Simanek EE. Nonviral vectors for gene delivery. Chem Rev, 2009, 109(2):259-302.

[48] Wasungu L, Hoekstra D. Cationic lipids, lipoplexes and intracellular delivery of genes. J Control Release, 2006, 116(2):255-264.

[49] Lonez C, Vandenbranden M, Ruysschaert JM. Cationic liposomal lipids: from gene carriers to cell signaling. Prog Lipid Res, 2008, 47(5):340-347.

[50] Patel S, Jung D, Yin PT, et al. NanoScript: a nanoparticle-based artificial transcription factor for effective gene regulation. ACS Nano, 2014, 8(9):8959-8967.

[51] Tiwari PM, Vig K, Dennis VA, et al. Functionalized gold nanoparticles and their biomedical applications. Nanomaterials (Basel), 2011, 1(1):31-63.

[52] Zhou J, Ralston J, Sedev R, et al. Functionalized gold nanoparticles: synthesis, structure and colloid stability. J Colloid Interface Sci, 2009, 331(2):251-262.

[53] Yin PT, Shah BP, Lee KB. Combined magnetic nanoparticle-based microRNA and hyperthermia therapy to enhance apoptosis in brain cancer cells. Small, 2014, 10(20):4106-4112.

[54] Kim J, Piao Y, Hyeon T. Multifunctional nanostructured materials for multimodal imaging, and simultaneous imaging and therapy. Chem Soc Rev, 2009, 38(2):372-390.

[55] Sapet C, Laurent N, de Chevigny A, et al. High transfection efficiency of neural stem cells with magnetofection. Biotechniques, 2011, 50(3):187-189.

[56] Jun YW, Lee JH, Cheon J. Chemical design of nanoparticle probes for high-performance magnetic resonance imaging. Angew Chem Int Ed Engl, 2008, 47(28):5122-5135.

[57] Jang JT, Nah H, Lee JH, et al. Critical enhancements of MRI contrast and hyperthermic effects by dopant-controlled magnetic nanoparticles. Angew Chem Int Ed Engl, 2009, 48(7):1234-1238.

[58] Shah BP, Pasquale N, De G, et al. Core-shell nanoparticle-based peptide therapeutics and combined hyperthermia for enhanced cancer cell apoptosis. ACS Nano, 2014, 8(9):9379-9387.

[59] Seo KW, Lee HW, Oh YI, et al. Anti-tumor effects of canine adipose tissue-derived mesenchymal stromal cell-based interferon-b gene therapy and cisplatin in a mouse melanoma model. Cytotherapy, 2011, 13(8):944-955.

[60] Ahn Jo, Lee Hw, Seo Kw, et al. Anti-tumor effect of adipose tissue derived-mesenchymal stem cells expressing interferon-β and treatment with cisplatin in a xenograft mouse model for canine melanoma. PLoS One, 2013, 8(9):e74897

[61] Petersen GF, Hilbert B, Trope G, et al. Efficient transduction of equine adipose-derived mesenchymal stem cells by VSV-G pseudotyped lentiviral vectors. Res Vet Sci, 2014, 97(3):616-622.

[62] Shimato S, Natsume A, Takeuchi H, et al. Human neural stem cells target and deliver therapeutic gene to experimental leptomeningeal medulloblastoma. Gene Ther, 2007, 14(15):1132-1142.

[63] Xin H, Kanehira M, Mizuguchi H, et al. Targeted delivery of CX3CL1 to multiple lung tumors by mesenchymal stem cells. Stem Cells, 2007, 25(7):1618-1626.

[64] Ren C, Kumar S, Chanda D, et al. Therapeutic potential of mesenchymal stem cells producing interferon-α in a mouse melanoma lung metastasis model. Stem Cells, 2008, 26(9):2332-2338.

[65] Studeny M, Marini FC, Dembinski JL, et al. Mesenchymal stem cells: potential precursors for tumor stroma and targeted-delivery vehicles for anticancer agents. J Natl Cancer Inst, 2004, 96(21):1593-1603.

[66] Kim SK, Cargioli TG, Machluf M, et al. PEX-producing human neural stem cells inhibit tumor growth in a mouse glioma model. Clin Cancer Res, 2005, 11(16):5965-5970.

[67] Katakowski M, Buller B, Zheng X, et al. Exosomes from marrow stromal cells expressing miR-146binhibit glioma growth. Cancer Lett, 2013, 335(1):201-204.

[68] Chen J, Li C, Chen L. The role of microvesicles derived from mesenchymal stem cells in lung diseases. Biomed Res Int, 2015, (2015):985814.

[69] Yan C, Li S, Li Z, et al. Human umbilical cord mesenchymal stem cells as vehicles of CD20-specific TRAIL fusion protein delivery: a double-target therapy against Non-Hodgkin's lymphoma. Mol Pharm, 2013, 10(1):142-151.

[70] Zhang Y, Wang J, Ren M, et al. Gene therapy of ovarian cancer using IL-21-secreting human umbilical cord mesenchymal stem cells in nude mice. J Ovarian Res, 2014, 7:8

[71] Dwyer RM, Ryan J, Havelin RJ, et al. Mesenchymal stem cell (MSC) mediated delivery of the sodium iodide symporter (NIS) supports radionuclide imaging and treatment of breast cancer. Stem Cells, 2011, 29(7):1149-1157.

[72] Amari A, Ebtekar M, Moazzeni SM, et al. *In vitro* generation of IL-35-expressing human Wharton's jelly-derived mesenchymal stem cells using lentiviral vector. Iran J Allergy Asthma Immunol, 2015, 14(4):416-426.

[73] Cucchiarini M, Madry H. Overexpression of human IGF-I via direct rAAV-mediated gene transfer improves the early repair of articular cartilage defects *in vivo*. Gene Therapy, 2014, 21(9):811-819.

[74] Yin X, Xu H, Jiang Y, et al. The effect of lentivirus-mediated PSPN genetic engineering bone marrow mesenchymal stem cells on Parkinson's disease rat model. PLoS One, 2014, 9(8):e105118.

[75] Wyse RD, Dunbar GL, Rossignol J. Use of genetically modified mesenchymal stem cells to treat neurodegenerative diseases. Int J Mol Sci, 2014, 15(2):1719-1745.

[76] Kim JH, Lee HJ, Song YS. Mesenchymal stem cell-based gene therapy for erectile dysfunction. Int J Impot Res, 2016, 28(3):81-87.

[77] Hu J, Yang Z, Wang J, et al. Infusion of Trx-1-overexpressing hucMSC prolongs the survival of acutely irradiated NOD/SCID mice by decreasing excessive inflammatory injury. PLoS One, 2013, 8(11):e78227.

[78] Hua J, He ZG, Qian DH, et al. Angiopoietin-1 gene-modifed human mesenchymal stem cells promote angiogenesis and reduce acute pancreatitis in rats. Int J Clin Exp Pathol, 2014, 7(7):3580-3595.

[79] Watanabe N, Ohashi K, Tatsumi K, et al. Genetically modified adipose tissue-derived stem/stromal cells, using simian immunodeficiency virus-based lentiviral vectors, in the treatment of hemophilia B. Hum Gene Ther, 2013, 24(3):283-294.

第二十章 间充质干细胞的分子影像学

第一节 干细胞的分子影像学

干细胞移植为治愈糖尿病、缺血性心脏病等退行性病变提供了可能。但在临床和实验研究中发现细胞移植的治疗效果并不理想，并且还有许多问题需要解决。其中，干细胞移植后在体内发生什么过程、干细胞的作用机制是旁分泌机制还是其增殖分化产生新的功能细胞仍是目前争论和研究的焦点。另外，干细胞移植后的长期存活率不高，也是制约干细胞应用的一个瓶颈。因而，需要进一步的关于机制临床和基础研究，对细胞移植中各重要环节进行更加统一的对比研究，包括：①何种干细胞移植效果最好；②如何追踪干细胞在活体内的行踪及结局；③干细胞移植的时机，包括急性期立即应用还是缓解以后应用；④干细胞移植的数量、移植途径；⑤干细胞修复损伤器官、改善功能的机制；⑥干细胞移植的安全性评价，能否在开展临床试验的同时充分保证患者安全。

干细胞移植后的监测一直存在问题，传统的实验方法需要在实验的终末期处死动物，进行免疫组化鉴定。然而，在实验过程中，移植细胞的动态变化难以被量化。利用分子影像技术长时间地动态观察干细胞在体内的归巢、迁移、分布、增殖、分化，以及其与其他细胞的相互作用可以解决上述难题。

第二节 分子影像学方法

分子影像学是医学影像技术和分子生物学、化学、物理学、放射医学、核医学及计算机科学相结合的一门新兴交叉学科。它是以体内特定分子为成像对比度源，利用现有的一些医学影像技术对人体内部生理或病理过程在分子水平上进行无损伤的、实时的成像。它将遗传基因信息、生物化学与新的成像探针进行综合，由精密的成像技术来检测，再通过一系列的图像后处理技术，达到显示活体组织在分子和细胞水平上的生物学过程的目的。分子成像主要是基于分子探针（molecular probe），生物体内的分子探针不同于基础研究的分子探针，在体内分子探针具有高度的亲和性、特异性和无毒性和超强的穿透生物屏障能力，并且便于检测。通常可根据对分子过程的靶向策略，将分子影像探针分为靶向性探针和可激活探针两类。目前常用的分子影像技术主要包括光学成像、核磁共振成像（MRI）、放射性核素成像[1]。

一、光学成像

光学成像技术是将光学过程与一定的分子性质相结合，用于分子和细胞生物学中的体外和体内应用。光学成像具有较高的空间分辨率，其原理是光在生物体内的迁徙规律。光学成像主要包括 3 个过程，即光吸收、光散射和荧光发射，不同的光学成像技术在光源、分子探测、成像对象等方面各有特点，在生物医学的应用也各有不同。

目前，常用的两种光学成像技术，即生物发光成像（bioluminescence imaging，BLI）和荧光成像（fluorescence imaging，FLI）技术已经被应用于活体中示踪移植间充质干细胞（MSC）[2-4]。BLI是通过探测荧光物质发出的荧光来进行成像的，与FLI相比，二者都存在荧光物质的电子从基态到激发态跃迁而发出光能的过程，但二者的主要区别在于电子是如何达到激发态的。在荧光发出过程中，激发电子的能量来自于光谱中近红外线、可见光、紫外线等的电磁辐射；而在生物发光中，激发电子的能量是由化学反应的热能提供的。目前，BLI的主要代表就是萤光素酶（luciferase）报告基因成像。常用到的一类酶为萤火虫萤光素酶和叩头虫萤光素酶，这两种酶的底物都是右旋萤光素（D-luciferin）[5-8]。BLI中还经常用到的一种酶类称为海肾萤光素酶和海洋桡角类萤光素酶，它们通常以腔肠素（coelenterazine）作为底物[5-8]。萤光素酶基因成像通常用于活体动物（如小鼠）实验中，萤光素脂溶性非常好，很容易随着血液循环全身，每注射一次萤光素酶底物能够使体内被标记细胞发光30min左右。由于萤光素酶催化底物反应每次只能产生一个光子，这种情况下人的肉眼是无法观察到的，但用一个高度灵敏的低温电荷耦合器件（charge coupled device，CCD）相机及特别设计的成像暗箱和成像软件，就可以观测并记录到这些光子。BLI具有灵敏度高、背景低及信号强度与细胞数量呈线性关系等优点，所以在基础医学研究中应用十分广泛。而FLI是利用特异性的荧光分子探针标记特定MSC，当采用特定波长的激发光激发荧光分子时就会产生荧光，所产生的荧光可以通过特定荧光显微镜观察[9,10]。用于标记的探针是生物学和医学研究中已经大量使用的荧光标志物，其空间分辨率能达到毫米级。FLI具有灵敏度高、时间分辨率高、简便快捷、费用低和通量相对高等诸多优点，所以近些年这项技术发展迅猛。

二、核磁共振成像

1973年，核磁共振成像（MRI）技术首次被提出[11]，随后被广泛应用于生物医学领域。MRI由于具有分辨率高、无放射性损伤和不受组织深度限制等优点，在分子影像学中应用具有其他影像学技术不可比拟的优越性，已成为分子影像学的重要方法和技术。目前，MRI常用的探针是钆（Gd）和磁性纳米颗粒（magnetic particle），如金钆螯合物和超顺磁性氧化铁（superparamagnetic iron oxide，SPIO）纳米颗粒[12]。根据不同的核磁共振探针类型，通常会采用不同的成像序列，以钆类为探针的对比剂，以测量组织的T1值为主，而以氧化铁为探针的对比剂，主要测量组织的T2值[13]。早期的MRI以钆类分子探针为主，这类探针依据抗原-抗体或配体-受体特异性结合原理。然而，钆类分子探针在成像时所需要的浓度比较高，当检测某些微量的靶点时，通常无法达到检测所需的钆类离子浓度。近年来，将钆类探针与高密度脂蛋白、脂质体、胶束等大分子物质结合，不仅能够延长探针的半衰期，而且具有更大的纵向弛豫率[14-16]。磁性纳米颗粒指由氧化铁核心或其他以铁和铁的化合物为主要磁性物质的纳米颗粒，比钆类探针具有更强的磁共振信号，而且磁性纳米颗粒也具有更高的弛豫率[17]。随着分子影像技术的快速发展，其他新型的探针也不断被引入到MRI中。例如，^{19}F对比剂在MRI中，它具有更高的灵敏度、较小的干扰信号、不易被代谢等优点[18]。全氟化碳（perfluorocarbon，PFC）是目前使用最广泛的磁共振探针，在现阶段的研究中，其他的含氟化合物如氟西汀和氟比洛芬等也可替代PFC作为MRI的探针[19]。然而，以^{19}F为探针的MRI技术仍然处于早期发展阶段，还存在信噪较低和成像时间长等缺点。广义的 MRI 还包括磁共振波谱成像和功能磁共振成像技术,这些 MRI 技术不需要相应的对比剂或特异性的分子探针,可以直接反映活体分子、细胞水平的变化。

三、放射性核素显像

放射性核素显像（radionuclide imaging）主要包括正电子发射断层成像（positron emission tomography, PET）和单光子发射计算机断层成像（singlephoton emission computed tomography, SPECT）。PET 主要是根据放射性示踪剂摄取变化来成像,18-氟脱氧葡萄糖（^{18}F-FDG）是最常用的放射性示踪剂,其检测原理基于病变葡萄糖有氧代谢活性的变化[20]。自 1998 年美国财政管理局（HCFA）将 ^{18}F-FDG PET 肿瘤显像纳入医保范围,^{18}F-FDG PET 已经成为敏感的、有效的、广为接受的成像技术[21]。在 SPECT 成像,通常用到的放射性同位素是 ^{99}Tc。由于 PET 和 SPECT 的灵敏度高达皮摩尔数量级,因此可以量化示踪剂放射性同位素,从而对活体中的生物进程进行研究。此外,许多报告基因已经被用于 PET 和 SPECT。例如,钠-碘同向载体作为一个报告基因已被用于心肌的显像[22]。报告基因在分子影像中有着它们自身的特点,而且将治疗基因连接于报告基因后,报告基因系统还能够同时监测 MSC 在体内的动态变化和基因治疗情况。

第三节　间充质干细胞的标记方法

使用多种成像模式对 MSC 示踪,依赖于分子影像技术的多功能性。目前,用于标记 MSC 的方法主要分两种,即直接标记法和间接标记法（图 20-1）[23]。

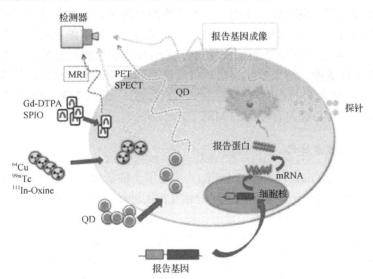

图 20-1　直接标记和间接标记的概括图（彩图请扫封底二维码）

左侧是直接标记常用的示踪剂;右侧是间接标记、报告基因和外源底物;MRI. 核磁共振成像,PET. 正电子发射断层成像,SPECT. 单光子发射计算机断层成像,Gd-DTPA. 二乙烯三胺五乙酸钆,SPIO. 超顺磁性氧化铁,QD. 量子点,^{64}Cu. 铜离子,^{99m}Tc. 放射性核素锝,^{111}In-Oxine. 放射性核素铟标记的 8-羟基喹啉

直接标记的方法主要依靠分子探针对细胞实施标记,该类探针主要包括用于 MRI 的

铁氧化物，^{18}F-hexafluorobenzene 或 ^{18}F-FDG 及 SPECT 的 ^{111}In-oxine 或 ^{111}In-tropolone[24-26]。在实验研究中，MRI 虽然在灵敏度和精确度上还有不足之处，但是这种非侵入性成像方式在很大程度上已经为 PET 临床研究提供了一种选择。MSC 的另一种标记方法是间接标记法，这种方法主要依赖于向 MSC 转入报告基因，利用报告基因的表达对 MSC 进行成像。

一、直接标记

MSC 的直接标记实质上是将特定的分子或物质直接引入细胞，将用这种方法标记的干细胞移植入体内，可以通过不同的成像技术追踪移植 MSC 的分布和去向，这是 MSC 标记最常用的方法。这种标记方法比较简单，通常不会涉及细胞基因修饰，但存在的缺点是在细胞分裂过程中标志物容易被稀释，从而导致单个细胞标记数量减少，同时标志物可能由于分布不对称而在子代细胞中丢失。因此，直接标记法目前在体内实验中主要用于观察 MSC 进入体内后的分布特点及归巢情况。现阶段经常采用磁性纳米颗粒（MNP）、放射性核素及纳米颗粒标记 MSC，并通过非侵入性成像技术进行体内的示踪。

（一）磁性颗粒标记

磁性颗粒标记（magnetic particle labeling）主要应用于 MRI 中，因为 MRI 具有较高的空间分辨率（全身临床扫描仪范围 50~300μm）和时间分辨率，所以广泛应用于临床前及临床研究中的干细胞示踪。其主要原理是当磁偶极子（如水或有机化合物中氢原子）置于磁场中时，它们能够有规律地排列成一行，这为 MSC 的示踪提供了方便。同时，在缺血性组织中，由于 MSC 富集了大量的造影剂，能够产生足够强的正、负信号，因此它们在体内能够被示踪。现阶段，使用广泛的造影剂有两种，一种包含了二乙烯三胺五乙酸钆（Gd-DTPA），另一种则包含了 SPIO[27]。如果进行短期干细胞跟踪，超顺磁性氧化铁应作为首选。将标记超顺磁性氧化铁菲立磁（Feridex）的骨髓间充质干细胞（BM-MSC）植入大鼠体内进行成像，结果显示移植后可以观察到移植细胞向缺血灶边缘迁移。采用 SPIO 标记 MSC，植入猪的心肌梗死模型[28]，结果表明，通过 MRI 可活体观察到移植细胞，组织学检查证实心脏内有存活的 MSC。

（二）放射性核素标记

放射性核素标记（radionuclide labeling）主要用于核素成像中，通过标记细胞内放射性强度，对生物体内的生化过程和细胞间的相互作用进行监测。核素成像技术主要包括正电子发射断层成像（PET）和单光子发射计算机断层成像（SPECT）。与 MRI 相比，PET 和 SPECT 具有较高的敏感性，因此有更多种类的显像剂可进行使用。随着空间分辨率的提高，核素成像已经特别适合进行细胞跟踪，无论是进行短期细胞跟踪，还是长期监测细胞活性和功能，核素成像都扮演着重要角色。用十六烷基 4-氟苯-18 标记大鼠 MSC 细胞膜上，研究 MSC 治疗大鼠慢性心肌梗死的效果，用显微 PET 造影显示经静脉注射到大鼠体内标记的 MSC[29]。结果显示，能够在心脏处发现大量放射性标记的 MSC。另一研究利用联合单光子发射电脑断层显像（SPECT/CT）扫描仪的高灵敏度，使用放射性造影剂和磁共振造影剂标记异体 BM-MSC，动态追踪其在急性心肌梗死模型体内的分布，静脉注射后 24~48h，肝、肾和脾均可观察到标记的 BM-MSC。因此，这种方法非常适合动态追踪间质干细胞在体内的分布[30]。然而，这种标记 MSC 的方法也存在自身的缺点，如放射性示踪物可能向非靶细胞转移，它们也可能对干细胞的活性、功能和分化能力造成不良影响，这些因素都会限制其应用范围。所以，放射性核素标记 MSC 所产生的效果有待进一步研究。

（三）纳米颗粒标记

纳米颗粒标记（nanoparticle labeling）早期主要使用量子点（quantum dot，QD）这种重要的荧光剂。发光胶体量子点是具有物理性质的无机半导体颗粒，它是能够发出 525~800nm 荧光的无机半导体颗粒。由于其特有的物理性质，量子点的总直径只有 2~10nm，这也是它量子性质的体现[31]。传统的有机标记剂通常只能发射单一波长光，并且具有窄的激发光谱和宽的发射光谱，而量子点具有单一波长的激发光和不同波长的发射光，这一特点使它成为多模态成像的理想探针[32,33]。由于量子点具有较强的亮度和耐光褪色性，较适用于活细胞成像，已经有研究成功利用量子点对 BM-MSC 进行标记，而且 MSC 的体外培养实验已经证明量子点不影响细胞的活性、增殖或分化。量子点可以标记 MSC 分化过程中的多个转录因子，因此利用量子点有助于系统、深入地研究 MSC 的分化机制及调控干细胞的分化[34]。然而，量子点对干细胞功能的长期影响尚不为人知，应该关注其金属芯（钳、镉和硒等）的暴露或溶解可能导致的中毒。

二、间接标记

间接标记是指在 MSC 移植前向细胞内引入报告基因，报告基因可以翻译成酶、受体、荧光或生物荧光蛋白。在注射特定的外源底物后，报告基因产物与底物特异性结合，采用光学成像或 PET 等方法就能够检测干细胞的存活情况和相关基因的表达情况。通常在报告基因表达稳定的情况下，可以观测到标记细胞的全部生命周期。但这种方法操作较为复杂，同时由 DNA 甲基化和组蛋白脱乙酰基导致的表观遗传基因沉默也可能影响报告基因的转录，对报告基因产生的信号具有抑制作用。报告基因标记（reporter gene labeling）通常应用于干细胞治疗的非侵入成像，主要研究植入干细胞的存活、转移和执行功能的效果。对体内基因表达的成像能够指导外源基因转移至器官、系统的细胞（转基因）或者内源基因中，当前大部分报告基因成像都用于指导转基因技术的使用。一般而言，报告基因编码某些易检验蛋白质的 DNA 序列，在成像情况下，报告基因编码的报告蛋白与一个特异性探针结合，产生分析信号，生成的信号能够被如 MRI、PET 和 SPECT 或光学电荷耦合器件（CCD）捕获和量化，所以报告蛋白就为研究基因活动提供了一个替代标记。

报告基因的稳定转染对于评价植入干细胞的生存状态十分重要，因为只要细胞有活性，报告基因就可以表达。在干细胞再生领域，一些传统的报告基因如绿色荧光蛋白（green fluorescent protein，GFP）、萤火虫萤光素酶（firefly luciferase，Fluc）、海肾萤光素酶（renilla luciferase，Rluc）和单纯疱疹病毒胸腺激酶（herpes simplex virus-thymidine kinase，HSV-TK）可以允许在小动物体内进行定位。然而，由于使用报告基因需要对细胞进行基因操作，这可能导致插入突变的发生，因此该技术存在一定的限制因素。在 MSC 移植方面有一个经典的实验[35]，使用三融合报告基因系统（Fluc-RFP-ttk）进行多模态成像，监测 MSC 移植到 NOD/SCID 小鼠后的情况。携带报告基因的 MSC 经过 BLI，信号十分明显，同时通过 PET，可对一个区域内细胞数量定量。在小动物模型中，运用这种 BLI 和 PET 技术对标记干细胞进行研究是相当可靠的方法。

第四节 间充质干细胞治疗的分子影像学

MSC 是干细胞的一个研究分支，干细胞是一类具有自我复制能力和多向分化潜能的细胞，它们可以不断地自我更新，并在特定条件下转变成为一种或多种构成人体组织或器官的细胞。干细胞是机体的起源细胞，是形成人体各种组织器官的始祖细胞。这类干细胞存在于多种组织（如骨髓、脐血和脐带、胎盘、脂肪等），具有多向分化潜能，是非造血干细胞（HSC）的成体干细胞。例如，它们具有向多种间充质系列细胞（如成骨、软骨及脂肪细胞等）或非间充质系列细胞分化的潜能，并具有独特的细胞因子分泌功能。MSC 临床应用于解决多种血液系统疾病、心血管疾病、肝硬化、神经系统疾病、膝关节半月板部分切除损伤修复、自身免疫系统疾病等，并取得了重大突破，挽救了许多病患的生命。2004 年报道了首例半相合异体干细胞移植，该研究是运用 MSC 治疗移植物抗宿主病（GVHD）并获得了成功[36]，其后又报道了异体配型不合的 MSC 移植治疗 GVHD 的有效性，并且认为应用 MSC 治疗 GVHD 时不需要严格的配型，其后又有多篇利用异体未经配型的 MSC 治疗 GVHD、促进造血重建的报道，MSC 来源涉及骨髓、脂肪、牙周等，所以 MSC 已经被广泛运用于组织再生领域。另外，值得注意的是，未分化的 MSC 对间充质来源的原发肿瘤具有抑制其增殖的作用。将 MSC 与组织工程学结合起来，研究者认为利用 MSC 携带一些治疗性的药物或者分子可扩大其在肿瘤治疗领域的运用。如今利用 MSC 携带抗癌分子到原发肿瘤以及转移性肿瘤的研究已经应用到大量的疾病模型中[37]。一些治疗性的分子，包括干扰素（IFN）、白介素（IL）、前药激活的酶、凋亡诱导子及溶瘤病毒已经被 MSC 成功携带并运送到肿瘤部位。这不仅能够选择性地杀伤肿瘤，而且能够有效地克服肿瘤细胞对传统药物产生的抗化疗作用[38]。

分子成像方法的多功能性可以允许使用单模或多模式成像对 MSC 进行示踪[39]，这些方法包括用于 MRI 的氧化铁颗粒进行移植干细胞的直接标记，用于正电子发射断层成像（PET）的 ^{18}F-六氟苯或 ^{18}F-FDG，用于单光子发射计算机断层成像（SPECT）的 ^{111}In-oxin 或 ^{111}In-tropolone 对干细胞标记。无创 MRI 正在迅速成为临床前和临床研究最流行的成像方法，其中使用超顺磁性氧化铁（SPIO）纳米颗粒作为 MRI 对比增强剂可能是示踪 MSC 存活、迁移和监测治疗效果的最佳选择。间接标记依赖于移植前转导入细胞的报告基因的表达。使用报告基因追踪 MSC 移植的典型例子是 Love 等[35]的研究，他们使用三重融合报告系统（Fluc-RFP-ttk）进行多模态成像，以监测移植到 NOD/SCID 小鼠中的人 MSC（hMSC）。来自有报告基因的 hMSC 的立方体信号在 3 个月内通过 BLI 技术监测，另外这些信号也可以通过 PET 技术来进行干细胞数量的估算。运用报告基因标记 MSC，再通过 BLI 和 PET 在小动物模型中示踪 MSC 移植，现已经成为实验研究的可靠方法。

一、MSC 在再生医学治疗中的分子影像

MSC 可以分化成中胚层谱系的细胞，如成骨、脂肪和软骨细胞，但也具有向内胚层和神经外胚层分化的潜能。临床前和临床研究支持在临床上使用从同种异体供体获得的

MSC，在临床前研究中，MSC 已经应用于组织再生，包括造血器官、心脏、中枢神经系统（CNS）、皮肤、肾、肝、肺、关节、眼睛、胰腺和肾小球。根据目前的数据，BM-MSC 首先被提出用于再生医学用治疗。

（一）MSC 联合分子影像治疗心脏疾病

在西方国家，心脏相关疾病是常见的一种致死疾病。尽管这方面的研究取得了一些进展，但仍然有大量的患者无法得到治愈。目前，临床上用于治疗心脏相关疾病的干细胞主要为 BM-MSC，在该领域用于追踪干细胞的成像手段有多种。^{18}F-FDG 探针标记的干细胞，在体内可以被 PET 系统追踪，而 ^{99m}Tc-HMPAO 标记技术可以用于 SPECT 系统的成像研究中，而这些标记技术已被证实不具有细胞毒效应，且降解产物对机体几乎无毒[40]。临床上治疗心肌梗死（MI）的干细胞（如 BM-MSC）主要是通过旁分泌作用促进 MI 区内血管新生来改善 MI 后心功能和抑制心室重构的，故利用分子影像技术检测干细胞移植后 MI 区内血管新生情况是评价干细胞治疗 MI 效果的重要参数之一。已有很多实验室报道了利用 PET/SPECT 或 MRI 技术实时观察体内血管新生的研究，在此类工作中，主要是通过显示内皮细胞中整联蛋白（$\alpha_v\beta_3$）的表达情况来间接反映血管新生进程。在最近的临床研究中，利用 ^{18}F-Galacto-RGD 作为 $\alpha_v\beta_3$ 整联蛋白的探针，通过 PET 系统观察到了心梗患者 MI 区内的血管新生进程[41]。而在另一篇报道中，利用 ^{123}I-Gluco-RGD 探针通过 PET 系统追踪了猪体内的血管新生过程[42]。因此，利用这些分子成像模式，如 $\alpha_v\beta_3$、暴露的磷脂酰丝氨酸（PS）、葡萄糖代谢和细胞外基质（ECM）重塑等，可以更好地了解干细胞治疗后的分子变化（图 20-2）[43]。另外，在小动物实验中，将转基因的脂肪间充质干细胞（AD-MSC$^{Fluc/GFP}$）移植入小鼠 MI 模型中，利用高灵敏度的 BLI 系统可以示踪移植细胞在体内的存活情况[44]，真正实现了连续、实时的观察，收集的数据更连续，更具有说服力。

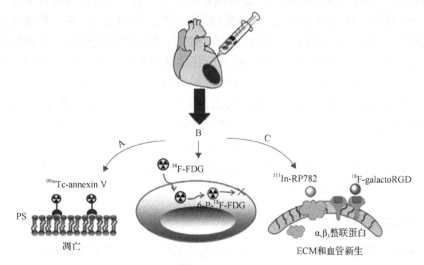

图 20-2 分子影像学方法示踪干细胞治疗后的作用机制（彩图请扫封底二维码）

A. 细胞凋亡可以应用探针 ^{99m}Tc-annexin V 进行 SPECT 成像；B. 细胞移植后的组织代谢可以利用 ^{18}F-FDG 进行 PET 成像；C. 细胞外基质的重构和血管新生可以应用分别靶向结合 MMP 和整合素 $\alpha_v\beta_3$ 的探针 ^{111}In-RP782、^{18}F-galactoRGD 进行成像；^{99m}Tc-annexin V. 放射性核素锝标记的膜联蛋白 V，PS. 磷脂酰丝氨酸，^{18}F-FDG. 18-氟脱氧葡萄糖，^{111}In-RP782. 放射性同位素的铟，^{18}F-galactoRGD. 18-氟标记的半乳糖多肽，ECM. 细胞外基质

(二) MSC 联合分子影像治疗急性肾损伤

尽管最近 10 年的研究使得人们对肾损伤及再生的理解不断加深，但急性肾损伤（acute kidney injury，AKI）患者的治疗手段仍然局限在支持性治疗。其他有效的方法，如肾透析和移植，受到医疗成本较高、并发症发生率高及器官短缺等因素限制。因此，迫切需要开发出新的治疗手段。干细胞具有自我更新和多向分化潜能，这些独特的性质使其成为过去 10 余年成体疾病治疗领域的热点。在与其他干细胞[如胚胎干细胞（ESC）和诱导多能干细胞（iPSC）]比较后，作为成体干细胞的 MSC 的应用研究最为广泛。MSC 最早从骨髓中获得，并应用于基础研究。已有很多研究显示，BM-MSC 具有促进肾脏修复的作用。在后来的研究工作中，人们最先从脂肪组织中分离出干细胞，并描述了其与 BM-MSC 相似的特性（包括形态、表面标志物、多向分化潜能）[45]。这一类细胞称作脂肪间充质干细胞（AD-MSC，也称为脂肪来源基质细胞）。近期的研究表明 AD-MSC 参与肾再生的机制主要与其旁分泌作用有关，通过分泌生长因子或趋化因子，AD-MSC 可以发挥免疫抑制、促增殖、抗凋亡及促进血管新生的功能。将转基因的 AD-MSC$^{Fluc/GFP}$ 移植入 AKI 模型中，利用分子影像学技术实时示踪 AD-MSC 的增殖、存活和迁移情况[46]。同时用 *VEGFR2-luc* 转基因小鼠建立 AKI 模型，探讨肾损伤修复过程中血管新生的机制[46]。

(三) MSC 联合分子影像治疗骨损伤

骨折后的修复过程类似于胚胎期的骨形成，它存在两种骨化形式，即软骨内成骨和膜内成骨。软骨内成骨先由 MSC 聚集并分化成软骨细胞，形成软骨组织，然后由骨组织取代软骨组织，最终实现成骨化[47]。膜内成骨过程中骨骼不需要形成软骨中介，而是由骨膜内 MSC 直接分化为成骨细胞，形成骨组织。显然，骨修复过程中，需要骨折局部聚集数量足够的 MSC。尽管骨折周围内也存在 MSC，但仅能满足轻度骨折或骨缺损的修复需要。对于严重或大块骨缺损，骨缺损部位往往缺乏足够活的 MSC。所以，向骨折或骨缺损部位引起足够的 MSC，是促进骨修复的一个重要策略。有试验研究将体外培养扩增的不同浓度 BM-MSC 通过静脉注入志愿者的体内，结果发现直至输入量达到 $5×10^7$cells/mL 均未发生免疫排斥反应，表明这类细胞可被机体良好地耐受[48]。而另一些实验表明，运用增强型绿色荧光蛋白（EGFP）标记 MSC，将其通过耳缘静脉注入骨骨折模型中，而后在骨断端组织中观察到了 EGFP 阳性细胞，证实了静脉途径给予的 MSC 可以迁移聚集到骨折区域的血肿组织内[49,50]。

二、MSC 在肿瘤治疗中的分子影像

有许多研究表明，MSC 可能抑制肿瘤细胞生长[51]。尤其在胶质瘤的研究中，不同来源的 MSC 表现出对肿瘤生长有抑制和浸润作用[52]。抑制胶质瘤的生长可以被 AD-MSC 分泌的高水平白介素（IL）所调节。联合 IFN-γ 及 G-CSF 能够诱导免疫炎症的级联反应。可是细胞因子之间的平衡是极其复杂的，在低水平 IL 的情况下，G-CSF 可能促进肿瘤增殖，这在黑色素瘤细胞系 A375 的实验中有所发现，并且 IL-6 对肿瘤的促进效应也得到

了讨论。MSC 分泌细胞因子模式的差异可能与激活其膜上不同的 Toll 样受体（TLR）有关，从而引起促炎症或者免疫抑制反应的应答。另外，MSC 已经显示出对原发性肿瘤或转移瘤具有趋向性[53,54]，因此 MSC 可以作为靶向治疗肿瘤的潜在载体。随着对 MSC 对肿瘤的特异性趋向性能力大量的研究，MSC 越来越被认为是用于向肿瘤部位递送治疗剂的良好载体。将含有单纯疱疹病毒胸苷激酶（HSV-TK）、海肾萤光素酶（Rluc）和红色荧光蛋白（red fluorescent protein，RFP）的三重融合基因的人脐带间充质干细胞（hUC-MSC）输注入预先建立的 GFP-Fluc 双融合 MDA-MB-231 乳腺癌模型中，通过 Fluc 和 Rluc 的 BLI 同时实时监测肿瘤细胞和 hUC-MSC 在体内的情况[38]。使用近红外（NIR）成像进一步证明 hUC-MSC 对体内肿瘤细胞的作用，结果表明，hUC-MSC 可以在一定程度上抑制肿瘤血管生成，并能够促进肿瘤细胞的凋亡。这些结果表明，分子成像是一个重要的工具，它不仅能够示踪 MSC 的移植情况，而且能实时监测 MSC 抑制肿瘤增殖等的治疗效果。因此，利用无侵袭性影像方法可以检测疾病的发展及移植细胞在体内的分布，这将有利于 MSC 的临床治疗应用。

第五节　间充质干细胞治疗的分子影像技术应用前景

MSC 的再生治疗为治愈疾病提供了一种可能，其中能够实时监测移植干细胞的命运是十分重要的。分子影像学作为一门飞速发展的学科，它可以对移植入体内细胞的存活、增殖和迁移情况进行一个详细的了解，进而为 MSC 对病灶的治疗提供了一种全新的模式。使用最佳的成像技术，可以更好地了解干细胞治疗的机制，也有助于在分子影像学发展中提出新的见解。而就细胞而言，需要选用一种最佳的探针对细胞进行标记，使得这些探针能够通过非侵入性的方式用于检测移植细胞的状态。目前，并无单一的最佳分子影像技术可示踪干细胞，因此最好能将灵敏度高、空间分辨率高和功能成像的各种分子影像技术高效有机地整合为一体，构建多模态成像。伴随实验技术的进步，再生医学对成像技术提出了更高的要求，也推进了成像技术的不断发展。未来分子影像技术应具有更高敏感度及特异度，更少免疫原性和致瘤性，以及更好的药代动力学特性。努力发展新的标记造影剂和多模态方法，增加对再生医学的深刻认识，促进 MSC 治疗及分子影像技术向临床转化是未来 MSC 研究领域的重中之重。

（张帅强　李宗金）

参 考 文 献

[1] Ray P. Multimodality molecular imaging of disease progression in living subjects. Journal of Biosciences, 2011, 36(3):499-504.

[2] Wu JC, Chen IY, Sundaresan G, et al. Molecular imaging of cardiac cell transplantation in living animals using optical bioluminescence and positron emission tomography. Circulation, 2003, 108(11):1302-1305.

[3] van der Bogt KEA, Hellingman AA, Lijkwan MA, et al. Molecular imaging of bone marrow mononuclear cell survival and homing in murine peripheral artery disease. JACC: Cardiovascular Imaging, 2012, 5(1):46-55.

[4] Jaffer FA, Libby P, Weissleder R. Optical and multimodality molecular imaging. Arteriosclerosis, Thrombosis, and Vascular Biology, 2009, 29(7):1017-1024.

[5] Li Z, Lee A, Huang M, et al. Imaging survival and function of transplanted cardiac resident stem cells. Journal of the American College of Cardiology, 2009, 53(14):1229-1240.

[6] Li Z, Wu JC, Sheikh AY, et al. Differentiation, survival, and function of embryonic stem cell–derived endothelial cells for ischemic heart disease. Circulation, 2007, 116(11 suppl):I46-I54.

[7] Xie X, Cao F, Sheikh AY, et al. Genetic modification of embryonic stem cells with VEGF enhances cell survival and improves cardiac function. Cloning and Stem Cells, 2007, 9(4):549-563.

[8] Pomper MG, Hammond H, Yu X, et al. Serial imaging of human embryonic stem-cell engraftment and teratoma formation in live mouse models. Cell Research, 2009, 19(3):370-379.

[9] Ly HQ, Hoshino K, Pomerantseva I, et al. *In vivo* myocardial distribution of multipotent progenitor cells following intracoronary delivery in a swine model of myocardial infarction. European Heart Journal, 2009, 30(23):2861-2868.

[10] Muller-Borer BJ, Collins MC, Gunst PR, et al. Quantum dot labeling of mesenchymal stem cells. Journal of Nanobiotechnology, 2007, 5(1):9.

[11] Lauterbur P C. Image formation by induced local interactions: examples employing nuclear magnetic resonance. Nature, 1973, 242(5394):190-191.

[12] Arbab AS, Janic B, Haller J, et al. *In vivo* cellular imaging for translational medical research. Current Medical Imaging Reviews, 2009, 5(1):19-38.

[13] James ML, Gambhir SS. A molecular imaging primer: modalities, imaging agents, and applications. Physiological Reviews, 2012, 92(2):897-965.

[14] Swanson SD, Kukowska-Latallo JF, Patri AK, et al. Targeted gadolinium-loaded dendrimer nanoparticles for tumor-specific magnetic resonance contrast enhancement. International Journal of Nanomedicine, 2008, 3(2):201.

[15] Mulder WJ, Strijkers GJ, van Tilborg GA, et al. Lipid-based nanoparticles for contrast-enhanced MRI and molecular imaging. NMR in Biomedicine, 2006, 19(1):142-164.

[16] Ghaghada KB, Ravoori M, Sabapathy D, et al. New dual mode gadolinium nanoparticle contrast agent for magnetic resonance imaging. PLoS One, 2009, 4(10):e7628.

[17] Straathof R, Strijkers GJ, Nicolay K. 2011. Target-specific paramagnetic and superparamagnetic micelles for molecular MR imaging. Methods in Molecular Biology, 2010, 771:691-715.

[18] Kadayakkara DK, Janjic JM, Pusateri LK, et al. *In vivo* observation of intracellular oximetry in perfluorocarbon-labeled glioma cells and chemotherapeutic response in the CNS using fluorine-19 MRI. Magnetic Resonance in Medicine, 2010, 64(5):1252-1259.

[19] Mizukami S, Takikawa R, Sugihara F, et al. Paramagnetic relaxation-based 19f MRI probe to detect protease activity. Journal of the American Chemical Society, 2008, 130(3):794-795.

[20] Musiek ES, Chen Y, Korczykowski M, et al. Direct comparison of fluorodeoxyglucose positron emission tomography and arterial spin labeling magnetic resonance imaging in Alzheimer's disease. Alzheimer's & Dementia: the Journal of the Alzheimer's Association, 2012, 8(1):51-59.

[21] Doyle B, Kemp BJ, Chareonthaitawee P, et al. Dynamic tracking during intracoronary injection of 18F-FDG-labeled progenitor cell therapy for acute myocardial infarction. Journal of Nuclear Medicine: Official Publication, Society of Nuclear Medicine, 2007, 48(10):1708-1714.

[22] Hoffman RM. Cellular and subcellular imaging in live mice using fluorescent proteins. Current Pharmaceutical Biotechnology, 2012, 13(4):537-544.

[23] Tong L, Zhao H, He Z, et al. Current perspectives on molecular imaging for tracking stem cell therapy. Medical Imaging in Clinical Practice: InTech, 2013, 1-17

[24] Hofmann M, Wollert KC, Meyer GP, et al. Monitoring of bone marrow cell homing into the infarcted human myocardium. Circulation, 2005, 111(17):2198-2202

[25] Rischpler C, Nekolla SG, Dregely I, et al. Hybrid PET/MR imaging of the heart: potential, initial experiences, and future prospects. Journal of nuclear medicine: official publication, Society of Nuclear Medicine, 2013, 54(3):402-415.

[26] Yamada M, Gurney PT, Chung J, et al. Manganese-guided cellular MRI of human embryonic stem cell and human bone marrow stromal cell viability. Magnetic Resonance in Medicine, 2009, 62(4):1047-1054.

[27] Yang JX, Tang WL, Wang XX. Superparamagnetic iron oxide nanoparticles may affect endothelial progenitor cell migration ability and adhesion capacity. Cytotherapy, 2010, 12(2):251-259.

[28] Arbab AS, Janic B, Haller J, et al. *In vivo* cellular imaging for translational medical research. Curr Med Imaging Rev, 2009, 5(1):19-38.

[29] Mazo M, Gavira JJ, Abizanda G, et al. Transplantation of mesenchymal stem cells exerts a greater long-term effect than bone marrow mononuclear cells in a chronic myocardial infarction model in rat. Cell Transplantation, 2010, 19(3):313-328.

[30] Graham JJ, Foltz WD, Vaags AK, et al. Long-term tracking of bone marrow progenitor cells following intracoronary injection post-myocardial infarction in swine using MRI. American Journal of Physiology Heart and Circulatory Physiology, 2010, 299(1):H125-H133.

[31] Li SC, Tachiki LM, Luo J, et al. A biological global positioning system: considerations for tracking stem cell behaviors in the whole body. Stem Cell Reviews, 2010, 6(2):317-333.

[32] Lin S, Xie X, Patel MR, et al. Quantum dot imaging for embryonic stem cells. BMC Biotechnology, 2007, 7(1):67.

[33] Pisanic TR, Blackwell JD, Shubayev VI, et al. Nanotoxicity of iron oxide nanoparticle internalization in growing neurons. Biomaterials, 2007, 28(16):2572-2581.

[34] Rosen AB, Kelly DJ, Schuldt AJ, et al. Finding fluorescent needles in the cardiac haystack: tracking human mesenchymal stem cells labeled with quantum dots for quantitative *in vivo* three-dimensional fluorescence analysis. Stem Cells, 2007, 25(8):2128-2138.

[35] Love Z, Wang F, Dennis J, et al. Imaging of mesenchymal stem cell transplant by bioluminescence and PET. Journal of Nuclear Medicine: Official Publication, Society of Nuclear Medicine, 2007, 48(12):2011-2020.

[36] Le Blanc K, Rasmusson I, Sundberg B, et al. Treatment of severe acute graft-versus-host disease with third party haploidentical mesenchymal stem cells. Lancet, 2004, 363(9419):1439-1441.

[37] Horwitz EM, Dominici M. How do mesenchymal stromal cells exert their therapeutic benefit? Cytotherapy, 2008, 10(8):771-774.

[38] Leng L, Wang Y, He N, et al. Molecular imaging for assessment of mesenchymal stem cells mediated breast cancer therapy. Biomaterials, 2014, 35(19):5162-5170.

[39] Psaltis PJ, Peterson KM, Xu R, et al. Noninvasive monitoring of oxidative stress in transplanted mesenchymal stromal cells. JACC Cardiovascular Imaging, 2013, 6(7):795-802.

[40] Haznedar R, Aki SZ, Akdemir OU, et al. Value of ^{18}F-fluorodeoxyglucose uptake in positron emission tomography/computed tomography in predicting survival in multiple myeloma. European Journal of Nuclear Medicine and Molecular Imaging, 2011, 38(6):1046-1053.

[41] Makowski M. ^{18}F-Galacto-RGD PET demonstrates elevated myocardial alpha_v-beta_3 expression in patients after myocardial infarction. European Heart Journal Supplements, 2009, 11(B):S21.

[42] Johnson LL, Schofield L, Donahay T, et al. Radiolabeled arginine-glycine-aspartic acid peptides to image angiogenesis in swine model of hibernating myocardium. JACC Cardiovascular Imaging, 2008, 1(4):500-510.

[43] Du W, Tao H, Zhao S, et al. Translational applications of molecular imaging in cardiovascular disease and stem cell therapy. Biochimie, 2015, 116:43-51.

[44] Yao X, Liu Y, Gao J, et al. Nitric oxide releasing hydrogel enhances the therapeutic efficacy of mesenchymal stem cells for myocardial infarction. Biomaterials, 2015, 60:130-140.

[45] Brzoska M, Geiger H, Gauer S, et al. Epithelial differentiation of human adipose tissue-derived adult stem cells. Biochemical and Biophysical Research Communications, 2005, 330(1):142-150.

[46] Feng G, Zhang J, Li Y, et al. IGF-1 C domain-modified hydrogel enhances cell therapy for AKI. Journal of the American Society of Nephrology, 2016, 27(8):2357-2669.

[47] Gerstenfeld LC, Cullinane DM, Barnes GL, et al. Fracture healing as a post-natal developmental process: molecular, spatial, and temporal aspects of its regulation. Journal of Cellular Biochemistry, 2003, 88(5):873-884.

[48] Lazarus H, Haynesworth SE, Gerson SL, et al. *Ex vivo* expansion and subsequent infusion of human bone marrow-derived stromal progenitor cells (mesenchymal progenitor cells): implications for therapeutic use. Bone Marrow Transplantation, 1995, 16(4):557-564.

[49] Blum JS, Temenoff JS, Park H, et al. Development and characterization of enhanced green fluorescent protein and luciferase expressing cell line for non-destructive evaluation of tissue engineering constructs. Biomaterials, 2004, 25(27):5809-5819.

[50] 段小军, 杨柳, 周跃, 等. 增强型绿色荧光蛋白标记技术对骨折后骨髓间充质干细胞的迁移示踪. 中国修复重建外科杂志, 2006, 20(2):102-106.

[51] Nakamura K, Ito Y, Kawano Y, et al. Antitumor effect of genetically engineered mesenchymal stem cells in a rat glioma model. Gene Therapy, 2004, 11(14):1155-1164.

[52] Dasari VR, Kaur K, Velpula KK, et al. Upregulation of PTEN in glioma cells by cord blood mesenchymal stem cells inhibits migration via downregulation of the PI3K/Akt pathway. PLoS One, 2010, 5(4):e10350.

[53] Yong RL, Shinojima N, Fueyo J, et al. Human bone marrow–derived mesenchymal stem cells for intravascular delivery of oncolytic adenovirus Δ24-RGD to human gliomas. Cancer Research, 2009, 69(23):8932-8940.

[54] Kidd S, Caldwell L, Dietrich M, et al. Mesenchymal stromal cells alone or expressing interferon-β suppress pancreatic tumors *in vivo*, an effect countered by anti-inflammatory treatment. Cytotherapy, 2010, 12(5):615-625.

第二十一章 基于拉曼散射的间充质干细胞无标记检测

第一节 引 言

干细胞作为生物活体被研究和应用，它们的功能特性（如自我更新、高度增殖、多向分化等）不仅取决于自身的生物组成及结构，还高度依赖于其生理状态（如细胞活力、衰老等）。在生物医学应用中，大量具有高活力状态的干细胞是干细胞治疗、基于干细胞的基因治疗及组织工程取得成功的基本条件。因此，实时准确地获取干细胞的生理状态信息对于干细胞的体外培养扩增、生物学研究，干细胞药物研究和开发及临床应用具有重要意义。

目前，生物医学上主要基于形态学、生物化学和免疫学方法对细胞进行检测。形态学方法方便、直接、可视性强，但一般用于定性分析，且分析结果的准确性在很大程度上依赖于实验人员的经验，人为因素干扰大。生化学方法和免疫学方法灵敏、准确、特异性强，且利用流式细胞仪、酶标仪、荧光测定仪、生物发光仪、分光光度计等可提供定量的检测结果。其中，流式细胞仪（flow cytometer，FCM）集成有激光技术、电子物理技术、光电测量技术、计算机技术、细胞荧光化学技术和单克隆抗体技术等，能对处于快速直线流动状态中的细胞或生物颗粒进行多参数、快速的定量分析和分选，是当前细胞检测的有力手段。然而，尽管流式细胞术发展成熟，但大多数研究仍然基于传统的生物化学和免疫学方法。这些方法往往是侵入性的，其中用到的生化反应、染色、标记、固定、细胞溶解等处理手段往往会改变细胞正常的生长环境和生理功能，甚至破坏细胞结构，不利于细胞的实时在线检测。因此，探索一种无外源性标记、非侵入性的细胞检测技术显得非常必要，已成为干细胞研究和产业化发展的重要课题。

光学技术由于具有无标记、非接触、微创甚至无创的特点而有望成为细胞检测的重要手段，成为传统生物化学方法的有益补充。特别是利用细胞本征的光学信号进行细胞生物学研究及相关检测已成为当前生物医学领域的研究热点。在光学技术中，拉曼光谱技术以其对分子结构的指示灵敏、受水干扰小、不依赖于激发光波长等优点日益受到人们的关注，特别是拉曼光谱与显微成像技术相结合，为在亚细胞水平上无标记地检测和研究细胞分子结构提供了新的有力工具，正被越来越多地应用于 MSC 的研究中。

第二节 拉曼散射的基本概念

一、光的散射

光散射是自然界中普遍存在的现象。当一束光通过浑浊的液体时，我们可以从侧面清晰地看到在溶液中传播的光束，这就是由光散射所导致的结果。即使在单色光入射的情况下，被物质散射的光也并非具有完全相同的频率。其中，大部分散射光的频率与入射光频率相同，但也有一部分散射光的频率发生了改变。根据散射光与入射光的频率是否一致，人们将散射分为弹性散射和非弹性散射，前者频率不发生改变，而后者的频率会出现红移（散射光频率小于入射光频率）或蓝移（散射光频率大于入射光频率）。

根据散射中心尺寸的不同，弹性散射可进一步分为瑞利散射和米氏散射。前者的散射中心（如分子）尺寸比入射光波长小很多，且散射光强度与光波长的四次方成反比，这就是著名的瑞利散射定律；后者由尺寸较大的散射中心（如尘埃颗粒）所引起，其散射光强度往往很大，并不存在与波长4次方成反比的关系。

在散射光频率发生改变的非弹性散射中，根据频率改变的大小，也可进一步分为布里渊散射和拉曼散射。在光散射研究中，能量通常以波数（cm^{-1}）为单位。波数与波长成反比，是指1cm的单位长度上所含的波的数目，在数学上表示为（$1/\lambda$），其中λ为波长，且换算为cm单位。布里渊散射所引起的波数变化约为$0.1cm^{-1}$，而拉曼散射的波数变化一般大于$1cm^{-1}$，基本上落在与分子的转动能级、振动能级和电子能级之间的跃迁相联系的范围内。因此，在利用非弹性散射研究分子结构时，一般不考虑难以与入射光和弹性散射区分的布里渊散射，而主要采用拉曼散射光谱技术。

二、拉曼散射

1928年，印度物理学家Raman在研究液体苯的散射光谱时，从实验中发现了一种非弹性散射[1]，人们称为拉曼散射。随后，人们以拉曼散射为基础，建立了拉曼光谱分析技术，曾一度成为研究分子结构的主要手段。然而，原始的拉曼光谱测量以汞弧灯为光源，存在激发光能量弱、曝光时间长、荧光干扰大等缺点，仅限于测量无色液体样品。这使得拉曼光谱的发展陷入了低谷。直到20世纪60年代，激光的问世为拉曼光谱的发展带来了新的契机。与早期的汞弧灯光源相比，激光输出功率大且能量集中，单色性和相干性好，在很大程度上弥补了拉曼光谱信号弱的不足，使拉曼光谱技术得到了快速发展。利用散射物质的拉曼频移可表征物质分子的振动和转动能级特性，因此不同的分子结构（如化学键、官能团等）都有其特征性的拉曼散射光谱，拉曼光谱也被形象地称为分子"指纹光谱"[2]。这一独特优势使其成为能够准确提供物质化学结构信息的重要工具，在物质成分鉴定及分子结构研究等领域发挥了重要作用。随着光学探测技术、成像技术、纳米材料技术等的进步，一些基于拉曼散射的新技术如显微拉曼光谱及成像技术[2]、共振拉曼光谱技术[3]、表面增强拉曼光谱技术[4]、相干反斯托克斯拉曼光谱技术[5]等得到了快速的发展，极大地扩展了拉曼光谱的应用领域。

拉曼散射的原理与我们所熟知的荧光、磷光等光致发光现象截然不同。荧光、磷光来源于分子从电子激发态向基态的辐射跃迁，而拉曼散射来源于分子对入射光的非弹性散射[6]。根据量子理论，分子在入射光的作用下，对应于光散射效应的跃迁矩（从n态到m态跃迁）矩阵元为

$$\hat{\mu}_{mu}(t) = \frac{1}{\hbar}\sum_{k}\left[\frac{(\hat{\mu}_{mk} \cdot \hat{E}_0)\hat{\mu}_{kn}}{\omega_{km} + \omega_0} + \frac{(\hat{\mu}_{kn} \cdot \hat{E}_0)\hat{\mu}_{mk}}{\omega_{kn} - \omega_0}\right]\exp[-i(\omega_0 + \omega_{mn})t]$$
$$+ \frac{1}{\hbar}\sum_{k}\left[\frac{(\hat{\mu}_{mk} \cdot \hat{E}_0)\hat{\mu}_{kn}}{\omega_{km} + \omega_0} + \frac{(\hat{\mu}_{kn} \cdot \hat{E}_0)\hat{\mu}_{mk}}{\omega_{kn} - \omega_0}\right]\exp[i(\omega_0 + \omega_{mn})t] \quad (21\text{-}1)$$

式中，ω_0 为入射光的圆频率。根据 Klein 的相关原则，一个复数的跃迁矩为 $\hat{\mu}_{ij}\exp(-i\omega_{ij}t)$。如果 $\omega_{ij}>0$，它就相当于一个实际的辐射跃迁矩,辐射的圆频率为 ω_{ij}。因此,对于式(21-1)的第一项来说，如果存在 $\omega_0-\omega_{mn}>0$，则它可引起光的散射，散射光的圆频率为 $(\omega_0-\omega_{mn})$。如果入射光是紫外光或可见光或近红外光，而分子的电子态不发生改变，只是振动态或转动态跃迁，则通常有 $\omega_0>\omega_{mn}$，即上述 $\omega_0-\omega_{mn}>0$ 的条件总是能满足。这时可以存在三种情况，如图 21-1 所示。

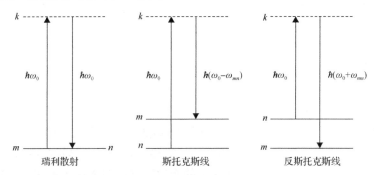

图 21-1 拉曼散射跃迁示意图

1) $\omega_{mn}=0$：即 $\omega_m=\omega_n$，跃迁的终态 m 的能量等于初始态 n 的能量，或终态与初始态为同一个态。这时散射光的圆频率等于入射光的圆频率 ω_0，这就是瑞利散射。

2) $\omega_{mn}>0$：即 $\omega_m>\omega_n$，跃迁的终态 m 的能级高于初始态 n 的能级，这时散射光的圆频率等于 $\omega_0-\omega_{mn}$，这就是斯托克斯线。

3) $\omega_{mn}<0$：即 $\omega_m<\omega_n$，跃迁的终态 m 的能级低于初始态 n 的能级，这时散射光的圆频率等于 $\omega_0-\omega_{mn}=\omega_0+\omega_{nm}$，这就是反斯托克斯线。

在式（21-1）中，k 态起着重要的作用，通常称为中间态，是虚态。我们可以把瑞利散射过程理解为分子吸收光子 $\hbar\omega_0$ 由 n 态跃迁到 k 态，随即由 k 态跃迁回 n 态而发射光子 $\hbar\omega_0$。相似的，拉曼散射过程可以理解为分子吸收光子 $\hbar\omega_0$，由 n 态跃迁到 k 态，随即由 k 态跃迁到 n 态而发射光子 $\hbar\omega_0(\omega_0-\omega_{nm})$。如果 n 态是下能级，得到的是斯托克斯线，如果 n 态是上能级，得到的就是反斯托克斯线。由于能级的热分布，下能级的分子数密度要比上能级大得多，因此斯托克斯线的强度要比反斯托克斯线强得多。

从上述拉曼散射的量子理论解释中可以看到，起着中间态作用的 k 态是虚态，这意味着在拉曼散射中并不存在分子对入射光的吸收过程，因此与荧光、磷光不同，拉曼散射不依赖于入射光波长。事实上，由于多数物质的拉曼散射较弱，容易受到荧光的干扰，因此在选择入射光波长时一般会避开分子的吸收波段。此外，拉曼散射光与入射光的圆频率差的绝对值 $|\omega_{mn}|$，取决于分子振动能级和转动能级的分布特性，与入射光的圆频率 ω_0 无关。因此，拉曼光谱图的横坐标不是一个绝对量（如散射光的波长、频率等），而是一个相对量（散射光与入射光的频率差，以波数表示），这个频率差称为拉曼位移。

三、拉曼光谱的预处理

拉曼光谱是指拉曼散射强度随拉曼位移的分布，一般表现为一系列窄频宽的拉曼峰。

拉曼峰可通过峰位（峰顶点对应的拉曼位移）、峰的半高宽（full width at half-maximum, FWHM）和峰高这三个参数进行表征。理想情况下，分子的拉曼光谱是基线为零的一系列拉曼峰的分布，但实际测量中，难以避免本底信号、杂散光和分子荧光等对拉曼散射信号的干扰，从而形成形状复杂且随机多变的基线，其上叠加着拉曼峰。对于定性分析，一般不需对光谱数据进行预处理，通过考察光谱中某个峰是否存在、峰宽和峰高是否发生明显改变等信息来反映物质组成或相应分子结构的变化。但对于定量分析，则一般需要进行光谱预处理，去除本底、基线等对拉曼峰的影响。常用的拉曼光谱预处理包括扣除本底、基线校正、强度归一化、多峰拟合等。

（一）扣除本底

本底信号主要来源于玻片、培养皿、培养液、缓冲液等细胞测量环境中其他物质的荧光或其对光的散射。作为扣除本底的示例，图 21-2 给出了未经处理的细胞原始拉曼散射信号（曲线 a）和来自测量环境中石英玻片和磷酸盐缓冲液（PBS）的本底信号（曲线 b）。可以看到，前者是后者和拉曼峰的叠加。需要注意的是，这里扣除本底并不是将曲线 a 和曲线 b 直接相减。由于基底（玻片）的光散射强度对入射光的焦点位置十分敏感，在每次测量中，入射光焦点到玻片的距离可能不尽相同，因此本底信号的强度并不完全一样。所以，扣除本底信号前首要明确本底信号的特征，如曲线 b 在 $800cm^{-1}$ 附近和 $1070cm^{-1}$ 附近有两个明显的突起，这两个特征也出现在细胞信号中，可以判断是受本底的影响。扣除本底时，需要将本底信号乘上一个系数，使本底信号 $800cm^{-1}$ 和 $1070cm^{-1}$ 附近的突起高度与细胞信号中相应的突起高度相当，然后进行相减。对于本底特征不明显的细胞拉曼数据，可以免去扣除本底处理。

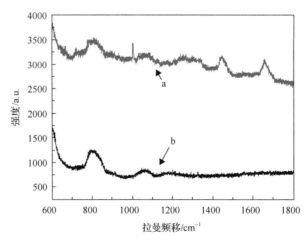

图 21-2　细胞原始拉曼散射信号（a）与来自石英玻片和 PBS 的本底信号（b）（彩图请扫封底二维码）

（二）基线校正

扣除本底后，拉曼数据中一般仍会存在非零的基线。由于细胞的拉曼散射信号较弱，其测量时间较长（一般需数十秒），因此极易受到基线的影响。基线的成因非常复杂，可能来源于样品的荧光、环境杂散光、测量系统基线漂移等多个因素，导致基线的强度和形状随机多变。如图21-3A所示，在同样的环境和条件下对细胞进行两次拉曼光谱测量，两者的基线表现出明显的差异，这将对光谱数据的定量分析造成严重的干扰。

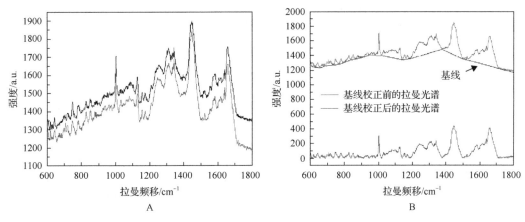

图21-3 细胞拉曼光谱的基线校正（彩图请扫封底二维码）
A. 未经基线校正的细胞拉曼光谱，B. 多点基线校正示意图

基线校正的方法很多，大体可分为两种类型：一种是采用特定函数（如高次多项式函数）对基线进行拟合；另一种是采用多点基线校正（multipoint baseline correction）方法，即根据基线的形状特征，选取若干波谷点作为校正点，将相邻校正点之间以直线连接，代表基线，如图21-3B所示。严格来说，所有的基线校正方法都只是近似地描述基线，也都存在误差。相对而言，多点基线校正方法由于以波谷作为校正点，在一定程度上避免了函数拟合引入的误差，在许多细胞拉曼光谱的研究工作中被采用[7,8]。

（三）强度归一化

细胞的拉曼散射强度不仅取决于细胞内的分子结构，还受到入射光强度、入射光焦点位置、探测器灵敏度等多方面因素的影响。因此，在定量分析时一般需要对光谱强度进行归一化处理，其关键在于内标峰的选择。目前采用较多的是位于$1448cm^{-1}$的拉曼峰[7-10]，归属于CH变形振动。由于这一分子振动存在于蛋白质、核酸、脂质和糖类这4类生物大分子中，因此该拉曼峰被认为是包含了细胞内所有生物大分子的贡献，从而可作为衡量细胞整体拉曼散射强度水平的内标峰，光谱强度以其为标准进行归一化。

（四）多峰拟合

对于由多个拉曼峰彼此叠加而形成的光谱包络，可以采用多峰拟合（multiple peak fitting）的方法将隐藏在包络下的各个峰的参数计算出来。多峰拟合的原理及过程如下：先通过数学方法确定隐藏在包络下的峰的个数及大致峰位。例如，对包络区域的强度分布进行二阶求导运算，原始曲线中明显的或隐藏的实际峰表现为二阶求导后得到的曲线中的极小值，如图21-4A所示。在此基础上，利用数学函数描述峰形（通常采用高斯函数），不断调整各个峰的位置、高度和宽度参数，使得各峰叠加后形成的曲线与实际曲线达到最好的符合程度。因此，可通过这一过程将叠加在一起的拉曼峰分解出来，从而求得各个峰的特征参数。图 21-4B 给出了拉曼光谱多峰拟合的一个示例。

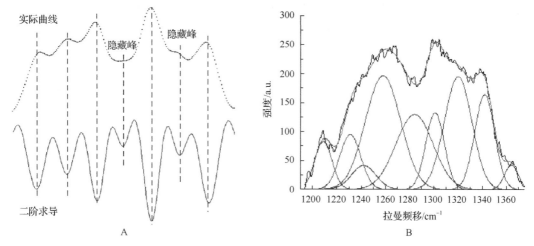

图 21-4　光谱的多峰拟合（彩图请扫封底二维码）
A. 实际曲线中明显或隐藏的峰表现为二阶导数曲线中的极小值；B. 拉曼光谱多峰拟合示意图

第三节　间充质干细胞的典型拉曼光谱

典型的 MSC 的拉曼光谱如图 21-5 所示。可以看到，有 4 个拉曼峰最为明显，它们分别是位于 1003cm^{-1} 的苯丙氨酸环对称呼吸振动，位于 1301cm^{-1} 的脂质 CH$_2$ 扭曲振动，位于 1448cm^{-1} 的来自蛋白质、核酸、脂质和糖类分子的 CH 变形振动，以及位于 1659cm^{-1} 的 α 螺旋蛋白分子酰胺 I 振动。此外，其余许多拉曼峰虽然强度较小，但也清晰可辨，这些拉曼峰包含了细胞中 4 类主要生物大分子的结构信息。各个拉曼峰的归属指认列于表 21-1。

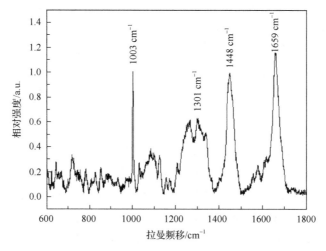

图 21-5　MSC 的典型拉曼光谱

表 21-1 MSC 的拉曼峰归属指认

拉曼位移/cm^{-1}	归属指认[7,9,11-15]
621	碱基腺嘌呤五元环变形振动
643	碱基胸腺嘧啶环角弯曲振动
667	碱基鸟嘌呤环呼吸振动
719	脂质 CN$^+$(CH$_3$)$_3$ 伸缩振动
732	碱基腺嘌呤环呼吸振动
744	碱基胸腺嘧啶 C=O 平面外弯曲振动
758	色氨酸环呼吸振动
784	DNA O—P—O 伸缩振动
811	RNA O—P—O 伸缩振动
827	酪氨酸环平面外呼吸振动
853	酪氨酸环呼吸振动
877	脂质 C—C 对称伸缩振动
900	脱氧核糖
934	蛋白质 C—C 伸缩振动
980	脂质=CH 弯曲振动
1003	苯丙氨酸环对称呼吸振动
1010	脱氧核糖 C—O 伸缩振动
1031	苯丙氨酸
1064	脂质链 C—C 伸缩振动
1083	蛋白质 C—N 伸缩振动，脂质链 C—C 伸缩振动，糖 C—O 伸缩振动
1097	DNA O—P—O 伸缩振动
1127	蛋白质 C—N，C—C 伸缩振动
1157	蛋白质 C—C，C—N 伸缩振动
1174	酪氨酸，苯丙氨酸
1209	苯丙氨酸 C—C$_6$H$_5$ 伸缩振动
1231	蛋白质酰胺Ⅲ，无规卷曲
1242	蛋白质酰胺Ⅲ，β 折叠
1258	蛋白质酰胺Ⅲ，β 折叠
1284	蛋白质酰胺Ⅲ，α 螺旋
1301	脂质 CH$_2$ 扭曲振动
1320	碱基鸟嘌呤，蛋白质 CH 变形振动
1342	蛋白质 CH 变形振动
1364	脂质 CH$_3$ 对称伸缩振动
1448	核酸、蛋白质、脂质、糖类 CH 变形振动
1659	蛋白质酰胺Ⅰ，α 螺旋

第四节 拉曼光谱在间充质干细胞检测中的应用

一、拉曼光谱用于 MSC 活力变化检测

人脐带间充质干细胞（hUC-MSC）活力下降引起的拉曼光谱变化主要出现在

617~690cm^{-1}、703~770cm^{-1}、840~945cm^{-1}、950~1020cm^{-1}、1190~1375cm^{-1} 频带及 1659cm^{-1} 拉曼峰，如图 21-6 所示[16]。高活力（活细胞率>90%）与低活力（活细胞率<20%）hUC-MSC 拉曼光谱之间，各拉曼峰的拉曼位移和半高宽并没有出现明显变化，但峰高表现出有显著差异，反映出高活力与低活力细胞之间重要的分子结构变化信息。

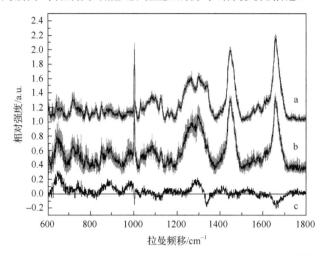

图 21-6 高活力与低活力间充质干细胞的拉曼光谱对比[16]

（a）高活力细胞的平均拉曼光谱；（b）低活力细胞的平均拉曼光谱；（c）将光谱 b 与光谱 a 相减得到的差谱；灰色阴影表示同组细胞光谱之间的标准差；所有光谱以 1448cm^{-1} 峰为内标进行强度归一划；为清楚起见，光谱在纵轴方向作了平移

除 1659cm^{-1} 拉曼峰外，其他几个频带没有出现单一的峰形，而是表现出由多个拉曼峰彼此叠加形成的光谱包络特征。这些频带包含了多个在拉曼频移上彼此接近的分子振动，集中反映了蛋白质、核酸、脂质和糖类的分子结构信息。通过多峰拟合及统计学分析，发现两组细胞之间有 13 个拉曼峰的强度存在显著差异（$P<0.05$），如图 21-7 所示[16]。

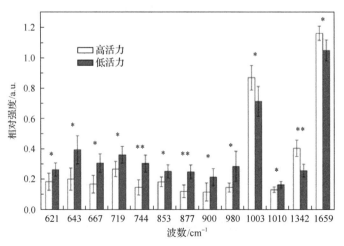

图 21-7 高活力与低活力细胞的拉曼峰强度对比[16]

*表示 $P<0.05$，**表示 $P<0.001$

当细胞活力下降，归属于核酸分子振动的 621cm^{-1}、643cm^{-1}、667cm^{-1}、744cm^{-1}、900cm^{-1} 和 1010cm^{-1} 拉曼峰强度上升。同时，归属于脂质分子振动的 719cm^{-1}、877cm^{-1} 和 980cm^{-1} 拉曼峰的强度也明显增强。然而，位于 1003cm^{-1}、1342cm^{-1} 和 1659cm^{-1} 归属于蛋白质分子振动的拉曼峰强度出现了下降。在这些拉曼峰中，最具有统计学意义（$P<0.001$）的显著差异体现在位于 744cm^{-1}、877cm^{-1} 和 1342cm^{-1} 的三个拉曼峰，它们分别归属于核酸碱基胸腺嘧啶的 C=O 平面外弯曲振动、脂质的 C—C 对称伸缩振动和蛋白质分子的 CH 变形振动。其中，在低活力细胞光谱中位于 744cm^{-1} 和 877cm^{-1} 的两个拉曼峰的强度均比高活力细胞中对应拉曼峰强度高出约一倍，而位于 1342cm^{-1} 的拉曼峰的强度则下降了约 30%，表明这三个分子振动对 hUC-MSC 活力下降最为敏感。此外，进一步研究发现这三个拉曼峰随细胞活力下降呈现单调变化[17]，因而有可能作为光谱标记对 hUC-MSC 的活力进行表征。

进一步的研究表明，低活力细胞内活性氧（ROS）水平出现了显著上升，且随细胞活力下降，细胞内相对的过量 ROS 产量与 1342cm^{-1}、877cm^{-1}、744cm^{-1} 三个拉曼峰的相对强度变化量之间存在高度的线性相关[17]。因此，hUC-MSC 活力下降过程中出现的分子振动变化可能是由过量 ROS 对生物大分子攻击所引起的。

位于 744cm^{-1}、667cm^{-1}、643cm^{-1} 和 621cm^{-1} 归属于 DNA 碱基分子振动的拉曼峰的强度变化可能源于 ROS 对 DNA 碱基堆积结构的破坏。正常情况下，DNA 分子呈现双螺旋结构，碱基在横向氢键结合力和纵向堆积力的作用下，形成高密度的压缩体。当堆积结构被破坏，碱基之间氢键断裂，DNA 双链解旋甚至断链，碱基分子将更多地暴露在光照之下。这相当于增大了碱基对激发光的散射截面，从而产生"增色"效应，导致碱基分子振动相对应的拉曼峰强度增加。

此外，ROS 可通过脂质过氧化反应对生物膜造成氧化损伤。这一过程不可避免地会导致脂类分子所处的微环境及分子空间构象发生变化，可能使得某些具有拉曼活性的化学键对激发光的散射截面增加，造成相应拉曼峰（如 980cm^{-1}、877cm^{-1} 和 719cm^{-1}）强度的增强。

蛋白质分子二级结构的构象变化可通过拉曼光谱清晰观察到。例如，位于 1659cm^{-1} 的拉曼峰强度降低，表明 α-螺旋结构减少。这可能是由于 ROS 的攻击导致分子中的氢键、二硫键和碳硫键发生改变甚至断裂，这些化学键在维持蛋白质空间结构方面起着重要作用。同时，蛋白质空间结构变化极有可能引起化学键（如 C—H 键）所处微环境的改变，从而影响其对激发光的散射。此外，ROS 能引起蛋白质降解。这些可能是造成归属于蛋白质分子振动的拉曼峰（如 1003cm^{-1}、1342cm^{-1} 和 1659cm^{-1}）强度下降的原因。

二、拉曼光谱用于 MSC 复制性衰老检测

hUC-MSC 传代培养，分别取第 5 代、第 10 代、第 20 代、第 30 代的细胞进行拉曼光谱检测，这些不同代次的细胞代表了不同的细胞衰老状态，其拉曼光谱[18]示于图 21-8。

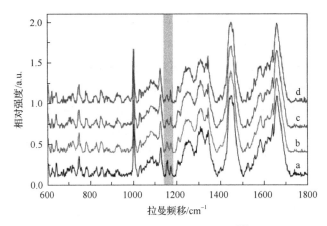

图 21-8 不同衰老程度的间充质干细胞的拉曼光谱对比[18]（彩图请扫封底二维码）

细胞传代数分别为 5 代（a）、10 代（b）、20 代（c）、30 代（d）

随着细胞传代数的增加，归属于蛋白质 C—C、C—N 伸缩振动的 1157cm^{-1} 拉曼峰和归属于芳香族氨基酸（酪氨酸、苯丙氨酸）的 1174cm^{-1} 拉曼峰的相对强度发生了规律性变化，两者的强度比值 I_{1157}/I_{1174} 随细胞衰老呈现单调下降趋势，且在不同代次的细胞之间表现出有统计学差异（$P<0.05$），如图 21-9 所示[18]。因此，该比值有可能作为一项光谱指标用于表征 hUC-MSC 的衰老。

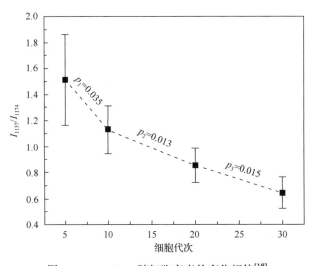

图 21-9 I_{1157}/I_{1174} 随细胞衰老的变化规律[18]

三、拉曼光谱用于 MSC 分化检测

（一）拉曼光谱检测 MSC 向成骨细胞的分化

MSC 向成骨细胞分化的一个重要光谱特征在于 960cm^{-1} 附近的拉曼峰，对应于磷酸酯基团的 P—O 对称伸缩振动，归属于羟基磷灰石（hydroxyapatite，HA）[19,20]。在未分化的 MSC 和诱导分化 7 天内的 MSC 的拉曼光谱中[图 21-10（A）]，未发现 960cm^{-1} 附

近存在拉曼峰；在诱导分化 7 天时，960cm^{-1} 拉曼峰已经清晰可辨[图 21-10（B）]，表明 MSC 开始矿化；此后，该峰的强度随分化时间的延长急剧上升[图 21-10（C），（D）]。因此，960cm^{-1} 拉曼峰可作为特异性的光谱标记用于监测 MSC 向成骨细胞的分化[19,20]，并已在一些研究工作中得到应用。Gao 等[21]利用 950cm^{-1} 附近的拉曼散射信号对细胞进行成像，成功开发了一种非侵入性和无标记的成像方法用于评价聚合物支架上 MSC 的成骨分化。Azrad 等[12]以 960cm^{-1} 拉曼峰作为光谱标志用于评估一种麋鹿茸粉提取物对 MSC 成骨分化的促进作用。

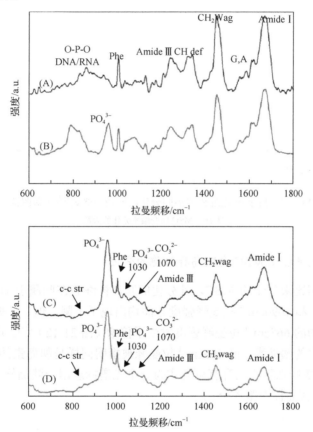

图 21-10　诱导 MSC 向成骨细胞分化，不同分化时间的 MSC 拉曼光谱[19]（彩图请扫封底二维码）
（A）7 天，（B）14 天，（C）21 天，（D）28 天

（二）拉曼光谱检测 MSC 向肝细胞的分化

未分化的 MSC 和原代肝细胞在拉曼光谱上的一个显著差异体现在 2800~3000cm^{-1} 频带（图 21-11）[22]，该频带包含位于 2852cm^{-1} 和 2898cm^{-1} 的两个拉曼峰，分别对应于脂褐素的 CH_2 对称伸缩振动和 CH 伸缩振动。在未分化的 MSC 中，该频带未出现明显的拉曼信号；随着诱导分化时间的延长，在分化 10 天后，已分化 MSC 的 2800~3000cm^{-1} 频带积分强度显著高于未分化的 MSC 对照组，表现出与原代肝细胞相似的拉曼光谱特征。此外，以位于 400~565cm^{-1} 的本底信号为参考，在诱导 MSC 分化的 7~17 天，

2800~3000cm^{-1} 频带积分强度随分化时间延长呈线性上升[22], 表明 MSC 分化过程中脂褐素形成。因此, 2800~3000cm^{-1} 频带的积分强度可作为一项拉曼光谱指标用于检测 MSC 向肝细胞的分化。

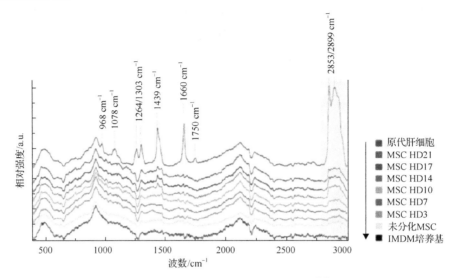

图 21-11　诱导 MSC 向肝细胞分化过程中不同分化时间拉曼光谱[22]（彩图请扫描封底二维码）
HD 表示 MSC 向肝细胞分化的天数

（三）拉曼光谱检测 MSC 向胰岛样细胞的分化

研究发现, 相比未分化的 MSC, 其分化的胰岛样细胞在归属于 DNA 的 623cm^{-1}、644cm^{-1}、784cm^{-1} 和 1090cm^{-1} 拉曼峰强度方面均出现了下降, 但, 一般认为归属于碱基鸟嘌呤环呼吸振动的 665cm^{-1} 拉曼峰强度发生了倍增（图 21-12）[23]。这一结果可能来源于胰岛样细胞产生的胰岛素, 而 665cm^{-1} 拉曼峰被指明同样是胰岛素的特征拉曼峰[23,24]。因此, 665cm^{-1} 拉曼峰强度的上升有可能作为一个光谱标志用于胰岛样细胞的鉴定、分选以及监测 MSC 向胰岛样细胞的分化。

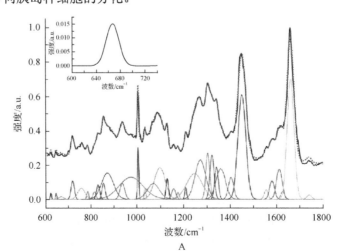

A

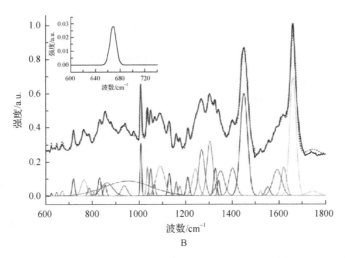

图 21-12　间充质干细胞及其分化的胰岛样细胞的拉曼光谱多峰拟合[23]（彩图请扫封底二维码）

A. 未分化 MSC 拉曼光谱的多峰拟合；B. 胰岛样细胞拉曼光谱的多峰拟合；两个小图是 665cm^{-1} 拉曼峰的放大图

综上所述，随着显微拉曼光谱技术的发展，其已经在 MSC 的生物学研究及无标记检测等方面展现出具有极大的研究价值和广阔的应用前景。由于具有"分子指纹光谱"特性，它不仅能提供特异性的光谱标记用于表征细胞的生物学功能及生理状态变化，也可从分子结构角度为 MSC 的生物学和临床应用研究提供新的有力手段。总的来说，当前虽然取得了许多可喜的研究成果，但 MSC 的拉曼光谱研究及应用仍处于起步阶段，未来的工作可从以下三个方面深入：①进一步提高 MSC 拉曼光谱研究的深度和广度，探索更多可特异性表征 MSC 生物学变化的光谱标记，为 MSC 研究提供更多无标记检测方法，同时为 MSC 研究提供新的视角和实验依据；②在现有成果的基础上开展工程应用研究，结合光学技术、电子技术、计算机及软件技术、智能信息处理技术等开发基于拉曼散射的 MSC 无标记自动检测分析系统，实现仪器的小型化、便携式、低成本、智能化，促进拉曼光谱在 MSC 研究和生产中的应用；③针对细胞拉曼散射强度较弱的问题开展拉曼增强技术研究，缩短测量时间，提高检测效率。相信在不远的将来，随着光电检测、纳米材料、拉曼增强、信号处理等相关技术的快速进步，细胞拉曼光谱技术将迎来长足的发展，在 MSC 研究和生产中得到广泛应用，成为 MSC 无标记检测的重要手段。

（白　华）

参 考 文 献

[1] Raman CV. A new radiation. Indian J Phys, 1928, 2:387-398.

[2] Swain RJ, Stevens MM. Raman microspectroscopy for non-invasive biochemical analysis of single cells. Biochemical Society Transactions, 2007, 35(3):544-549.

[3] Wood BR, Langford SJ, Cooke BM, et al. Raman imaging of hemozoin within the food vacuole of Plasmodium falciparum trophozoites. FEBS Letters, 2003, 554(3):247-252.

[4] CW Shi, Cao XW, Chen X, et al. Intracellular surface-enhanced Raman scattering probes based on TAT peptide-conjugated Au nanostars for distinguishing the differentiation of lung resident mesenchymal stem cells. Biomaterials, 2015, 58:10-25.

[5] Rodriguez LG, Lockett SJ, Holtom GR. Coherent anti-stokes raman scattering microscopy: a biological review. Cytometry Part A, 2006, 69A:779-791.

[6] 张树霖. 拉曼光谱学及其在纳米结构中的应用（上册）-拉曼光谱学基础. 许应瑛译, 北京: 北京大学出版社, 2017.

[7] Notingher I, Verrier S, Haque S, et al. Spectroscopic study of human lung epithelial cells (A549) in culture: living cells versus dead cells. Biopolymers, 2003, 72:230-240.

[8] Verrier S, Notingher I, Polak JM, et al. *In situ* monitoring of cell death using raman microspectroscopy. Biopolymers, 2004, 74:157-162.

[9] Chan JW, Lieu DK, Huser T, et al. Label-free separation of human embryonic stem cells and their cardiac derivatives using Raman spectroscopy. Anal Chem, 2009, 81:1324-1331.

[10] Short KW, Carpenter S, Freyer JP, et al. Raman spectroscopy detects biochemical changes due to proliferation in mammalian cell cultures. Biophysical Journal, 2005, 88:4274-4288.

[11] Notingher I, Bisson I, Polak JM, et al. *In situ* spectroscopic study of nucleic acids in differentiating embryonic stem cells. Vib Spectros, 2004, 35:199-203.

[12] Azrad E, Zahor D, Vago R, et al. Probing the effect of an extract of elk velvet antler powder on mesenchymal stem cells using raman microspectroscopy: enhanced differentiation toward osteogenic fate. J Raman Spectrosc, 2006, 37:480-486.

[13] Mohamed TA, Shabaan IA, Zoghaib WM, et al. Tautomerism, normal coordinate analysis, vibrational assignments, calculated IR, raman and NMR spectra of adenine. J Mol Struct, 2009, 938:263-276.

[14] Singh JS. FTIR and Raman spectra and fundamental frequencies of biomolecule: 5-methyluracil (thymine). J Mol Struct, 2008, 876:127-133.

[15] Nishimura Y, Tsuboi M, Nakano T, et al. Raman diagnosis of nucleic acid structure: sugar-puckering and glycosidic conformation in the guanosine moiety. Nucleic Acids Res, 1983, 11:1579-1588.

[16] Bai H, Chen P, Fang H, et al. Detecting viability transitions of umbilical cord mesenchymal stem cells by Raman micro-spectroscopy. Laser Phys Lett, 2011, 8(1):78-84.

[17] Bai H, Chen P, Tang GQ, et al. Relations between reactive oxygen species and raman spectral variations of human umbilical cord mesenchymal stem cells with different viability. Laser Physics, 2011, 21(6):1122-1129.

[18] Bai H, Li HY, Han ZB, et al. Label-free assessment of replicative senescence in mesenchymal stem cells by raman microspectroscopy. Biomedical Optics Express, 2015, 6(11):4493-4500.

[19] McManus LL, Burke GA, McCafferty MM, et al. Raman spectroscopic monitoring of the osteogenic differentiation of human mesenchymal stem cells. Analyst, 2011, 136:2471-2481.

[20] Chiang HK, Peng FY, Hung SC, et al. *In situ* raman spectroscopic monitoring of hydroxyapatite as humanmesenchymal stem cells differentiate into osteoblasts. J Raman Spectrosc, 2009, 40:546-549.

[21] Gao Y, Xu CJ, Wang LH. Non-invasive monitoring of the osteogenic differentiation of human mesenchymal stem cells on a polycaprolactone scaffold using Raman imaging. RSC Adv, 2016, 6:61771-61776.

[22] Wu HH, Ho JH, Lee OK. Detection of hepatic maturation by Raman spectroscopy in mesenchymal stromal cells undergoing hepatic differentiation. Stem Cell Research & Therapy, 2016, 7:6.

[23] Su X, Fang SY, Zhang DS, et al. Quantitative raman spectral changes of the differentiation of mesenchymal stem cells into islet-like cells by biochemical component analysis and multiple peak fitting. J Biomed Opt, 2015, 20(12):125002.

[24] Hilderink J, Otto C, Slump C, et al. Label-free detection of insulin and glucagon within human islets of langerhans using raman spectroscopy. PLoS One, 2013, 8(10):e78148.

第二十二章　间充质干细胞外泌体

第一节　间充质干细胞和外泌体

间充质干细胞（MSC）是一种具有自我更新能力的成体干细胞，广泛存在于骨髓和其他组织中，包括脂肪组织、外周血和围产期组织。MSC对多种疾病具有治疗作用，许多MSC产品已进入临床试验阶段，为多种疾病的治愈带来了新的希望。据统计，2006~2012年，通过美国食品药品监督管理局（FDA）允许进入临床试验阶段的MSC产品就有大约500种。随着越来越多的临床应用，MSC的功能机制已经成为当前研究的热点。体内和体外的研究显示，MSC具有免疫调节、促血管新生和组织再生的能力，但是其作用机制还不清楚。越来越多的证据表明，MSC治疗过程中其旁分泌作用起了主导作用。研究显示，MSC通过释放小分子物质，如生长因子、细胞因子和趋化因子，发挥其治疗作用。除了释放这些可溶性小分子物质外，MSC也释放大量的外泌体（exosome），这些外泌体可以携带大量的MSC来源的生物活性因子在细胞间交流，从而在正常生理和疾病病理过程中发挥作用[1]。

大量的研究发现，MSC来源外泌体具有与其来源MSC相似的治疗作用。现有研究已证明MSC分泌的外泌体在减少心肌损伤范围[2,3]、促进组织损伤修复[4,5]、促进急性肾小管损伤修复[6]、促进神经再生、减少肺损伤和免疫系统调节[7-9]等方面有重要作用。

第二节　外泌体的研究

越来越多的研究证明，细胞与其邻近细胞能够通过非经典的分泌胞外囊泡的方式进行交流。细胞外囊泡（extracellular vesicle，EV）最初被认为是细胞碎片或细胞排泄不需要或有毒物质的途径。但它们古老的进化起源和保守的生成机制表明，EV在细胞间通讯中具有重要作用[10,11]。EV有很多的类型，包括外泌体（30~150nm）[12]、微囊泡（microvesicle，MV）（100~350nm）[13]和凋亡小体（apoptotic body）（500~1000nm）[14]，其中外泌体被研究的最为广泛[12,15]。外泌体通过在细胞间传递信息物质，在生理和病理过程中起到非常重要的作用。

一、外泌体的形成过程

外泌体是由细胞经过"内吞-融合-外排"等一系列调控过程而形成的[16]。首先多囊泡胞内体（multivesicular endosome，MVE）以向内出芽方式形成较小的内囊泡（internal vesicle），内囊泡内含有蛋白质、微RNA（miRNA）及mRNA，接下来当MVE与细胞膜融合时，内囊泡作为外泌体被释放到细胞外，或者MVE与溶酶体融合从而降解MVE中包含的物质（图22-1）。释放出来的外泌体能够运行到距离较远的组织而影响细胞的行为和生理功能。外泌体的目的地是由外泌体表面的特异性配体的结合方式决定的。因此外泌体进入靶细胞的途径可以有两种：第一种是通过靶细胞的胞吞作用被摄入到细胞内；第二种是通过膜融合的方式与靶细胞膜融合，进而直接释放其含有的物质到目的细胞。

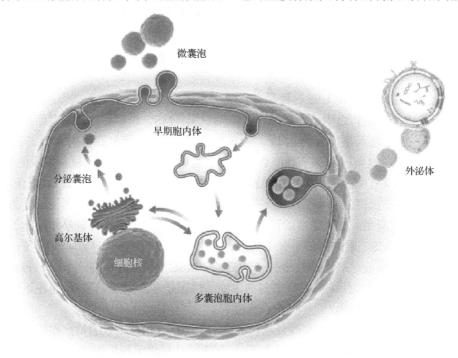

图 22-1　囊泡的产生和类型[17]（彩图请扫封底二维码）

二、外泌体的生物学特性

外泌体是一种直径为 30~150nm，密度为 1.10~1.18g/mL 的囊泡样小体，因为外泌体的半径较小，普通显微镜无法对其进行观察，所以现在主要用透射电镜来观察外泌体的形态，应用动态光散射仪分析外泌体的粒径[12]。外泌体内含有与其细胞来源相关的蛋白质、miRNA 及 mRNA 等物质。另外，不同细胞来源的外泌体具有相同的膜蛋白分布，如四跨膜蛋白超家族（CD63、CD9 和 CD81）、热休克蛋白（HSP70 和 HSP90）及主要组织相容性复合体（MHC）分子[18]。通过应用流式细胞术、蛋白免疫印迹（Western Blot）和酶联免疫吸附试验（enzyme-linked immunosorbent assay，ELISA）可以检测这些蛋白，对外泌体进行鉴定。

三、外泌体的组成和功能

外泌体生物学功能的发挥依赖于其所运载的物质被靶细胞通过胞吞作用内在化[1,19]。外泌体运载的大量功能性分子（图 22-2），我们可以在 ExoCarta 网站（http://www.exocarta.org/）中查找[20]。外泌体在靶细胞内的作用方式较复杂，外泌体内载有的 mRNA 可在靶细胞内翻译出相应蛋白质，miRNA 可通过降解靶细胞的 mRNA 或抑制 mRNA 翻译来调控目的蛋白的表达，而干扰小 RNA（siRNA）则可直接敲除靶细胞中的目标基因发挥基因沉默作用[21]。

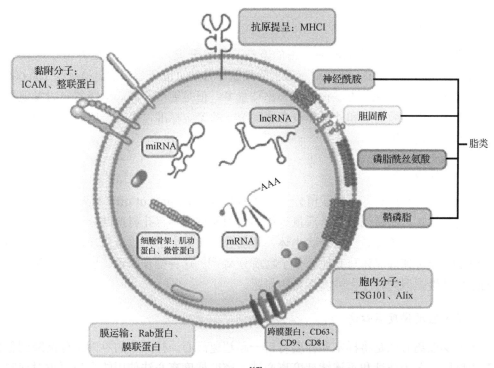

图 22-2 外泌体的组成成分[17]（彩图请扫封底二维码）

外泌体中含有蛋白质、mRNA、miRNA 和脂类等生物活性成分；ICAM. 细胞间黏附分子；lncRNA. 长链非编码 RNA

尽管外泌体中的分子是细胞依赖性的，但是一些分子进入外泌体并不是随机的。在生理或者是病理条件改变的情况下，外泌体所含组分是可以改变的。这说明微环境的改变，可以调节细胞分泌外泌体所运载的组分。这是由于微环境改变，调节细胞内成分改变，这种改变进一步转移到外泌体中。有研究比较了几种细胞产生的外泌体，发现一些 miRNA（如 miR-150、miR-142-3p、miR-451）易于进入外泌体[22]。另外的研究发现，与肺癌、直肠癌相比较，胃癌细胞分泌的外泌体中 let-7 家族成员表达水平更高[23]。

外泌体既可以通过质膜受体直接激活受体细胞，也可以转运蛋白、mRNA、miRNA 甚至细胞器进入受体细胞内，同时可以携带不同病理状态下细胞所含有的特异性物质，从而在生理学和病理学上都发挥着重要作用[16]。外泌体的质膜含有丰富的胆固醇、鞘磷脂、神经酰胺、脂筏及磷脂酰丝氨酸，外泌体不仅可以保护这些生物活性物质在细胞外环境中不被降解和稀释，也可以促进这些物质通过组织液或血液进行远距离输送，同时外泌体膜上特异性的表面配体允许它与受体细胞高效率结合[24]。

四、外泌体的分离方法

外泌体分离纯化是进行外泌体研究的重要前提。只有得到纯化的具有生物活性的外泌体，才能够对外泌体的作用及治疗效果进行更好的研究。自从 Thery 等[12]应用超速离心法提取外泌体后，产生了很多分离纯化外泌体的方法。但是，还没有一种方法能同时保证外泌体的含量、纯度、生物活性。

（一）超速离心法

超速离心法是外泌体提取最常用的方法。简单来说，收集细胞培养液以后依次在 $300×g$、$2\ 000×g$、$10\ 000×g$ 条件下离心去除细胞碎片和大分子蛋白质，最后经 $100\ 000×g$ 离心得到外泌体[12]。此种方法得到的外泌体量多，但是纯度不高，电镜鉴定时发现外泌体聚集成块，由于微囊泡和外泌体没有非常统一的鉴定标准，也有一些研究认为此种方法得到的不全是外泌体。

（二）过滤离心

过滤离心是利用具有不同截留分子量（molecular weight cut-off，MWCO）的超滤膜离心分离外泌体。截留相对分子质量是指能自由通过某种有孔材料的分子中最大分子的相对分子质量。外泌体是一个囊状小体，相对分子质量大于一般蛋白质，因此选择不同的 MWCO 膜可使外泌体与其他大分子物质分离[25]。这种操作简单、省时，不影响外泌体的生物活性，但同样存在纯度不高的问题。

（三）密度梯度离心法

密度梯度离心法是将样本和梯度材料一起超速离心，样品中的不同组分沉降到各自的等密度区，分为连续和不连续梯度离心法。密度梯度离心法使用的介质要求对细胞无毒，在高浓度时黏度不高且易将 pH 调至中性。实验中常用蔗糖密度梯度离心法[8,26]，在离心的基础上，预先将两种浓度蔗糖溶液（如 2.5mol/L 和 0.25mol/L）配成连续梯度体系置于超速离心管中，样本铺在蔗糖溶液上，$100\ 000×g$ 离心 16h，外泌体会沉降到等密度区（1.10~1.18g/mL）。用此种方法分离到的外泌体纯度高，但是前期准备工作繁杂、耗时，得到的量少。

（四）免疫磁珠法

免疫磁珠是包被有单克隆抗体的球形磁性微粒，可特异性地与靶物质结合。同样，在离心的基础上，预先使磁珠包被针对外泌体相关抗原的抗体（如 CD9、CD63 和 Alix），之后与外泌体共同孵育，蒸馏水冲洗后，重悬于磷酸盐缓冲液（PBS）中[27]。这种方法可以保证外泌体形态的完整，特异性高、操作简单、不需要昂贵的仪器设备，但是非中性 pH 和非生理性盐浓度会影响外泌体的生物活性，不便进行下一步的实验。

（五）化学沉淀法

聚合物如聚乙二醇（PEG）通常用于沉淀病毒和其他小颗粒。实验的原理可能是根据外泌体在不同溶剂中溶解性存在差异来从实验样品中沉淀外泌体。沉淀物可以采用低速离心或过滤分离获取。Biosciences 有一款名为 ExoQuick 的试剂盒（www.systembio.com），可以从细胞条件培养基、血清或尿液中沉淀外泌体。美国生命技术公司

（www.lifetechnologies.com/exosomes）推出 5 种提取试剂，从而可从不同样本类型中快速地回收所有的外泌体，适用于细胞培养液，血清、血浆、尿液等体液（唾液、乳汁、脑脊液、腹水和羊膜流体）。PEG 通过捆绑水分子，促使溶解度低的外泌体从水中沉淀出来。2016 年，Rider 等[28]发表的一篇关于外泌体提取方法的文献介绍，通过终浓度为 8%的 PEG6000 与生物体液共孵育 14h 后，10 000×g 离心 1h，可将直径在 30~150nm，且包含丰富 RNA 的外泌体沉淀下来。这种方法简单快速、产品回收率高。缺点是耗时，PEG 可沉淀蛋白质和其他脂质体，需要后续处理除去 PEG。

（六）色谱法

色谱法是根据凝胶孔隙的孔径大小与样品分子尺寸的相对关系而对溶质进行分离、分析的方法。样品中大分子不能进入凝胶孔，只能沿多孔凝胶粒子之间的空隙通过色谱柱，首先被流动相洗脱出来；小分子可进入凝胶中绝大部分孔洞，在柱中长时间滞留，更慢地被洗脱出。分离到的外泌体在电镜下大小均一，但是需要特殊的设备，应用不广泛[29]。

五、外泌体的分析方法

鉴定外泌体的方法很多，主要有扫描电子显微镜法、原子力显微镜法、纳米追踪分析法、流式细胞仪检测法、蛋白免疫印迹、酶联免疫吸附试验等。通常采用 2 或 3 种方法联合鉴定。

（一）电子显微镜检查法

电子显微镜按结构和用途可分为透射式电子显微镜、扫描式电子显微镜、反射式电子显微镜和发射式电子显微镜等。透射式电子显微镜常用于观察那些用普通显微镜所不能分辨的细微物质结构；扫描式电子显微镜主要用于观察固体表面的形貌，也能与 X 射线衍射仪或电子能谱仪相结合，构成电子微探针，用于物质成分分析；发射式电子显微镜用于自发射电子表面的研究。电子显微技术（electron microscopy，EM）可以检测到较小的颗粒，但是过程烦琐，操作技术难度大，需要专业人士操作，不适合高通量分析。

（二）动态光散射技术

动态光散射（dynamic light scattering，DLS）技术是在纳米颗粒做无规则布朗运动时，观测散射光随时间的波动性得到颗粒的速度，并通过斯托克斯-爱因斯坦方程测定颗粒粒径及粒度分布[30]。

（三）纳米颗粒追踪分析技术

纳米颗粒跟踪分析（nanoparticle tracking analysis，NTA）技术可以直接并实时观测纳米颗粒，通过光学显微镜收集纳米颗粒的散射光信号，拍摄其在溶液中做布朗运动的影像，再对每个颗粒的布朗运动进行追踪和分析，从而计算出纳米颗粒的流体力学半径和浓度。NTA 技术对外泌体进行单独检测，免受复杂样本（如血清、尿液等）环境的影响。NTA 技术只是检测颗粒的尺寸不能分析颗粒的组成[31]。DLS 和 NTA 技术可以检测到较小的颗粒，但是特异性差，往往高估颗粒的数量，对于复杂样品（大小不均一）计数误差大。

（四）生物免疫学分析技术

生物免疫学分析技术常用的有 ELISA 和 Western Blot。生物公司有专门检测 CD63 和 CD81 表达量的 ELISA 试剂盒。ELISA 操作简单，适合高通量分析，但是无法知道颗粒的数量和大小。Western Blot 操作烦琐，需要大量的样本。

（五）流式细胞仪检测法

多数外泌体具有独特的蛋白和脂成分，携带细胞膜所表达的蛋白-微管蛋白和肌动蛋白等细胞骨架蛋白、四跨膜蛋白超家族（CD63、CD9、CD81、CD82）、热休克蛋白 hsp70 和 hsp90、TSG101、Alix、浮舰蛋白-1（flotillin-1）、葡萄糖新陈代谢酶、信号转导蛋白（如异三聚体 G 蛋白）、MHC 分子、网格蛋白、转运和融合蛋白（如膜联蛋白）及翻译延伸因子[26]。因此，常用特定外泌体的荧光抗体（如 CD63、CD9、CD81）进行流式细胞仪检测。

另外，外泌体是一种脂质体，而 PKH67 荧光细胞标记试剂可在细胞膜的脂质双分子层中稳定结合绿色荧光染料，因此可以标记含有脂质双分子层膜的外泌体。捷克科学家利用 Apogee 超灵敏纳米流式细胞仪研发出检测胞外囊泡和外泌体的流式新方法[27]。这方法可以快速地在细胞培养液和复杂的人体样本中定量与检测泌外体及胞外囊泡，而且不需要用特别试剂，如外泌体磁珠试剂盒。整个过程主要用亲脂性荧光染料 PKH67、针对特定外泌体和胞外囊泡的荧光抗体及超灵敏纳米流式细胞仪 A50-Micro。这个新的技术可应用于日后快速和常规的胞外囊泡与外泌体检测，在临床医学领域有很大的优势。

第三节　间充质干细胞来源外泌体的治疗作用

MSC 的功能机制已经成为当前研究的热点。体内和体外的研究显示，MSC 具有免疫调节、促血管新生和组织再生的能力，但是其作用机制还不清楚。越来越多的证据表明 MSC 治疗过程中其旁分泌作用起了主导作用。研究显示，MSC 通过释放小分子物质，如生长因子、细胞因子和趋化因子，发挥其治疗作用。除了释放这些可溶性小分子物质外，MSC 也释放大量的外泌体，这些外泌体可以在细胞间交流，从而在正常的生理过程和疾病的病理过程中发挥作用[1]。外泌体的特性与其来源细胞有关，现有研究已证明 MSC 分泌的外泌体在减少心肌损伤范围[2,3]、促进组织损伤修复[4,5]、促进急性肾小管损伤修复[6]、促进神经再生、减少肺损伤和免疫系统调节[7-9]等方面有重要作用。当前，大量的文献报道指出 MSC 来源的外泌体具有显著的促进血管新生的作用[32,33]。

一、MSC 来源外泌体与心血管疾病

心血管疾病（CVD）已经成为世界范围内最主要的致死病因。研究者投入了巨大的精力来研发相应的治疗手段用于改善这些疾病的预后[34]。在过去的 10 年里，以干细胞为

基础的治疗技术已取得重要进展。不同类型的干细胞、MSC 及其分泌的因子对治疗 CVD 具有非常重要的潜在价值[35]。许多研究发现，MSC 条件培养基可以改善小鼠和猪心肌梗死损伤，减少心肌梗死面积[36]。后来证明，这些治疗效果主要来源于条件培养基中的外泌体[37]。

最近研究发现，MSC 来源外泌体对 CVD 具有治疗效果是由于外泌体具有促血管新生的能力[38]。干细胞来源的外泌体由于含有 20S 蛋白酶体、合成 ATP 所需的各种酶、调控血管新生的 miRNA 及血管内皮生长因子（VEGF）等物质，在抗缺血/再灌注损伤、促进血管新生及缺血器官功能恢复中具有重要作用[39]。正常人骨髓间充质干细胞（hBM-MSC）分泌的外泌体可抑制外周血单个核细胞（PBMC）分泌干扰素-γ（IFN-γ），内含免疫相关 miRNA，如 miR301、miR22 和 miR-let-7a 等，能促进人脐静脉内皮细胞（HUVEC）网状结构形成和血管形成[40]。血管新生是内皮细胞增殖和游走形成小血管的过程，K562 细胞来源的外泌体携带有 miR-92，可促进 HUVEC 游走和管状结构形成，从而促进血管新生[41]。Bian 等[42]的研究发现 BM-MSC 在缺氧条件下产生的细胞外囊泡（EV）能够促进血管形成，保护心肌组织免受缺血损害。另外，Teng 等[43]也证实 BM-MSC 来源的外泌体能够刺激血管新生，从而改善大鼠心肌缺血模型的心肌功能。其他研究也发现，脐带间充质干细胞（UC-MSC）来源的外泌体也可以通过促进血管新生来改善缺血心肌功能[44]。同时，不同来源的 MSC 产生的外泌体具有相同的促进血管新生的作用。

二、MSC 来源外泌体与免疫系统间作用

外泌体作为细胞间交流信号的传导者的和分子转移的运载体，能促进蛋白质、脂类、miRNA 和 RNA 的细胞间传递。外泌体的免疫活性影响免疫调节机制，包括免疫激活、免疫抑制、免疫监视和细胞间的沟通。除了免疫细胞，肿瘤细胞外泌体活性也会影响生理和病理过程。据观察，免疫细胞外泌体如树突细胞（DC）外泌体调节免疫反应，这些微囊泡执行免疫调节的方式使之成为潜在的免疫治疗试剂。事实上，肿瘤和免疫细胞外泌体已被证实可携带肿瘤抗原并促进免疫力，从而清除促进肿瘤产生的 $CD4^+$ T 细胞，直接抑制肿瘤生长和恶性肿瘤发展。进一步了解这些外泌体的生物学功能，特别是参与免疫细胞定位、交换和调节的分子机制，有可能为研究免疫调节和治疗干预提供重要的手段[45]。

最近几年的研究发现，脂多糖（LPS）预处理的 MSC 旁分泌作用显著增强，包括营养支持作用的增强和再生与修复能力的提高。MSC 可能分泌大量的外泌体影响细胞与细胞之间的交流并为组织损伤修复维持一个动态和自我平衡的内环境。韩卫东等[9]研究评价了 LPS 预处理的 MSC 来源的外泌体对慢性炎症和伤口修复的治疗效果与机制。研究结果显示，LPS pre-Exo 可上调抗炎因子的表达和促进 M2 型巨噬细胞的激活。miRNA 芯片筛选结果显示，LPS pre-Exo 中有独特的 let-7b 表达，而 let-7b/TLR4 通路与巨噬细胞极化和炎症消除有关。进一步研究结果显示，TLR4/NF-κB/STAT3/Akt 调节通路在 LPS pre-Exo 调节炎症反应中巨噬细胞的极化上发挥着重要作用。研究结论：LPS 预处理的 MSC 通过外泌体传递 let-7b 调节炎症反应中巨噬细胞的极化，这些外泌体具有为伤口修复提供免疫治疗的潜能。

三、MSC 来源外泌体与神经系统疾病

Xin 等[25]第一次利用多能 MSC 的外泌体来系统性治疗脑卒中后的小鼠，外泌体促进了神经与血管的构建及其功能的恢复。对成年雄性 Wistar 鼠进行持续 2h 的大脑中动脉阻塞，24h 后从尾静脉注入含 100mg 蛋白的外泌体沉淀或同等体积的磷酸盐缓冲液（PBS）。处死脑卒中后 28 天的小鼠，利用组织病理学和免疫组化来确认神经突重建、神经发生和血管生成。研究结果表明，静脉注射外泌体可促进功能恢复和神经突重建、神经发生和血管生成。外泌体来源的 miRNA 和其他信号分子一样，调节着生物信号网络，参与多种生理过程。MSC 来源的外泌体通过转移 miR-133b 到神经细胞，可促进神经轴突的生长[46]。

四、MSC 来源外泌体与肿瘤血管新生

血管新生是肿瘤增长的重要环节。肿瘤细胞通过过度表达血管新生相关的生长因子[如血管内皮生长因子（VEGF）]来促进血管新生。由于 MSC 具有天然的肿瘤趋向能力，因此许多研究开始探讨 MSC 在肿瘤血管新生中的作用。这些研究产生了有争议的结果：一些研究阐述了 MSC 对肿瘤血管新生具有支持作用[47-49]；另外一些研究发现 MSC 具有广泛的抑制肿瘤血管新生的作用[50,51]。与这些结果相同，MSC 来源的外泌体在肿瘤血管新生中的作用也产生了分歧。有报道显示，MSC 来源的外泌体通过上调肿瘤细胞的 VEGF 表达，促进肿瘤血管新生；而另外的一个研究结果恰恰相反，表明 MSC 来源的外泌体通过下调肿瘤细胞的 VEGF 表达，抑制肿瘤血管新生[52,53]。这种分歧还不能很好地解释，可能 MSC 及其外泌体对不同的肿瘤细胞具有不同的效果。另外，MSC 的异质性也导致其产生的外泌体携带不同的物质，产生不同的治疗结果。

有研究显示，鼠来源 BM-MSC 分泌的外泌体可以显著地下调鼠乳腺癌细胞（4T1）的 VEGF 表达水平，抑制内皮细胞的增殖和迁移，进而抑制血管新生，这种抑制效果是浓度依赖的。进一步研究显示，这种外泌体含有 miRNA-16，miRNA-16 可以抑制 VEGF 的表达。因此，这种外泌体转移 miRNA-16 进入肿瘤细胞，抑制 VEGF 表达[53]。相反的，另一项研究显示 hBM-MSC 来源的外泌体通过活化 ERK1/2 和 p38 MAPK 途径，促进胃癌细胞（SGC-7901）的 VEGF 表达，这种促进效果也是剂量依赖的。在这个研究中，作者抑制了 ERK1/2 活化，从而逆转了 VEGF 表达升高的趋势[52]。

五、MSC 来源外泌体与组织修复

血管新生在组织修复中起到重要的作用，是肉芽组织形成所必需的过程。在创伤愈合过程中，血管以出芽的形式侵入到创面的纤维组织中，形成新的毛细血管网。以前的研究显示，MSC 的条件培养基能够促进创伤愈合、组织修复。现在很多的研究已经证实，外泌体在这个过程中起到了至关重要的作用。然而，其作用机制还不是很清楚。

研究发现，人脐带间充质干细胞（hUC-MSC）分泌的外泌体介导 WNT4 信号通路，参与皮肤创伤修复。作者发现在小鼠体内，hUC-MSC 介导的 Wnt/b-连环蛋白活性在上皮形成和细胞增殖中发挥重要作用。当 hUC-MSC 敲除 Wnt4 后，b-连环蛋白活性消失，

其分泌的外泌体在体内的皮肤修复疗效明显下降[4]。Shabbir 等[54]的研究结果显示，BM-MSC 分泌的外泌体可以提高正常创伤和慢性创伤中成纤维细胞的增殖与迁移能力，作者阐述了 BM-MSC 来源的外泌体可运载具有转录活性的 STAT3，进入靶细胞后，可以促进血管内皮生长因子（VEGF）、肝细胞生长因子（HGF）和白介素 6（IL-6）表达。STAT3 在创伤愈合中扮演着重要角色，包括促进功能性细胞迁移、增殖，促进血管新生和生长因子产生[55]。因此，BM-MSC 来源的外泌体能够激活 Akt、ERK1/2 和 STAT3，这些信号通路都与调节血管新生相关。诱导多能干细胞（iPSC）来源的 MSC 分泌的外泌体也能对创伤产生很好的治疗效果。研究发现，这种治疗效果是通过外泌体在创伤部位显著提高胶原合成和血管新生实现的[54]。

从临床转化的角度来看，MSC 来源的外泌体在各种临床疾病的模型中取得了令人鼓舞的治疗效果。这些研究结果显示，外泌体可以有效地运载重要的具有治疗效果的分子，从而刺激不同细胞的增殖，促进组织修复和血管新生。同时，外泌体是可以持续再生的，用外泌体代替干细胞可以成为一种更有效的治疗策略。虽然外泌体起作用的机制尚不明确，某些现有结果也存在一些争议。但是与细胞相比，外泌体具有更高的稳定性、更小的免疫排斥反应和更少的微环境的影响。目前已被应用于多种疾病模型中，并起到了一定的疗效。深入研究 MSC 来源的外泌体的功能，将其开发成为一种既具有 MSC 特点而又能克服其缺陷的新型治疗方式，是今后具有重要探讨价值的研究方向。

第四节　提高间充质干细胞来源外泌体生物活性的研究

寻找提高 MSC 来源外泌体促进血管新生能力的方法，就是寻找一种能够使外泌体携带更多促进血管新生因子的方法[19]。不同细胞来源的外泌体具有不同的治疗潜能。例如，内皮祖细胞（EPC）来源的外泌体具有促进血管新生的潜能，可以解释为 EPC 来源的外泌体在缺血的病理条件下具有促进血管新生的功能。实验也证实了这种说法，EPC 来源的外泌体可以加速大鼠动脉内皮损伤处的再内皮化；这种外泌体也能刺激内皮细胞分泌更多的生长因子[56]。一些前期研究结果显示，不同胎盘组织来源的 MSC 有不同的能力，可以推断出不同来源的 MSC 产生的外泌体的治疗作用也会各不相同[57]。另外，除了细胞来源，细胞所处环境变化和细胞被人为修饰，也已经被考虑作为产生具有更好治疗作用外泌的方法。目前，有几种用于给外泌体装载更多的有益物质的方法已经被研究，目的是调节外泌体的治疗潜能。

一、外泌体在纳米医学中的应用

从再生医学的角度讲，外泌体的治疗作用完全符合干细胞的治疗作用的逻辑过程。干细胞治疗一度被誉为具有神奇效果的治疗方法。同样，装载了大量干细胞因子（SCF）的外泌体也被发现具有与干细胞相似的治疗效果。这引发了大量研究者深入调查单独应用外泌体是否可以提供较好的药理作用，甚至替代干细胞成为新的治疗手段。这项研究工作已得到初步成果：外泌体已经被证明能够抑制细胞凋亡和促进细胞增殖，诱导血管新生，调节炎症和免疫反应，诱发凝血机制，影响分化通路，并且能够增强细胞的植入[58]。

外泌体来源广泛，MSC 来源的外泌体已经成为研究的热点。在动物模型研究中，如心肌缺血、肾缺血、胰岛移植、肺动脉高压、骨软骨缺损、关节炎、烧伤、移植物抗宿主病（GVHD）和炎症，MSC 来源的外泌体已经显示出了具有巨大的组织再生和损伤修复潜力[58]。除了外泌体自身表现出来的效果，外泌体已被用来作为在体内释放药物和寡核苷酸的载体，来源于 DC 和巨噬细胞的外泌体已被用来作为传染病和癌症疫苗[59,60]。早期临床研究显示，外泌体作为脑膜炎、癌症和 GVHD 的疫苗是安全和有效的。

以外泌体为基础的疗法规避了细胞治疗一些难以解决的问题，如应激反应引起的坏死或异常分化。外泌体的体积小，相比于整个细胞应用于治疗更具有优势，包括降低腹腔巨噬细胞的吞噬功能和避免血管栓塞，容易注射，可提高肿瘤血管通过性。虽然合成的纳米级小载体（如脂质体、纳米）可以达到类似的效果，但是外泌体具有以细胞为基础的生物结构和功能，在治疗应用中具有更大的优势。例如，外泌体可提供天然的生物相容性，更高的化学稳定性，更远距离的细胞间交流和固有的细胞间通信、融合和传递能力[58]。一些研究也表明，外泌体具有细胞选择性融合和组织特异性取向，以及穿透血脑屏障等严密组织结构的能力[61]。另外，脂质体和纳米颗粒系统在试剂的选择、制备流程和表面功能化方面具有高度的方法灵活性。这种纳米级颗粒的合成系统可以装载更多的仿生材料，如靶向性的抗体或趋化性的配体，以及非生物单位，如造影剂或光热材料。

二、外泌体的改造技术

修饰外泌体的许多方法在以前已被用于细胞的功能化。细胞修饰一般是通过特定内源性物质的生物合成或直接的外源性物质的加载来实现的。这两种方法都可以用来操纵细胞分泌来改进外泌体功能，而后一种方法也可以用来直接纯化外泌体。这一研究领域越来越受到研究者的重视，主要是研究新的外泌体功能化技术，并使外泌体具有更高的治疗效果。

（一）通过细胞修饰改造外泌体

几十年来，研究人员将非天然物质引入细胞以增强其治疗功能[62]。现在这些技术极有可能应用到外泌体的构建中。例如，装载到细胞膜的生物材料必然会被传递到外泌体膜上，而内在的生物活性材料可以被包装进入外泌体。利用这些方案，细胞修饰和细胞工程技术将适用于外泌体的功能化。在本节中，我们将讨论如何应用这些技术如基因工程、代谢标记和外源性装载等，来改进外泌体的成分，进而促进外泌体的治疗效果。

1. 通过细胞基因修饰改造外泌体

毫无疑问，基因工程是应用最多的细胞工程化策略，这种方法也可以用来调节泌体在治疗中的应用。有研究显示，mRNA 转染入细胞可以转载进入外泌体，然后在靶细胞可以检测到相应蛋白的表达[63,64]。同样，基因工程可以将丰富的非编码 RNA 序列转载入外泌体，如 miRNA 或干扰小 RNA（siRNA）[65,66]。这种研究方法利用了外泌体与生俱来的细胞结合能力和防止 RNA 被酶降解的功能。但是，在这些实验中需要解决几个问题。例如，研究表明，观察到的效果不一定完全来源于外泌体所包裹的 RNA，实际上可能有一部分是大分子蛋白复合物、脂蛋白或蛋白寡核苷酸复合物携带的 RNA。因此，对外泌体纯化对于研究外泌体的功能是非常必要的。

一些研究发现,应用超速离心法从瞬时转染质粒 DNA 或 mRNA 的 HEK293FT 细胞中分离外泌体,这些外泌体装载有这些转染质粒 DNA 或 mRNA 进入靶细胞。结果显示,质粒 DNA 可以在靶细胞表达相应的蛋白,但 mRNA 不能在靶细胞表达相应的蛋白[67]。这可能是因为外泌体传递的 mRNA 在受体细胞的溶酶体中快速降解,这一过程阻止了功能蛋白的表达。

以上这些研究都是使用寡核苷酸来诱导或调节靶细胞中基因的表达。另一种方法是诱导母细胞基因表达蛋白,直接分离装载有这种蛋白的外泌体应用于临床治疗。一个非常有效的做法是将编码的蛋白质与外泌体中富含的蛋白质构成融合表达基因,从而确保目的蛋白定位到外泌体。这种方法需要仔细的设计和对外泌体蛋白分子生物学的充分了解。Stickney 等[68]用这种方法,将荧光蛋白报告基因和萤光素酶报告基因与外泌体富含的 CD63 蛋白基因制成融合基因。这个体系可以以可视化图像的方式来研究外泌体的生物合成、细胞之间的传递和外泌体输入后在体内的分布。最近,Lai 等[37]报道了更先进的可视化系统。他们应用棕榈酰化信号与 RNA 结合序列融合,转录后棕榈酰化的 mRNA 结合序列靶向人胚胎肾(human embryonic kidney,HEK)细胞膜,它被包装进入外泌体。在外泌体中,RNA 结合序列结合表达 GFP 的噬菌体外壳蛋白,从而可以直接可视化 mRNA 包装进入外泌体。

2. 通过代谢标记改造外泌体

代谢标记是一个行之有效的细胞改造策略,规避了许多遗传操作改变细胞功能的问题。这种方法利用的是细胞利用培养基中的营养成分,如氨基酸、脂类、核酸或糖进行新陈代谢的过程。这些营养物质由细胞通过合成代谢,分别整合到基因组、蛋白质组、磷脂组和糖组。这些营养物质被修饰后整合到细胞中,可以用于对细胞生物学过程进行研究。代谢标记是一种不加区分修改整个细胞的生物分子技术,往往替换细胞质膜中的成分。因此,外泌体必将包含经过代谢标记的成分,如胞内蛋白或胞膜脂质。Wang 等[69]最近应用非天然氨基酸 L-azidohomoalanine 作为甲硫氨酸替代物,将叠氮基团整合到黑色素细胞来源外泌体的蛋白质组中。应用代谢标记细胞进而标记外泌体的研究还很少,Wang 等的研究是一个非常好的应用代谢标记改造外泌体的例子。

3. 通过细胞摄取外来物质改造外泌体成分

遗传修饰和代谢标记策略是利用细胞生物合成产生内源性生物活性物质装载入外泌体。另一种方法是将外源性物质直接装载入细胞,进而被外泌体所携带。这种外泌体能够更好地将药物运输到靶细胞,起到更好的治疗作用。许多研究已经采用这种方法得到较好的效果。在一个研究模型中,研究者将卟啉类化合物作为一个药物模型转入细胞中,这种细胞分泌的外泌体装载有这种卟啉类化合物,作用于癌细胞模型。结果显示,含有卟啉类化合物的外泌体显示出具有更强的杀伤肿瘤细胞的作用[70]。用相似的方法,对细胞或外泌体加载 VEGF 等促进血管新生的因子,可以促进这些因子被靶细胞内吞,起到更好地促进血管新生的作用。这些研究需要注意的,要对加载的蛋白因子的含量进行控制,适当的因子含量才能起到更好地促进血管新生的作用。

外泌体装载外源性物质的多少通常依赖于细胞中这种物质的装载量，而这又是由物质-细胞相互作用的强度决定的。例如，很小或根本没有细胞结合能力的纳米颗粒将会与细胞质膜发生很弱的、非特异性的相互作用。在这种情况下，高的纳米颗粒浓度和延长孵育时间能够最大限度地促进纳米颗粒与细胞结合，并达到足够的细胞装载量。例如，Neubert 和 Glumm[71]认为用 0.5mm 的超顺磁性氧化铁纳米颗粒（SPION）经神经元的原代细胞培养产生加载的外泌体，需要培养 24h。应用巨噬细胞产生外泌体可以更好地避免这种情况的发生，因为巨噬细胞具有吞噬能力，可以通过吞噬作用积极地内化大量外源性物质。Silva 等[72]用氧化铁纳米颗粒和小分子光敏剂与巨噬细胞共孵育，产生具有磁性和光学反应性的外泌体。这些外泌体称为 theranosome，具有磁力靶向性，用于核磁共振成像（MRI）和光动力疗法。然而，这种方法的一个关键限制因素是它依赖于细胞的吞噬作用。如何增加非吞噬细胞的负载能力，使细胞含有更多的外源物质是研究人员所面临的一个挑战。

增加细胞结合能力的一种方法是利用外源性材料和细胞质膜间的疏水反应。除了利用小的疏水分子外，另外常用的方法是使用脂质体系统，脂质体载体可以直接插入脂质双分子层膜与细胞膜融合，而被封装在内的亲水基团可以释放入细胞。这种方法已经由 Lee 等证明。Lee 等[73]用膜融合脂质体技术将疏水性成分加载到细胞膜，将亲水性成分加载到细胞液。有趣的是，作者观察到加载到细胞膜脂质的量与纳入外泌体的脂质量不同。这说明这种方法明显缺乏控制可控性。以脂质体为基础的加载策略的另一个限制是加载外源性物质的效率低下，这也正是制约这种方法发展的关键所在。

（二）直接改造外泌体

以改造细胞为基础的外泌体修饰策略仅仅能将细胞内的小部分生物活性因子装载入外泌体。这样的结果是花费很高，但是效果较小。相反，直接对纯化的外泌体进行功能性的修饰可以解决这一问题。外泌体的直接改造主要有以下几种方法。

1. 外泌体膜的修饰

对外泌体膜的修饰可分为共价修饰和非共价修饰。外泌体膜可以被修饰，这是外泌体与活细胞相比的优势所在。例如，外泌体可以经受住不能用于活细胞的膜修饰操作过程，如过高的温度、压力和低渗或者高渗的溶液。Smyth 等[74]应用共价修饰外泌体膜的方法，将含有叠氮化物的荧光团以共价连接的方式结合到外泌体膜上。除了以共价结合的方式修饰外泌体外，还有几种以非共价结合方式修饰外泌体的方法，如多加电位结合、受体-配体结合和疏水作用插入。但是，对外泌体膜的修饰也能够引起外泌体囊泡的聚集，很可能会影响外泌的膜结合能力，从而影响外泌体将携带的生物活性物质运送至靶细胞。

2. 外泌体直接加载活性物质

为了直接利用外泌体的细胞结合能力和保护 RNA 不被 RNA 酶降解，一些研究应用不同的方法直接为外泌体加载所需的生物活性物质，如生长因子和药物治疗分子等。Cooper 等[75]将 α-Syn siRNA 应用电穿孔的方法加载进入外泌体，可以减少模型小鼠脑内的 α-突触核蛋白聚集，从而减轻帕金森病的症状。Tian 等[76]分离提取了小鼠不成熟树突细胞（iDC）分泌的外泌体，将化疗药物阿霉素装载进入这种外泌体，对肿瘤起到了很好的抑制效果。这种对外泌体直接加载生物活性分子的方法可以保证将目的分子直接装载进入外泌体。但是，这种方法对外泌体表型和细胞结合能力的影响还需要进一步研究。

三、微环境改变诱导增强外泌体的促血管新生作用

最近研究发现，微环境改变的刺激可以延长 MSC 移植后的存活时间、提高 MSC 的分泌能力，从而使 MSC 具有更好的治疗作用。那么同理，这样的环境改变同样可改变外泌体的分子组成。Anderson 等[77]研究发现，MSC 分泌的外泌体会随着环境的改变而改变。在这项研究中，研究人员将 MSC 暴露在外周动脉疾病（PAD）的环境中培养，产生的外泌体所含有的表皮生长因子（EGF）、成纤维细胞生长因子（FGF）和血小板源生长因子（PDGF）显著升高，从而提高了外泌体的促血管新生作用。

（一）应激反应诱导增强外泌体的促血管新生作用

外泌体不是细胞的缩小版，它的分子组成不是缩小化的细胞分子组成。外泌体只含有丰富的特殊的 RNA 和蛋白质，而缺少其他物质。这说明有相应的机制控制外泌体装载其所能含有的物质。另外，外泌体的组成成分不是一成不变的，它的成分能够因细胞所处微环境的改变而改变。这说明外泌体装载物质也是一个可调节的过程。但是，调节物质进入外泌体的机制还不清楚。机体应激状态，如缺氧、缺血、放射都能改变外泌体的内含物含量，从而影响外泌体的功能[19]。

组织损伤后首先使机体处于应激状态，机体首先需要克服这个状态才能进行有效的组织再生治疗。各种应激反应可以改变外泌体介导的信号通路。应激诱导能使外泌体中的小 RNA 和蛋白质组成发生变化，从而为受损的靶细胞提供保护作用。Borges 等[78]的研究结果表明，缺氧损伤能够使内皮细胞产生更多的外泌体，并且改变了外泌体的组成，这是通过转化生长因子-β（TGF-β）介导的通路来实现的。另有研究发现，缺氧条件下脑神经胶质瘤细胞分泌的外泌体与正常状态下分泌的外泌体相比，含有更多的促血管新生的生长因子和细胞因子。缺氧刺激也可以使 MSC 来源的外泌体更快地被内皮细胞内吞，起到更好地促进心肌缺血模型血管新生的作用[79]。

（二）生物活性因子刺激诱导增强外泌体的促血管新生作用

研究发现，一些生物活性因子的刺激也能够增强 MSC 分泌的外泌体的促血管新生作用。Song 等[80]的研究发现，应用 IL-1β 预刺激 MSC，可以使 MSC 分泌的外泌体中 miR-146a 水平增加，从而提高 MSC 对败血症的治疗作用。Anderson 等[77]应用血小板源生长因子（PDGF）刺激 MSC，实验结果显示，经过 PDGF 刺激后，MSC 能够产生更多的外泌体，而且这种外泌体含有促血管新生分子 c-Kit 和它的配体 SCF，一些抑制血管新生的因子在这些外泌体中显著减少。SCF/c-Kit 信号通路可以促进内皮细胞的增殖、迁移和管状形成能力，并且可以招募自身的 MSC。所以，PDGF 刺激 MSC 产生的运载 c-Kit 和它的配体 SCF 的外泌体被内皮细胞吞后，可以提高内皮细胞的促血管新生作用[81]。

增加 MSC 来源外泌体中促血管新生因子的含量，可以直接或间接地通过以上叙述的

方法来完成。但是，每种方法都有其优点和局限性。所以，对其有效性还需要更多的比较研究来确定[19]。

（杜　为　李宗金）

参 考 文 献

[1] Ranganath SH, Levy O, Inamdar MS, et al. Harnessing the mesenchymal stem cell secretome for the treatment of cardiovascular disease. Cell Stem Cell, 2012, 10(3):244-258.

[2] Ma J, Zhao Y, Sun L, et al. Exosomes derived from Akt-modified human umbilical cord mesenchymal stem cells improve cardiac regeneration and promote angiogenesis via activating platelet-derived growth factor D. Stem Cells Transl Med, 2017, 6(1):51-59.

[3] Sahoo S, Losordo DW. Exosomes and cardiac repair after myocardial infarction. Circ Res, 2014, 114(2):333-344.

[4] Zhang B, Wang M, Gong A, et al. HucMSC-exosome mediated-Wnt4 signaling is required for cutaneous wound healing. Stem Cells, 2015, 33(7):2158-2168.

[5] Zhang B, Wu X, Zhang X, et al. Human umbilical cord mesenchymal stem cell exosomes enhance angiogenesis through the Wnt4/beta-catenin pathway. Stem Cells Transl Med, 2015, 4(5):513-522.

[6] Bruno S, Grange C, Deregibus MC, et al. Mesenchymal stem cell-derived microvesicles protect against acute tubular injury. J Am Soc Nephrol, 2009, 20(5):1053-1067.

[7] Drommelschmidt K, Serdar M, Bendix I, et al. Mesenchymal stem cell-derived extracellular vesicles ameliorate inflammation-induced preterm brain injury. Brain Behav Immun, 2017, 60:220-232.

[8] Lee C, Mitsialis SA, Aslam M, et al. Exosomes mediate the cytoprotective action of mesenchymal stromal cells on hypoxia-induced pulmonary hypertension. Circulation, 2012, 126(22):2601-2611.

[9] Ti D, Hao H, Tong C, et al. LPS-preconditioned mesenchymal stromal cells modify macrophage polarization for resolution of chronic inflammation via exosome-shuttled let-7b. J Transl Med, 2015, 13:308.

[10] Raposo G, Stoorvogel W. Extracellular vesicles: exosomes, microvesicles, and friends. J Cell Biol, 2013, 200(4):373-383.

[11] Gould SJ, Raposo G. As we wait: coping with an imperfect nomenclature for extracellular vesicles. J Extracell Vesicles, 2013, 2(1):20389.

[12] Thery C, Zitvogel L, Amigorena S. Exosomes: composition, biogenesis and function. Nat Rev Immunol, 2002, 2(8):569-579.

[13] Gyorgy B, Szabo TG, Pasztoi M, et al. Membrane vesicles, current state-of-the-art: emerging role of extracellular vesicles. Cell Mol Life Sci, 2011, 68(16):2667-2688.

[14] Andaloussi SEL, Mager I, Breakefield XO, et al. Extracellular vesicles: biology and emerging therapeutic opportunities. Nat Rev Drug Discov, 2013, 12(5):347-357.

[15] Thery C. Exosomes: secreted vesicles and intercellular communications. F1000 Biol Rep, 2011, 3:15.

[16] Desrochers LM, Antonyak MA, Cerione RA. Extracellular vesicles: satellites of information transfer in cancer and stem cell biology. Dev Cell, 2016, 37(4): 301-309.

[17] Riazifar M, Pone EJ, Lotvall J, et al. Stem cell extracellular vesicles: extended messages of regeneration. Annu Rev Pharmacol Toxicol, 2017, 57:125-154.

[18] Katsuda T, Kosaka N, Takeshita F, et al. The therapeutic potential of mesenchymal stem cell-derived extracellular vesicles. Proteomics, 2013, 13(10-11):1637-1653.

[19] Alcayaga-Miranda F, Varas-Godoy M, Khoury M. Harnessing the angiogenic potential of stem cell-derived exosomes for vascular regeneration. Stem Cells Int, 2016, (2016):3409169.

[20] Keerthikumar S, Chisanga D, Ariyaratne D, et al. ExoCarta: a web-based compendium of exosomal cargo. J Mol Biol, 2016, 428(4):688-692.

[21] Corrado C, Raimondo S, Chiesi A, et al. Exosomes as intercellular signaling organelles involved in health and disease: basic science and clinical applications. Int J Mol Sci, 2013, 14(3):5338-5366.

[22] Guduric-Fuchs J, O'Connor A, Camp B, et al. Selective extracellular vesicle-mediated export of an overlapping set of microRNAs from multiple cell types. BMC Genomics, 2012, 13:357.

[23] Ohshima K, Inoue K, Fujiwara A, et al. Let-7 microRNA family is selectively secreted into the extracellular environment via exosomes in a metastatic gastric cancer cell line. PLoS One, 2010, 5(10):e13247.

[24] Rani S, Ryan AE, Griffin MD, et al. Mesenchymal stem cell-derived extracellular vesicles: toward cell-free therapeutic applications. Mol Ther, 2015, 23(5):812-823.

[25] Xin H, Li Y, Cui Y, et al. Systemic administration of exosomes released from mesenchymal stromal cells promote functional recovery and neurovascular plasticity after stroke in rats. J Cereb Blood Flow Metab, 2013, 33(11):1711-1715.

[26] Pospichalova V, Svoboda J, Dave Z, et al. Simplified protocol for flow cytometry analysis of fluorescently labeled exosomes and microvesicles using dedicated flow cytometer. J Extracell Vesicles, 2015, 4:25530.

[27] Chen TS, Arslan F, Yin Y, et al. Enabling a robust scalable manufacturing process for therapeutic exosomes through oncogenic immortalization of human ESC-derived MSCs. J Transl Med, 2011, 9:47.

[28] Rider MA, Hurwitz SN, Meckes DG Jr. ExtraPEG: a polyethylene glycol-based method for enrichment of extracellular vesicles. Sci Rep, 2016, 6:23978.

[29] Boing AN, van der Pol E, Grootemaat AE, et al. Single-step isolation of extracellular vesicles by size-exclusion chromatography. J Extracell Vesicles, 2014, (3):23430.

[30] Im H, Shao H, Park YI, et al. Label-free detection and molecular profiling of exosomes with a nano-plasmonic sensor. Nat Biotechnol, 2014, 32(5):490-485.

[31] Im H, Shao H, Weissleder R, et al. Nano-plasmonic exosome diagnostics. Expert Rev Mol Diagn, 2015, 15(6):725-733.

[32] Chen J, Liu Z, Hong MM, et al. Proangiogenic compositions of microvesicles derived from human umbilical cord mesenchymal stem cells. PLoS One, 2014, 9(12):1-16.

[33] Merino-Gonzalez C, Zuniga FA, Escudero C, et al. Mesenchymal stem cell-derived extracellular vesicles promote angiogenesis: potencial clinical application. Front Physiol, 2016, 7:24.

[34] Hirsch AT, Duval S. The global pandemic of peripheral artery disease. Lancet, 2013, 382(9901):1312-1314.

[35] Nguyen PK, Riegler J, Wu JC. Stem cell imaging: from bench to bedside. Cell Stem Cell, 2014, 14(4):431-444.

[36] Timmers L, Lim SK, Arslan F, et al. Reduction of myocardial infarct size by human mesenchymal stem cell conditioned medium. Stem Cell Res, 2007, 1(2):129-137.

[37] Lai RC, Arslan F, Lee MM, et al. Exosome secreted by MSC reduces myocardial ischemia/reperfusion injury. Stem Cell Res, 2010, 4(3):214-222.

[38] Waldenstrom A, Ronquist G. Role of exosomes in myocardial remodeling. Circ Res, 2014, 114(2):315-324.

[39] Penfornis P, Vallabhaneni KC, Whitt J, et al. Extracellular vesicles as carriers of microRNA, proteins and lipids in tumor microenvironment. Int J Cancer, 2016, 138(1):14-21.

[40] Jansen F, Yang X, Proebsting S, et al. MicroRNA expression in circulating microvesicles predicts cardiovascular events in patients with coronary artery disease. J Am Heart Assoc, 2014, 3(6):e001249.

[41] Umezu T, Ohyashiki K, Kuroda M, et al. Leukemia cell to endothelial cell communication via exosomal miRNAs. Oncogene, 2013, 32(22):2747-2755.

[42] Bian S, Zhang L, Duan L, et al. Extracellular vesicles derived from human bone marrow mesenchymal stem cells promote angiogenesis in a rat myocardial infarction model. J Mol Med (Berl), 2014, 92(4):387-397.

[43] Teng X, Chen L, Chen W, et al. Mesenchymal stem cell-derived exosomes improve the microenvironment of infarcted myocardium contributing to angiogenesis and anti-inflammation. Cell Physiol Biochem, 2015, 37(6):2415-2424.

[44] Zhao Y, Sun X, Cao W, et al. Exosomes derived from human umbilical cord mesenchymal stem cells relieve acute myocardial ischemic injury. Stem Cells Int, 2015, 2015(1):761643.

[45] Greening DW, Gopal SK, Xu R, et al. Exosomes and their roles in immune regulation and cancer. Semin Cell Dev Biol, 2015, 40:72-81.

[46] Xin H, Li Y, Buller B, et al. Exosome-mediated transfer of miR-133b from multipotent mesenchymal stromal cells to neural cells contributes to neurite outgrowth. Stem Cells, 2012, 30(7):1556-1564.

[47] Zhang T, Lee YW, Rui YF, et al. Bone marrow-derived mesenchymal stem cells promote growth and angiogenesis of breast and prostate tumors. Stem Cell Res Ther, 2013, 4(3):70.

[48] Beckermann BM, Kallifatidis G, Groth A, et al. VEGF expression by mesenchymal stem cells contributes to angiogenesis in pancreatic carcinoma. Br J Cancer, 2008, 99(4):622-631.

[49] Orecchioni S, Gregato G, Martin-Padura I, et al. Complementary populations of human adipose CD34$^+$ progenitor cells promote growth, angiogenesis, and metastasis of breast cancer. Cancer Res, 2013, 73(19):5880-5891.

[50] Otsu K, Das S, Houser SD, et al. Concentration-dependent inhibition of angiogenesis by mesenchymal stem cells. Blood, 2009, 113(18):4197-4205.

[51] Ho IA, Toh HC, Ng WH, et al. Human bone marrow-derived mesenchymal stem cells suppress human glioma growth through inhibition of angiogenesis. Stem Cells, 2013, 31(1):146-155.

[52] Zhu W, Huang L, Li Y, et al. Exosomes derived from human bone marrow mesenchymal stem cells promote tumor growth *in vivo*. Cancer Lett, 2012, 315(1):28-37.

[53] Lee JK, Park SR, Jung BK, et al. Exosomes derived from mesenchymal stem cells suppress angiogenesis by down-regulating VEGF expression in breast cancer cells. PLoS One, 2013, 8(12):e84256.

[54] Shabbir A, Cox A, Rodriguez-Menocal L, et al. Mesenchymal stem cell exosomes induce proliferation and migration of normal and chronic wound fibroblasts, and enhance angiogenesis *in vitro*. Stem Cells Dev, 2015, 24(14):1635-1647.

[55] Sano S, Chan KS, DiGiovanni J. Impact of Stat3 activation upon skin biology: a dichotomy of its role between homeostasis and diseases. J Dermatol Sci, 2008, 50(1):1-14.

[56] Li X, Chen C, Wei L, et al. Exosomes derived from endothelial progenitor cells attenuate vascular repair and accelerate reendothelialization by enhancing endothelial function. Cytotherapy, 2016, 18(2):253-262.

[57] Gonzalez PL, Carvajal C, Cuenca J, et al. Chorion mesenchymal stem cells show superior differentiation, immunosuppressive, and angiogenic potentials in comparison with haploidentical maternal placental cells. Stem Cells Transl Med, 2015, 4(10):1109-1121.

[58] Armstrong JP, Holme MN, Stevens MM. Re-engineering extracellular vesicles as smart nanoscale therapeutics. ACS Nano, 2017, 11(1):69-83.

[59] Beauvillain C, Juste MO, Dion S, et al. Exosomes are an effective vaccine against congenital toxoplasmosis in mice. Vaccine, 2009, 27(11):1750-1757.

[60] Del Cacho E, Gallego M, Lee SH, et al. Induction of protective immunity against Eimeria tenella infection using antigen-loaded dendritic cells (DC) and DC-derived exosomes. Vaccine, 2011, 29(21):3818-3825.

[61] Wood MJ, O'Loughlin AJ, Samira L. Exosomes and the blood-brain barrier: implications for neurological diseases. Ther Deliv, 2011, 2(9):1095-1099.

[62] Armstrong JP, Shakur R, Horne JP, et al. Artificial membrane-binding proteins stimulate oxygenation of stem cells during engineering of large cartilage tissue. Nat Commun, 2015, 6:7405.

[63] Zomer A, Maynard C, Verweij FJ, et al. *In vivo* imaging reveals extracellular vesicle-mediated phenocopying of metastatic behavior. Cell, 2015, 161(5):1046-1057.

[64] Ridder K, Keller S, Dams M, et al. Extracellular vesicle-mediated transfer of genetic information between the hematopoietic system and the brain in response to inflammation. PLoS Biol, 2014, 12(6):e1001874.

[65] Luo SS, Ishibashi O, Ishikawa G, et al. Human villous trophoblasts express and secrete placenta-specific microRNAs into maternal circulation via exosomes. Biol Reprod, 2009, 81(4):717-729.

[66] Buck AH, Coakley G, Simbari F, et al. Exosomes secreted by nematode parasites transfer small RNAs to mammalian cells and modulate innate immunity. Nat Commun, 2014, 5:5488.

[67] Kanada M, Bachmann MH, Hardy JW, et al. Differential fates of biomolecules delivered to target cells via extracellular vesicles. Proc Natl Acad Sci USA, 2015, 112(12):E1433- E1442.

[68] Stickney Z, Losacco J, McDevitt S, et al. Development of exosome surface display technology in living human cells. Biochem Biophys Res Commun, 2016, 472(1):53-59.

[69] Wang M, Altinoglu S, Takeda YS, et al. Integrating protein engineering and bioorthogonal click conjugation for extracellular vesicle modulation and intracellular delivery. PLoS One, 2015, 10(11):e0141860.

[70] Fuhrmann G, Serio A, Mazo M, et al. Active loading into extracellular vesicles significantly improves the cellular uptake and photodynamic effect of porphyrins. J Control Release, 2015, 205:35-44.

[71] Neubert J, Glumm J. Promoting neuronal regeneration using extracellular vesicles loaded with superparamagnetic iron oxide nanoparticles. Neural Regen Res, 2016, 11(1):61-63.

[72] Silva AK, Luciani N, Gazeau F, et al. Combining magnetic nanoparticles with cell derived microvesicles for drug loading and targeting. Nanomedicine, 2015, 11(3): 645-655.

[73] Lee J, Lee H, Goh U, et al. Cellular engineering with membrane fusogenic liposomes to produce functionalized extracellular vesicles. ACS Appl Mater Interfaces, 2016, 8(11):6790-6795.

[74] Smyth T, Petrova K, Payton NM, et al. Surface functionalization of exosomes using click chemistry. Bioconjug Chem, 2014, 25(10):1777-1784.

[75] Cooper JM, Wiklander PB, Nordin JZ, et al. Systemic exosomal siRNA delivery reduced alpha-synuclein aggregates in brains of transgenic mice. Mov Disord, 2014, 29(12):1476-1485.

[76] Tian Y, Li S, Song J, et al. A doxorubicin delivery platform using engineered natural membrane vesicle exosomes for targeted tumor therapy. Biomaterials, 2014, 35(7):2383-2390.

[77] Anderson JD, Johansson HJ, Graham CS, et al. Comprehensive proteomic analysis of mesenchymal stem cell exosomes reveals modulation of angiogenesis via nuclear factor-kappaB signaling. Stem Cells, 2016, 34(3):601-613.

[78] Borges FT, Melo SA, Ozdemir BC, et al. TGF-beta1-containing exosomes from injured epithelial cells activate fibroblasts to initiate tissue regenerative responses and fibrosis. J Am Soc Nephrol, 2013, 24(3):385-392.

[79] Kucharzewska P, Christianson HC, Welch JE, et al. Exosomes reflect the hypoxic status of glioma cells and mediate hypoxia-dependent activation of vascular cells during tumor development. Proc Natl Acad Sci USA, 2013, 110(18):7312-7317.

[80] Song Y, Dou H, Li X, et al. Exosomal miR-146a contributes to the enhanced therapeutic efficacy of interleukin-1beta-primed mesenchymal stem cells against sepsis. Stem Cells, 2017, 35(5).

[81] Lopatina T, Bruno S, Tetta C, et al. Platelet-derived growth factor regulates the secretion of extracellular vesicles by adipose mesenchymal stem cells and enhances their angiogenic potential. Cell Commun Signal, 2014, 12:1-12.

第二十三章 间充质干细胞药物开发策略

随着技术的进步和医疗条件的改善，人类历史上的大规模传染性疾病和急性疾病已经得到了有效控制，因而人类对生命质量的期望不断提高。同时，我国已经步入老龄化社会，与老龄化相关的重大、难以治愈的疾病如糖尿病、心血管疾病、癌症和阿尔茨海默病等发病率不断攀升。以分子药物和手术治疗为支柱的现当代医学显得力不从心，逐渐遭遇玻璃天花板。以干细胞技术为核心、被科学界誉为第三次医学革命的再生医学已是大势所趋，有望通过对疾病组织细胞再生、替代与修复，从根本上改变医疗困局，为人类面临的创伤性、神经性、免疫性、代谢性、退行性及肿瘤性等难治性重大疾病提供崭新的治疗方案，从而成为生物医药领域的新支柱。

间充质干细胞（MSC）是干细胞家族的重要成员，其具有多向分化潜能、造血支持和促进干细胞植入、免疫调节和自我复制等特点，具有有较高的安全性，而且不论是自体还是同种异源使用一般都不会引起宿主的免疫反应，被认为是继造血干细胞（HSC）之后最接近大规模临床应用的成体干细胞，有希望用于治疗多种疾病。其中，胎盘和脐带来源的 MSC 具有分化潜力大、增殖能力强、免疫原性低、取材方便、无道德伦理问题限制、易于工业化制备等特征，成为药物开发理想的研究材料。本章以围产期组织来源的 MSC 为例，探讨干细胞药物开发基本流程及策略。

第一节 干细胞临床转化探索之路

一、国际干细胞临床研究竞争激烈

干细胞研究和再生医学的发展为人类面临的创伤性、免疫性、代谢性、退行性及肿瘤性等难治性重大疾病将提供革命性的治疗方案。世界各国纷纷投入大量的人力、物力和财力抢占干细胞领域的制高点。截至 2018 年 3 月，美国国立卫生研究院（NIH）临床试验网站（ClinicalTrials.gov）注册的全球干细胞相关临床试验超过 6504 项，Ⅰ期、Ⅱ期、Ⅲ期、Ⅳ期临床研究分别为 2185 项、3147 项、652 项和 183 项。其中，HSC 研究 2293 项，约占 35%；MSC 研究 807 项，约占 12%，主要涉及的疾病类型包括脑卒中、肺结节病、外周动脉疾病（PAD）、类风湿关节炎（RA）、特发性肺纤维化（IPF）及免疫系统疾病等；胚胎干细胞（ESC）/诱导多能干细胞（iPSC）研究 115 项，约占 2%，主要集中在眼睛、胰腺和多种神经退行性疾病或损伤[如帕金森病、肌萎缩侧索硬化（amyotrophic lateral sclerosis，ALS）和脊髓损伤]治疗领域；神经干细胞（NSC）研究 55 项，约占 1%，主要是旨在修复受损的中枢神经系统（CNS）。

目前全球共有将近 300 种干细胞相关药物正在研发，迄今，美国、加拿大、欧盟、韩国、日本、印度、新西兰等国家和地区的药品监督管理部门已经批准了 17 个干细胞制剂上市（表 23-1）。2016 年 2 月，TEMCELL®（即 Prochymal®）干细胞制剂在日本上市销售用于治疗移植物抗宿主病（GVHD），成为首个得到完全批准的大规模生产的干细胞产品。

表 23-1　国际上已获批准上市的干细胞产品[1]

商品名	开发机构	细胞来源	适应证	批准机构	时间	类型
Prochymal	Osiris Therapeutics	骨髓间充质干细胞	移植物抗宿主病 肠道克罗恩病 Ⅰ型糖尿病	美国 FDA	2005-12 2009-12 2010-05	孤儿药
MPC	Mesoblast	自体间充质前体细胞	受损骨组织修复	澳大利亚 TGA	2010-10	医疗产品
Hearticellgram-AMI	FCB-Pharmicell	自体骨髓间质干细胞	急性心肌梗死	韩国 KFDA	2011-07（已撤销）	药品，条件批准 Ⅲ期临床失败
Hemacord Ducord Allocord Clevecord	New York Blood Center、Duke University、SSM、Cleveland 等 6 家	脐血来源造血祖细胞	血液及免疫系统疾病	美国 FDA	2011-11 2012-10 2013-05 2016-09	生物制品
Cartistem	Medipost	脐血间充质干细胞	退行性关节炎和膝关节软骨损伤	韩国 KFDA	2012-01	药品
Cuepistem	Anterogen	自体脂肪间充质干细胞	复杂性克罗恩病并发肛瘘	韩国 KFDA	2012-01	药品
Prochymal	Osiris Therapeutics	骨髓间充质干细胞	儿童移植物抗宿主病	加拿大卫生部	2012-05	药品
MultiStem	America Stem Cell	骨髓等来源多能成体祖细胞	赫尔勒综合征	美国 FDA	2012-07	孤儿药
Holoclar	Chiesi Farmaceutici	自体角膜缘干细胞	灼伤引起的角膜缘干细胞缺陷症	欧洲 EMA	2015-02	先进医疗产品
Temcell	Mesoblast	骨髓间充质干细胞	移植物抗宿主病	日本 MPDA	2016-02	药品，完全批准
Stempeucel	Stempeutics	骨髓混合间充质干细胞	血栓闭塞性脉管炎（Buerger 病）	印度 DCGI	2016-06	药品，条件批准
Strimvelis	GSK	基因修饰自体造血干细胞	儿童 ADA-SCID 免疫缺陷症	欧洲 EMA	2016-08	先进医疗产品

从数据上可以看出，已经有相当多的投资用于干细胞治疗临床前研究和临床试验中，但只有为数不多的品种最终获准临床应用。Geron、ACT、StemCells、Osiris 公司等干细胞领域的先驱从 20 世纪 90 年代涉足干细胞治疗产品的开发，所开展的 ESC、NSC、MSC 项目遭遇诸多波折而被迫中止，分别卖给了 BioTime、Astellas、MicroBot、Mesoblast 公司。

国际上已获准上市的干细胞药品的前景并不太乐观，Osiris 公司在探索 Prochymal 用于治疗成人 GVHD、克罗恩病和糖尿病等病症时屡屡受挫，2013 年 11 月该公司将 Prochymal 转让给澳大利亚 Mesoblast 公司。2013 年 11 月，FCB-Pharmicell 公司用于急性心梗治疗的干细胞药物 Hearticellgram-AMI 第 Ⅱ/Ⅲ 期的临床试验结果显示失败。

二、中国干细胞临床监管滞后、有花无果

20 世纪 90 年代以来，我国开始大力支持干细胞临床研究及应用，原卫生部药政管理局于 1993 年颁布的《人的体细胞治疗及基因治疗临床研究质控要点》[2]，2003 年国家

食品药品监督管理总局发布《人体细胞治疗研究和制剂质量控制技术指导原则》[3]中，干细胞作为体细胞制剂纳入药品管理，2009年卫生部发文《医疗技术临床应用管理办法》[4]，将干细胞移植归为第三类医疗技术，引发了干细胞治疗究竟是医疗技术还是药品的长期争论。这一时期，我国成为世界上接受干细胞治疗的患者中最多的国家，其安全性问题受到广泛关注[5-7]，2012年1月卫生部发文《关于开展干细胞临床研究和应用自查自纠工作的通知》[8]，叫停了未经批准的干细胞治疗临床研究。

国家卫生和计划生育委员会与食品药品监督管理总局在2015年8月联合发布《临床干细胞临床研究管理办法（试行）》[9]与《干细胞制剂质量控制及临床前研究指导原则（试行）》[10]，新办法为开放干细胞临床研究设立了七道门槛，实行备案管理制，为探索干细胞临床转化迈出积极的一步。

我国在干细胞及其转化应用研究领域起步较早，基础研究几乎与发达国家同步。科技部门通过973计划、863计划、国家重大科学研究计划等对干细胞的基础研究、关键技术和资源平台建设给予了大力支持，取得了一批标志性成果。目前，我国干细胞领域的论文数量排名国际第2位；一批研究机构进入了国际研究机构前20位，专利数量已经排名国际第3位。

国家食品药品监督管理总局官网的数据显示，2004年以来，药品审评中心共受理了10项干细胞新药临床申请（表23-2），其中"骨髓原始间充质干细胞""自体骨髓间充质干细胞注射液""间充质干细胞心梗注射液""脐带血巨核系祖细胞注射液"获准开展临床试验，但截至2015年3月，所有这些干细胞药物均在不同阶段夭折[1]。长期以来干细胞临床研究监管滞后、相关标准缺失，到目前为止，我国除HSC外尚无任何一种干细胞治疗技术或产品获准临床应用，真正意义上实现创新的干细胞技术尚未形成。

表23-2 国家药品审评中心受理的干细胞制剂

受理号	药品名称	来源	承办日期	备注
CXSL1300090	注射用人脐带间充质干细胞	异体	2014-3-14	受理
CXSL1300091	注射用人脐带间充质干细胞	异体	2014-3-14	受理
CXSL1200056	脐带间充质干细胞抗肝纤维化注射液	异体	2013-8-2	受理
CXSL1000057	脐带间充质干细胞抗肝纤维化注射液	异体	2011-10-25	受理
CXSL1600082	人牙髓间充质干细胞注射液	异体	2016-9-26	撤回
CXSL1300001	人脐带间充质干细胞注射液	异体	2013-3-7	受理
CXSL0600068	脐带间充质干细胞注射液	异体	2007-2-7	受理
CXSL0500073	间充质干细胞肝纤维化注射液	异体	2005-12-23	受理
X0408234	间充质干细胞心梗注射液	异体	2005-1-5	Ⅰ/Ⅱ临床
X0400586	骨髓原始间充质干细胞/GVHD	异体	2004-10-10	Ⅰ/Ⅱ临床
X0407487	自体骨髓间充质干细胞注射液/心肌缺血	异体	2004-12-11	Ⅰ/Ⅱ临床
X0404119	脐带血红系祖细胞注射液/造血支持	异体	2004-8-1	Ⅰ/Ⅱ临床
X0404120	脐带血巨核系祖细胞注射液/血小板减少症	异体	2004-7-29	Ⅰ/Ⅱ临床

注：信息源自国家药品审评中心官网

三、干细胞临床监管问题复杂、面临挑战

国际上,一项新的医疗技术在被大规模应用于临床治疗之前,至少需要经过临床前研究和临床试验两个阶段,只要有足够充分的证据证明新疗法优于既有的疗法,且在安全性、有效性、质量可控性及医学伦理上均通过专业评估后新疗法才可以应用于临床。

干细胞从获得到移植使用的过程复杂,由于干细胞组织来源多样,组织分化能力不同,供者遗传背景复杂,干细胞体外制备过程较长且要引入多种外来成分,还可能污染致病性外源因子,特别是新鲜制备细胞,一般要求 24h 内回输,无法完成质量检查,带来急性毒性、异常毒性和病原体感染等安全风险;体外操作过程可能会带来干细胞的生物学特性发生改变而导致免疫毒性和致瘤风险;在保存、运输、复苏、配制过程中细胞制剂处在动态变化之中。如若对干细胞制剂缺乏有效的质量控制手段和标准,缺乏规范的临床前评价数据,则无法保证细胞临床应用的安全性和有效性。

由于机构间缺乏需遵循的统一规范,各机构对细胞来源和细胞制剂无统一的衡量标准,因此临床结果无法相互认可和形成公认的结论,阻碍了干细胞研究的进程;同时,缺乏安全性、有效性依据和质量标准的细胞临床使用,有可能对人体造成伤害甚至危及患者生命。眼下风靡全球的"转化医学"倡导"从实验室到病床",这是一个连续、双向、开放的研究过程,意味着它可以直接将干细胞治疗送入一个"灰色地带"。

美国约有 570 个干细胞治疗诊所,且数量还在不断增多。2012 年 2 月 1 日美国食品药品监督管理局(FDA)对位于科罗拉多州的"再生科学"公司的干细胞产品 Regenexx 的临床应用行为发出禁令,FDA 声明人体中的干细胞属于药物,应由其负责管辖,并和美国司法部一道提起诉讼。7 月 23 日,美国干细胞疗法诉讼暂告一段落,华盛顿特区地方法院承认了美国 FDA 拥有管理干细胞疗法的权力。《自然》杂志在线报道称,这一判决主要取决于法院是否同意 FDA 关于干细胞是药物的论断[11]。2014 年和 2015 年,FDA 公布了一系列提议,拟对大量干细胞诊所加以管理,这些诊所宣称能进行未被证明的干细胞疗法。这些提议遭到了干细胞诊所和某些患者的强烈谴责,这些患者不想等到这些干细胞疗法被证明有效才开始接受治疗。

国际上干细胞临床转化普遍实行分级分类监管模式,根据干细胞来源(自体/异体)、特性(胚胎/成体)和体外处理(扩增/诱导/修饰)方式分类,在分级管理方面根据安全性进行分级,临床转化研究根据风险级别和类型按照相关规范进行评估。世界卫生组织(WHO)、国际细胞治疗协会(ISCT)、国际干细胞研究学会(ISSCR)、细胞治疗认证基金会(FACT)等国际组织长期致力于规范临床细胞治疗,建立了一系列细胞制剂、药品临床应用指南、质量评估体系及认证标准[12-14]。世界各国对于细胞临床转化监管并无统一认识,甚至是按技术管理还是按药物管理都有很大分歧,各国制定的监管模式与质量评估体系也不尽相同[15]。

美国 FDA 对细胞及干细胞治疗采取了分级分类管理模式,按干细胞种类和风险级别管理。FDA 针对细胞和基因治疗产品(CGT)发布了《人体细胞治疗和基因治疗指南》[13]、《人类细胞、组织以及细胞的与基于组织的产品机构的现行良好组织规范》(GTP)、《细胞和基因治疗产品的效力试验指南》等一系列法规指南[16-19]。

欧洲药品管理局（EMA）设有专门的先进治疗委员会，对先进医学治疗产品（ATMP）制定了类似美国的法规，如《高级医学治疗产品》《人类使用医学产品的临床试验》等[20,21]。

韩国作为国家战略在大力推进干细胞转化应用，韩国 KFDA 成立了专门机构来修订相关的法规和指南，在整个生物制品里面单列出细胞及干细胞治疗产品的《药品生产质量管理规范》（GMP）指南，把干细胞作为生物制品来管理，并批准上市了全球首个干细胞治疗新药。

日本采用双轨制对干细胞临床转化进行管理，厚生劳动省既接纳干细胞药品类注册申报，也受理以先进治疗技术启动的临床研究[22,23]。

干细胞治疗究竟是临床技术还是药物？我国之前没有明确的界定。由于伦理、宗教环境和人口优势，以及法规和标准的滞后或缺位，"神奇的干细胞疗法"不失时机地钻了空子。由于监管不力，一些还没有进行正规临床试验证明其有效性和安全性的细胞疗法几乎在全国的三甲医院遍地开花。2016年的"魏则西事件"引起舆论关注之后，国家卫生和计划生育委员会再度明令叫停当时的细胞疗法，要求干细胞、免疫细胞治疗技术研究按临床规定执行。

综上所述，干细胞为医药史上最为复杂的治疗性产品，研发者与监管者均缺乏经验。在缺乏系统的工程化、标准化开发前提下，难以保证细胞制剂质量可控性及临床应用的安全性与有效性。

四、药物开发是干细胞临床转化的必由之路

相当长一段时期干细胞治疗被作为治疗技术来对待，而"技术"本身就存在着探索性和不确定性，受制于人员、技能、场所、时间等条件要求，难以满足不断增加的临床需求。

一项治疗手段实现临床转化需要满足安全有效、质量可控、经济可行等基本条件。在缺乏系统的工程化、标准化开发前提下，无法实现细胞制剂的质量可控性，更无从保证干细胞临床应用的安全性与有效性。标准化是实现质量可控性前提，自动化可大幅度降低人为因素干扰，为标准化提供保障，而规模化制备则为实现标准化、自动化提供经济可行的条件。

经过数十年的发展，国际上已建立起包括 GLP、GCP、GMP、GSP 在内的相对完善的药品评价与管理体系，参照药品监管和评价体系开发干细胞药物有利于建立起统一的规范和标准，促进社会各专业化分工合作，突破干细胞临床转化瓶颈。

药品必须具有安全性、有效性与质量可控性（稳定性与均一性）三个基本要素，干细胞制剂也不应例外。安全性：只有在患者的获益大于不良反应，或可解除、缓解不良反应的情况下才可按规定的适应证和用法、用量使用。有效性：适应证明确、疗效肯定，是药品的根本属性。稳定性：在规定的条件下保持其有效性和安全性的能力，是药品的重要质量特征。均一性：每一单位产品的物理化学分布一致，均符合有效性、安全性的规定要求，是体现药品标准的质量特性。

干细胞与化学药物、抗生素、蛋白质药物等化学成分明确的分子药物的研发理念与规律有很大不同，干细胞制剂的动态性、异质性问题从根本上挑战了传统药物均一性、稳定性的基本质量要求。干细胞从获得到移植使用面临一系列技术障碍；人类干细胞移植的历史不长，诸多的科学问题悬而未决；干细胞临床转化需要从业者有足够的谨慎和耐心。

2016年12月16日，国家食品药品监督管理总局药品审评中心对外发布了关于《细胞制剂研究与评价技术指导原则（征求意见稿）》的通知[24]，经过一年的反复讨论与修订，于2017年12月22日颁布《细胞治疗产品研究与评价技术指导原则（试行）》[25]，该原则基本结束了长期以来细胞治疗到底是药品还是临床技术的争论，度过了"第三类医疗技术"向"药品"监管的过渡期，产品上市审批道路逐渐打通，为细胞治疗健康规范发展带来了新的机遇。

第一，指导原则明确了大方向，无论是自体还是异体细胞，无论是干细胞还是免疫细胞，无论是诱导分化还是基因修饰的细胞，无论是游离的或是与辅助材料结合的细胞，除生殖细胞及其相关干细胞外，只要经过体外培养和操作制成的活细胞产品，均按照药品进行注册和监管，结束了"放开细胞治疗技术"等不切实际的期待，避免了对产业政策走向的误读、误判和误导。

第二，指导原则有机地将细胞制剂纳入了现有药品评价体系，基本上涵盖了安全性、有效性及质量可控性方面的重大关切问题，对伦理、知情权、非预期变化应对、风险管控等给予了充分关注。同时，突出了细胞治疗产品种类繁多，其治疗原理、体内生物学行为、临床应用的利弊存在巨大差别和不确定性，以及与小分子药物、大分子生物药物和医疗器械之间的不同，遵循"具体情况具体分析"原则保留了一定灵活处理的空间，结束了过往细胞学家与药审专家"鸡同鸭讲"、各说各话的局面。

第三，指导原则从客观上提高了准入门槛，药品开发将会是细胞治疗技术临床转化的唯一出口，可预期未来细胞治疗技术会成为生物医药行业的一个细分领域，而非一个独立的行业。在这个框架下产业回到同一起跑线上公平竞争，但是细胞制剂开发过程占用大量资源，风险巨大，迫使企业在进行战略决策时将更加谨慎，不可心存一丝侥幸，需要踏踏实实走好每一步，从而避免一些企业急功近利的投机行为。

第四，指导原则的实施将结束学术专家主导"细胞行业"的时代，在新药创制框架下建立起统一的规范和标准，促进社会各专业分工合作，提高执业药师与工程技术人员的话语权，促进培养与引进熟悉药物开发和细胞技术的综合型人才，共同突破细胞制剂临床转化的技术瓶颈。

指导原则紧跟国际前沿，跳出传统药物监管的思维，坚持以临床价值为导向，管理与研发并肩前行，本着审慎研发、审评、管理的同时，也为探究与开发者留出空间，共同促进和保护公众的健康。指导原则将会是细胞制剂评审的一个基本原则，虽充分考虑了专业性与合理性，但细胞治疗产品涉及范围广，同时具有个性化要求，一些灵活性条款落实在具体操作上会让人十分茫然，干细胞药物开发的未知性、高风险对我国干细胞药物开发从业人员及药审专家在客观上提出了巨大的挑战。

中国干细胞研究几乎与发达国家同步，但面对干细胞产业快速发展，我国政府显得始料未及。其实，在现代生物医药创新领域我国整体落后于西方国家数十年。客观地说，过去60多年来我们的研发机构、企业和政府只有"仿制"的经验，尚未曾建立"新药创制"体系。从新药注册的层面上讲，只要是我国市场上没有出现过的药品就算是新药。数据显示，目前我国企业生产的化学药品97%为仿制药，其余基本都是剂型转换、分子结构修饰或抢仿的所谓"新药"。国际上研发成功一个新药，需要8~13亿美元、12年左右时间。而国内从早期研究到临床试验到新药证书取得，所需费用约在1亿元上下，去除劳动力、原材料成本因素外，新药研究规范与国际不接轨及缺乏创新投入是我国造不出真正意义上的原创新药的根本原因。至今我国仅有青蒿素被国际认可为原创新药，但其知识产权被国外制药企业所掌控。

"技术优先，科学随后"是世界各国技术和科学发展的一般规律，而我国干细胞技术发展规划、计划、重大技术创新项目的选择和评价等历来是科学家的天下，企业家、工程师很少有发言权，形成了政府资助项目跟风"重磅突破"而轻视"技术积累"的局面。同时，干细胞技术开发和临床应用研究还涉及诸多伦理学问题、规范和标准问题、工程技术问题，涉及多学科学术交叉、多技术综合运用，是一项空前复杂的系统工程，其所面临的挑战不是当下干细胞主战场上的细胞学家和医师能够考虑周详的，亦非单一机构或部门能够独立解决，需要社会化、专业化的分工协作。干细胞产业决策中科学家"越位"、工程师"缺位"亦制约我国干细胞产业链的形成和发展。

近年来，CFDA开展了仿制药一致性评价、新药临床数据打假等专项行动，两年来发布了一系列里程碑式的政策和措施，取消认证使GMP/GLP/GCP成为起步，重新定义"新药"，施行药品上市许可持有人制度，建立药品专利链接，完善药品医疗器械审评制度，创新药短缺药等优先审评审批，拓展性同情使用临床试验，加入人用药物注册技术要求国际协调会议（ICH）组织降低国外新药进入中国的政策门槛，改革的力度和效率前所未有，信号已非常明确，那就是鼓励创新药满足临床未被满足的临床需求，监管与国际接轨，实现中国新药研发和上市与全球同步，干细胞新药创制迎来历史最好机遇。

第二节　间充质干细胞药品开发流程

干细胞药物开发的目的是从"一对一"的个体化干细胞临床应用转向"一对多"的临床治疗方案，为临床提供标准化的干细胞制剂。药物的开发需要遵循常规药物开发的流程，包括药学研究（CMC）、非临床研究、临床试验申请（IND）、早期临床试验、确证性临床试验、新药申请（NDA）和批准上市等过程。

MSC制剂的药学研究主要包括工艺开发与质量评价。①干细胞库（MCB）与工作细胞库（WCB）建立：供者筛查、组织采集，细胞分离、纯化、培养、保藏、鉴别、效力检测，以及生物学特性、遗传学稳定性研究，干细胞库技术标准及工作标准建立。②干细胞制剂工艺开发与药学研究：大规模细胞扩增、细胞制备工艺、剂型选择、包装选择、处方筛选、制品冻存/复苏工艺，体外操作对干细胞生物学特性的影响，过程质量控制，干细胞药物放行标准建立。③干细胞制剂冷链传输系统及稳定性研究：产品冻存与复苏、冷链传输技术、临床快速检验、稳定性研究等，稳定性研究包括冻存条件下制剂影响因素实验、长期实验、模拟临床应用条件的实验等，依据这些实验获得的药品稳定性信息来确定药物的储存运输条件、包装及有效期。

干细胞制剂制备过程服从 GMP，但应充分考虑活药物与传统药物的区别，供者细胞的合理性及筛查；生产材料应考虑供者细胞、生产过程细胞分类分级管理，生产用原材料的风险评估，外源因子去除，限制动物及人来源材料，基因修饰/改造按高风险管控，辅料使用考虑必要性、安全性和合理性；制备工艺与过程控制应证明可行性与稳健性，生产工艺的设计应避免细胞发生非预期的或异常的变化，全过程监控、持续改进，用连续自动化、全封闭手段减少污染；强调过程控制与制品放行互补，高风险操作评估，细胞制剂与回输制剂转换验证；质量研究取代表批次及生产阶段细胞的特性、功能性、纯度和安全性等分析；质量控制基于过程理解，兼顾认知，逐步完善，确证性与商业化保持一致；临床质量核准，方法验证与《中国药典》适用性验证，快速微量方法相互验证；稳定性研究应采取连续工艺，适用包装密封性研究、冷冻储存适应性研究等，运输条件，模拟使用；应对直接容器和过程容器进行安全性评估和相容性研究，次级包装应考察遮光性、密封性和抗击机械压力。

非临床研究包括临床前药效学、药物的作用机制、一般药理、毒理、药物相互作用、药代动力学等研究，是在动物身上进行的研究，目的是捕捉药物的安全性和有效性信号，了解药物在动物体内的吸收分布代谢排泄（ADME）及药物药代动力学（PK）/药效学（PD）。相关动物种群中，对药物的局部和全身毒性反应进行识别、鉴定、量化，识别其毒性的靶器官和部位，可逆性（急性或慢性毒性），量效关系，人体应用时的建议安全剂量及剂量递增方案，毒性/活性的潜在靶器官的鉴定，识别临床监测参数，患者的入组标准，终止潜在的不成功的开发计划，优化给药方案/种类合理性。干细胞制剂的药理毒理学研究内容基本包括：①干细胞药物分布和代谢，同种/异种干细胞在动物体内的分布、定植、分化和转归等；②干细胞药物临床前有效性评价，异种/同种干细胞在动物（鼠及大动物等）疾病模型体内有效性评价；③干细胞药物临床前安全性评价，一般毒性、特殊毒性（致瘤、促瘤、免疫毒性、生殖毒性等）、长期毒性评价试验；由新药开发机构进行实验室和动物研究，以观察 MSC 针对目标疾病的量效关系，同时对 MSC 进行安全性评估，动物体内安全性评估必须在 GLP 实验室完成。

新药临床试验申请：在临床前试验完成后，新药注册受理前召开药品审评中心与注册申请人沟通会议，然后向国家食品药品监督管理总局药品审评中心（CDE）提请一份 IND。提出的 IND 需包括以下内容：先期的体外实验结果，体内的作用机制，药理、药效研究结果，安全性研究结果，后续研究的适应证和临床方案等，临床评价包括安全性、有效性评价，临床研究风险及应急控制预案。同时根据需要，所用围产期干细胞种子库、半成品及成品在相关检定机构进行复核检验。IND 需得到审核部门的审核和批准，审评机构自受理之日起 60 个工作日后，没有给出否定或质疑的审查意见即视为同意，申请人可按照递交的方案开展临床试验。新药临床研究需要遵循赫尔辛基宣言伦理原则、《新药审批办法》《新药审批办法》《药品注册管理办法》中国/WHO/ICH GCP 指导原则、CDE 注册要求/新药临床研究指导原则。

临床试验是针对人类进行的、在健康志愿者和患病患者身上进行的试验，目的是在人类中测定药物实际的安全性和有效性，细胞制剂的临床研究分为早期临床试验和确证性临床试验，取代传统药物的Ⅰ~Ⅲ期临床试验。早期临床试验主要目的为安全性，次要目的为有效性，药代动力学研究单次与多次给药后细胞活力、增殖、分化、分布、迁移，预期存活时间及生物学功能，是否长期存在或持久作用等；概念验证试验（proof-of-concept，PoC）应考虑短期效应与长期结局，剂量探索（dose finding）试验包括起始剂量、有效剂量、最大耐受剂量等；然后是确证性临床试验，每期试验由于研究目的和研究问题不同而可能包含多个试验项目，考虑细胞活性持续时间，迟发安全性要开展长期随访，持续监测安全性与药理活性，考虑长期有效性及充分暴露安全性，长期安全性监测失效、感染、免疫原性、免疫抑制、恶性转化等风险。通过两个阶段的临床试验，新药开发机构将分析所有的试验数据，如果数据能够成功证明药物的安全性和有效性，机构将向药品监督机构提出新药申请。一旦药品监督机构批准了干细胞新药申请，此种干细胞新药就可以被医师用于处方。新药上市与上市后再评价：GMP设施建设，产品生产验证，新药临床应用安全性数据追踪与分析。

MSC药物是需要进行体外扩增的制品，由于需要较多的体外操作，对其安全性评价将是重点，到目前为止，国内尚无被批准的MSC药物。以下为需要扩增的MSC药物开发流程（图23-1）。

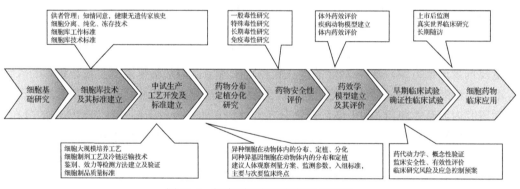

图23-1 间充质干细胞药物开发流程

一、临床级 MSC 库及其标准建立

参照国际干细胞指南和国家相关标准规范，对胎盘、脐带的供者筛查、样本采集、分离、纯化、诱导、入库筛选及所涉及的人员、设备设施、原辅材料、环境、操作程序进行规范，建立干细胞库技术标准，为干细胞供试品制备提供临床级种子来源。参考质控指标包括细胞鉴别、个体DNA指纹图谱、人白细胞抗原（HLA）分型、多向分化潜能、细胞纯度、染色体核型分析、端粒酶活性和致瘤性。生物活性检测包括总细胞数和活细胞数，细胞活力和效力检测。外来因子检测细菌、真菌、支原体、病毒[人类免疫缺陷病毒-1（HIV-1）、乙型肝炎病毒（HBV）、丙型肝炎病毒（HCV）、巨细胞病毒（CMV）、EB病毒（EBV）等]，以及相关的逆转录病毒，若使用抗生素有残留也应进行测试。

二、标准 MSC 供试品制备

建立干细胞的大规模培养、诱导分化、制剂制备、冷冻贮存、冷链传输、临床配制等工艺，依照 GMP 要求，质量保证将贯穿细胞制备的整个流程，制定标准操作程序（SOP），对所涉及的人员、设备设施、原辅材料、环境、操作程序进行规范。工艺稳定后开展工艺验证，制备连续 10 批以上符合标准的供试品。供试品制备能力为每批数十个临床剂量，为细胞制剂质量评价提供均一、稳定的供试品。

三、标准 MSC 供试品质量评价

针对 MSC 治疗目标疾病开展分子水平、细胞水平、动物水平研究，设置从供者筛查到临床回输放行全过程的质量控制点、安全性评价指标、功能性检测指标等，实现对细胞形态、生物标志物、细胞纯度、细胞活率、倍增时间、细胞世代、传代稳定性、迁移能力、分化能力、异常免疫反应、成瘤能力、生物学效力开展评估，建立分析测试方法并进行验证。评估体外过程对生物学功能、免疫原性、基因组稳定性等方面的影响，确定限定传代代次。参照《中国药典》（第三部）形式制定《供临床用间充质干细胞制造与检定规程（草案）》[26]，实现干细胞供者筛查、入库筛查、过程质量控制、中间品检测、半成品检测、细胞制剂检测及原辅材料检测功能。

四、临床前实验动物评价

根据干细胞制剂类型、适应证和给药途径建立动物评价模型，结合体外细胞学研究数据开展低、高代次和限定代次的干细胞制剂的组织分布、迁移、归巢、定植和分化、转归研究，致/促/抑瘤实验、免疫原性和免疫毒性实验、溶血实验、过敏实验、急/长毒性实验、单次给药毒性实验，重复给药毒性实验，生殖毒性实验，遗传毒性实验，特殊安全性实验，其他毒性实验，有效性实验等，对安全性指标、生物学效力参数进行验证和评估。相关动物种群中，对局部和全身毒性反应进行识别、鉴定、量化，识别其毒性/活性的靶器官和部位，确认急性或慢性毒性的可逆性及量效关系；为人体应用时的建议安全剂量及剂量递增方案，识别临床监测参数，建议患者的入组标准，建议主要与次要临床终点。

五、临床试验

早期临床试验是首次在人体进行的试验研究目的是了解药物的安全性，确定有效的剂量范围/最小有效剂量（minimal effective dose，MED），确定药物的适应证/最佳剂量，对药物的疗效进行初步评价，并证明与安慰剂或者对照相比的有效性和安全性，观察人体对新药的耐受程度及研究药物代谢动力学，为制定给药方案提供依据。早期临床试验最低病例数是 20~30 例，总体一般不超过 100 例，通常采用药物浓度递增的方法，估计达到患者不能接受毒性前的最大耐受剂量（maximum tolerated dose，MTD）。确证性临床试验通常是随机、双盲、对照（安慰剂/阳性对照）的多中心、大规模的验证性研究，以确认新药对目标适应证患者的疗效和安全性，必须提供有统计学意义的疗效证据。产品上市后，仍需要在"真实世界"环境下继续评价干细胞新药的风险和安全性及其风险效益。

新药从实验室进入到临床试验阶段通常称为新药的"死亡之谷"。新药开发是一个长期复杂的过程，资金、人才密集，成功率低，风险巨大。我国目前还未有干细胞药物被批准上市，主要是干细胞药物评审相对滞后，以及研发机构针对干细胞药物安全性的评价未能说服审批机构。尽管如此，还是有一些机构已经申报或即将申报干细胞药物。随着《临床干细胞临床研究管理办法（试行）》与《干细胞制剂质量控制及临床前研究指导原则（试行）》等文件出台，国家干细胞临床研究专家委员会成立，干细胞临床研究机构及研究项目启动备案工作。规范干细胞临床研究有助于为干细胞药物适应证选择提供概念验证试验（PoC），减小药物开发风险。

第三节　制约干细胞药物开发的技术瓶颈

干细胞植入体内后我们不知道它们去了哪里，做了些什么。干细胞药物挑战了传统医疗技术和药物开发的一些基本思想体系，但人们有能力从伦理学、安全性、有效性、质量可控性和经济可行性几个方面入手，在将干细胞植入受者体内前对其进行筛选、观察、操作、模拟及控制，从而突破制约干细胞药物开发的技术瓶颈。

一、合理的组织来源是干细胞药物开发的起点

许多实验室已发布自己在干细胞领域的研究成果，而事实上多数发表在顶级刊物上的研究成果难以重现，排除试剂、装备、操作等方面的因素，干细胞作为研究主体，其来源的差异对研究结果判断的影响不容忽视。

（一）干细胞供者遗传背景复杂

家族史、既往病史、性别、年龄、种族、血型、组织相容性抗原、身体条件（心理、生理和疾病等）、生活习性（饮食习惯、工作习惯、运动习惯、用药习惯等）、生存环境（大气环境、水环境、居住环境、工作环境、交通环境等）受各种物理、化学、生物因素的动态影响，世界上不存在两株遗传背景一致的干细胞。

（二）不同组织来源干细胞具有不同的生物学特性

人体260多种组织均由干细胞分化而来，干细胞无处不在，有细胞活动的地方就有干细胞的身影，干细胞的组织来源非常丰富，包括胚胎、成体组织（骨髓、牙髓、毛囊、上皮、肌肉、血管、脂肪、神经等）、围产期组织（脐带、羊水、羊膜、胎盘、脐血等）中均含有干细胞或细胞。以MSC为例，除了骨髓、脐带来源外，脂肪、胎盘、宫内膜、牙髓、羊水、羊膜等均含有MSC，但是不同组织来源的MSC的生物学特性具有异质性。

（三）干细胞为动态活细胞

目前临床上包括抗生素、化学小分子、生物大分子在内的药物等均为结构和成分明确并相对稳定的体系。而干细胞属于活细胞，具有异质性，其大小、形态或多或少具有一定差异，在功能、行为、状态方面也可能有不同。同种干细胞在不同微环境中可以发挥不同作用。

供临床用干细胞种类（组织来源）的选择可考虑以下几方面：①无伦理学障碍；②材料有稳定的来源；③具有明确筛选指标和理论依据；④符合规模化制备要求；⑤体外过程可控；⑥安全性在可接受范围内。

围产期组织来源干细胞为 0 岁细胞，更健康、更富有活力，免疫原性低，为废物组织再利用、无伦理学障碍，易于标准化制备，很好地解决了干细胞药物开发的所需种子细胞的组织来源问题。目前，国内外正在开发多个围产期干细胞药物，以色列 Pluristem 治疗公司利用胎盘贴壁细胞开发的干细胞药物 PLX-PAD，获得了 EMA 和 FDA 批准用于治疗下肢 PAD 人体临床试验，已进入Ⅲ期临床阶段。天津昂赛细胞基因工程有限公司开发的注射用 MSC（脐带）根据新药研究规范和干细胞的特性建立了质量标准，并获得了完备、规范的临床前药学、安全性和有效性评价数据，正在开发治疗 GVHD 的 MSC 制剂[27,28]。

二、有效的评价技术是干细胞药物开发的关键

（一）干细胞药物的质量源于设计（QbD）

干细胞制剂的质量高度取决于工艺路线与产品形态的选择，再深刻理解干细胞生物学特性和目标疾病治疗机制基础上，针对干细胞"异质性""动态性"与临床制品要求的"均一性""稳定性"之间的突出矛盾，以"标准化、自动化、规模化"为手段，预先定义好质量目标及工艺控制点，进行系统的开发。例如，细胞制剂工艺设计需要考虑：机械法还是酶法分离细胞，开放操作还是选择封闭工作站，选择手动传代还是连续自动扩增，选择 2D 还是 3D 培养，低分种率高代次还是限定低代次，考虑异源血清还是无动物源性培养基，制剂是游离细胞形式还是与基质结合，产品是新鲜细胞还是冻存制品，采用 DMSO 保护还是新型冻存剂，等等。总之，新药开发需要考虑未来 30 年的技术替代与竞争，整个新药开发过程风险守恒的特征，将前置的风险降低失败与反复带来总体资金与时间成本，有人指出，风险每后置一步其危害将放大 20 倍。

（二）质量可控性是干细胞药物安全性和有效性的基础和前提

干细胞组织来源多样，供者遗传背景复杂，体外分离、纯化、扩增、诱导和细胞制剂制备过程漫长且要引入抗生素、生长因子、抗体、胶原酶、蛋白酶、动物血清等多种外来成分；还可能污染细菌、支原体、病毒等外源因子；体外操作过程可能会引起干细胞的增殖潜能、分化潜能、生物学效力、组织相容性抗原、端粒酶、核型、原癌和抑癌基因等发生改变而带来风险；在保存、运输、复苏、配制过程中，新鲜活细胞制剂的存活率、生物学效力等均可能发生变化。如果细胞制备过程及细胞制剂缺乏有效的质量控制方法和相应的标准，干细胞临床应用的安全性和有效性无法实现。

(三) 干细胞制剂有效性评价指标[29]

目前临床上一般采用活细胞计数方法决定输注干细胞数量，在某些患者身上干细胞治疗显现出了"令人惊讶"的效果，而有部分患者则反映"没什么效果"。事实上，干细胞的体外操作是一个动态过程，活细胞数量并不能有效反映干细胞生理状况的改变，有人甚至将处死的干细胞输注到动物体内获得了与输注活细胞相当的实验结果。干细胞的有效性与其分化潜能、倍增时间、分泌因子、世代数、归巢特性等如何相关联？这里需要引入一个可量化的干细胞的定义单位——生物学效力（biological potency）。

生物学效力或称为"效价"，是一个评价生物药物有效性的重要指标，用来测量药物产生特定生理功能的浓度或数量；生物药物的物理数量（质量）往往无法直接代表其生物学效力，如蛋白质失活后含量不变效力下降，一般需要重新定义效价单位作为金标准来衡量其生物学效力。例如，通常用抑菌实验的半数抑制微生物生长的最高稀释度测量抗生素效价单位，用细胞病变（cytopathogenic effect，CPE）抑制法测量能抑制50%细胞病变或50%病毒空斑形成的最高稀释度来测量干扰素（IFN）活性单位。

干细胞的生物学效力的要求：①效力指标设定需要找出与临床治疗效应相关的关键机制通路，能够真实反映量效关系，要考虑干细胞是多靶点、多通路的；②设定需要效力指标能够代表细胞制剂在制备过程中与临床使用时具有一致性；③效力指标设定应是优化的，最好用体外实验来替代动物模型，要求操作便捷、能够成为制品放行标准；④效力指标设定应通过分析方法学验证：线性、范围、准确度、精密度（包括重复性和重现性）、检测限、定量限和耐用性等。

细胞表面标志物虽可以用来鉴别干细胞的种类和评价干细胞的纯度，但无法直接反映细胞的生物学功能，甚至无法区分细胞死活；诱导分化试验较难定量，而且周期长、变异性大；微RNA（miRNA）及因子表达直接测定难以反映体内状况。

干细胞生物学效力测定方法的建立可以从以下方面入手：①考虑将干细胞置于"激活"状态，而非简单的体外培养或保存条件下，如模拟组织损伤微环境；②观测"激活"前后干细胞行为差异，研究这些差异与干细胞分化、抗炎、抗凋亡潜能之间的关系；③建立用于评估生物学活性的刺激系统，筛选与目标疾病治疗相关的关键作用因子；④最终开发出干细胞生物学效力测定方法、建立放行标准。

实验动物模型困扰干细胞临床前安全性和有效性的评价：实验动物常用来评价新药的体内分布、代谢、有效性、急性毒性、特殊毒性、长期毒性等，为新药人体临床观察提供基础数据。动物模型建立的一般原则：①在动物身上复制的模型应尽可能近似于人类疾病的情况；②物模型应该是可重复的，甚至是可以标准化的；③复制的模型可特异地、可靠地反映疾病机能、代谢、结构变化和主要症状和体征；④复制动物模型时应尽量考虑到今后临床应用和便于控制其疾病发展；⑤复制动物模型时所采用的方法应尽量做到容易执行和合乎经济原则。

但是，动物模型是否能够为临床研究提供有效参考仍然值得考虑：①干细胞不符合一般药物代谢规律，异种使用涉及排异问题；②干细胞具有种属特异性，同种植入供试品又与未来临床试验不一致；③标记后活细胞示踪实验可能造成干细胞生物特征发生改变，定植、分化、转归实验追踪困难；干细胞进入体内后组织归巢、病理性归巢的规律如何？是如何定植、分化的？④动物生命周期不足以体现人类干细胞长期植入后果。

三、风险控制预案是干细胞药物临床研究的保障

干细胞药物的开发与化学药物、抗生素、蛋白质药物等这些化学结构明确的分子有着很大的不同，挑战了传统药物开发的一些基本理念和规律，人类对干细胞的认识尚处在相对初级的阶段。

（一）作用机制

干细胞直接定植分化参与组织修复，还是通过分泌细胞因子改变微环境发挥机体调节作用？激活了宿主内源性干细胞进行组织修复，还是与宿主细胞相互作用改变机体细胞状态？干细胞体外培养特性与体内生物学行为是否一致？动物试结果能否用于预测干细胞在人体内的作用？自体干细胞与异体干细胞植入的区别？

（二）细胞使用后体内命运

以 ESC 为例，干细胞进入体内后组织归巢、病理性归巢的规律如何，是如何定植、分化的，还是被免疫清除？如果干细胞在体内非靶器官、靶组织"归巢"，非目的性"分化"，产生非需要生物学效力，引起持续不良后果如何终止？但目前认为 MSC 不能在体内长期存活，目前临床试验显示安全性方面表现良好。

（三）有效性

干细胞治疗/移植疗效与患者的病种、病程、身体状况、组织配型、年龄、种族、性别、习惯、经历，以及移植细胞数量或输注途径等之间存在怎样的关系，如何确定量效关系？

（四）干细胞治疗/移植可能引起的不良反应

目前报道最多的细胞使用后出现的不良反应是一过性发热、寒战，绝大多数不需要特殊处理；也有报道少数人会出现嗜睡、兴奋、恶心、呕吐等症状；极个别的患者可能出现皮疹等过敏反应。造血干细胞移植（HSCT）后会出现 GVHD，应用 MSC 可以治疗 GVHD。应用 iPSC 或 ESC 分化后的神经等细胞可能会出现肿瘤或囊肿，也有报道提到使用流产胚胎来源的细胞治疗，也导致了肿瘤形成[30]。

（五）干细胞治疗病种和人群选择建议

包括：①无有效手段控制疾病进程者；②有突破现有手段局限性的理论依据；③干细胞治疗带来的可能利益明显大于潜在风险；④肿瘤患者或肿瘤高危人群；⑤是否明确组织 HLA 配型（相合、半相合）；⑥孕妇等特殊人群；⑦能否做到长期随访。

干细胞技术让一些传统医疗手段束手无策的患者重新燃起希望，但干细胞治疗犹如开启了"潘多拉魔盒"，人们充满了好奇、期待和恐惧。恐惧源自于结果的不确定性，只

有我们对最差的状况有了充分的估计和应对准备，把风险控制在可以接受的范围内，人们才能充分享受新技术和产品带来的福祉。

必须承认的是，某些特殊情况下，干细胞技术的临床应用可能先于严格的新药开发程序完成，就是所谓的"医疗新技术"，医学史上有不少这样的例子。但是，对于公众和潜在的消费者而言，在干细胞科学研究尚在进行、临床应用条件尚未成熟及管理规则还有待制定的情况下，如何采取必要的措施保护自身的健康与经济利益，如何配合科学界与管理机构监督干细胞治疗的规范性，是一个值得认真考虑的问题。同时，我们必须清醒地认识到，从干细胞研究到实际应用需要经过漫长艰苦的发展过程。面对具有无限潜力的干细胞研究和其应用前景，科学界、临床工作者、产业界及广大消费公众都需要有一个现实的态度，需要耐心与理解，需要承担各自的责任，互相配合，以保证科学研究与临床应用能够在最有利的环境中进行。

第四节　国外已上市的干细胞药物及生物制品概要

一、Prochymal——FDA 批准的首个干细胞孤儿药

药品名称：Prochymal（孤儿药，图 23-2）
开发机构：美国 Osiris Therapeutics（www.osiris.com）
活性成分：骨髓源间充质干细胞
适 应 证：1 型糖尿病
批准时间：2010 年 5 月 4 日
批准机构：美国食品药品监督管理局

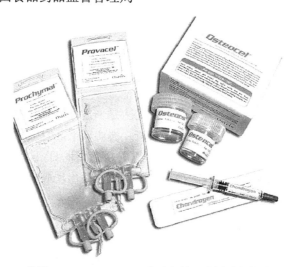

图 23-2　美国 Osiris Therapeutics 公司产品（彩图请扫封底二维码）

二、Hearticellgram-AMI——全球首个获准上市的干细胞药物

药品名称：Hearticellgram-AMI（图 23-3）

开发机构：韩国 FCB-Pharmicell（http://www.pharmicell.com/ch/index.html）
活性成分：自身骨髓间质干细胞
适 应 证：急性心肌梗死
批准时间：2011 年 7 月 1 日
批准机构：韩国食品与药品管理局

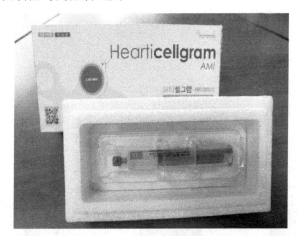

图 23-3　Hearticellgram-AMI 外观（彩图请扫封底二维码）

三、Hemacord——FDA 批准的首个干细胞制剂

药品名称：Hemacord（生物制品，图 23-4）
开发机构：美国纽约血液中心（www.nybc.org）
活性成分：脐血造血祖细胞
适 应 证：重建因先天性、后天性与化疗引起的血液与免疫系统缺失疾病，如血液恶性病、贺勒氏疾病（MPS-I）、脑白质萎缩症（Krabble disease）、X 性联性肾上腺脑白质失养症、免疫不全症、骨髓衰竭与乙型地中海型贫血等
批准时间：2011 年 11 月 10 日
批准机构：美国食品药品监督管理局

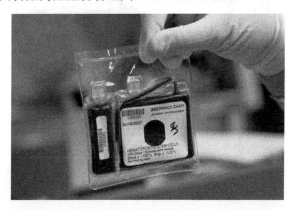

图 23-4　Hemacord 外观（彩图请扫封底二维码）

四、Cartistem——全球首个异基因干细胞药物

药品名称：Cartistem（图 23-5）
开发机构：韩国 Medipost Inc（www.medi-post.com）
活性成分：新生儿脐血源间充质干细胞
适 应 证：退行性关节炎和受损膝盖软骨
批准时间：2012 年 1 月 19 日
批准机构：韩国食品与药品管理局

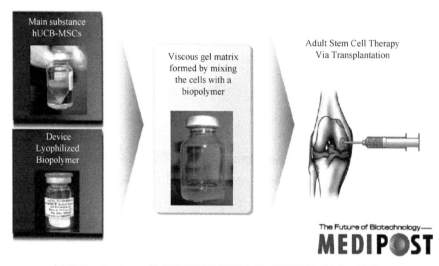

图 23-5　Cartistem 的产品外观及使用途径（彩图请扫封底二维码）

五、Prochymal——欧美首个异基因干细胞药物

药品名称：Prochymal（图 23-6）

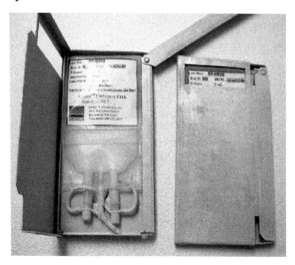

图 23-6　Prochymal 外观（彩图请扫封底二维码）

开发机构：美国 Osiris Therapeutics（www.osiris.com）
活性成分：骨髓间充质干细胞
适 应 证：儿童移植物抗宿主病（GVHD）
批准时间：2012 年 5 月 17 日
批准机构：加拿大卫生部

六、MultiStem——FDA 批准的第二个干细胞孤儿药

药品名称：MultiStem（孤儿药，图 23-7）
开发机构：America Stem Cell, Inc.（www.athersys.com）
活性成分：多能成体祖细胞（骨髓或其他器官来源）
适 应 证：赫尔勒氏综合征（Hurler's Syndrome），即 I 型黏多糖贮积症（MPS-I）
批准时间：2012 年 7 月 10 日
批准机构：美国食品药品监督管理局

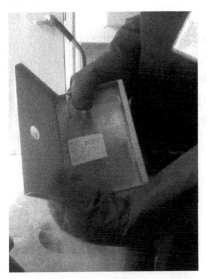

图 23-7　MultiStem 外观（彩图请扫封底二维码）

七、Holoclar——欧盟首个干细胞先进医疗产品

药品名称：Holoclar（图 23-8）
开发机构：Chiesi Farmaceutici（www.chiesigroup.com）
活性成分：自体角膜缘干细胞
适 应 证：灼伤引起的角膜缘干细胞缺陷症
批准时间：2015 年 2 月 17 日
批准机构：欧洲药品管理局

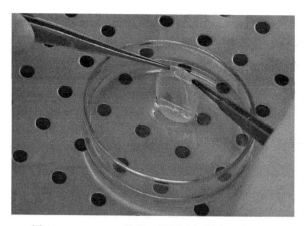

图 23-8　Holoclar 外观（彩图请扫封底二维码）

八、Stempeucel——印度首个干细胞药品

药品名称：Stempeucel（图 23-9）
开发机构：Stempeutics（www.stempeutics.com）
活性成分：骨髓混合间充质干细胞
适 应 证：血栓闭塞性脉管炎（Buerger 病）
批准时间：2016 年 6 月 25 日
批准机构：印度药品管理总局

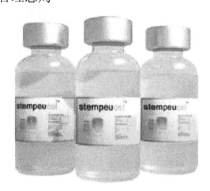

图 23-9　Stempeucel 外观（彩图请扫封底二维码）

九、Strimvelis——全球首个干细胞基因治疗产品

药品名称：Strimvelis（图 23-10）
开发机构：GlaxoSmithKline（www.gsk.com）
活性成分：腺苷脱氨酶基因修饰自体造血干细胞
适 应 证：儿童腺苷脱氨酶缺乏性重度联合免疫缺陷症（ADA-SCID）
批准时间：2016 年 5 月 27 日
批准机构：欧洲药品管理局

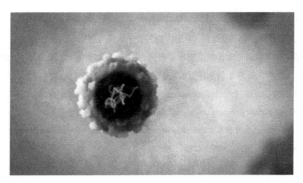

图 23-10　Strimvelis 概念图（彩图请扫封底二维码）

（张　磊　韩之波）

参 考 文 献

[1] 张磊. 干细胞创新如何跨越"死亡之谷". 中国医药生物技术, 2015, 10(5):385-391.
[2] 原国家卫生部. 人的体细胞治疗及基因治疗临床研究质控要点. 1993.
[3] 人体细胞治疗研究和制剂质量控制技术指导原则. 国药监注[2003]109 号.
[4] 医疗技术临床应用管理办法. 卫医政发[2009]18 号.
[5] None. Collective responsibilities. Nature, 2009, 457:935.
[6] Listed N. Stem-cell laws in China fall short. Nature, 2010, 467:633.
[7] Cyranoski D. China's stem-cell rules go unheeded. Nature, 2012, 484:149-150.
[8] 关于开展干细胞临床研究和应用自查自纠工作的通知. 卫办科教函[2011]1177 号.
[9] 中华人民共和国卫生计生委, 国家食品药品监督管理总局. 干细胞制剂质量控制及临床前研究指导原则(试行). 2015.
[10] 中华人民共和国卫生计生委, 国家食品药品监督管理总局. 干细胞临床研究管理办法(试行). 2015.
[11] David C. FDA's claims over stem cells upheld. Nature, 2012, 488:14.
[12] WHO. Recommendations for the evaluation of animal cell cultures as substrates for the manufacture of biological medicinal products and for the characterization of cell banks. 2010.
[13] International standards for hematopoietic cellular therapy product collection, processing, and administration. FACT-JACIE International Standards Sixth Edition, 2015.
[14] ISSCR Guidelines for clinical translation of stem cells. 2008.
[15] Daniel C. Stem cells take root in drug development. Nature, 2012.
[16] FDA Center for Biologics Evaluation and Research. Guidance for Industry Potency Tests for Cellular and Gene Therapy Products. 2011.
[17] FDA Guidance for human somatic cell therapy and gene therapy. 1998.
[18] FDA Guidance-Content and review of CMC information for human somatic cell therapy IND application. 2008.
[19] FDA Guidance for Industry-Current Good Tissue Practice (CGTP) and Additional Requirements for Manufactures of Human Cells, Tissues, and Cellular and Tissue-Based Products (HCT/Ps). 2009.
[20] EMA Guideline on human cell-based medicinal products. 2007.
[21] The European Parliament and The Council, Clinical trials on medicinal products for human use. 2014.
[22] Japan. The Safety of Regenerative Medicine. 2013.
[23] Japan. Pharmaceuticals and Medical Devices Act. 2014.
[24] 国家食品药品监督管理总局. 细胞制品研究与评价技术指导原则（征求意见稿）. 2016.

[25] 国家食品药品监督管理总局. 细胞治疗产品研究与评价技术指导原则（试行）. 2017.
[26] 国家药典委员会. 中华人民共和国药典. 北京: 中国医药科技出版社, 2015.
[27] 耿洁, 张磊, 韩之波, 等. 临床用间充质干细胞的质量研究. 中国医药生物技术, 2013, 8(3):225-230.
[28] 耿洁, 张磊, 王斌, 等. 人脐带间充质干细胞注射液的长期稳定性研究. 中国医药生物技术, 2013, 8(3):230-234.
[29] Galipeau J, Mauro K, John B, et al. International society for cellular therapy perspective on immune functional assays for mesenchymal stromal cells as potency release criterion for advanced phase clinical trials. Cytotherapy, 2016, 18(2):151-159.
[30] Amariglio N, Hirshberg A, Scheithauer BW, et al. Donor-derived brain tumor following neural stem cell transplantation in an ataxia telangiectasia patient. PLoS Med, 2009, 6(2):e1000029.